Revision

- Have you stated clearly and specifically the purpose of the report?
- Have you put into the report everything required? Do you have sufficient supporting evidence? Have you stated the implications of your information clearly?
- Are all your facts and numbers accurate?
- Have you answered the questions your readers are likely to have?
- Does the report contain anything that you would do well to cut out?
- Does your organization suit the needs of your content and your audience?
- Are your paragraphs clear, well organized, and of reasonable length? Are there suitable transitions from one point to the next?
- Is your prose style clear and readable?
- Is your tone appropriate to your audience?
- Are all your statements ethical? For example, have you avoided making ambiguous statements or statements that deliberately lead the reader to faulty inferences?
- Are your graphs clear and accurate? Are they well placed? Do they present your information honestly?
- Is your document readable, accessible, and visually effective?
- Are there people you should share your draft with—for example, members of the target audience—before going on to a final draft?

Editing

- Have you checked thoroughly for misspellings and other mechanical errors?
- Have you included all the formal elements your report needs?
- Are format features such as headings, margins, spacing, typefaces, and documentation consistent throughout the draft?
- Are your headings and titles clear, properly worded, and parallel? Do your headings in the text match those in the table of contents?
- Is your documentation system the one required? Have you documented wherever appropriate? Do the numbers in the text match those in the notes?
- Have you keyed the tables and figures into your text and have you sufficiently discussed them?
- Are all parts and pages of the manuscript in the correct order?
- Will the format of the typed or printed report be functional, clear, and attractive?
- Does your manuscript satisfy stylebook specifications governing it?
- Have you included required notices, distribution lists, and identifying code numbers?
- Do you have written permission to reproduce extended quotations or other matter under copyright? (Necessary only when your work is to be published or copyrighted.)
- While you were composing the manuscript, did you have any doubts or misgivings that you should now check out?
- Have you edited your manuscript for matters both large and small?
- What remains to be done, such as proofreading final copy?

REPORTING TECHNICAL INFORMATION

REPORTING TECHNICAL INFORMATION

8th Edition

Kenneth W. Houp
Late, The Pennsylvania State University

Thomas E. Pearsall
Emeritus, University of Minnesota

Elizabeth Tebeaux
Texas A&M University

Contributing Author: Janice C. Redish
President, Redish & Associates, Inc.
Bethesda, Maryland

ALLYN AND BACON
Boston London Toronto Sydney Tokyo Singapore

Editor: Eben W. Ludlow
Production Supervision: P. M. Gordon Associates
Production Manager: Nick Sklitsis
Art Director: Patricia Smythe
Text and Cover Designer: Patricia Smythe
Cover Illustration: Salem Krieger
Photo Researcher: Chris Migdol
Illustrations: TopDesk Publishers' Group
Compositor: TopDesk Publishers' Group

A Simon & Schuster Company
Needham Heights, Massachusetts 02194

Library of Congress Cataloging-in-Publication Data

Houp, Kenneth W., 1913–
Reporting technical information / Kenneth W. Houp, Thomas E. Pearsall, Elizabeth Tebeaux.— 8th ed.
p. cm.
Includes bibliographical references (p.) and index.
ISBN 0-02-393351-8
1. Technical writing. I. Pearsall, Thomas E. II. Tebeaux. Elizabeth. III. Title.
T11. H59 1995
808´. 0666-dc20 94-23097
CIP

Printed in the United States of America

10 9 8 7 6 5 4 3 2 1 99 98 97 96 95 94

For Eben W. Ludlow—editor and friend

PREFACE

With this eighth edition, we bring aboard a new member of the *Reporting Technical Information* team: Elizabeth Tebeaux, Professor of English at Texas A&M and President of the Association of Teachers of Technical Writing. Professor Tebeaux has long been active in the teaching, scholarship, and administration of technical communication. She agrees thoroughly with the long-standing objectives of *Reporting Technical Information:*

> Our readers should be able to analyze a writing situation correctly; to find and organize material appropriate to audience, purpose, and situation; to design a functional report or letter that answers the needs of both writers and readers; and to write that report or letter correctly, clearly, and persuasively.

Janice C. Redish, President of Redish & Associates, Inc., serves once again as a contributing author.

Plan of the Book

This is what you will find in the eighth edition.

Chapter 1, "An Overview of Technical Writing." Chapter 1 defines technical writing and describes the world of workplace writing—the forms that technical reporting takes and the problems that writers of technical information encounter.

Part I: Basics. In Part I, emphasis falls on the composing process. Chapter 2, "Composing," discusses how to analyze a writing situation; how to discover, arrange, write, revise, and edit the information you will need; and how to deal with ethical concerns. With this chapter begins the emphasis on audience that is central to this book. Chapter 3, "Writing Collaboratively," deals with how to compose technical writing in collaboration with others. Chapter 4, "Writing for Your Readers," discusses some of the ways you can adapt your writing to various audiences. Finally, Chapter 5, "Achieving a Readable Style," shows you how paying attention to elements of your style at the paragraph, sentence, and language levels can make your writing more readable.

Part II: Techniques. In Part II we build on the basic concepts of Part I. The three chapters in Part II describe and demonstrate the techniques you will need to inform, describe, define, and argue. With these techniques you can arrange, draft, and revise much of the writing you will do as a professional in your field. As in Part I, we emphasize the need to consider audience and purpose no matter what technique you are using. With the techniques you master in Part II, you will have many of the skills you will need to apply in writing documents such as instructions, proposals, and feasibility reports, the subjects of Part IV, "Applications."

Part III: Document Design. Part III deals with document design and graphics. Good design—creating a format that helps readers find information and read selectively—is vitally important in technical writing. Chapter 9, "Document Design," deals primarily with the format and appearance of the document. Chapter 10, "Design Elements of Reports," tells you how to construct all those elements that full reports need, such as covers, tables of contents, abstracts, introductions, discussions, summaries, and notes. Chapter 11, "Graphical Elements," tells you how to use tables, graphs, drawings, and photographs to inform your reader about such things as concepts, processes, trends, relationships, and objects. Technical writing is marked by an extensive use of graphics.

Part IV: Applications. The first six chapters of Part IV put all the basics, techniques, design features, and graphics of the first three parts to work. The chapters of Part IV discuss correspondence, the job hunt, and reports such as feasibility reports, instructions, progress reports, and proposals. Chapters 14 and 15, in particular, show you how to deal with any kind of report, no matter what it is called. Chapter 18 discusses how to put the principles and techniques of this book to work in oral reports. In short, Part IV covers most of the kinds of reports, written and spoken, that professionals in every field have to deliver.

Part V: Handbook. Any living language is a growing, flexible instrument with rules that are constantly changing by virtue of the way it is used by its live, independent speakers and writers. Only the rules of a dead language are unalterably fixed. Nevertheless, at any point in a language's development, certain conventions of usage are in force. Certain constructions are considered errors that mark the person who uses them as uneducated. It is with these conventions and errors that this handbook primarily deals. We also include sections on outlining and avoiding sexist language.

Appendixes. Appendix A guides you to technical reference sources in your own field. Appendix B is a bibliography that leads you to other sources for the many subjects covered in *Reporting Technical Information.*

Major Changes in the Eighth Edition

- Ethical considerations are now introduced in Chapter 2, "Composing." Technical writing often has consequences for large numbers of people; therefore, technical writers have to learn to deal with ethical considerations as a part of the composing process. This new section makes this important point.
- Chapter 4, "Writing for Your Readers," has been totally reorganized. In earlier editions, the emphasis in the chapter was on four kinds of audiences: lay, executive, technicians, and experts. These audiences are still present in the chapter, but the chapter is no longer organized around them. Rather the chapter is organized around universal considerations such as point of view, providing background, and style. This change should provide more flexibility for users of this chapter.
- Chapter 9, "Document Design," now contains more examples that show differences in documents when principles of good design are omitted and when they are used. Also, exercises at the end of the chapter provide practice in redesigning for better reader accessibility to reports.
- Chapter 14, "Development of Reports," and Chapter 15, "Development of Analytical Reports," are new chapters that emphasize that effective reports respond to context and that graphics and page design must be considered in report development. The new chapters make clear that while generic reports such as feasibility reports do exist, technical writers must also learn to develop reports that do not fit cleanly into some generic category.

Besides these major changes, numerous small changes in style and substance have been made. We have added fresh examples from the many fields that users of this text represent. In particular, you will find many fresh examples in Chapter 11, "Graphical Elements." The four-color graphics insert in this chapter contains all new examples. We have updated Appendix A, "Technical Reference Books and Guides," particularly the section on computerized information retrieval. In short, we have followed our touchstone that all writing is subject to infinite improvement. We invite teachers and students to use our E-mail address—tpearsall@aol.com—to send us comments and suggestions.

Teaching Ancillaries

With this eighth edition, we continue to offer extensive teaching aids.

- An expanded Instructor's Manual with transparency masters and tests. The revised manual, co-authored with Professor Richard Raymond of Armstrong State College, includes advice on course planning, sample syllabi, suggested examination questions, and, for each chapter, teaching objectives, teaching hints, a discussion of exercises and assignments, a chapter quiz, transparency masters, and exercise solutions.
- A set of fifty two-color transparency acetates. The transparencies display both existing text figures and new student examples.

Acknowledgments

Detailed acknowledgments to the many sources we have drawn upon for this book are found in our "Chapter Notes." We thank the colleagues who took the time to review our work and make so many useful suggestions: Virginia A. Book, University of Nebraska, Lincoln; David E. Boudreaux, Nicholls State University; Jerry Harris, DeVry Institute of Technology, Chicago; Timothy C. Kennedy, Oregon State University; Andrew Linder, DeVry Institute of Technology, Toronto; Paul Meyer, New Mexico State University; Joe Nickell, University of Kentucky; Anita C. Nordbrock, Embry-Riddle Aeronautical University–Prescott Campus; Sandra Manoogian Pearce, Moorhead State University; Carolyn Plumb, University of Washington; Brenda R. Sims, University of North Texas; Valerie J. Vance, Oregon Institute of Technology.

Donald J. Barrett, Chief Reference Librarian, United States Air Force Academy, has once again revised Appendix A, "Technical Reference Books and Guides," for us, and Professor James Connolly, University of Minnesota, has again contributed the section on visuals found in Chapter 18, "Oral Reports."

Reporting Technical Information has had the good fortune to have two of college publishing's best editors, Madalyn Stone and Eben W. Ludlow, to aid in the development of this book. Pat Smythe has once again furnished a design that is both efficient and attractive. Finally, we express our love and gratitude to our spouses, Anne and William Jene, for their loving and loyal support.

Thomas E. Pearsall
Elizabeth Tebeaux

BRIEF CONTENTS

CONTENTS

CHAPTER 1

AN OVERVIEW OF TECHNICAL WRITING

This first chapter is purely introductory. It is intended to give you the broadest possible view of technical writing. Beginning with Chapter 2, we go into details, but in order to be meaningful, these details must be seen against the background given here.

Some Matters of Definition

As you work your way through this book, you will see that technical writing is essentially a problem-solving process that involves the following elements at one or more stages of the process:

- A technical subject matter that is peculiar to or characteristic of a particular art, science, trade, technology, or profession.
- A recognition and accurate definition of the communication problem involved.
- The beginning of the solution through the establishment of the role of the communicator and the purpose and audience (or audiences) of the communication.
- Discovery of the accurate, precise information needed for the solution of the problem through thinking, study, investigation, observation, analysis, experimentation, and measurement.
- The arrangement and presentation of the information thus gained so that it achieves the writer's purpose and is clear, useful, and persuasive.

The final product of this problem-solving process is a piece of technical writing that may range in size and complexity from a simple memorandum to a stack of books. To expand our overview of technical writing, we discuss it under these five headings:

The Substance of Technical Writing
The Nature of Technical Writing
The Attributes of Good Technical Writers
The Qualities of Good Technical Writing
A Day in the Life of Two Technical Writers

The Substance of Technical Writing

Organizations produce technical writing for internal and external use. Internally, documents such as feasibility reports, technical notes, and memorandums go from superiors to subordinates, from subordinates to superiors, and between colleagues at the same level. If documents move in more than one direction, they may have to be drafted in more than one version. Company policy, tact, and the need to know are important considerations for intracompany paperwork.

Many examples come to mind. The director of information services studies and reports on the feasibility of providing middle management with personal computers. The research department reports the results of tests on new products. The personnel department instructs new employees about company policies and procedures. In fact, the outsider cannot imagine the amount and variety of paperwork a company generates simply to keep its internal affairs in order. Survey research indicates that college-educated employees spend about 20% of their time on the job writing.[1] In fact, most college-educated workers rank the ability to write well as very important or critically important to their job performance.[2]

Externally, letters and reports of many kinds go to other companies, the government, and to the users of the company's products. Let us cite a few of the many possibilities: A computer company prepares instructional manuals to accompany its computers. A university department prepares a proposal to a state government offering to provide research services. An architectural firm prepares progress reports to inform clients of the status of contracted building programs. An insurance company writes letters accepting or denying claims by its policyholders.

The manufacture of information has become a major industry in its own right. Much of that information is research related. Many government agencies, scientific laboratories, and commercial companies make research their principal business. They may undertake this research to satisfy their internal needs or the needs of related organizations. The people who conduct the research may include social scientists, computer scientists, chemists, physicists, mathematicians, psychologists—the whole array of professional specialists. They record and transmit much of this research via reports. The clients for such research may be government agencies or other institutions that are not equipped to do their own research. Reports may, in fact, be the only products of some companies and laboratories.

Much technical writing goes on at universities and colleges. Professors have a personal or professional curiosity that entices them into research. If they believe that their findings are important, they publicize the information in various ways—books, journal articles, papers for professional societies. Students assigned research problems present what they have done and learned in laboratory reports, monographs, and theses.

Many reports are prepared for public use. For example, a state department of natural resources is entrusted not only with conserving our woodlands, wetlands, and wildlife but also with making the public aware of these resources. State and federally supported agricultural extension services have as a major responsibility the preparation and dissemination of agricultural information for interested users. Profit-earning companies have to create and improve their public image and also attract customers and employees. Airlines, railroads, distributors of goods and services, all have to keep in the public view. Pamphlets, posted notices, and radio and television announcements are commonly used to meet these needs.

Myriad applications such as these—company memos and reports, government publications, research reports, public relations releases—create a great flood of paperwork. Some of it is only of passing interest; some of it makes history. Some of it is prepared by full-time professional writers, but most of it is prepared by professionals in a technical field who are writing about their own work.

The Nature of Technical Writing

Technical writing, whether done by professional writers or professionals in a technical field, is a specialty within the field of writing as a whole. It requires a working knowledge of the technical subject matter and terminology. People working with technical documents need to learn about document design, standards for abbreviations, the rules that govern the writing of numbers, the uses of tables and graphs, and the needs and expectations of people who use technical documents.

And yet a broad and sound foundation in other writing is a tremendous asset for those who write technical documents, for it gives them versatility both on and off the job. They can write a good letter, prepare a brochure, compose a report. In this comprehensive sense, they are simply *writers.* The same writing skills that are important in a college classroom are important on the job. Surveys show that workers rank writing skills in this order of importance:[3]

1. Clarity
2. Conciseness
3. Organization
4. Grammar

They understand, too, that not all writing is done in the same tone and style. As writers, they have not one style but a battery of them:

<table>
<tr><td>...the very nice plant my mother had on her table in the front hall.</td><td>Everyday, homey diction; much depends on the reader's imagination</td></tr>
<tr><td>...in a shaft of yellow sunlight, a white-flowering begonia in a red clay pot.</td><td>Pictorial, vivid, sensory; shows, rather than tells</td></tr>
<tr><td>...a 12-inch begonia propagated from a 3-inch cutting; age, 42 days.</td><td>Specific, technical, factually informative</td></tr>
</table>

As someone who writes technical documents, whether part-time or full-time, you may have to use all of these styles, for your job will be to convey your message to your intended readers. By playing the right tune with these styles in different combinations, and by adding other writing skills in generous measure, you can produce leaflets, proposals, brochures, sales literature, reports to stockholders, and a great variety of letters.

In writing intended for your professional colleagues, you will be nearer to the third begonia example than to the first two. Your diction will be objective and accurate. By relying on this style, you can produce operating manuals, feasibility reports, research reports, progress reports, and similar materials. When your audience and purpose are appropriate for this style, your writing is likely to have these characteristics:

- Your purpose is usually spelled out in the opening paragraph or two. All included information bears upon the accomplishment of the stated purpose. For example, a technical paper on smoke detectors may set forth only one major objective: to determine the relative effectiveness of photoelectric and ionization chamber types in detecting smoldering fires, flaming fires, and high temperatures. Other major topics would be reserved for other papers.
- The vocabulary tends to be specialized. Some of the terms may not appear in general dictionaries. If the audience shares the writer's specialization, such terms may not be defined within the text, on the assumption that professional colleagues will be familiar with them. At other times, the terms may be listed and defined in accompanying glossaries.
- Sentences are highly specific and fact filled.
- When appropriate to the material, numbers and dimensions are numerous.
- Signs, symbols, and formulas may pepper the text.
- Graphs and tables may substitute for prose or reinforce and expand upon the surrounding prose. Figures and illustrations of all sorts are widely used, sometimes to supplement prose, sometimes to replace it.
- Documentation and credits appear in notes and bibliographies.

As this list makes clear, audience analysis is tremendously important to successful technical writing. What is appropriate for your professional colleagues may be inappropriate for the general public. In matters of definition, for example, terms are not normally defined if the audience is expected to know them. But the indispensable corollary to that proposition is that terms *have to be defined* when your audience, for whatever reason, cannot be expected to know them. Sentences can be fact filled when the audience is highly professional and highly motivated. However, when your readers do not share your motivation, profession, and enthusiasm, you should slow your pace and make your prose less dense. In technical writing, you have to know your audience as well as your objectives and adapt your style and material to both.

The Attributes of Good Technical Writers

To write clear and effective reports, you build upon the natural talents you have in communicating ideas to others. How can you build successfully? What skills, characteristics, and attitudes are of most value to the technical writer? From experience, we can summarize some of the major attributes that will stand you in good stead:

- Be reasonably methodical and painstaking. Plan your work for the day and for the rest of the week. Look up from time to time to take stock of what you and others are doing, so that you do not squander your time and energy on minor tasks that should be put off or dispensed with altogether. File your correspondence. Keep at your desk the supplies you need to do your work. Keep a clear head about ways and means for accomplishing your purpose.
- Be objective. Try not to get emotionally attached to anything you have written; be ready to chuck any or all of it into the wastebasket. While reading your own prose or that of your colleagues, do not ask whether you or they are to be pleased but whether the intended audience will be pleased, informed, satisfied, and persuaded.
- In your research, keep in mind that most of what you do will eventually have to be presented in writing. Do your work so that it will be honestly and effectively reportable. Keep a notebook, a computer journal, or a deck of note cards. Record what you do and learn.
- Remind yourself frequently that *clarity* is your most important attribute. Until the sense of a piece of writing is made indisputably clear, until the intended reader can understand it, nothing else can profitably be done with it.
- As someone who writes, understand that writing is something that can be learned, even as chemistry, physics, and mathematics can be. The rules of writing are not as exact as those of science, but they can never be thrown overboard if you are to bring your substance home to your reader.

One writer, who knew well the nature and substance of technical writing, summed up the way to be successful with three imperatives that underlie much of this book:

1. Know your reader.
2. Know your objective.
3. Be simple, direct, and concise.[4]

The Qualities of Good Technical Writing

Because the qualities of good technical writing vary, depending upon audience and objective, we cannot offer you a list that applies equally to everything you write. However, some qualities are apparent in good technical writing:

Good technical writing...

- Arrives by the date it is due.
- Is well designed. It makes a good impression when it is picked up, handled, and flipped through.
- Has the necessary preliminary or front matter to characterize the report and disclose its purpose and scope.
- Has a body that provides essential information and that is written clearly without jargon or padding.
- When appropriate, uses tables and graphs to present and clarify its content.
- Has, when needed, a summary or set of conclusions to reveal the results obtained.
- Has been so designed that it can be read selectively: for instance, by some users, only the summary; by other users, only the introduction and conclusions; by still other users, the entire report.
- Has a rational and readily discernible plan, such as may be revealed by the table of contents and a series of headings throughout the report.
- Reads coherently and cumulatively from beginning to end.
- Answers readers' questions as these questions arise in their minds.
- Conveys an overall impression of authority, thoroughness, soundness, and honest work.

Beyond all these basic characteristics, good technical writing is free from typographical errors, grammatical slips, and misspelled words. Little flaws distract attention from the writer's main points.

A Day in the Life of Two Technical Writers

To summarize, let us describe two representative writers, whom we shall identify as Marie Enderson and Ted Freedman.

Marie Enderson: Computer Specialist and Occasional Technical Writer

Marie has a bachelor's degree in engineering technology. She works in the information services division of a small electronics company that employs some 400 people. Marie has been with the company for a little over a year. Since her childhood, she has been recognized as a whiz at mathematics. In college, she was drawn to the use and design of computing systems. Her major responsibility is to provide technical support for computer systems users in the company.

Marie's first project with the company was a design for an automated system for the shipping department. She interviewed the supervisors and workers in the department to establish the department's needs. She then matched the needs to available off-the-shelf equipment and programs and designed a system to automate much of the department's work. After finishing her design, she had to prepare a written report and oral briefing describing it for the shipping department and her boss. She had a ghastly time the next two weeks. She found, as do many novices at writing, that she knew what she wanted to say but not where or how to say it. The 10-page report did somehow get written and, after a thorough overhaul by Ted Freedman (whom we'll meet next), was presented. Her oral report was a summary of the written report, and it was well received. Her system design was accepted and will be implemented in several months.

Marie's first experience with on-the-job technical writing taught her four important things: (1) An engineer is not simply a person whose only product is a new design or a gadget that works; (2) things that go on in your head and hands are lost unless they are recorded; (3) writing about what you have thought and done is a recurring necessity; and (4) technical writing, strange and difficult as it may seem at first, is something that can be learned by anyone of reasonable intelligence and perseverance.

Marie's present project is a set of instructions for the accounting department to help them use an automated system that was installed over a year ago. Marie's predecessor had installed the equipment and furnished the accountants a set of the manuals produced by the computer and program manufacturers. The manuals are well written, but because they are written by different manufacturers for a general audience, they do not integrate the components of the system in a way meaningful to the accountants. Marie has studied the system and interviewed the users to determine their needs. She has drafted a 20-page booklet that supplements the manufacturers' manuals and shows the accountants how to use the new equipment and programs in their work. She has sent the draft to Ted Freedman for his comments.

Ted Freedman: Technical Writer and Company Editor

Ted Freedman was hired three years ago by the company as a technical writer–editor. He holds a bachelor's degree in technical communication. His office is a sparsely furnished cubicle down the hall from the publications and mailing departments. His office possessions include an old typewriter, a brand-new personal computer and printer, a four-foot shelf of dictionaries and reference manuals, and an extra-large wastepaper basket.

At 8:45 this morning Ted is scheduled for a project review session in the company auditorium. He arrives at the auditorium with five minutes to spare. For the next hour he studies flip charts, slide projections on the huge screen, chalk-and-blackboard plans for company reorganization (minor), and staffing proposals for three new projects totaling $578,400. From the platform, Chief Scientist Muldoon requests that Ted develop research timetables and preview reporting needs.

At 10:20 he meets with a commercial printer to examine the artwork and layout for a plush report the company is preparing for a state commission. The work looks good but needs a little typographical variety, he suggests.

At 12:55, back from lunch in the company cafeteria, Ted glances over the memos that collected on his desk during the morning—nothing urgent. Then he opens the manila envelope lying in his mail rack. In it is a computer disk that contains Marie's instructional booklet and a printout of the booklet.

At 1:30 he calls Marie and arranges for a meeting at 3:00 so that they can run through the draft together. In the meantime he looks over the printout. He notices some computer jargon. He is pretty sure the accountants would not have a clear idea of the distinction between Standard Generalized Markup Language (SGML) and American Standard Code for Information Interchange (ASCII). He circles both phrases. Reading on, he finds a spot where the text should be supported by a graphic. He makes a note of it in the margin. He realizes that the booklet would be more accessible to the reader with more headings in it. He puts Marie's disk into his word processor and, scrolling through her text on the screen, inserts headings that fit her arrangement and material. Thus, the afternoon wears on.

At 3:00, Marie arrives and the two confer, make changes, and plan later alterations in the draft. As before, they work amicably together. They intersperse their writing and editing with an occasional trip to the water cooler, a chat with a department head, and a trip to the library to consult a specialized reference work.

Ted is good at his work and considered to have a great future with the company.

Marie and Ted are roughly representative of many thousands of technical writer–editors, most of them, like Marie, part-time as the need arises. To gain a more rounded understanding of their duties and behavior, we would have to pay them many additional visits; however, certain things are evident even from this brief visit. Like most writers on the job, they work in collaboration with others. Also, much of the time they are not writing at all, in the popular sense. Some of the time they are simply listening hard to what people are saying to one another—trying to clarify, simplify, and translate into other terms. A generous portion of their time is spent on tasks that have little direct connection with writing but eventually provide grist for the writing mill. The techniques, tools, and processes that writers such as Ted and Marie need to accomplish their work are the subject matter of this book.

Exercises

1. As your instructor directs, bring to class one or more documents that you believe to be technical. In what respects is the writing technical? Subject matter? Purpose? Tone? Specialized vocabulary? How has the writer used numbers, formulas, tables, and graphs? Are there headings and transitional features that guide readers through the document? Is it easy to scan the document and select certain parts of it for more intensive reading?
2. Rewrite a brief paragraph of technical prose (perhaps a document submitted in Exercise 1) to substantially lower its technical level. Explain what you have done and why.
3. With the help of *Ulrich's International Periodicals Directory* and a professor in your field, find several periodicals in that field. Examine one or more copies of them. In what ways and to what extent does your examination of such periodicals confirm or change your first impressions of technical writing?
4. On a two-column page, list your present assets and limitations as a technical writer.

5. Turn to the job advertisements section of a large metropolitan newspaper such as the Sunday *New York Times.* What advertisements for technical writers do you find? What qualifications are demanded of them?
6. Talk with a professional person—if possible, one in a field you would like to enter. Ask how much writing he or she does and what kinds. Ask how much importance is attached to good writing on the job. Write a short report of what you learn for your instructor.

PART I
BASICS

In Part I, the emphasis falls on the composing process. Chapter 2, "Composing," discusses how to analyze a writing situation, and how to discover the information you will need, arrange it, write it, revise it, and edit it. With this chapter begins the emphasis on audience that is central to this book. Chapter 3, "Writing Collaboratively," deals with how to compose technical writing in collaboration with others. Chapter 4, "Writing for Your Readers," discusses some of the ways you can adapt your writing to various audiences. Finally, Chapter 5, "Achieving a Readable Style," shows you how paying attention to elements of your style at the paragraph, sentence, and language level can make your writing more readable.

CHAPTER 2 COMPOSING

The composing process is similar to all high-level reasoning processes in that we do not understand it completely. Most researchers in artificial intelligence—the use of computers to replicate human reasoning—have concluded that no general rule of reasoning works for all problems. As one of them put it, "The human brain ... has an incredibly large processing capacity, much greater than several Cray computers [one of the new generation of supercomputers], and it is beyond our understanding in its ability to connect, recall, make judgments, and act. Thus, all experiments to discover a generalized problem solving system were, in a practical sense, failures."[1] Another researcher put it more bluntly: "There is no reason but hubris to believe that we are any closer to understanding intelligence than the alchemists were to the secrets of nuclear physics." [2]

Just how complex is the human brain that is "much greater than several Cray computers"? As one authority points out, "There are more neurons in the human brain than stars in the Milky Way—educated estimates put the number of neurons at about 10^{12} or one trillion. Each of those cells can 'talk' to as many as 1,000 other cells, making 10^{15} connections."[3] Given that level of complexity and those kinds of numbers, no one can map out completely how any complex, high-level, problem-solving process works. And the composing process is precisely that: a complex, high-level, problem-solving process.

Since classical times we have understood some things about the composing process. Aristotle, for example, recognized the wisdom of taking one's audience into account. In recent years, empirical research has revealed additional useful facts about the process. What we tell you in this chapter is based upon those classical concepts that have stood the test of time and modern research. We don't pretend to have all the answers, or even that all our answers are right for you. But we can say that the process we describe draws upon the actual practices of experienced writers, and it works for them.

For most skilled and experienced writers, the composing process breaks up into roughly five parts. The first part involves **situational analysis,** that time when you're trying to bring a thought from nowhere to somewhere. It's a time when you think about such things as your audience, your topic, and your purpose. In the second stage, you "discover" the material you need to satisfy your purpose and your audience. That **discovery** process may go on completely within the trillion cells of your brain or, as is often the case in technical writing, in libraries, laboratories, and workplaces as well.

When the discovery stage is almost complete, you pass into a stage where you **arrange** your material. That is, before writing a draft, you may rough out a plan for it or even a fairly complete outline.

With your arrangement in hand, you are ready for the fourth part of the composing process, the **drafting** and **revising** of your document. For many competent writers, drafting and revising are separate steps; for others, they are almost concurrent.

In the final stage of the writing process, you **edit** your work to satisfy the requirements of standard English and proper format.

Time spent on these five parts is usually not equal. Situational analysis, discovery, and arrangement for a complicated piece of work may take 80% or more of the time you spend on the project. For an easy piece of routine writing, these first three stages may take a few minutes, and drafting and revising may take up the bulk of the time. Some situations call for careful, scrupulous editing; others do not.

The process is often not linear. If the drafting bogs down, you may have to return to the situational analysis stage to resolve the problem. Drafting and revising may alternate as you write for a while, then stop to read and revise. But, in rough outline, what we have described for you is the competent writer's composing process. Throughout this book, we frequently deal with the process. We remind you again and again of the needs of your audience and provide you ways to discover material to satisfy different purposes and topics. In the rest of this chapter we provide some strategies you can use to develop a competent writing process of your own. Because any part of the process can be done in cooperation with others, we provide information on how to write collaboratively (Chapter 3). Also, because technical writers must be ethical, we provide a section in this chapter on **ethical considerations.**

Situational Analysis

In this section we discuss situational analysis, dealing first with topic and purpose, and then with audience and persona.

Topic and Purpose

The topics and purposes of technical writing are found in the situations of technical writing. The topics are many. You may have a mechanism or process to explain; that is your topic. You may have to define a term or explain a procedure. You may have to report the results and conclusions of a scientific experiment or a comparison shopping study. New research has to be proposed. Delayed work has to be explained. All these and many more are the topics of technical writing.

Although the topics of technical writing are varied, the purposes are more limited. Generally, your purpose is either to inform or to argue. Most topics can be handled in one of these two ways, depending upon the situation. Often, you are simply informing. For example, the situation may call for you to describe a mechanism so that someone can understand it. As you will see in Chapter 7, "Defining and Describing," mechanism description will often call for you to divide the mechanism into its component parts and then describe these parts, perhaps as to size, shape, material, and purpose. As another example, you may have to define a term from your discipline. In your definition, you may tell what category the thing defined belongs to and what distinguishes it from other members of the same category. You may give an example of the thing described.

On the other hand, when dealing with your mechanism or definition, you may really be mounting an argument. You may not merely be describing a mechanism; you may be attempting to demonstrate its superiority to other mechanisms of the same type. To do so, you'll need to argue, perhaps by showing how your mechanism is more economical and easier to maintain than other mechanisms. In the same way, you may not simply be defining a term; you may be arguing that your definition is more comprehensive or more correct than previous definitions of the same term.

Be sure to have your topic and purpose in hand before you proceed on in your writing project. It's good practice to write them down, something like this:

> I will define alcoholism in a way that reflects recent research. Further, I will demonstrate that my definition, which includes the genetic causes of alcoholism as well as the environmental ones, is more complete and accurate than definitions that deal with environmental causes alone.

Will the topic and purpose change as you proceed with your project? That depends on the situation. Frequently, the situation may call for you to stick closely to a narrow topic and purpose: *We have to explain to our clients our progress (or lack of progress) in installing the air conditioning system in their new plant.* Or, in another typical situation, *We have to provide instructions for the bank tellers who will use the computer consoles we have installed at their stations.* Although the way you handle such topics and purposes is subject to change as you explore them, the topics and purposes themselves really are not subject to change. On the other hand, the situation may call for you to explore a topic, perhaps the effect that the rising age of the American population may have on the restaurant business. Although you have defined your topic well enough to begin your exploration, the precise topic and the purpose may have to wait until you discover more information about your subject.

Audience and Persona

Writers make important decisions about content and style based upon consideration of the audience and the persona the writer wants to project. **Persona** refers to the role the writer has or assumes when writing. It relates to, among other things, the position of the writer and his or her relationship to the audience and the situation. For example, a bank lending officer might assume one persona when writing to a loan applicant and a different persona when writing to a supervisor to justify a loan that has been made.

Professional people consider both audience and persona seriously when composing, as this quote from a hydrology consultant at an engineering firm indicates:

> We write about a wide range of subject matters. Some things are familiar to a lay audience. Most people can understand a study about floods. They can understand a study that defines a 100-year flood plain. They can imagine, say, water covering a street familiar to them. But other subjects are very difficult to communicate. We work with three-dimensional models of water currents, for example, that are based upon very recondite hydrolic movements. We also have a wide audience range. Some of our reports are read by citizen groups. Sometimes we write for a client who has a technical problem of some sort and is only interested in what to do about it. And sometimes we write for audiences with high technical expertise like the Army Corps of Engineers. Audiences like the Army Corps expect a report to be written in a scientific journal style, and they even want the data so they can re-analyze it. A lot of times the audience is mixed. A regulatory agency may know little about the subject of one of our reports, but they may have a technically trained person on the staff who does. In any case, we must understand what it is that the client wants, and we must be aware of what he knows about the subject. We must convince clients that we know what we are doing. We depend upon return business and word-of-mouth reputation, and we must make a good impression the first time. Much of the technical reputation of this company rides on how we present ourselves in our technical reports.[4]

Here are some questions you need to ask about your audience and persona when you are preparing to write.

What Is the Level of Knowledge and Experience of Your Readers? In technical writing, the knowledge and experience your readers possess are key factors. Do your readers understand your professional and technical language? If they do, your task is easier than if they do not. When they do not, you have to be particularly alert to your word choice, choosing simpler terms when possible, defining terms when simpler choices are not possible. It goes beyond word choice. There are whole concepts that a lay audience may not have. Geologists, for example, thoroughly understand the concept of plate tectonics and can assume that geologists in their audience understand it equally well. When addressing a lay audience, however, the geologist writer would be wise to assume little understanding of the concept. If the geologist wishes to use the concept, he or she will have to take time to explain it in a way that the audience can grasp.

What Is the Reader's Point of View? Point of view relates to the reader's purpose and concerns. Suppose that you are writing about a procedure. People may read about procedures for many reasons. In one case, the reader may wish to perform the procedure. In another, the reader may have to make a decision about whether to adopt the procedure. In yet another, readers may simply want generalized information about the procedure, perhaps because they find it interesting.

Each case calls for a different selection of content and a different style. Readers wishing to perform the procedure need a complete set of step-by-step instructions. The decision maker needs to know by what criteria the procedure has been evaluated and why, under these criteria, it is a better choice than other procedures. Readers for interest want the general concept of the procedure explained in language they can understand.

What Is Your Relationship to the Reader? Are the readers your bosses, clients, subordinates, peers, or students? If you are a public employee, are you writing to a taxpayer who contributes to your salary? Writers in the workplace, when interviewed about how they write, reveal that they pay a good deal of attention to the effect of such relationships on tone, as these quotations demonstrate:[5]

- Writing to my boss, I try to pinpoint things a little more.
- When you have something as personal as a phone call or a conversation back and forth ... I feel free to use "I" rather than "we."
- We always want them to realize they can call on us if they have any questions.
- This [referring to a statement] is a bit more on a personal level ... The other [statement] is much too formal.
- Just to say "Send his address," would, I think, be a little too authoritarian.

The roles writers find themselves in also affect their choice of content. Imagine the difference in approach between a Chevrolet sales representative trying to sell a fleet of Chevrolets to a company, and a young executive of the same company reporting to his or her superiors that the results of a feasibility study demonstrate Chevrolets to be the best purchase. In the first instance, the sales representative is likely to be more enthusiastic about Chevrolets than other makes. The decision makers would expect and understand such enthusiasm and would allow for it. In the second instance, the decision makers will expect a more balanced approach from the young executive.

What Is Your Reader's Attitude about What You Are Going to Say? Audiences can be suspicious and hostile. They may be apathetic. Of course, they may be friendly and interested. Their attitude should affect how you approach them. If you have an unfriendly audience, you must take particular care to explain your position carefully in language that is understandable but not patronizing. You may need more examples than you would with a friendly audience. A friendly audience may be persuaded with less information. With a friendly audience, you may present your conclusions first and then support them. With an unfriendly audience, it's a sound idea to present your support first and then your conclusions.

Readers may have attitudes about the language you use. For example, public health officials have had a difficult time expressing how to avoid exposure to AIDS. Such advice, to be effective, must refer very explicitly to sexual practices. Newspapers have had to change their usual practice to allow such language to be printed, and some readers have found the language offensive. In most cases, the interest in AIDS prevention has won out over reader sensibilities, but the problem illustrates well the social context of audience analysis.

What Persona Do You Wish to Project? If you have read many scientific journals, you have probably noticed that they have a certain tone about them, a tone to which words such as *objective, formal,* and *restrained* readily apply. Scientists, to find acceptance in such journals, must adopt such a tone. A breezy, light journalistic style, though it might be just as clear, would not be acceptable. In the same way, bankers must present themselves in a careful, formal way. We're not likely to give our money for safekeeping into the hands of someone who comes on like a television used-car dealer. Young executives writing to their bosses are likely to be a bit deferential. The bosses, in turn, want to sound firm but reasonable and not authoritarian. What has come to be called "corporate culture" plays a role in the persona a writer may adopt. In writing, you must project the values and attitudes of the organization you work for. To do so, you may look over past correspondence and reports to see what practices have been used, what sort of tone writers in the organization have adopted.

Taking on a persona when you write is something like taking on a persona when you dress. The student who exchanges his blue jeans and running shoes for a business suit and leather shoes when he reports for a job interview is slipping out of one persona into another. The teacher who exchanges a comfortable sweater and skirt for a businesslike dress when she leaves the classroom to consult in industry is exchanging one persona for another. It's a common enough experience in life, and you should not be surprised to find such experiences in writing situations. Both dressing and writing have their own rhetorics. However, don't misinterpret anything we have said as a rationale for being obscure or jargony. You should be clear no matter what persona you adopt.

Discovery

At some point in your writing process you must "discover" the material you will use in your writing. Discovery is teasing out of your mind the information you will use and modify to meet the needs of your topic, purpose, audience, and persona. Discovery is making connections. It's the putting together of two pieces of information to create a third piece that didn't exist before the connection. A mind that is well stocked with information will probably be successful at discovery. Those trillion neurons need something to work with; the more you read, observe, and experience, the better writer you are likely to be.

Of course, all the material you need may not be in your mind when you begin. Discovery includes using libraries and laboratories to fill in the gaps in your knowledge. You may also use interviews, on-site inspections, letters of inquiry, and the like to gather information. The techniques we discuss here will enable you to explore your own mind.

Brainstorming

In brainstorming you uncritically jot down every idea about a subject that pops into your head, without thought of organization. The key to successful brainstorming is that you do not attempt to evaluate or arrange your material at the first stage. These processes come later. Evaluation or arrangement at the first critical stage may cause you to discard an idea that could prove valuable in the context of all the ideas that the brainstorming session produces. Also, avoiding evaluation at this point prevents the self-censorship that often blocks a writer.

Because brainstorming is a fairly painless process, it's frequently a good device to break down the normal resistance most of us have to hard thinking. It can result in your writing down a good deal more information than you ever thought you possessed. It can quickly reveal holes in your knowledge, which can be filled with information you gather later.

Using Arrangement Patterns for Discovery

Although you do not arrange your material in the discovery stage, you can use familiar arrangement patterns as aids in discovering your material.[6] For example, suppose your purpose is to describe a procedure for a reader who wishes to perform the procedure. If you were familiar with writing instructions, as you will be after reading Chapter 16, "Instructions," you would know that a set of instructions often lists and sometimes describes the tools that must be used to perform the procedure. Furthermore, instructions describe the steps of the procedure, normally in chronological order. Knowing what is normally required for a set of instructions, you can brainstorm your material in a more guided way.

You can begin by writing down the tools that will be needed for the procedure. Think about what you know about your audience. Are they experienced with the tools needed? If so, simply list the tools. If they are not experienced, jot down some information they'll need to use the tools properly.

When you are done with the tools, write down the steps of the procedure. Keep your readers in mind. Are there some steps so unfamiliar to your readers that you need to provide additional information to help them perform the steps? If so, list what that information might be.

As in brainstorming, in very little time you can get information out where you can see it. Also, as in brainstorming, if there are gaps in your knowledge, you can discover them early enough to fill them.

Another task frequently encountered in technical writing is arguing to support an opinion. In discovering an argument, you can begin by stating that opinion clearly, perhaps something like "Women should get equal pay for equal work." Next, you can turn your attention to the subarguments that might support such an opinion. For example, first, you would have to establish that in many instances women are not getting equal pay for equal work, and, therefore, a problem really does exist. Then you might think of a philosophical argument: Ethically, women have a right to equal pay for equal work. You might think of an economic argument: Women's needs to support themselves and their families equal those of men. And so forth. As you think about subarguments, you will begin to think about the information you will need to support them. Some of it you may have; some you may need to research. The very form and needs of your argument serve as powerful tools to help you discover your material.

Other Successful Discovery Techniques

Most experienced writers develop their own discovery techniques. In the workplace, writers often use past documents of a similar nature to jog their minds. Many professional writers keep journals that they can mine for ideas and data. Scientists keep laboratory notebooks that can be invaluable when it's time to write up the research. In the workplace, people talk to each other to discover and refine ideas.

Asking questions, particularly from the reader's point of view, is a powerful discovery technique. Suppose you were describing the use of computers for word processing. What questions might the reader have? *Do I have to be an expert typist to use a word processor? What are the advantages of word processing? The disadvantages? How do I judge the effectiveness of a word processor? What do word processors cost? Will word processing make writing and revising easier?* As you ask and answer such questions, you are discovering your material.

When you have established your topic and purpose, analyzed your audience, and discovered your material, it's time to think about arrangement.

Arrangement

When you begin your arrangement, you should have a good deal of material to work with. You should have notes on your audience, purpose, and persona. Your discovered material may take various forms. It may be a series of notes produced by brainstorming or other discovery techniques. You may have cards filled with notes taken during library research or notebooks filled with jottings made during laboratory research. You may have previous reports and correspondence on the topic you are writing about. You may have ideas for graphs and tables to use in presenting your material. In fact, you may have so much material that you do not know where to begin.

You can save yourself much initial chaos and frustration if you remember that certain kinds of reports (and sections of reports) have fairly standard arrangements. The same arrangement patterns that helped you discover your material can now serve you as models of arrangement. For instance, you might divide your subject into a series of topics, as we have done with the chapters of this book. If you're describing a procedure and know that your readers wish to perform that procedure, you may use a standard instruction arrangement: introduction, tool list and description, and steps of the procedure in chronological order. If you are arranging an argument, you have your major opinion, often called the major thesis, and your subarguments, often called minor theses. You'll probably want to consider the strength of your minor theses when you arrange your argument. Generally, you want to start and finish your argument with strong minor theses. Weaker minor theses you'll place in the middle of your argument.

Documents such as progress reports, proposals, and empirical research reports have fairly definite arrangements that we describe for you in Part IV, "Applications." Not all the arrangements described will fit your needs exactly. You must be creative and imaginative when using them. But they do exist. Use them when they are appropriate.

How thorough you are at this stage depends upon such things as the complexity of the material you are working with and your own working habits. Simple material does not require complicated outlines. Perhaps nowhere else in the writing process does personality play such a prominent role as it does at the arrangement stage. Some people prepare fairly complete arrangement patterns; others do not.

An article by Blaine McKee, "Do Professional Writers Use an Outline When They Write?" explores the outlining practices of professional technical writers.[7] Dr. McKee found that only 5% of the writers used no outline at all, and only 5% reported using an elaborate sentence outline. Most reported using some form of topic outline or a mixture of words, phrases, and sentences. Most kept their outlines flexible and informal, warning against getting tied to a rigid outline too early. But all but 5%

did feel the need to think through their material and to get some sort of arrangement pattern down on paper before beginning the first draft of a report.

Most experienced writers are usually thorough but informal in writing down their arrangement patterns. However, if you need a formal outline and need instruction in preparing one, look in Part V under "Outlining."

In technical writing, graphs and tables are important techniques for presenting material. It's not too early to think about them while you are arranging your material. For help in planning and selecting graphs and tables, see Chapter 11, "Graphical Elements."

Drafting and Revising

When you have finished arranging your material, you are ready to draft and revise your report. Keep in mind that writing is not an easy mechanical job. But we do give you suggestions that should make a tough job easier.

The Rough Draft

Writing a rough draft is a very personal thing. Few writers do it exactly alike. As you have seen, most write from a plan of some sort; a few do not. Some write at a fever pitch; others write slowly. Some writers leave revision entirely for a separate step. Some revise for style and even edit for mechanics as they go along, working slowly, trying to get it right the first time. All we can do is describe in general the practices of most professional writers. Take our suggestions and apply them to your own practices. Use the ones that make the job easier for you and revise or discard the rest.

Probably our most important suggestion is to begin writing as soon after the prewriting stage as possible. Writing is hard work. Most people, even professionals, procrastinate. Almost anything can serve as an excuse to put the job off: one more book to read, a movie that has to be seen, anything. The following column by Art Buchwald describes the problem of getting started in a manner that most writers would agree is only mildly exaggerated.

> MARTHA'S VINEYARD—There are many great places where you can't write a book, but as far as I'm concerned none compares to Martha's Vineyard.
>
> This is how I managed not to write a book and I pass it on to fledgling authors as well as old-timers who have vowed to produce a great work of art this summer.
>
> The first thing you need is lots of paper, a solid typewriter, preferably electric, and a quiet spot in the house overlooking the water.

You get up at 6 in the morning and go for a dip in the sea. Then you come back and make yourself a hearty breakfast.

By 7 A.M. you are ready to begin Page 1, Chapter 1. You insert a piece of paper in the typewriter and start to type "It was the best of times ..." Then you look out the window and you see a sea gull diving for a fish. This is not an ordinary sea gull. It seems to have a broken wing and you get up from the desk to observe it on the off chance that somewhere in the book you may want to insert a scene of a sea gull with a broken wing trying to dive for a fish. (It would make a great shot when the book is sold to the movies and the lovers are in bed.)

It is now 8 A.M. and the sounds of people getting up distract you. There is no sense trying to work with everyone crashing around the house. So you write a letter to your editor telling him how well the book is going and that you're even more optimistic about this one than the last one which the publisher never advertised.

It is now 9 A.M. and you go into the kitchen and scream at your wife, "How am I going to get any work done around here if the kids are making all that racket? It doesn't mean anything in this family that I have to make a living."

Your wife kicks all the kids out of the house and you go back to your desk ... You look out the window again and you see a sailboat in trouble. You take your binoculars and study the situation carefully. If it gets worse you may have to call the Coast Guard. But after a half-hour of struggling they seem to have things under control.

Then you remember you were supposed to receive a check from the *Saturday Review* so you walk to the post office, pause at the drugstore for newspapers, and stop at the hardware store for rubber cement to repair your daughter's raft.

You're back at your desk at 1 P.M. when you remember you haven't had lunch. So you fix yourself a tuna fish sandwich and read the newspapers.

It is now 2:30 P.M. and you are about to hit the keys when Bill Styron calls. He announces they have just received a load of lobsters at Menemsha and he's driving over to get some before they're all gone. Well, you say to yourself, you can always write a book on the Vineyard, but how often can you get fresh lobster?

So you agree to go with Styron for just an hour.

Two hours later with the thought of fresh lobster as inspiration, you sit down at the typewriter. The doorbell rings and Norma Brustein is standing there in her tennis togs looking for a fourth for doubles.

You don't want to hurt Norma's feelings so you get your racket and for the next hour play a fierce game of tennis, which is the only opportunity you have had all day of taking your mind off your book.

It is now 6 P.M. and the kids are back in the house, so there is no sense trying to get work done any more for that day.

So you put the cover on the typewriter with a secure feeling that no matter how ambitious you are about working there will always be somebody on the Vineyard ready and eager to save you.

But you must begin and the sooner the better. Find a quiet place to work, one with few distractions. Choose a time of day when you feel like working, and go to work. What writing tools should you use? Our tools for the first draft of almost any piece of writing are a yellow, legal-sized pad and a can filled with pencils of 2.5 hardness. Other writers compose on typewriters or word processors. One writer we know insists upon using white paper and a fountain pen filled with blue ink. The moral of all this is that the tools used are a matter of individual choice. There is something a bit ritualistic about writing, and most competent writers insist upon their own rituals.

Where should you begin? Usually, it's a good strategy to begin not with the beginning but with the section that you think will be the easiest to write. If you do so, the whole task will seem less overwhelming. As you write one section, ideas for handling others will pop into your mind. When you finish an easy section, go on to a tougher one. In effect, you are writing a series of short, easily handled reports rather than one long one. Think of a 1,500-word report as three short, connected, 500-word reports. You will be amazed at how much easier this attitude makes the job. We should point out that some writers do prefer to begin with their introductions and even to write their summaries, conclusions, and recommendations (if any) first. They feel this sets their purpose, plan, and final goals firmly in their minds. If you like to work that way, fine. Do remember, though, to check such elements after you have written the discussion to see whether they still fit.

How fast should you write? Again, this is a personal thing, but most professional writers write rapidly. We advise you not to worry overmuch about phraseology or spelling in a rough draft. Proceed as swiftly as you can to get your ideas on paper. Later, you can smooth out your phrasing and check your spelling, either with your dictionary or your word processor's spelling checker. However, if you do get stalled, reading over what you have written and tinkering with it a bit is a good way to get the flow going again. In fact, two researchers of the writing process found that their subjects spent up to a third of their time pausing. Generally, the pauses occurred at the ends of paragraphs or when the writer was searching for examples to illustrate an abstraction.[8]

Do not write for more than two hours at a stretch. This time span is one reason why you want to begin writing a long, important report at least a week before it is due. A report written in one long five- or six-hour stretch reflects the writer's exhaustion. Break at a point where you are sure of the next paragraph or two. When you come back to the writing, read over the previous few paragraphs to help you collect your thoughts and then begin at once.

Make your rough draft very full. You will find it easier to delete material later than to add it. Nonprofessional writers often write thin discussions because they think in terms of the writing time-span rather than the reading time-span. They have been writing on a subject for perhaps an hour and have grown a little bored with it. They feel that if they add details for another half-hour they will bore the reader. Remember this: At 250 words a minute, average readers can read an hour's writing output in several minutes. Spending less time with the material than the writer must, they will not get bored. Rather than wanting less detail, they may want more. Don't infer from this advice that you should pad your report. Brevity is a virtue in professional reports. But the report should include enough detail to demonstrate to the reader that you know what you're talking about. The path between conciseness on one hand and completeness on the other is often something of a tightrope.

As you write your rough draft, indicate where your references will go. Be alert for paragraphs full of numbers and statistics and consider presenting such information in tables and graphs. Be alert for places where you will need headings and other transitional devices to guide your readers through the report. (See Part III, "Document Design.")

Whether your planning has been detailed or casual, keep in mind that writing is a creative process. Discovery does not stop when you begin to write. The reverse is usually true. For most people, writing stimulates discovery. Writing clarifies your thoughts, refines your ideas, and leads you to new connections. Therefore, be flexible. Be willing to revise your plan to accommodate new insights as they occur.

Revision

Some writers revise while they are writing. For them, revision as a separate step is little more than minor editing, checking for misspellings and awkward phrases. For other writers, particularly those who write in a headlong flight, revising is truly rewriting and sometimes even rearranging the rough draft. Naturally, there are many gradations between these two extremes. We, for example, do some revising as we write, but we save most of it for a separate step. Most of our revising occurs when we type from our yellow legal pad manuscript into our word processor. At this stage we pause and ponder and rework our language, content, and arrangement.

Whether you revise while you write or in a separate step, you should be concerned about arrangement, content, logic, style, graphics, and document design. In some situations you may want to show your work to others and seek their advice.

Arrangement and Content In checking your arrangement and content, try to put yourself in your reader's place. Does your discussion take too much for granted? Are questions left unanswered that the reader will want answered? Are links of thought missing? Have you provided smooth transitions from section to section, paragraph to paragraph? Do some paragraphs need to be split, others combined? Is some vital thought buried deep in the discussion when it should be put into an emphatic position at the beginning or end? Have you avoided irrelevant material or unwanted repetitions?

In checking content, be sure that you have been specific enough. Have you quantified when necessary? Have you stated that "In 1994, 52% of the workers took at least 12 days of sick leave," rather than, "In a previous year, a majority of the workers took a large amount of sick leave"? Have you given enough examples, facts, and numbers to support your generalizations? Conversely, have you generalized enough to unify your ideas and to put them into the sharpest possible focus? Have you adapted your material to your audience?

Is your information accurate? Don't rely on even a good memory for facts and figures that you are not totally sure of. Follow up any gut feeling you have that anything you have written seems inaccurate, even if it means a trip back to the library or laboratory. Check and double-check your math and equations. You can destroy an argument (or a piece of machinery) with a misplaced decimal point.

Logic Be rigorous in your logic. Can you really claim that A caused B? Have you sufficiently taken into account other contributing factors? Examine your discussion for every conceivable weakness of arrangement and content and be ready to pull it apart. All writers find it difficult to be harshly critical of their own work, but a critical eye is essential.

Style After you have revised your draft for arrangement and content, read it over for style. (We treat this as a separate step, which it is. But, of course, if you find a clumsy sentence while revising for arrangement and content, rewrite it immediately.) Use Chapter 5, "Achieving a Readable Style," to help you. Rewrite unneeded passive voice sentences. Cut out words that add nothing to your thought. Cross out the pretentious words and substitute simpler ones. If you find a cliché, try to express the same idea in different words. Simplify; cut out the artificiality and the jargon. Be sure the diction and sentence structures are suitable to the occasion and the audience. Remember that you are trying to write understandably, not impressively. The final product should carry your ideas to the reader's brain by the shortest, simplest path.

Graphics Much technical information is presented in tables and graphs. When dealing with content that has visual components, you should probably present at least some of that content graphically. When you have numerous statistics, particularly statistics that you are comparing to each other, you probably should display them in tables or graphs. (For help in such matters, see Chapter 11, "Graphical Elements.")

Document Design Good document design—the use of tables of content, headings, the right typeface, proper spacing, and so forth—are integral to good technical writing. We offer detailed guidance in this area in Chapter 9, "Document Design" and Chapter 10, "Design Elements."

Sharing Your Work In actual workplace situations, writers often share their drafts with colleagues and ask for their opinions. Often, someone who is not as close to the material as the writer can spot flaws far more quickly than can the writer. As you'll see when we discuss revising and editing with word processing, you may also share your work with your personal computer.

When you are writing instructions, it's an excellent idea to share an early draft with people who are similar in aptitude and knowledge to the people for whom the instructions are intended. See if they can follow the instructions. Ask them to tell you where they had trouble carrying out your instructions or where poor vocabulary choice or insufficient content threw them off track.

Editing

Editing is a separate step that follows drafting and revising. It's the next-to-final step before you release your report to its intended audience. When drafting and revising a manuscript, you may have to backtrack to the discovery or the arranging stage of the process. But when you are editing, it's either because you are satisfied with your draft or because you have run out of time. In the editing stage you make sure your report is as mechanically perfect as possible, that it meets the requirements of standard English and whatever format requirements your situation calls for. If you are working for a large organization or the government, you may have to concern yourself with things such as stylebook specifications, distribution lists, and code numbers.

Checking Mechanics

Begin by checking your mechanics. Are you a poor speller? Check every word that looks the least bit doubtful. Some particularly poor spellers read their draft backwards to be sure that they catch all misspelled words. Develop a healthy sense of doubt and use a good dictionary or the spelling checker of your word processing program. Do you have trouble with subject-verb agreement? Be particularly alert for such errors. In Part V, we have provided you with a handbook that covers some of the more common mechanical problems. A word processor can help by providing a check for some of the errors a computer program is able to detect.

Checking Documentation

When you are satisfied with your mechanics, check your documentation. Be sure that all notes and numbers match. Be sure that you have used the same style throughout for your notes. For help in documentation see Chapter 10, "Design Elements."

Checking Graphics

Check your graphics for accuracy, and be sure you have mentioned them at the appropriate place in the report. Are your graphics well placed? If they are numbered, be sure that their numbers and the numbers you use in referring to them match.

Checking Document Design

In your drafting and revision, you should have made sure that your document's design made your document readable and accessible for your readers. In your editing, check for more mundane but nevertheless important things. For instance, is your table of contents complete? Is it accurate? Does it match the headings you have used? Do your headers and footers accurately describe your material?

Again, word processing makes all these tedious little tasks easier. With the word processor you don't have the agony of retyping an entire page to correct for one or two errors. When you are satisfied you have done all that needs to be done, print or type your final draft. Whether you, a typist, or a word processor types your final draft, proofread it one more time before you turn it over to your audience. The author of a report is responsible for all errors, no matter how the report is prepared.

Revising and Editing with Word Processing

Word processing has eased revision considerably since the time when revision called for writing between the lines and in the margins and even cutting up reports and splicing them together in new arrangements. Word processing offers, in effect, unlimited space for revision, allowing the writer to insert new sentences, paragraphs, and headings at any point in the report. Sentences and paragraphs can be easily moved from one place to another. Writers can get a cleanly typed printout at any stage of the writing and revising process. Because a clean text is easier to read than one full of arrows and written-over sentences, writers working with word processing can check their arrangement, content, logic, and style more easily than ever before. Because the machine does the retyping that even small changes sometimes entail, writers are far less reluctant to make needed changes than in the past.

To further help the writer with editing and revising, spelling checkers and grammar and style checkers are available.

Spelling Checkers

You may have a spelling checker in your word processing program or one that you have bought separately. Spelling checkers compare every word in your text with the dictionary included in the program. Most spelling checkers have no sense of grammar or usage. They will stop on a correctly spelled word if that word does not happen to be in the program's dictionary, as is true of many technical terms. More importantly, they will not stop at a word that is in the program's dictionary when that word is used incorrectly in context.

A spelling checker won't catch errors like these:

> The student's all came to class today.
>
> They wanted to here your speech.
>
> They wanted to hear you speech.

For a list of words that sound alike but have different spellings and meanings, such as *weather* and *whether*, see the entry for "Spelling Error" in Part V, "Handbook." Use a spelling checker first, but be sure to proofread as well.

Grammar and Style Checkers

You can get programs that check your work for grammar, punctuation, and style. Some will flag passive sentences, long sentences, wordy phrases, double words (such as *and and*), unmatched pairs of quotation marks, and other problems. Some also flag problems with subject–verb agreement, incorrect possessives, and other grammatical faults. Some give your text a "readability rating" according to one or more readability formulas.

Grammar and style checkers can be helpful. They can make you more aware of your writing style. If you tend to write in the passive voice, they'll press you to change to active voice. If you tend to use wordy phrases or unnecessarily long words, you'll see shorter, crisper alternatives.

Use grammar and style checkers with great caution, however. Some current text-analysis programs are too rule-bound to be flexible, and some of the rules may be of doubtful validity. As one authority says, "Syntax writers that can truly evaluate a user's writing style await future breakthroughs in artificial intelligence."[9]

Think about the advice they give you in light of the purpose and audience for your document. Not every passive voice sentence should be rewritten as an active sentence. Not every sentence more than 22 words is too long. Grammar and style checkers work only at the sentence and word level, but the most serious problems with many documents are in their content and overall organization. If you change words and sentences here and there without considering larger issues of content and arrangement, you may actually be making your document less useful and understandable.

Ethical Considerations

Because technical writing often has consequences for large numbers of people, ethical considerations frequently play a role in the writing process.[10] For example, it is sometimes tempting in a feasibility report to soft-pedal results that do not support the recommendation the writer wishes to make. It may seem advantageous in a proposal to exaggerate an organization's ability to do a certain kind of research. A scientist may be too willing to ignore results that do not fit his or her theory and report those that do. Each of these acts is unethical.

Understanding Ethical Behavior

What makes an act unethical? Why should we be ethical? Let us briefly answer those two questions and then offer a few suggestions about how to behave ethically.

What Makes an Act Unethical? Most of us carry around ethical rules in our heads. Most of us, no doubt, would agree that it is unethical to lie, cheat, and steal. Further extended, we would likely agree that it is wrong to make promises we don't intend to keep. Where do such ethical rules come from? In part, they are rules learned at home or in a religion or simply in the rough and tumble of growing up. The loss of friends who catch one in a lie can be a lasting ethical lesson. Philosophers have long attempted theories to support ethical behavior. Most embrace either logic, consequences, or some combination of the two.

Logically, as the 18th-century German philosopher Immanuel Kant proposed, we should not act in a way that we cannot will to be universal behavior. For example, if you make a promise that you have no intention of keeping, you cannot will that to be universal behavior. For if you did, all promises would be worthless, and it would be pointless to make a promise, false or otherwise.

Another group of philosophers, the utilitarians, make consequences their test for ethical behavior. An act should do the greatest good for the greatest number of people or, conversely, create the least amount of evil for the fewest people. For example, causing an industrial plant to clean up its smokestack emissions may be an economic evil for the company and its stockholders but be the greatest good for the many people who must breathe those emissions. Medical scientists who fudge their data may win promotion for themselves but injure unsuspecting people who are mistreated as a result of the deception.

Most of us have sufficient ethical knowledge to act ethically if we want to. Why should we want to?

Why Should We Act Ethically? We don't have to act ethically. We don't have to will that our acts become universal behavior as Kant would have us do. Despite the utilitarians, we don't have to worry about the greatest good for the greatest number of people. We can act in our own selfish interests if we want to. If acting ethically is a voluntary act, why bother?

We can list some pragmatic, nonaltruistic reasons for acting ethically. For one thing, some unethical acts are also illegal. You can end up in prison for stealing or otherwise bilking people of money. Furthermore, organizations that intend to prosper in the long term need to build a reputation for ethical behavior. Professionals must act with integrity to survive in their work environments. Unethical acts can help an organization or an individual initially, but in the long run, they usually do more harm than good.

However, perhaps the most important justification for acting ethically is less obvious than these individual and organizational reasons. Acting ethically is a price we pay for living in a free, civilized society. A nonethical society would either be barbaric or totalitarian. A world without ethics would be a world in which anything goes: murder, theft, rape, pillage, lying, and cheating in all their forms. It would be a society unfit to live in. Conversely, when ethics are lacking, the state, in order to maintain a civilization, would have to have laws restricting all kinds of unethical behavior. In part, because we have unethical people, we live in such a society right now. We do have laws, for example, condemning theft, murder, and insider trading. We would not need environmental laws if every company voluntarily acted in the best interests of the people. But a state that attempted to control all nonethical behavior would be a totalitarian state, in its own way as bad as a barbaric one.

Perhaps, then, the best motivation for acting ethically is that it allows us to live in a civilized society without the heavy hand of government constantly on us.

Recognizing Unethical Writing

Perhaps the first step in writing ethically is to recognize the ways in which people can be unethical when they write.

Using Ambiguous Language In Chapter 5, "Achieving a Readable Style," we discuss ways to write clearly and help your readers to understand you. We urge you to write with precision and to avoid ambiguous language. Unclear writing usually results from a faulty style, but not always. It can result from a deliberate attempt to mislead or manipulate the reader by hiding unfavorable information.

The principal meaning of *ambiguous* is "capable of having two meanings." The word *majority*, for instance, means "the greater part of something" and "a number more than half of the total." Imagine the writer of a feasibility report who wishes to recommend a change in company policy. He takes a survey of all the workers in the company and finds that 50.1% of the 20% who returned his survey favor the change. In his report he writes "A majority of those who returned the survey favor the change." By counting on the ambiguity of *majority*, he makes a stronger case for change than if he reported the actual bare majority the survey revealed. In addition, by not revealing that this "majority" represents only 10% of the company's workers, he further strengthens what is actually rather weak support for his case. He has not lied, but through ambiguity he has certainly misled his audience.

Making False Implications Writers can imply that things are better than they are by manipulating language. For example, a writer answering an inquiry about her company's voltage generator could reply, "Our voltage generator is designed to operate from the heat of Saudi Arabian deserts to the frozen tundra of Greenland." It may be true that the generator was *designed* that way, but if it only *operates* properly between Atlanta and Toronto, the writer has made a false implication without telling an outright lie.

For another example, imagine a mutual fund that led its market in returns for 10 years. In the eleventh year, the original fund manager retires and a new manager takes over. In that year and the next, the fund drops to the bottom tenth of its market in returns. The writer of an advertising brochure for the fund writes the following: "Our fund has led the market for ten of the last twelve years." The writer has avoided an outright lie, but clearly has made an unethical statement.

Manipulating Data In *Honor in Science*, Sigma Xi, the Scientific Research Society, lists three ways scientists can present their results unethically:

- *Trimming:* the smoothing of irregularities to make the data look extremely accurate and precise.
- *Cooking:* retaining only those results that fit the theory and discarding others.
- *Forging:* inventing some or all of the research data that are reported, and even reporting experiments that were never performed.

Only the last of these three manipulations is clearly a lie, but all misrepresent the data and are unethical.

Writing Ethically

We are probably most tempted to write unethically when our own interests or the interests of our organization are at stake. For example, you may be writing a proposal for your research laboratory to do a significant and costly piece of research for a large government agency. A proposal is a sales document, after all. It's sensible practice to cast your laboratory in its best light. But the temptation to go too far is always present. You may be tempted to exaggerate the expertise of the scientists who will carry out the job. Through imprecise language you may hide the deficiencies of your laboratory or overstate its attributes.

On the other hand, you may write unethically simply by not recognizing the consequences of what you have written. A way to bring the consequences of your writing to the foreground is to construct a fault tree diagram at the point in your planning or writing at which you recognize that there are various options open to you. As you construct your fault tree, draw each of your possible options as a branch. List the consequences for each branch, or option. If any of the listed consequences lead to another consequence, draw another branch showing that consequence, and so on, until you have exhausted all reasonable options. Let us illustrate.

Imagine yourself to be a newly graduated civil engineer. You are hired by a land developer to develop plans for streets and sewage disposal for a large parcel of land on which he plans to build 45 houses. In walking the parcel, you discover that about half of it is a waste dump filled with trees and other vegetation covered with several feet of soil. When you draw this to the developer's attention, he tells you that while building other housing developments, he used this parcel of land as a dump. Upon further questioning, he reveals that he has never sought a permit for the dump from the county, so you are dealing with an unauthorized dump. You realize that a dump filled with vegetation may create substantial amounts of highly explosive methane gas. You recognize three possible options you can recommend to the developer:

1. Proceed with the development as planned.
2. Delay the building until the contents of the dump have been removed.
3. Cancel the development plans.

To help yourself sort out the consequences of the actions, you develop the fault tree shown in Figure 2-1.

Your fault tree makes it clear that you cannot ethically recommend option 1. Option 2 looks good, despite some possible negative consequences. You realize you'll need some further work to determine the cost of removing the dump. Option 3 is ethical but probably not cost effective. If the developer chooses option 2 or 3, you have fulfilled your

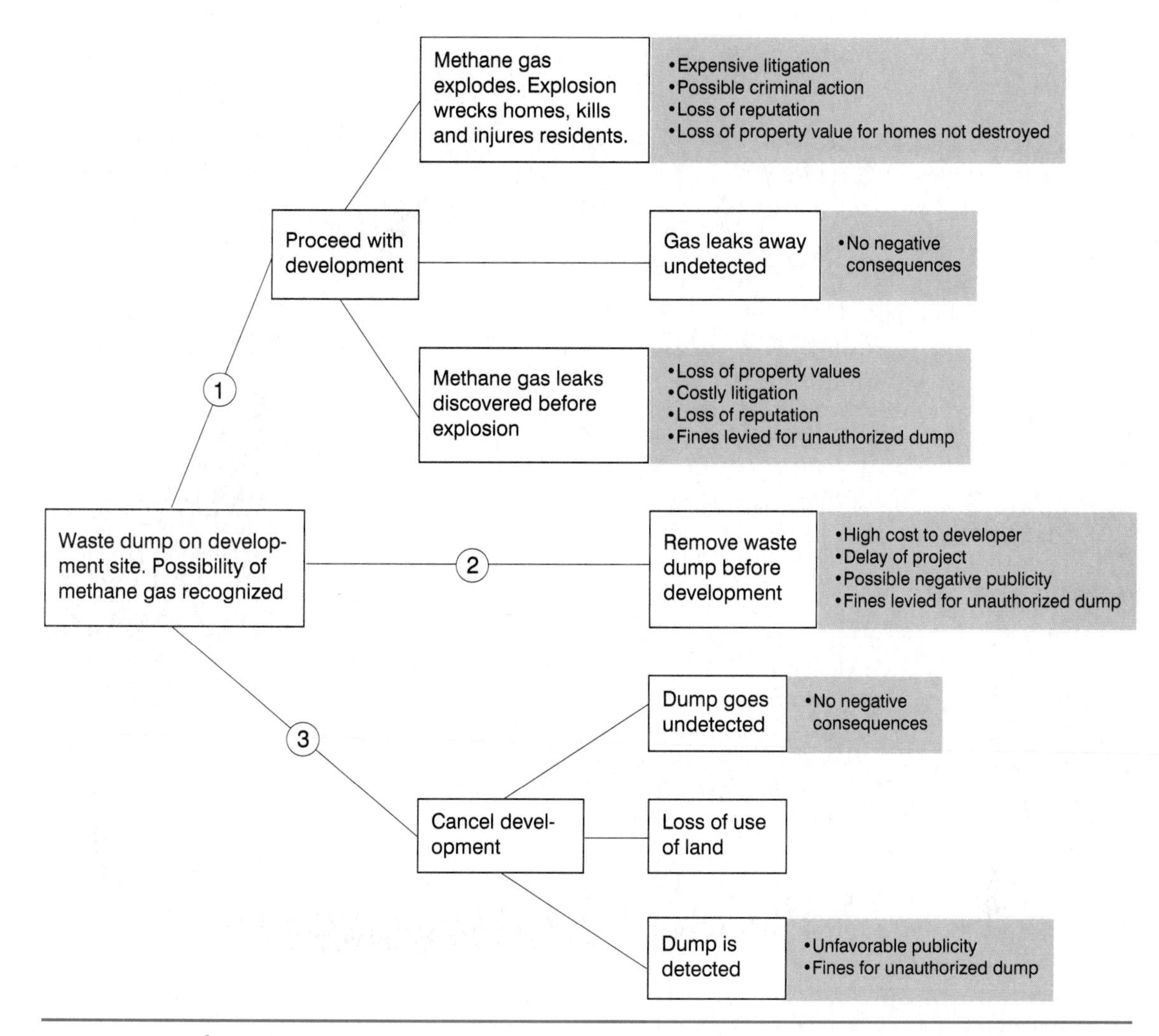

Figure 2–1 Fault Tree

ethical duty. If the developer decides to go ahead with the development, you have another ethical choice. Should you keep a copy of your report to protect yourself but remain quiet, or should you blow the whistle on the developer? Given the possible cost in human misery if the developer goes ahead, there seems to be little choice; you'll have to blow the whistle.

In this section, we have made you aware of some of the ethical difficulties you may encounter when writing and given you ways to deal with them. However, no amount of reading about ethics can make you ethical. To be ethical, you must have a good will, a moral sense, and, frequently, courage.

Planning and Revision Checklists

The following questions are a summary of the key points in Chapter 2, and they provide a checklist when you are composing.

SITUATIONAL ANALYSIS

- What is your topic?
- Why are you writing about this topic? What is your purpose (or purposes)?
- What are your readers' educational levels? What knowledge and experience do they have in the subject area?
- What will your readers do with the information? What is their purpose?
- Do your readers have any expectations as to style and tone? Serious? Light? Formal?
- What is your relationship to your readers? How will this relationship affect your approach to them?
- What are your readers' attitudes about what you are going to say?

DISCOVERY

- What discovery approach can you use? Brainstorming? Using arrangement patterns? Other?
- Are there documents similar to the one you are planning that would help you?
- Do you have notes or journal entries available?
- What questions are your readers likely to want answered?
- Do you have all the information you need? If not, where can you find it? People? The library? Laboratory research?
- What tables, graphs, diagrams, or other graphic aids will you need?

ARRANGEMENT

- Are there standard arrangement patterns that would help you, such as instructions, arguments, or proposals?
- Will you need to modify a standard pattern to suit your needs?
- Do you need a formal outline?
- When completed, does your organizational plan fit your topic, material, purpose, and audience?
- What headings and subheadings will you use to reveal your organization and content to your readers?
- Is everything in your plan relevant to your topic, purpose, and audience?
- If you have a formal outline, does it follow outlining conventions? Are entries grammatically parallel? Is each section divided into at least two parts? Is the capitalization correct? Are the entries substantive?

DRAFTING

- Do you have a comfortable place to work?
- Where in your organizational plan can you begin confidently?
- Where will your graphical elements be placed?

REVISION

- Have you stated clearly and specifically the purpose of the report?
- Have you put into the report everything required? Do you have sufficient supporting evidence? Have you stated the implications of your information clearly?
- Are all your facts and numbers accurate?
- Have you answered the questions your readers are likely to have?
- Does the report contain anything that you should cut out?
- Does your organization suit the needs of your content and your audience?
- Are your paragraphs clear, well organized, and of reasonable length? Are there suitable transitions from one point to the next?
- Is your prose style clear and readable?
- Is your tone appropriate to your audience?
- Are all your statements ethical? Have you avoided making misleading ambiguous statements or statements that deliberately lead the reader to faulty inferences?
- Are your graphs and tables clear and accurate? Are they well placed? Do they present your information honestly?
- Is your document readable, accessible, and visually effective?
- Are there people with whom you should share your draft—for example, members of the target audience—before going on to a final draft?

EDITING

- Have you checked thoroughly for misspellings and other mechanical errors?
- Have you included all the formal elements that your report needs?
- Are design elements such as headings, margins, spacing, typefaces, and documentation consistent throughout the draft?
- Are your headings and titles clear, properly worded, and parallel? Do the headings in the text match those in the table of contents?
- Is your documentation system appropriate and complete? Have you documented wherever appropriate? Do the numbers in the text match those in the notes?
- Have you keyed the tables and figures into your text and have you sufficiently discussed them?
- Are all parts and pages of the manuscript in the correct order?
- Will the format of the typed or printed report be functional, clear, and attractive?
- Does your manuscript satisfy stylebook specifications governing it?
- Have you included required notices, distribution lists, and identifying code numbers?
- Do you have written permission to reproduce extended quotations or other matter under copyright? (Necessary only when your work is to be published or copyrighted.)
- While you were composing the manuscript did you have any doubts or misgivings that you should now check out?
- Have you edited your manuscript for matters both large and small?
- What remains to be done, such as proofreading final copy?

Exercises

1. Describe accurately and completely your current writing process. Be prepared to discuss your description in class.
2. Interview someone who has to write frequently (such as one of your professors). Ask about the person's writing process. Base your questions upon the process described in this chapter; that is, ask about situational analysis and arranging, drafting, revising, and editing techniques. Take good notes during the interview, and write a report describing the interviewee's writing process.
3. Choose some technical or semitechnical topic you can write about with little research—perhaps a topic related to a hobby or some school subject you enjoy. Decide upon a purpose and audience for writing about that topic. For example, you could instruct high school seniors in some laboratory technique. You could explain some technical concept or term to someone who doesn't understand it—to one of your parents, perhaps. Analyze your audience and persona following the suggestions in this chapter. With your purpose, audience, and persona in mind, brainstorm your topic. After you complete the brainstorming, examine and evaluate what you have. Reexamine your topic and purpose to see if information you have thought of during the brainstorming has changed them. Keeping your specific topic, purpose, audience, and persona in mind, arrange your brainstorming notes into a rough outline. Do not worry overmuch about outline format, such as roman numerals, parallel headings, and so forth.
4. Turn the informal outline you constructed for Exercise 3 into a formal outline. (See "Outlining" in Part V, the "Handbook.")
5. Write a rough draft of the report you planned in Exercises 3 and 4. Allow several classmates to read it and comment upon it. Revise and edit the rough draft into a final, well-written and well-typed draft. Submit all your outlines and drafts to your instructor.
6. Read the following case and write a memo as discussed at the end of the case. For information on writing memos, see Chapter 12, "Correspondence."

Radon is an odorless radioactive gas produced by the breakdown of uranium in the soil. Exposure to it can cause lung cancer. The U.S. Surgeon General considers radon to be second only to smoking as a cause of lung cancer in the United States. Imagine that you live in an area where radon is a potential health threat in people's houses. Because of the threat, people frequently hire radon removal contractors to test for radon levels in their houses and, when necessary, to install radon removal systems.

You obtain summer employment with one such contractor. His name is John May and his firm is called May Radon Removal. Typically, the contractor tests the house for radon and then presents a proposal to the householder detailing the work to be done and setting a price for the work. To obtain more information about radon and its reduction, you read a government booklet titled *Consumer's Guide to Radon Reduction.*[11] In the booklet, you learn that the most expensive radon reduction systems are needed for houses that are built on concrete slabs or with basements. Systems for such houses can run as high as $2,500. In houses built over crawl spaces, homeowners can reduce radon to safe levels simply by increasing the ventilation of the crawl space. This seldom costs more than $500. You realize that you have been helping Mr. May install expensive systems, suitable for houses with basements and concrete slabs, in houses built over crawl spaces.

You look at a proposal being presented to a person who owns a house with a crawl space. In the proposal, you find that Mr. May has recommended suction depressurization, a system normally used under basements or slabs. It requires an expensive installation of pipes and fans in the soil under the house to trap and suck away radon. He offers no alternatives to this system. In the proposal, Mr. May justifies the suction depressurization system with this statement: "Suction depressurization is the most common and usually the most reliable radon reduction method." From your research on radon, you know this is a true statement.

Is the contractor being unethical? If so, in what way? Construct a fault tree to answer these two questions. When you have arrived at an answer, write a memo to your instructor stating your opinion and a justification for that opinion. If you found the contractor to be unethical, propose how you plan to handle the problem. Include a copy of your fault tree with your memo.

CHAPTER 3 WRITING COLLABORATIVELY

As we point out to you in Chapter 2, you can write collaboratively as well as individually. Organizations conduct a good deal of their business through group conferences. In a group conference, people gather, usually in a comfortable setting, to share information, ideas, and opinions. Organizations use group conferences for planning, disseminating information, and, most of all, for problem solving. As a problem-solving activity, writing lends itself particularly well to conferencing techniques. In fact, collaborative writing is common in the workplace.[1]

In this chapter, we discuss some of the ways that people can collaborate on a piece of writing. We conclude with a brief discussion of group conferencing skills, skills that are useful not only for collaborative writing, but for any conference situation you are likely to find yourself in.

People cooperate in many ways in the workplace. One of the ways they cooperate is by sharing their writing with one another. Someone writing a report may pass it to a co-worker and ask for general

Planning	Drafting	Revising	Editing
• Keep all discussions objective. • Record the discussion. • Analyze situation and audience. • Establish purpose. • Discover content. • Organize content. • Agree on style and tone. • Agree on format. • Choose coordinator. • Seek opinion on plan from outside group.	• Choose a drafting plan: – Divide the work. – Draft in collaboration. – Choose a lead writer. • Consult with group when needed. • Stick to deadlines.	• Revise for content, organization, style, and tone. • Be concerned with accuracy and ethics. • Make criterion-based comments. • Make reader-based comments. • Be objective in discussion, not personal. • Remember, people get attached to their writing. • Accept criticism gracefully. • Don't avoid debate, but keep discussions as friendly and positive as possible. • Know when to quit revising. • Seek an opinion from outside the group.	• Edit for format and standard usage. • Check and double-check for inconsistencies in such things as margins, typeface, documentation, and headings.

Figure 3–1 The Collaboration Process

comments or for specific feedback on the report's style, tone, accuracy, or even grammar.

However, the collaborative writing we discuss in this chapter is more complex than a simple sharing. Rather, it is the working together of a group over an extended period of time to produce a document. In producing the document, the group shares the responsibility for the document, the decision making, and the work. Figure 3–1 shows the major steps of the collaborative process. Collaborative writing can be two people working together or five or six. Writing groups with more than seven members are likely to be unwieldy. In any case, all the elements of composing—situational analysis, discovery, arrangement, drafting and revising, and editing—generally benefit by having more than one person work on them.

Groups sometimes digress and wander off the point of the discussion. Therefore, it helps to have set procedures that guide discussion without stifling it. To that end, we have provided planning and revision checklists at the end of this chapter and many others. The checklist provided on the front endpapers of this book combines the Chapter 2, "Composing," checklist with key elements from the checklist that follows this chapter. These checklists raise questions about topic, purpose, and audience that will keep the individual or the group on track.

Planning

The advantage of working in a group is that you are likely to hit upon key elements that working alone you might overlook. The collaborative process greatly enhances situational analysis and discovery. Shared information about audience is often more accurate and complete than individual knowledge. By hammering out a purpose statement that satisfies all its members, a group heightens the probability that the purpose statement will be on target.

The flow of ideas in a group situational analysis and in a discovery brainstorming session will come so rapidly that you risk losing some of them. One or two people in the group should serve as recorders to capture the thoughts before they are lost. It helps if the recording is done so that all can see—on a blackboard, a pad on an easel, or a computer screen. During the brainstorming, remember to accept all ideas, no matter how outlandish they may appear. Evaluation and selection will follow.

The group can take one of the more organized approaches to discovery. For instance, if instructions are clearly called for, the group can use the arrangement pattern of instructions to guide discovery. If discovery includes gathering information, working in a group can speed up the process. The group can divide the work to be done, assigning portions of the work according to the expertise of each group member.

When the brainstorming and other discovery techniques are finished, the group must evaluate the results. This is a time when trouble can occur. When everyone is brainstorming, it's fun to listen to the flow. There is a synergy working that helps to produce more ideas than any one individual is likely to develop working alone. When the time comes to evaluate and select ideas, however, some ideas will be rejected, and tension in the group may result. Feelings may be ruffled. Keep the discussion as open but as objective as you can. Divorce as much as possible the ideas from those who offered them. Evaluate the ideas on their merits—on how well they fit the purpose and the intended audience. Whatever you do, don't attack people for their ideas. Again, someone should keep track of the discussion in a way that the group can follow.

In collaborative writing, a good way of evaluating the ideas and information you are working with is to arrange them into an organizational plan. The act of arranging will highlight those ideas that work without shining too bright a spotlight on those that don't. A formal outline is not always necessary, but a group usually needs a tighter, more detailed organizational plan than does an individual. (See "Outlining" in Part V, "Handbook.")

Do not be in a hurry at this stage (or any other stage) to reach agreement. Collaborative groups should not be afraid of argument and disagreement. Objective discussion about such elements as purpose, content, style, and tone are absolutely necessary if all members of the group are to visualize the report in the same way. A failure to get a true consensus on how the report is to meet its purpose and how it should be written can lead to serious difficulties later in the process.

While in the planning stage, a group should take three other steps that can save a lot of hassle and bother later on.

First, they should agree on as many format features as possible, such as spacing, typefaces, table and graph design, and the form of headings and footnotes. Part III, "Document Design," will help you with this step.

Second, the group should set deadlines for completed work and stick to them. The deadlines should allow ample time for the revising stage and for the delays that seem inevitable in writing projects.

Third, the group should choose a coordinator from among themselves. The group should give the coordinator the authority to enforce deadlines, call meetings, and otherwise shepherd the group through the collaborative process. Unless the coordinator abuses his or her authority, the group should give the coordinator full cooperation.

When the planning is finished, you may want to take one more step. Collaborative writing, like individual writing, can profit from networking with individuals or groups outside your immediate working group. You may want to seek comments about your content and organizational plan from people with particular knowledge of the subject area. If you're writing in a large organization, it might pay to seek advice from people senior to you who may see political implications your group has overlooked. In writing instructions, you would be wise to discuss your plan with several members of the group to be instructed. Be ready to go back to the drawing board if your networking reveals serious flaws in your plan.

Drafting

In the actual drafting of a document, a group can choose one of several possible approaches.

Dividing the Work

For lengthy documents, perhaps the most common procedure is to divide the drafting among the group. Each member of the group takes responsibility for a segment of the organizational plan and writes a draft based upon the group plan. It's always possible, even likely, that each writer will alter the plan to some degree. If the alterations are slight enough that they do not cause major problems for group members working on other segments of the plan, such alterations are appropriate. However, if such changes will cause problems for others, the people affected should be consulted.

Allow generous deadlines when you divide the work. Even when a group has agreed on the design features, there will be many stylistic differences in the first drafts. A group that divides the work must be prepared to spend a good deal of time revising and polishing to get a final product in which all the segments fit together smoothly.

Drafting in Collaboration

In a second method of collaboration, a group may want to draft the document in collaboration, rather than dividing up the work. Word processing, in particular, makes such close collaboration possible. Two or three people sitting before a keyboard and a screen will find that they can write together. Generally, one person will control the keyboard, but all collaborators can read the screen and provide immediate feedback as changes are made to the document. Although such close collaboration is possible, it is a method seldom used in the workplace, probably because it is time consuming and, therefore, costly. Its use is most often reserved for short, important documents where the writers must weigh every word and nuance.

One Person Doing the Drafting

The third method is to have one person draft the entire document. This produces a uniformity of style but has the obvious disadvantage in the classroom that not everyone will get needed writing experience. An alternative approach is to divide the work but then appoint a lead writer to put the segments together, blending the parts into a stylistic whole. The group may even give the lead writer the authority to make editorial decisions in cases where the group can not reach agreement on its own. In large organizations you will find all of these methods or combinations of them in use.

Revising and Editing

Collaboration works particularly well in revising and editing. People working in a group frequently will see problems in a draft, and solutions to those problems, that a person working alone will not see.

Revising

In revising, concern yourself primarily with content, organization, style, and tone. Be concerned with how well a draft fits purpose and organization. When the group can work together in the same location, everyone should have a copy of the draft, either on paper or on the computer screen. Comments about the draft should be both criterion based and reader based.[2]

Criterion-Based Comments Criterion-based comments measure the draft against some standard. For example, the sentences may violate stylistic standards by being too long or by containing pretentious language. (See Chapter 5, "Achieving a Readable Style.") Perhaps in classifying information, the writer has not followed good classification procedures. (See Chapter 6, "Informing.") The group should hold the draft to strict standards of ethics and accuracy. Whatever the problem may be, approach it in a positive manner. Say something like "The content in this sentence is good. It says what needs to be said, but maybe it would work better if we divided it into two sentences. A sixty-word sentence may be more than our audience can handle."

Reader-Based Comments Reader-based comments are simply your reaction as a reader to what is before you. Compliment the draft whenever you can: "This is good. You really helped me understand this point." Or you can express something that troubles you: "This paragraph has good factual content, but perhaps it could explain the implications of the facts more clearly. At this point, I'm asking what does it all mean. Can we provide an answer to the 'so-what' question here?"

Word Processing Word processing offers an attractive technique for revising, particularly when geography or conflicting schedules keep group members apart. Each member can do a draft on a disk and then send a copy of the disk to one or more co-authors. The co-author can make suggested revisions on the disk and send it back to the original author. The revisions should be highlighted in some way, perhaps with asterisks or brackets; many word processors include a redlining feature for this very purpose. If a printer is available, a redlining program such as CompareRite or DocuComp may also be used. The original author can react to the changes in a way he or she thinks appropriate. Collaborators can use electronic mail in a similar way. If the collaborators can get together, they can slip the disk into the word processor and work on it side by side.

Comments from Outside the Group As with the organizational plan, you should consider seeking comments on your drafts from people outside the group. People senior to you in your organization can help you to ensure that the tone and content of your work reflect the values and attitudes of the organization.

Problems in the Group Although effective, collaborative revision can cause problems in the group. We all get attached to what we write. Criticism of our work can sting as much as adverse comments about our personality or habits. Therefore, all members of the group should be particularly careful at this stage. Support other members of the group with compliments whenever possible. Try to begin any discussion by saying something good about a draft. As in discussing the plan, keep comments objective and not personal. Be positive rather than negative. Show how a suggested change will make the segment you are discussing stronger—for instance, by making it fit audience and purpose better.

If you are the writer whose work is being discussed, be open to criticism. Do not take criticism personally. Be ready to support your position, but also be ready to listen to opposing arguments. Really *listen.* Remember that the group is working toward a common goal—a successful document. You don't have to be a pushover for the opinions of others, but be open enough to recognize when the comments you hear are accurate and valid. If you are convinced that revision is necessary, make the changes gracefully and move on to the next point. If you react angrily and defensively to criticism, you poison the well. Other group members will feel unable to work with you and may find it necessary to isolate you and work around you. Harmony in a group is important to its success. Debate is appropriate and necessary, but all discussions should be kept as friendly and positive as possible.

Know when to quit revising. As we have said in every preface to every edition of this book, "All writing is subject to infinite improvement." However, none of us has infinity in which to do our work. When the group agrees that the document satisfies the situation and purpose for which it is being written, it's time to move on to editing.

Editing

Make editing a separate process from revision. In editing, your major concerns are format and standard usage. Editing by a group is more easily accomplished than is revision. Whether a sentence is too long may be debatable. If a subject and verb are not in agreement, that's a fact. Use the handbook (Part V) of this text to help you to find and correct errors. Final editing should also include making the format consistent throughout the document. This is a particularly important step when the work of drafting has been divided among the group. Even if the group agrees beforehand about format, inconsistencies will crop up. Be alert for them. All the equal headings should look alike. Margins and spacing should be consistent. Footnotes should all be in the same style, and so forth. See Part III, "Document Design" for help in this important area.

The final product should be seamless. That is, no one should be able to tell where Mary's work leaves off and John's begins. To help you reach such a goal, we provide you with some principles of conferencing.

Collaboration in the Workplace

The collaboration process we have described in this chapter, or one very much like it, is the one you will probably use in a classroom setting. It is also the one you are likely to use in the workplace when a group voluntarily comes together to produce a piece of work. As such, it is a fairly democratic process. However, in the workplace, collaboration may be assigned by management rather than a voluntary decision made by members of a group. In such a case there may be several significant differences from what we have described.

In an assigned collaboration, people may be placed in the group because they can provide technical knowledge and assistance the group may need to carry out its assignment. For example, within a state department of transportation, a group might be assigned to produce an environmental impact statement in preparation for building a new highway. The group might include a wildlife biologist, a civil engineer, a social scientist, and an archaeologist. Furthermore, a professional writer may be assigned to the group to help with the composing process from planning to editing.

Rather than the group choosing a leader or a coordinator, management may assign someone to be the leader. Good leadership encourages democratic process and collaboration and enables people to do what they do best. However, there are times in the workplace when an assigned leader may be arbitrary, for example, about work assignments and deadlines.

Finally, in the workplace, there is often a prescribed process for reviewing the collaborative results. This process may involve senior executives and people with special knowledge, such as attorneys and accountants. The reviewers may demand changes in the document. The group may have some right of appeal, but, in general, the wishes of the review panel are likely to prevail.

Group Conferences

Collaborative writing is valuable as a means of writing and learning to write. In a school setting, collaborative writing is doubly valuable because it also gives you experience in group conferencing. You will find group conferencing skills necessary in the workplace. Most organizations use the group conference for training, problem solving, and other tasks. In this section we briefly describe good conference behavior and summarize the useful roles conferees can play. You'll find these principles useful in any conference and certainly in collaborative writing.

Conference Behavior

A good group conference is a pleasure to observe. A bad conference distresses conferees and observers alike. In a bad group conference, the climate is defensive. Conferees feel insecure, constantly fearing a personal attack and preparing to defend themselves. The leader of a bad conference can't talk without pontificating; advice is given as though from on high. The group punishes members who deviate from the majority will. As a result, ideas offered are tired and trite. Creative ideas are rejected. People compete for status and control, and they consider the rejection of their ideas a personal insult. They attack those who reject their contributions. Everyone goes on the defensive, and energy that should be focused on the group's task flows needlessly in endless debate. As a rule, the leader ends up dictating the solutions, perhaps what was wanted all along.

In a good group conference, the climate is permissive and supportive. Members truly listen to one another. People assert their own ideas, but they do not censure the opinions of others. The general attitude is, "We have a task to do; let's get on with it." Members reward each other with compliments for good ideas and do not reject ideas because they are new and strange. When members do reject an idea, they do it gently with no hint of a personal attack on its originator. People feel free to operate in such a climate. They come forward with more and better ideas. They drop the defensive postures that waste so much energy and put the energy instead into the group's task.

How do members of a group arrive at such a supportive climate? To simplify things, we present a list of **dos** and **don'ts.** Our principles cannot guarantee a good conference, but if they are followed they can help contribute to a successful outcome.

Dos

- Do be considerate of others. Stimulate people to act rather than pressuring them.
- Do be loyal to the conference leader without saying yes to everything. Do assert yourself when you have a contribution to make or when you disagree.
- Do support the other members of the group with compliments and friendliness.
- Do be aware that other people have feelings. Remember that conferees with hurt feelings will drag their feet or actively disrupt a conference.
- Do have empathy for the other conferees. See their point of view. Do not assume you know what they are saying or are going to say. Really listen and hear what they are saying.
- Do conclude contributions you make to a group by inviting

criticism of them. Detach yourself from your ideas and see them objectively as you hope others will. Be ready to criticize your own ideas.

- Do understand that communication often breaks down. Do not be shocked when you are misunderstood or when you misunderstand others.
- Do feel free to disagree with the ideas of other group members, but never attack people personally for their ideas.
- Do remember that most ideas that are not obvious seem strange at first, yet they may be the best ideas.

Don'ts

- Don't try to monopolize or dominate a conference. The confident person feels secure and is willing to listen to the ideas of others. Confident people are not afraid to adopt the ideas of others in preference to their own, giving full credit when they do so.
- Don't continually play the expert. You will annoy other conferees with constant advice and criticism based upon your expertise.
- Don't pressure people to accept your views.
- Don't make people pay for past mistakes with continuing punishment. Instead, change the situation to prevent future mistakes.
- Don't let personal arguments foul a meeting. Stop arguments before they reach the personal stage by rephrasing them in an objective way.

Perhaps the rule "Do unto others as you would have them do unto you" best summarizes all these *dos* and *don'ts.* When you speak you want to be listened to. Listen to others.

Group Roles

You can play many roles in a group conference. Sometimes you bring new ideas before the group and urge their acceptance. Perhaps at other times you serve as information giver and at still others as harmonizer, resolving differences and smoothing ruffled egos. We describe these useful roles that you as a conference leader or member can play. We purposely do not distinguish between leader and member roles. In a well-run conference, an observer would have difficulty knowing who the leader is. We divide the roles into two groups: **task roles,** which move the group toward the accomplishment of its task; and **group maintenance roles,** which maintain the group in a harmonious working condition.

Task Roles When you play a task role, you help the group accomplish its set task. Some people play one or two of these roles almost exclusively, but most people slide easily in and out of most of them.

- **Initiators** are the idea givers, the starters. They move the group toward its task, perhaps by proposing or defining the task or by suggesting a solution to a problem or a way of arriving at the solution.
- **Information seekers** see where needed facts are thin or missing. They solicit the group for facts relevant to the task at hand.
- **Information givers** provide data and evidence relevant to the task. They may do so on their own or in response to the information seekers.
- **Opinion seekers** canvass group members for their beliefs and opinions concerning a problem. They might encourage the group to state the value judgments that form the basis for the criteria of a problem solution.
- **Opinion givers** volunteer their beliefs, judgments, and opinions to the group or respond readily to the opinion seekers. They help set the criteria for a problem solution.
- **Clarifiers** act when they see the group is confused about a conferee's contribution. They attempt to clear away the confusion by restating the contribution or by supplying additional relevant information, opinion, or interpretation.
- **Elaborators** further develop the contributions of others. They give examples, analogies, and additional information. They might carry a proposed solution to a problem into the future and speculate about how it would work.
- **Summarizers** draw together the ideas, opinions, and facts of the group into a coherent whole. They may state the criteria that a group has set or the solution to the problem agreed upon. Often, after a summary, they may call for the group to move on to the next phase of work.

Group Maintenance Roles When you play a group maintenance role, you help to build and maintain the supportive group climate. Some people are so task oriented that they ignore the feelings of others as they push forward to complete the task. Without the proper climate in a group, the members will often fail to complete their task.

- **Encouragers** respond warmly to the contributions of others. They express appreciation for ideas and reward conferees by complimenting them. They go out of their way to encourage and reward the reticent members of the group when they do contribute.

- **Feeling expressers** sound out the group for its feelings. They sense when some members of the group are unhappy and get their feelings out in the open. They may do so by expressing the unhappiness as their own and thus encourage the others to come into the discussion.
- **Harmonizers** step between warring members of the group. They smooth ruffled egos and attempt to lift conflicts from the personality level and objectify them. With a neutral digression, they may lead the group away from conflict long enough for tempers to cool, allowing people to see the conflict objectively.
- **Compromisers** voluntarily withdraw their ideas or solutions in order to maintain group harmony. They freely admit error. With such actions, they build a climate in which conferees do not think their status is riding on their every contribution.
- **Gatekeepers** are alert for blocked-out members of the group. They subtly swing the discussion away from the forceful members to the quiet ones and give them a chance to contribute.

Planning and Revision Checklists

The following questions are a summary of the key points in this chapter, and they provide a checklist for composing collaboratively. To be most effective, the questions in this checklist should be combined with the checklist questions following Chapter 2, "Composing." To help you use the two checklists together, we have combined Chapter 2 questions with the key questions from this list and printed them in the front endpapers of this book.

PLANNING

- Is the group using appropriate checklists to guide discussion?
- Has the group appointed a recorder to capture the group's ideas during the planning process?
- When planning is completed, does the group have an organizational plan sufficiently complete to serve as a basis for evaluation?
- How will the group approach the drafting stage?
  - By dividing the work among different writers?
  - By writing together as a group?
  - By assigning the work to one person?
- Has the group agreed on format elements such as spacing, typography, table and graph design, headings, and documentation?
- Has the group set deadlines for the work to be completed?
- Should the group appoint a coordinator for the project?
- Are there people you should share your draft with? Supervisors? Peers? Members of the target audience?

REVISION

- Are format elements such as headings, margins, spacing, typefaces, and documentation consistent throughout the group's documents?
- Does the group have criteria with which to measure the effectiveness of the draft?
- Is the document accurate and ethical?
- Do people phrase their criticisms in an objective, positive way, avoiding personal and negative comments?
- Are the writers open to criticism of their work?
- Is the climate in the group supportive and permissive? Do members of the group play group maintenance roles as well as task roles, encouraging one another to express their opinions?

Exercises

By following the techniques outlined in this chapter, groups could do most of the writing exercises in this book as a collaborative exercise. For a warm-up exercise in working collaboratively, work the following problem:

- Divide into groups of three to five people. Consider each group to be a small consulting firm. An executive in a client company has requested a definition of a technical term used in a document the firm has prepared for that company.
- The group plans, drafts, revises, and edits an extended definition (see Chapter 7, "Defining and Describing") in a memo format (see Chapter 12, "Correspondence") for the client.
- Following the completion of the memo, the group critiques its own performance. Before beginning the critique, the group must appoint a recorder to summarize the critique.

  How well did the members operate as a group?

  What methods did the group use to work together to analyze purpose and audience and to discover its material?

  What technique did the group use to draft its memo?

  Was the group successful in maintaining harmony while carrying out its task?

  What trouble spots emerged?

  What conclusions did the group reach that will help future collaborative efforts?
- The recorders report to the class the summaries of the groups. Using the summaries as a starting point, the class discusses collaborative writing.

CHAPTER 4

WRITING FOR YOUR READERS

In Chapter 2, "Composing," we tell you how and why to analyze your readers. In this chapter we provide you with some ways to use that analysis to plan strategies for communicating successfully with your readers. Consider what their point of view is: What will be their concerns and interests when they read what you write? What do your readers already know or not know? Are they highly knowledgeable in your subject matter area or not? Do they already have a good grasp of the technical vocabulary and concepts you will use or not? What are their reading habits? When you have answers to such questions, you will be better prepared to consider matters such as these:

- Satisfying your readers' points of view
- Providing needed background
- Helping readers through your document
- Choosing an effective style
- Choosing appropriate graphics
- Discourse communities
- Writing for combined audiences

Point of View

One of the most important audience characteristics is their point of view. Why are they reading your document? What is their purpose? What are their expectations? Using point of view as your criterion, you can break audiences down into four convenient categories: lay people, executives, experts, and technicians. Do understand, though, that these categories, while a convenient starting point, are something of a fiction. As you will see as you read the chapter, some executives are close to being lay people, others closer to experts, and so forth. But thinking in terms of these four categories will help you analyze any audience.

Lay People

Who are **lay people**? They are fourth-graders learning how the moon causes solar eclipses. They are the bank clerk reading a Sunday newspaper story about genetic engineering and the biologist reading an article in *Scientific American* titled "The Nature of Metals." In short, we are all lay people when we are outside our own particular fields of specialization. Most lay people have at least a high school diploma. In 1991, 78.4% of the U.S. population over 25 years of age had at least four years of high school; 21.4% had at least four years of college.[1] Despite these encouraging statistics, some studies indicate that as many as 27 million adult Americans read at only marginal levels. They have difficulty reading things such as newspapers, bus schedules, and catalogs.[2] Many high school graduates and even college graduates have only a smattering of mathematics and science and are a little vague about both subjects.

Lay people represent a wide range of educational levels. Regardless of that difference, when most people are reading as lay people, they have a similar point of view. They read for interest. They read to understand the world in which they live. In these days of environmental concern and consumerism, they may be reading as a prelude to action.

Lay people are generally much more concerned with what things do than how they work. Their interest is personal. What impact will this new development have on them? They are more interested in the fact that widespread computer networks may invade their privacy than the fact that computers work on a binary number system. They are more interested in the safety, efficiency, and cost of nuclear waste disposal than in the technical details of such disposal.

Executives

In general, executives are the managers, supervisors, administrators, and decision makers of an organization. Legislators, granting agency reviewers, high school principals, and farmers, as well as chief executive officers, function as executives. Although most executives have college degrees and many have technical experience, they represent many disciplines, not necessarily including the one you are writing about. Some executives may have training in management, accounting, a social science, or the humanities, but little or no technical background. What are their concerns and interests?

Executives want to know how technological developments will affect the development of their companies. Although executives resemble lay readers in many ways, there is also a significant difference between them. What lay people read influences their lives and their decisions, but they only occasionally have to act directly upon it. Executives, however, must often make decisions based upon what they read. People and profits figure largely in executive decisions.

Executives must also consider the social, economic, and environmental effects of their decisions upon the community. Aesthetics, public health and safety, and conservation are key factors, and few executives would consider a report complete if it did not deal with them. All this means that executives are usually more interested in the implications of data than in the data themselves.

What questions do executives want you to answer in a report written for them? They want to know how a new process or piece of equipment

Problems
What is it?
Why undertaken?
Magnitude and importance?
What is being done? By whom?
Approaches used?
Thorough and complete?
Suggested solution? Best? Consider others?
What now?
Who does it?
Time factors?

New Projects and Products
Potential?
Risks?
Scope of application?
Commercial implications?
Competition?
Importance to Company?
More work to be done? Any problems?
Required manpower, facilities, and equipment?
Relative importance to other projects or products?
Life of project or product line?
Effect on Westinghouse technical position?
Priorities required?
Proposed schedule?
Target date?

Tests and Experiments
What tested or investigated?
Why? How?
What did it show?
Better ways?
Conclusions? Recommendations?
Implications to Company?

Materials and Processes
Properties, characteristics, capabilities?
Limitations?
Use requirements and environment?
Areas and scope of application?
Cost factors?
Availability and sources?
What else will do it?
Problems in using?
Significance of application to Company?

Field Troubles and Special Design Problems
Specific equipment involved?
What trouble developed? Any trouble history?
How much involved?
Responsibility? Others? Westinghouse?
What is needed?
Special requirements and environment?
Who does it? Time factors?
Most practical solution? Recommended action?
Suggested product design changes?

Figure 4–1 What Managers Want to Know
Source: James W. Souther, "What to Report," *IEEE Transactions on Professional Communication* PC-28 (1985):6.

can be used. What new markets will it open up? What will it cost, and why is the cost justified? What are the alternatives?

Why did you choose the new equipment over the other alternatives? Give some information about the also-rans. Convince the executive that you have explored the problem thoroughly. For all the alternatives, include comments on cost, size of the project, time to completion, future costs in upkeep and replacement, and the effects on productivity, efficiency, and profits. Consider such aspects as new staffing, competition, experimental results, and potential problems. What are the risks involved? What environmental impact will this new development have? Figure 4–1, "What Managers Want to Know," lists the information executives feel they need in a report.

Experts, in particular, often find writing for executives a difficult task. Experts are often most interested in methodology and theory. As Professor Mary Coney points out, executives are more interested in function. Excerpts from a salmon study done for the Alaskan Fish and Game Department illustrate the frame of mind the expert researcher should have while writing for the executive. In the introduction, the researcher poses the questions that will be answered in the report. They are the questions an executive would ask:

> Why have they gone? Can the runs be restored to any significant degree? Is it reasonable to base a large industry on the harvest cycle of a wild resource? What should be done? What should be done now?[3]

The stated purpose of the report further reassures the executive that the researcher is on the right track:

> Our approach has been first to gather and understand as much relevant information as could reasonably be found; and then to organize, interpret, and project toward the goal of defining a conceptual framework for successful actions by the State of Alaska through its Department of Fish and Game.[4]

Here it is obvious that scientific findings will be wedded to executive needs. Function—successful action—lies at the heart of the report.

Experts

Who are experts? For our purposes we will define **experts** as people with either a master's degree or a doctorate in their fields or a bachelor's degree and years of experience, such as college professors, industrial researchers, and engineers who design and build. They know their fields intimately. Experts are very concerned with how and why things work. They want to see the theoretical calculations and the results of basic research. They want your observations, your facts—what you have seen, what you have measured. They expect you to work your way through your data and your interpretation of those data to your conclusions. In reporting such things, be as complete as time, space, and human patience allow. Modern technology and science depend upon many people cooperatively accumulating facts, many of which seem trivial standing alone.

Technicians

Technicians are at the heart of any operation. They are the people who build, maintain, and operate equipment. Technicians' educational levels vary widely. Most typically, technicians will range anywhere from a high school to a college graduate. They may have been trained in a vocational school. The high school graduate may have a great deal of on-the-job training and experience. The college graduate may be better educated about theory but have less practical experience.

Because technicians build, maintain, and operate equipment, their major concern is with how-to information. How do I use this word processing program? How do I operate this lathe? How do I replace the fuel pump in this engine? And so forth.

Although technical audiences care more about how-to information and more about the practical application of a theory than about the theory itself, they will appreciate some theory. How much theory you give, and how complex you make it, depends on their education and their point of view. College-educated technicians, or those with extensive experience, border on being experts, and you can treat them much as you would an expert. Less expert technicians or people operating as technicians in areas outside their major interest require much less theory. For example, a college professor installing a new washer in her kitchen sink faucet is acting as a technician but would probably not desire much background theory about plumbing.

Providing Needed Background

Without sufficient background, readers will not be able to comprehend and absorb your material. For example, most North American readers of this book know enough about baseball to comprehend a sentence such as, "Casey hit Cohen's high hard one down the right field line, moving Morrisey from first to third." North American readers have what reading experts call a *schema* upon which to hang the sentence. That is, they can visualize the baseball field. They know that Casey has to be the batter at home plate and Cohen has to be the pitcher. They know where the right field line is and that it is far enough away from third base to allow Morrisey to gain a few steps on the throw from right field. All that is a lot to know, but readers with the appropriate baseball schema can easily organize and integrate the new information in the sentence into their knowledge. Readers without the appropriate schema can make little sense of the sentence and will not absorb the information in it.

Consider the schema your audience possesses. To give any audience the schema they need, you may have to define terms and explain concepts.

Defining Terms

Although we learn our professional language long after we have acquired our common, everyday language, it becomes such a part of our life that we often forget that others don't share it with us. We forget that we are, in effect, bilingual, possessing both a common language that we share with others and a professional language that we share with a much smaller group. In reaching out to your audience, you must remember that you may need to define specialized terms. You are the host when you write. You have invited your readers to come to you. You owe them every courtesy, and defining difficult terms is a courtesy. If you force your readers to the dictionary every fourth line, their interest will soon flag.

Depending upon the needs of the audience, terms can be defined either briefly, often by the substitution of more familiar terms, or at length. The following definition of *ground water* combines both techniques. Within a paragraph definition of *ground water*, the writer has used brief definitions consisting of a few familiar terms. Notice that in some brief definitions the word defined precedes the familiar term and in others the familiar term precedes the word defined. Notice that the paragraph definition proceeds from familiar things such as *precipitation* and *plants* to such unfamiliar things as the *saturation zone* and *artesian pressure:*

Ground Water

> Precipitation may seep into the soil. This water replenishes the soil moisture or is used in growing plants and returned to the atmosphere by transpiration (water vapor released to the air by plants). Water that seeps downward (percolates) below the root zone reaches a level at which all the openings or voids in the ground are filled with water. This zone is known as the "saturation zone." Water in the saturation zone is referred to as "ground water." The upper surface of the saturation zone, if not restricted by an impermeable layer, is called the "water table." When the ground formation over the saturation zone keeps the ground water at a pressure greater than atmospheric pressure, the ground water is under "artesian pressure."[5]

Be careful not to distort the true meaning of terms if you substitute more common terms for technical language. One researcher, for example, felt his work was distorted by this lead in a newspaper story:

> A research group reported Friday that marijuana causes chimpanzees to overestimate the passage of time, and a single dose can keep them befuddled for up to three days.

The researcher commented:

> The term "befuddle" was not employed in our scientific report, and the statement in the news article "and a single dose can keep them befuddled for up to three days" is erroneous and misleading. Three days were required to recover normal baseline performance following administration of high doses.[6]

Scientists choose words very precisely and for good reason. Although scientific findings must be interpreted for nonscientific readers, to distort or to sensationalize their work is a disservice to them and to the reader.

The need for definition varies from audience to audience. Audiences with a weak schema may need many words defined. Those with a strong schema need few definitions. To define a word that a reader already knows would be a mistake, but scientific and technical vocabularies are constantly growing. If you are using words that you think any audience may not know, even one that shares your professional knowledge, don't hesitate to define. (See Chapter 7, "Defining and Describing.")

Explaining Concepts

The scientific world is full of complex concepts that your readers may or may not comprehend. For example, expert geologists understand the concept of plate tectonics. Nongeologists may not; for them, a writer would have to explain that the land masses of our planet float on plates that are slowly but inexorably moving. One plate crushing against another can cause earthquakes and raise mountains. The subcontinent of India, in crushing against Asia, raised the Himalayan Mountains. Plate action along the west coast of North America causes earthquakes and volcanic activity.

One way to help a reader visualize a concept is through analogy. In analogy you move from the familiar to the unfamiliar. For example, to help readers understand the slowness with which lithospheric plates move, you could tell them that the plates move at about the same speed that fingernails grow, a few inches a year. For readers who have difficulty grasping the enormousness of geologic time compared to a human life span, you can provide an analogy like this one: If the time that the earth has existed is seen as the height of the Empire State Building, then the time of human existence would be comparable to a dime on top of the building.

Analogy can be useful even when writing for the technical audience. Analogy bridges the gap between a reader's general information and the particular object or theory you are trying to explain. An excerpt from the *Bell Laboratories Record* article on waveguides illustrates the principle:

> Every electron orbiting about an atomic nucleus gives rise to magnetic fields. In some materials the field comes largely from the motion of the electron around the nucleus, but in ferromagnetic materials it depends more on the spin associated with the electron itself. The spin creates a small magnetic movement that is precisely aligned with the axis of spin (this can be visualized as similar to the alignment of the earth's magnetic field between the north and south poles.)[7]

As with definitions, some audiences need many concepts explained in order to develop the necessary schema; others need few or none explained. Your audience analysis should tell you how much explanation you have to give any audience.

Helping Readers Through Your Report

Readers need help in getting through written reports—even very good readers. The more new information a document provides, the tougher it is for the reader. You can help readers by being directive, by providing an appropriate context for your material, and by organizing around your audience's reading habits.[8]

Be Directive

Direct your readers through your documents. That is, provide them with a road map by including sufficient introductions, transitions, summaries, and the like. A longer document needs a set of headings consistently applied and probably a table of contents. Design your document so that your reader comes to know where certain information is presented and how it is presented. For example, the planning and revision checklists in this book are all bulleted lists that come just before the exercises at the end of the chapters. After reading a few chapters, you have learned where to look for them.

You will find instruction about designing documents in Chapter 9, "Document Design." In Chapter 10, "Design Elements," you'll find the information you need to prepare introductions, summaries, conclusions, headings, and the like.

Provide an Appropriate Context

Readers don't read words or sentences in isolation. They understand what they are reading by relating the passage they are reading to passages that have gone before or that come later. That is, they understand what they are reading by putting it in context. Two ways you can help readers put a passage into an appropriate context are by moving from the familiar to the unfamiliar and, particularly in instructions, by providing the reason for an action before describing the action.

In the following example, the writer begins with glacial deposition, a form of deposition familiar to his readers. The writer then uses glacial deposition to help the readers understand beach deposition. In the fourth paragraph the writer uses beach deposition, now familiar to his readers, to introduce river and delta deposition:

> From the nature and distribution of glacial deposits, geologists have formed a picture of what the earth looked like during a glacial event.
>
> In the same manner, geologists have recognized rocks that were once ancient beach deposits because most beaches are composed of well-sorted sand. The action of waves along the shores of ancient seas washed out the silt and clay and left behind rounded grains of sand, just as those along present shorelines.

Offshore, where the bottom waters are calmer, the finer sediment settles as mud. In a general way, the size of the sediment grains shows the direction of slope on the sea floor, because the sandy sediment will be near the shore and the mud will be offshore. The same principle of sorting also applies to ancient rocks; marine sandstones and conglomerates were formed closer to shore than were the finer textured marine shales and siltstones.

The same concept of sorting can be used with nonmarine formations, such as river and delta deposits.[9]

When reading instructions, readers expect the reason for an action to be given before the action. Because the following passage inverts that order, it frustrates the reader's expectations:

Drag the image of the disk into the trash to eject the disk.

The following is the order the reader expects:

To eject the disk, drag the image of the disk into the trash.

In Chapter 5, "Achieving a Readable Style," we discuss other ways to make a document more readable.

Organize Around Audience Reading Habits

Readability authorities have classified the way people read into five methods:[10]

- **Skimming.** Going through a document very quickly, mainly to get a general idea of its nature and contents
- **Scanning.** Reading rapidly to find specific needed facts or conclusions, such as looking only for financial information in a report
- **Search reading.** Scanning, but slowed down so as to pick up more of the content
- **Receptive reading.** Reading at whatever speed is necessary for high comprehension
- **Critical reading.** Reading to evaluate the document and its contents

Analyze your own reading habits. When you are reading a newspaper, you probably skim, scan, or search. When you find a story that interests you, you switch over to receptive reading. When you have as much information as you need or want from the article, you are likely to scan or search the rest of it quickly to see if you need or want anything else from it. News stories are structured to allow you to do precisely that. They begin with the journalist's *who, what, when, where,* and *why*. The rest of the story provides additional detail. When you have all the detail you want, you can quit reading and still have the gist of the story.

Because writers for popular magazines such as *Discover* and *National Geographic* know that people scan the publication looking for articles that interest them, they often include an interest-catching introduction like this one from an article about space satellite remote sensing instruments:

> The hike through the harsh jungle of northern Guatemala's Peten region had been long and arduous, at times even dangerous. Now at El Mirador, the prehistoric Mayan temples seemed to spring out from under the thick forest canopy and touch the sky. Tom Sever climbed the steep staircase of the 51-meter high Danta temple, the largest in Peten, and scaled its limestone walls. With labored breath, he eased his way carefully along onto the narrow ledge. The magnitude and splendor of the lush rain forest stretched out before him. Could there be a more beautiful sight on Earth, he wondered.
>
> As an archaeologist and anthropologist, Sever was even more interested in the people who had built the temple. How had they gotten there? Where were the paths leading from the temple to the now vanished Mayan cities that had been here more than a thousand years ago?
>
> As Sever scanned the countryside, he saw no sign of the ancient roadways in the dense forest. He knew they were there, though. He had already seen them—in pictures taken from above.[11]

If the writer can grab your interest in the introduction, you are more likely to continue reading.

Professor James Souther, who researched the reading habits of executives, reported that executive reading habits are surprisingly similar.[12] Professor Souther's research seems to indicate that executives most often skim, scan, and search read. They are likely to read receptively or critically only in such sections as introductions, summaries, and conclusions. Professor Souther reports that "All managers said they read the *abstract*; most said they read the *introduction* and *background* sections as well as the sections containing *conclusions* and *recommendations* to gain a better perspective of the material being reported and to find an answer to that all-important question—'What do we do next?'"[13]

Organize your executive reports so that your readers can easily find the sections they want. Be directive, providing good introductions, transitions, headings, and summaries. Executives expect to see conclusions and recommendations early in a report with the justification following. If you feel you must report a large amount of technical data, put the data in an appendix where executives can read them if they want to (or assign experts to read them). Managers seldom read the body and appendixes of reports. When they do, it is because, as Professor Souther

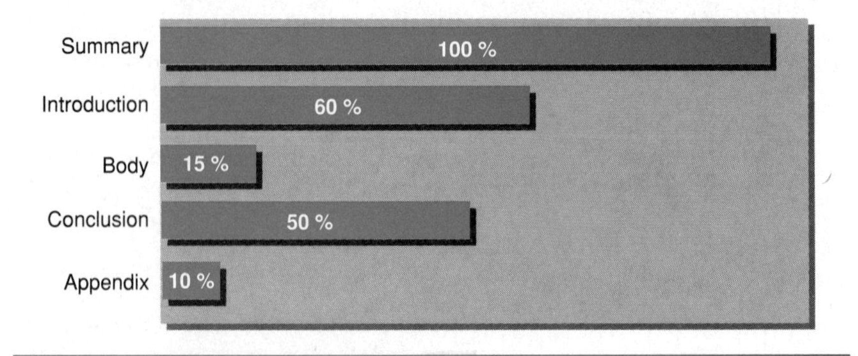

Figure 4–2 How Managers Read Reports
Source: James W. Souther, "What to Report," *IEEE Transactions on Professional Communication* PC-28 (1985): 6.

reports, they are "especially interested ... deeply involved ... forced to read by the urgency of the problem," or "skeptical" of the writer's conclusions.[14] Figure 4–2 illustrates how Westinghouse managers read reports.

When writing for executives, you should give your conclusions and recommendations clearly. In writing a report for an executive, interpret your material and present its implications—don't just give the facts. James Souther points out that in executive reports, the professional judgment of the writer should be the focal point. "True," Professor Souther writes, "it is judgment based on objective study and evaluation of the evidence; but it is judgment nevertheless."[15] The researcher who amasses huge amounts of detail but neglects to state the implications, conclusions, and recommendations that follow from the facts has failed to do the complete job.

People reading in fields where they have expert or technical knowledge have a greater tolerance for data than do people reading as lay people or executives, and are more inclined to read receptively. However, because of the enormous amounts of information available in most fields, experts and technicians have also become scanners and search readers. Research indicates that experts review reports quickly to see whether the reports contain needed information. In their review, the experts depend upon, in order of importance, the *summary, conclusions, abstract, title page,* and *introduction.*[16] Therefore, these components must be carefully written in expert reports. As in executive reports, make information accessible through the use of headings.

Technicians scan instruction manuals looking for needed information, rather than reading the manual in sequence. Therefore, manual writers provide directive devices such as tables of contents and headings to allow their readers to read selectively.

Style

Choose a writing style that matches the reading ability and technical knowledge of your audience. For instance, when writing for a nontechnical audience, avoid the more complex sentence structures and technical vocabulary that might be appropriate for an audience that shares your knowledge. Whoever your audience is, you should be no more complicated than necessary.

Plain Language

Use plain language. Some scientific specialties are loaded with mathematics. Others, such as biochemistry, are full of formulas, complicated charts, and diagrams incomprehensible to lay people. Equations, formulas, and diagrams are useful shorthand expressions for experts. They convey information with a precision impossible to obtain in any other way. But what experts sometimes forget are the years spent learning how to handle such precise tools. The average person, lacking those years of training, often cannot understand them. When you write for people who don't share your knowledge, you must express your ideas in plain language.

In Chapter 5, "Achieving a Readable Style," we discuss how to achieve a style that aids rather than impedes understanding. The example that follows illustrates that style. It is from a Public Health Service pamphlet that explains living cells. To give background information about cells, the writer uses a vocabulary suited to her audience—lay people with a good general reading ability—and well-constructed sentences of a reasonable length. She adds a measure of human interest and closes with an interesting analogy:

> All cells—whether from a bacterium, plant, mouse or human—are made of the same basic materials: nucleic acids, proteins, carbohydrates, water, fats, and salts.
>
> "The uniformity of the earth's life, more astonishing than its diversity, is accountable by the high probability that we derived, originally, from a single cell," notes physician Lewis Thomas in The Lives of a Cell. "It is from the progeny of this parent cell that we take our looks; we still share genes around, and the resemblance of the enzymes of grasses to those of whales is a family resemblance."
>
> The genetic material in all these cells is deoxyribonucleic acid (DNA), a large molecule that directs the making of duplicate cells. DNA also directs the building of proteins according to a complex code. Even the simplest living cells—the mycoplasma—contain a relatively large amount of DNA, enough to code for up to a thousand different proteins. Every human cell has about 6 feet of very tightly wound DNA strands contained within its nucleus. Every adult carries about 100 billion miles of ultrathin DNA strands in his or her body—a distance greater than the diameter of the solar system.[17]

Human Interest

Human interest serves two purposes when you are writing for lay people: It motivates people to read and it seems to help them retain more of what they read.

You often have to motivate lay people to read. Most of us, at any educational level, have an interest in other human beings and in human personalities. Most writers for lay audiences recognize this interest and use it to gain acceptability for their subject matter. For example, an article in *Time* about farm problems will give us statistical information about the number of farmers losing their farms. But the writer of the article knows that many of us do not relate very well to bare, abstract statistics. Therefore, the writer will also usually introduce people into the article. Perhaps Bill and Mary Ellen Clark and their two children, a "typical" farm family living in Iowa, will be described. We'll learn what effect losing their farm has had on their lives. We are interested in learning about what happens to real people, and through such knowledge we can better understand the farm problem.

In the following example, notice how the introduction to an article on hurricane prediction uses human drama to capture the reader's interest:

> Today, some 45 million Americans live in coastal areas vulnerable to hurricanes. Many of them have seen merely the fringes of a hurricane and believe it capable of nothing more dangerous than tearing off shingles or flattening large trees. Only a fraction have endured the unbelievable fury at the storm's center.
>
> How bad, in fact, can a hurricane be? Bob Sheets, a forecaster with the National Hurricane Center, presents visitors with two photographs of an apartment house in Pass Christian, Mississippi. The first shows a solid, three-story brick structure, separated from the Gulf of Mexico by an eight-foot seawall, a four-lane highway, a row of substantial oak trees, a generous front lawn, and a swimming pool. When Hurricane Camille was predicted in August 1969, 25 people felt confident enough to stage a "hurricane party" there. The second photo shows the same site, after Camille passed through. Only the swimming pool remained. Of the partygoers, one managed to cling to the upper branches of a tree. Another was swept out to sea and cast back 12 hours later and four miles down the beach, semi-conscious, but somehow still alive. The other 23 died.[18]

The technique involved in this excerpt is explained by the editor of a science magazine designed for a lay audience:

> We take out some of the content whenever it gets in the way of the story telling. We stress readability and the quality and freshness of the writing over content because we think the first imperative is that people actually read the article—not a trivial task when you are trying to interest two million very different people in the complexities of cosmology or molecular biology. We also think that what most people carry away from a popular

> magazine article is a rough sense of the subject, not the details. So we emphasize the cultural context, the human impact, the anecdotal example, precisely because they contribute more to that lingering impression and are more important to the lives of our readers than the detailed physics or the viral mechanisms, however scientifically elegant.[19]

In addition to motivating people to read, human interest also seems to improve readers' retention of what they read. In one experiment, *Time-Life* editors were asked to rewrite a passage from a high school history text. They introduced what they called "nuggets" into the passage, that is, anecdotes and stories about historical people. The result: student recall rose from 20% on the original passage to 60% on the rewritten one. Other research indicates that examples, questions, elaborations, and summaries all improve reader recall.[20]

While providing human interest and perhaps even human drama, be careful not to exaggerate scientific achievements. Journalists who write stories about scientific achievements sometimes forget this need for caution, to the dismay of the scientists involved. A newspaper story concerning research on the skin ailment psoriasis carried this headline: "Psoriasis Cure Breakthrough Seen." The lead of the story announced, "Scientists Wednesday announced a breakthrough in treating psoriasis, the skin disease which causes misery for about 6 million Americans." The scientist involved criticized the story, saying that nowhere in the report presented by the scientists was the word *breakthrough* used. He concluded by saying, "The last sentence of our write-up said cure of psoriasis is probably 50 years away. Yet the title of this article you sent says 'Psoriasis Cure Breakthrough Seen.' All I can say is [censored]!"[21]

Technical Shorthand

When you are writing for people who share your technical knowledge, you may use any shorthand methods such as abbreviations, mathematical equations, chemical formulas, and scientific terms that you are sure your audience will comprehend. Complicated formulas and equations needed to support the conclusions, but not essential for understanding them, are often placed in an appendix rather than the body of the report.

Where possible, use the standard abbreviations and symbols for your field, as defined by the authorities in that field. Nonstandard symbols should be defined as in this example:

> From Einstein's mass–energy equation, one can write the relations:
>
> E(Btu) = m (lb) x 3.9 x 10
>
> m being the loss of mass.[22]

The writer does not define *E*, *Btu*, and *lb* because these are standard symbols and abbreviations for *energy*, *British thermal units*, and *pounds*. He does define *m* because it is not standard. Writers who do not define nonstandard symbols cannot expect readers to know what they are. In fact, in a few years, returning to their own reports, *they* may not know what the symbols mean. Symbols and abbreviations may be defined as they are used, or if there are many of them, they can be defined in a glossary at the front of the report. (See Chapter 10, "Design Elements.")

Qualifications

Technical reports are, of course, not merely calculations and facts. Technical people as well as executives want your inferences and conclusions. When you draw inferences from your facts and observations, be sure to make no unwarranted leaps. Stay within the bounds of the scientific method. In presenting your conclusions, be careful that your language shows where you are certain and where you are in doubt. In the interests of scientific honesty and caution, most expert discussions and conclusions contain qualifications as well as positive statements. The following excerpt, reporting research to see if time of day affects the adaptive response to exercise, is an example of the balance between positive statements and qualified statements that is typical of scientific style (we have removed the footnotes and have underlined the qualifiers):

> Our data indicate that the training effect is affected by time of day.
>
> The afternoon training was most effective in increasing VO_{2max} and adaptive response of heart rate among the three groups, and also showed an improved blood lactate response to exercise. Although there have been many studies of circadian rhythm or diurnal variations in response to exercise, this is the first report that shows the existence of diurnal variation in the training effect.
>
> It is difficult to explain why afternoon training was most effective because the training effect, represented by an increase in VO_{2max}, is the sum of numerous physiological processes. Diurnal variations in responses to exercise have been identified in measurements of body temperature as it relates to metabolism and in measurements of products of the neurohormonal system such as cortisol, renin, and catecholamine. There is a possibility that these factors together with other unknown factors may play different roles in the diurnal variation of training effects. In addition, our results suggest a possibility that heart rate and blood lactate levels may exhibit different patterns of responses to exercise that depend on the time of day. Most adaptation in heart rate occurred in the afternoon and morning groups, while most adaptation in blood lactate levels occurred in the evening followed by afternoon groups.[23]

Some people object to the amount of qualification in a scientific style. We agree that some writers of scientific reports are too timid in stating their conclusions. Nevertheless, scientists know that certainty in

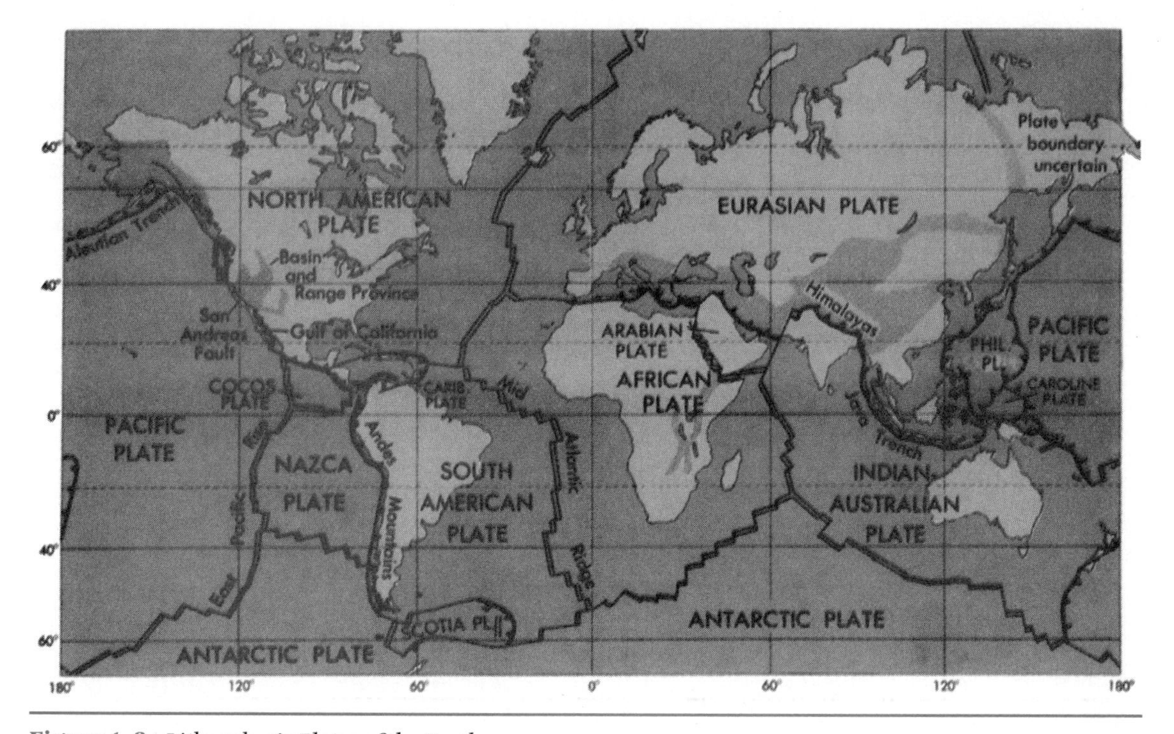

Figure 4–3 Lithospheric Plates of the Earth
Source: U.S. Geological Survey, *Our Changing Continent* (Washington, DC: GPO, 1991) 22–23.

science is a hard-won achievement. They are content with probabilities until thorough experimentation and observation remove all reasonable doubt. Their style reflects their basic caution and honesty.

Graphics

Graphs and tables are essential to most technical reports. Earlier we explained the concept of plate tectonics. To help you visualize that concept, we can provide a visual, as in Figure 4–3, that shows the lithospheric plates of the world and their relationship to one another.

You can use graphics to illustrate concepts and processes, show trends and relationships, and summarize material. You can use graphics to interest your readers as well as to inform and explain. You can use bar charts in place of equations to explain mathematical concepts or in place of formulas for chemical concepts. Or you can combine tables and pictographs that establish facts, formulas, or definitions quickly, clearly, and in a way that interests readers.

Percentage of Aged Receiving Income from Various Sources: 1962–1988

Income source	Percent 1962	Percent 1988
Social Security benefits	69	92
Private pensions	9	26
Government employee pensions	5	14
Assets	54	68
Earnings	36	22

Figure 4–4 A Simple Table
Source: Social Security Administration, *Fast Facts and Figures About Social Security* (Washington, DC: U.S. Department of Health and Human Services, 1990) 5.

Be sure to use graphics suited to your audience. For readers not expert in reading graphics, use simple tables and graphs, as in Figures 4–4 and 4–5. For more expert readers of graphics, you can use more complex tables and graphs, as in Figures 4–6 and 4–7. See Chapter 11, "Graphical Elements," for help in choosing and designing graphics-suitable to your audience.

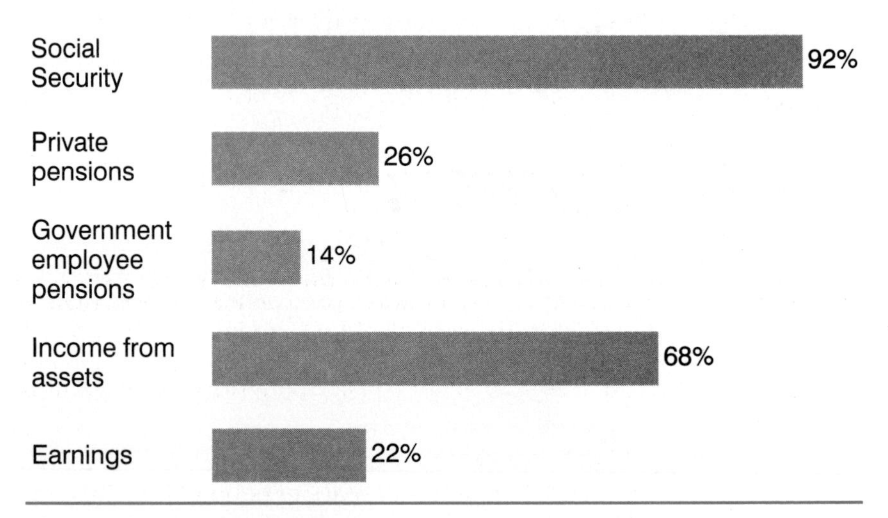

Figure 4–5 A Simple Graph
Source: Social Security Administration, *Fast Facts and Figures About Social Security* (Washington, DC: U.S. Department of Health and Human Services, 1990) 5.

Table 3. World Oil Consumption
(Million Barrels per Day)

			Projection Ranges								
	History		1995			2000			2010		
Region/Country	1989	1990	Base	Low	High	Base	Low	High	Base	Low	High
Market Economies											
OECD											
United States[a]	17.3	17.0	17.8	17.5	18.3	18.4	17.9	19.5	20.1	19.3	22.4
Canada	1.7	1.7	1.8	1.6	2.0	1.9	1.4	2.2	1.9	1.3	2.4
Japan	5.0	5.2	5.9	5.2	6.4	6.3	5.1	7.1	6.3	4.7	7.7
Europe	12.5	12.9	13.7	12.9	14.3	13.8	12.5	14.9	14.0	12.2	15.7
United Kingdom	1.7	1.7	1.9	1.7	2.1	1.9	1.5	2.3	1.9	1.5	2.6
France	1.9	1.8	2.0	1.8	2.2	2.1	1.6	2.3	2.1	1.5	2.5
Germany[b]	2.3	2.7	2.5	2.2	2.8	2.5	2.0	3.0	2.5	1.9	3.4
Italy	1.9	1.8	2.1	1.9	2.3	2.1	1.8	2.4	2.2	1.7	2.7
Netherlands	0.7	0.7	0.8	0.7	0.9	0.8	0.7	0.9	0.8	0.6	1.0
Other Europe	4.0	4.1	4.4	3.8	4.8	4.4	3.4	5.2	4.5	3.1	5.6
Other OECD	0.9	1.0	1.1	1.0	1.3	1.2	0.9	1.5	1.3	0.9	1.7
Total	37.6	37.8	40.3	39.2	41.2	41.6	39.7	43.4	43.6	41.0	46.8
OPEC	4.1	4.4	4.9	4.7	5.3	5.5	4.8	6.3	7.0	5.9	8.3
Other Developing Countries	10.5	11.1	13.2	12.0	14.3	15.0	12.5	17.1	18.1	14.3	21.6
Total Market Economies	52.2	53.3	58.5	56.8	60.0	62.1	58.9	65.0	68.7	64.0	73.7
Centrally Planned Economies[c]											
China	2.3	2.3	2.7	2.5	2.9	3.1	2.7	3.5	3.7	3.0	4.4
Former Soviet Union	8.7	8.4	6.5	5.0	8.0	7.5	6.0	9.0	8.9	6.3	11.4
Other CPE	2.4	2.0	2.0	1.7	2.1	2.3	2.0	2.5	3.0	2.4	3.6
Total	13.5	12.6	11.1	9.6	12.6	12.9	11.3	14.5	15.6	12.8	18.3
World Total	65.7	65.9	69.6	67.3	71.7	75.0	71.4	78.3	84.3	78.8	89.9

[a]Includes the 50 States and the District of Columbia. U.S. Territories are included in "Other OECD."
[b]The 1989 amount is for West Germany.
[c]Includes former, evolving, and current Centrally Planned Economies.
OECD = Organization for Economic Cooperation and Development.
OPEC = Organization of Petroleum Exporting Countries.

Notes: High and low range values for Europe and the 4 regional totals are not equal to the sum of the component countries or country groups but consist of the base value adjusted by the quantity: the square root of the sum of the squared deviations of the respective component countries or country groups from their base value. Other totals may not equal sum of components because of independent rounding.

Sources: **History**: Energy Information Administration, *International Energy Annual 1990*, DOE/EIA-0219(90) (1992) and *Monthly Energy Review*, DOE/EIA-0035(91/12) (1991). **Projections**: Energy Information Administration, *Annual Energy Outlook 1992*, DOE/EIA-0383(92) (1992); *Oil Market Simulation Model User's Manual* DOE/EIA-MO28(92) (1992); and World Energy Projection System Spreadsheet, 1992.

Figure 4–6 A Complex Table
Source: Energy Information Administration, *International Energy Outlook 1992* (Washington, DC: U.S. Department of Energy, 1992) 11.

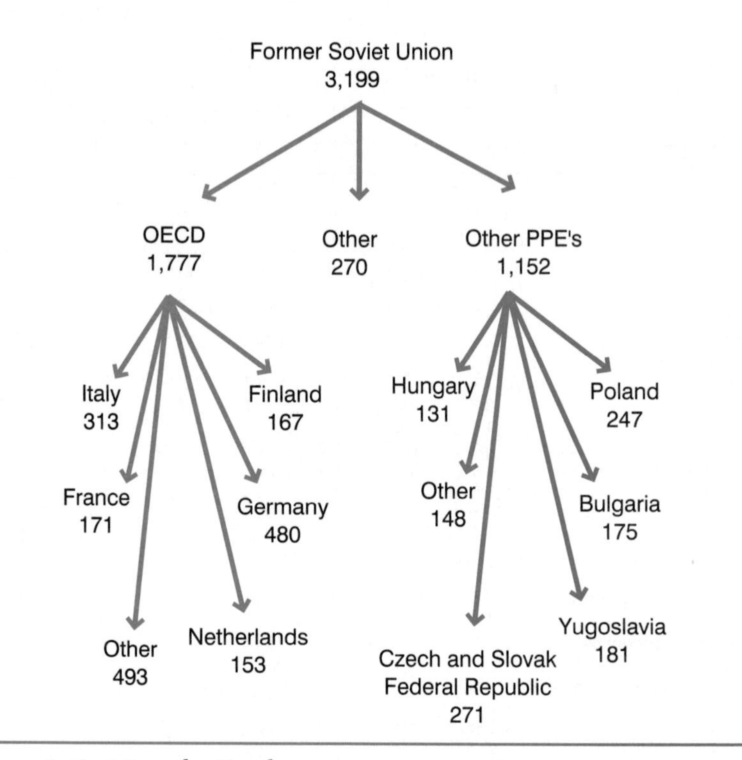

Figure 4–7 A Complex Graph
Source: Energy Information Administration, *International Energy Outlook 1992* (Washington, DC: U.S. Department of Energy, 1992) 36.

Discourse Communities

Discourse communities are communities in which people hold common ideas of what good discourse is. Within a given discourse community, members agree to a large extent about what information is important to them, what style is most acceptable, what lines of argument are most persuasive.[24] The organization for which you work may have a corporate style. As a member of that organization's discourse community, you will be expected to use that style. Various scientific disciplines are discourse communities in themselves. Some are rather plain spoken; others tend toward wordiness and excessive formality. For better or worse, scientists in these disciplines are expected to understand and use the prevailing style.

Most people belong to more than one discourse community. For example, an accountant who works for an insurance company may also in her spare time help to keep the books for her church. At the insurance company, she may find it necessary to cultivate a courteous but brisk style. The emphasis in all her reports will be on a straightforward statement of data and the implications of those data. At church, she may find it necessary to use a more indirect, less businesslike approach, perhaps more value oriented than money oriented.

In going from business to church, the accountant has crossed cultural boundaries. Not surprisingly, cultural boundaries exist between countries and regions of the world. Much of what we tell you in this book about plain style and writing directly is appropriate in North America and, perhaps with some modification, in Western Europe. Should you find yourself writing for readers in a discourse community outside those cultures, you might have to modify our advice.

Some experts on intercultural communication divide the cultures of the world into individualistic cultures and collectivistic cultures.[25] In individualistic cultures, people prize individual success. They value independent behavior highly. In a collectivistic culture, maintaining harmony and protecting the well-being of the group are more important than individual achievement.

To maintain harmony, people in a collectivistic culture tend to be indirect in communication—they talk around a problem to avoid hurting others. In an individualistic culture, people tend to be direct—"to say what they mean" and "to come to the point," as popular expressions have it. This is not to say that every member of a culture acts precisely like everyone else in the culture. However, people within a culture tend to communicate in similar ways.

North American and Western European cultures tend to be individualistic, whereas Asian cultures tend to be collectivistic. Audience adaptation has to take these significant differences into account. Businesspeople who have not done so have seen sales and opportunities slip away. Be forewarned: If you communicate outside your cultural boundaries, seek advice from experts who understand cultural differences.

The Combined Audience

Some writing situations call for you to write to people who clearly belong to one particular audience category. Many more situations do not. Sometimes, one person can have characteristics of several audiences. For instance, someone may be reading about stereo speaker systems that can be installed in his house. He may be a lay person in that he doesn't understand all the electronic jargon. He may be an executive in the sense that he may make a decision about buying such equipment. He may be a technician in that he may install the equipment himself.

Your audience may be a team of executives, experts, and technicians. Most organizational documents have more than one reader. In this age of photocopiers and fax machines, many reports, letters, and memos are copied for people other than the primary reader. Also, reports and correspondence may have a longer life than you anticipate. They go into filing cabinets, where they become a part of organizational history. They may be read years later, perhaps as background to a new development or even as evidence in a trial. Thus, your life is frequently complicated by the need to consider several audiences at once. As in any communication situation, the strategy you choose must suit your purpose, audience, and material.

Illustrative Example

Imagine that you are a computer expert working for a large investment company. You have just completed a study to decide which of two computer systems to install at the company. Let's call them Brand X and Brand Y. The system is to be purchased for the use of the research division of the company. The information systems division (ISD), for which you work, will buy, install, and maintain the system.

You have determined that Brand X, even though it costs slightly more, is the system to buy. You have to write a report stating your conclusions and recommendations and justifying them. You have a combined audience of three people, all of them senior to you:

Jack Anderson. A vice president of the company. The research division is one of his responsibilities. You have heard that his major concern is always cost, and that he doesn't like to read long reports. He is the decision maker in this matter, but he is a team player. He will certainly confer with the other people concerned before making his decision.

Sally Kroger. Head of the ISD. She has a Ph.D. in computer science. She has been with ISD for 10 years and head for the last 5. She is a stickler for detail and will want full justification for your conclusions and recommendations. She doesn't mind long reports; she even seems to prefer them.

Tony Martinez. Head of the research division. He has been complaining that his staff do not have adequate computer support for their research activities. It is because of his complaints that you conducted your study. He has a master's degree in business administration. You talked to him a good deal when you were assessing the division's computer needs. You know that he uses computers and has a practical interest in them but that he has no interest in computer theory.

How to satisfy this diverse audience? You have stacks of information in front of you: masses of statistics concerning input, output, power, remote consoles, ease of use, available programs, and cost. You have brainstormed a series of pros and cons for each system. The pros and cons are helpful, but they don't lend themselves to a coherent report. Also, they get you into more detail than Jack Anderson or Tony Martinez is likely to want.

You analyze the situation. You think about your audience and their true concerns. Jack Anderson will want to be sure that Tony Martinez's complaints are taken care of. Also, he will need to be persuaded that the more expensive Brand X is the better choice. Tony Martinez's major concern is that his needs will be satisfied. He will not care that the cost is higher for Brand X. Sally Kroger will be interested in cost, but it's not a vital factor for her. She will want to see that Martinez's needs are satisfied. She also will be quite interested in a comparison of the maintenance requirements of the two systems.

Now you're getting somewhere. The major concerns for your audience seem to be cost, needs satisfaction, and maintenance. Because Brand X meets the needs of the research division so well and because of its superior maintenance record, you have chosen it over Brand Y despite its higher cost.

The organization of your report begins to fall into place. You decide that your report will have three sections: *cost, needs satisfaction,* and *maintenance.* The cost section can be straightforward and short, probably based largely on some comparison tables. The needs section will show how you assessed the needs of the research division and how Brand X satisfies those needs far better than does Brand Y. You have some good quotes from Tony Martinez, which you can include in this section to show that he has been consulted. You remind yourself to keep this section free of computer jargon so that Tony Martinez and Jack Anderson can read it.

The maintenance section is of major concern to Sally Kroger and will probably not even be read by Jack Anderson or Tony Martinez. It's the section in which you can let yourself go with technical computer language and satisfy Sally's penchant for detail.

When the three sections are complete, you'll top them off with an *executive summary* that states your conclusions and recommendations and succinctly supports them. (See Chapter 10, "Design Elements.") You suspect that the summary may be the only section Jack Anderson reads, so you remind yourself to make sure that the support for buying the more expensive Brand X is adequate to meet his skepticism. You will also direct your readers through all parts of the report with a good introduction, transitions, and headings.

Step back for a moment and see what you have done. As shown in Figure 4–8, you have considered the major concern of each reader and

Reader	Main Concern	Style
Jack Anderson (Decision maker)	Cost	Cost section: straightforward and short, based on comparison tables
Tony Martinez (User)	Needs satisfaction	Needs satisfaction section: free of jargon, includes quotes
Sally Kroger (Information systems expert)	Maintenance	Maintenance section: detailed with technical language
Anderson, Martinez, Kroger	Satisfying Jack Anderson	Executive summary: conclusions, recommendations, with supporting evidence

Figure 4–8 Writing for a Combined Audience

made sure that you have addressed it. You have satisfied the needs of *the decision maker*, Jack Anderson; *the user*, Tony Martinez; and *the expert*, Sally Kroger. Because you know that executives scan reports and read them very selectively, you have organized and will format your report to make scanning easy.

Other Situations

Other situations may be more or less complex than the one we have described for you. For instance, you might need to write a short memo to a primary reader, but you know there will be several secondary readers. To satisfy the primary reader, you would have to adjust the level of technicality to the appropriate level. If the level of technicality is low, you run the risk of boring more technically competent secondary readers, but at least you won't confuse them. If the level of technicality is high, you might have to offer the secondary readers some help in the way of definitions, analogies, and so forth. Often, such help can be kept out of the way of a more technically competent primary reader by putting it in a graphic or an attachment. Because the graphic and attachment stand a little apart from the body of the memo, the primary reader can ignore them.

Sometimes, needed definitions can be placed in footnotes or glossaries where, again, the more expert reader can bypass them. We have much more to say about such matters in Chapters 7, 9, 10, and 11, where we discuss definition, document design, and graphics. Remember that all such devices and techniques are used to satisfy the needs of your readers. Therefore, always begin by considering your audience and consider it at every stage of the writing process, from preliminary research to final formatted document.

Planning and Revision Checklists

The following questions are a summary of the key points in this chapter, and they provide a checklist for planning and revising any document for your readers.

PLANNING

- How many readers will you have?
- Is one person (or any one group) your primary reader? Are there secondary readers? **Consider both primary and secondary readers when answering the questions that follow.**
- What is your relationship to the readers?
- What do you hope to accomplish with your document? What is your primary purpose? Secondary? Other?
- Why will the readers read your document? For enjoyment and interest? To learn a skill? To perform a task? To make a decision? To gain knowledge in the subject area? Other?
- What are the readers' interest, knowledge, and experience in the subject of your document? Do they have the necessary schema and background to assimilate your material easily? If not, what can you provide to help them? Explanations? Definitions? Graphics? Analogies? Examples? Anecdotes?
- What are the readers' attitudes toward your subject matter?
- What are your readers' attitudes toward you?
- If your document includes conclusions and recommendations, how will your readers feel about them? Enthusiastic? Friendly? Hostile? Skeptical? Indifferent?
- Is self-interest involved in the readers' attitudes? For instance, will the readers benefit or lose as a result of your document? Will the readers feel threatened or be made angry by the document?
- If the readers have negative attitudes (hostile, skeptical, indifferent), what can you do to overcome those attitudes? Provide more support? Point out long-term benefits? Soften your language? Make your presentation more interesting?
- Are you writing within a recognizable discourse community such as a corporation or a professional group? If so, are you familiar and comfortable with the customs and voice of that community? Is there a prevailing style and tone? What arguments are most persuasive? Do you have access to models of similar documents you can use to help you?
- Are you writing across cultural boundaries? Do you need to adjust your tone or style accordingly?
- How are the primary readers likely to read the report? Skim? Scan? Search read? Read for comprehension? Read critically? How will the secondary readers read? Given how your readers will read your document, what can you do to help your readers get the most from it? Will some of the following help: Table of contents? Graphics? Questions? Introductions and transitions? Summaries? Checklists?

REVISION

- Have you written your text at a level appropriate for your readers, both primary and secondary? Do you need more or fewer definitions, examples, descriptions, and explanations? Have you provided analogies and metaphors when they would be useful? Have you moved from the familiar to the unfamiliar?
- Have you met your objectives for each audience addressed?
- Will your readers feel that their objectives have been met?
- Is your purpose clearly stated in your introduction?
- Are your conclusions and recommendations clearly stated? Can your readers easily find them?
- Does your format allow your reader to read selectively? Do your introductions and transitions forecast what is to come? Have you furnished a table of contents if needed? Do you have enough headings and subheadings to guide your readers? Would review questions, summaries, or checklists be useful?
- Have you furnished good graphic support? Have you helped your reader when possible by using pictures, flowcharts, and diagrams?
- Have you achieved the proper tone? Have you inadvertently said anything rude or misleading? Have you taken cultural differences into account?
- Does your language distort or sensationalize your material in any way?

Exercises

1. Think about some concept in your discipline that you understand thoroughly. Write a letter explaining that concept to someone you know who does not understand it: a good friend, a parent, a sister, or a brother. In your planning, use the checklist provided in this chapter. Include graphics if you think they will help. Your goal is to interest your reader and teach him or her something about a world you understand and enjoy.
2. Take a journal article that discusses some development or concept in your field that may have practical value. For example, it may suggest a new method of producing a product, a new market for an old product, or a service that could be performed for a fee. Imagine an executive who might find a discussion of this concept or development useful. Assume this executive is intelligent and educated but that he or she knows little about the subject matter involved. Write that executive a memorandum in which you explain the concept or development, and fully discuss its implications to the executive's organization. Use graphics if they will help. Use the memorandum format shown in Chapter 12, "Correspondence."
3. Compare and contrast an article or report written for one kind of audience with another written for a different audience—perhaps an article

from *Science* with one from *Forbes.* Get articles on similar subjects and, if possible, in your discipline. Ask and answer some of these questions about the articles.

- What stylistic similarities do you see between the two articles? What dissimilarities? Which article has greater complexity? What indicates this complexity? What is the average sentence length in each article? Paragraph length? Which article defines the most terms?
- What is the author's major concern in each article? How do these concerns contrast?
- What similarities and dissimilarities of format and arrangement do you see?
- Are there differences in the kinds of graphics used in the two articles?
- Which article presents the most detail? Why? What kind of detail is presented in each article?
- How much background information are you given in each article? Does either article refer you to other books or articles?

4. Perhaps you are already planning a long paper for your writing course. Using the Planning Checklist in this chapter, write an analysis of the anticipated audience. Prepare your analysis as a memorandum addressed to your instructor. Use the memorandum format shown in Chapter 12, "Correspondence."

CHAPTER 5 ACHIEVING A READABLE STYLE

A readable text is one that an intended reader can comprehend without difficulty. Many things can make a text difficult to read. For example, the content may include unexplained concepts that the reader does not understand. Material that is new to the reader may not be explained in terms of material already familiar to the reader. The material may not be arranged or formatted in a way to make it accessible to the reader. We cover such aspects of readability elsewhere, notably in Chapter 2, "Composing"; Chapter 4, "Writing for Your Readers"; and Chapter 9, "Document Design." In this chapter, we deal with style elements at the paragraph, sentence, and word level that can make your text clearer and more readable.

Examples of unclear writing style are all too easy to find, even in places where we would hope to find clear, forceful prose. Read the following sentence:

> While determination of specific space needs and access cannot be accomplished until after a programmatic configuration is developed, it is apparent that physical space is excessive and that all appropriate means should be pursued to assure that the entire physical plant is utilized as fully as feasible.

This murky sentence comes from a report issued by a state higher education coordinating board. Actually, it's better than many examples we could show you. Although difficult, the sentence is probably readable. Others are simply indecipherable. When you have finished this chapter, you should be able to analyze a passage like the one just cited and show why it is so unclear. You should also know how to keep your own writing clear, concise, and vigorous. We discuss paragraphs, lists, clear sentence structure, specific words, and pomposity. We have broken our subject into five parts for simplicity's sake, but all the parts are closely related. All have one aim: readability.

If there is a style checker in your word processor, it will incorporate many of the principles we discuss in this chapter. Nevertheless, use it with great care. Style checkers used without understanding the principles involved in good style can be highly misleading. (We discuss style checkers and problems with them more fully in Chapter 2, "Composing."

The Paragraph

In Chapters 6, 7, and 8, we discuss ways to inform, define and describe, and argue. These strategies may be used not only to develop reports but also to develop paragraphs within reports. Thus, paragraphs will vary greatly in arrangement and length, depending upon their purpose.

The Central Statement

In technical writing, the **central statement** of a paragraph more often than not appears at the beginning of the paragraph. This placement provides the clarity of statement that good technical writing must have.

In a paragraph aimed at persuasion, however, the central statement may appear at the close, where it provides a suitable climax for the argument. Wherever you place the central statement, you can achieve unity by relating all the other details of the paragraph to the statement, as in this paragraph of inference and speculation:

We have underlined the central statement concerning the electronic library. The rest of the paragraph presents facts and speculation in support of the central statement.

> <u>The electronic library is coming.</u> Publishers are preparing for this eventuality. Already electronic storage of text is more economical and more compact than storage of paper. Whereas the conventional library expends much effort in managing the whereabouts of physical objects—books, maps, serials, and so forth—the electronic library will have no such objects to keep track of. Circulation control will be a thing of the past because the electronic book will always be available to users. Books will still be available for convenient reading and study; they will not become obsolete. They will, however, be printed mainly on demand and on the user's electronic printer. I expect to see specialty firms offering leather-bound volumes for scholars to order, and in this way all the familiar book formats will be preserved. Like the paperback publisher, the electronic publisher will distribute much more information than was previously available, and the information will be far less expensive to distribute and maintain.[1]

Paragraph Length

Examination of well-edited magazines such as *Scientific American* reveals that their paragraphs seldom average more than 100 words in length. Magazine editors know that paragraphs are for the reader. Paragraphing breaks the material into related subdivisions to enhance the reader's understanding. When paragraphs are too long, the central statements that provide the generalizations needed for reader understanding are either missing or hidden in the mass of supporting details.

In addition to considering the readers' need for clarifying generalizations, editors also consider the psychological effect of their pages. They know that large blocks of unbroken print have a forbidding appearance that intimidates the reader. If you follow the practice of experienced editors, you will break your paragraphs whenever your presentation definitely takes a new turn. As a general rule, paragraphs in reports and articles should average 100 words or fewer. In letters and memorandums, because of their page layout, you should probably hold average paragraph length to fewer than 60 words.

Transitions

Generally, a paragraph presents a further development in a continuing sequence of thought. In such a paragraph, the opening central statement will be so closely related to the preceding paragraph that it usually provides a sufficient transition. When a major transition between ideas is called for, consider using a short paragraph to guide the reader from one idea to the next.

The following four paragraphs provide an excellent example of paragraph development and transition:

Fossil is the key word. Fossil or fossils appears six times. Its repetition provides a major transitional device. The first sentence expresses the central theme of the four paragraphs: fossils provide clues to the past. The rest of the paragraph explains how geologists can use fossils to distinguish ancient seas from ancient land.

> The distribution of fossils (skeletons, shells, leaf impressions, footprints and dinosaur eggs) in rocks of a certain age tells something about the ancient distribution of lands and seas on the Earth's surface. The remains of coral and clamshells found in the very old limestones in parts of Pennsylvania and New York indicate that this region was once covered by a shallow sea. Similarly, the remains of ancestral horses and camels in rocks of South Dakota show that the area was then dry land or that land was nearby.

The central statement of the second paragraph shows its relationship to the central theme through the use of the word fossil. The phrase "will tell even more" leads into the subject of this paragraph, that fossils not only identify ancient areas of sea and land but also help define the shoreline of the sea.

> A closer look at these fossils will tell even more. Their distribution identifies the ancient areas of land and sea and also determines the approximate shoreline. The distribution of living forms shows that thick-shelled fossil animals once lived in shallow seas close to shore, where their shells were built to withstand the surging and pounding of waves. Thin-shelled, delicate fossil animals probably lived in deeper, calmer water offshore.

The phrase "In addition" marks the transition here. The central statement in this paragraph looks briefly back to the preceding paragraph and then announces the new subject: fossils provide information about the water temperature of the ancient seas. Many central statements provide transition by looking both forward and backward.

> In addition to providing a measure of water depth, fossils can also be used to indicate the former temperature of water. In order to survive, certain types of present-day coral must live in warm and shallow tropical saltwaters, such as the seas around Florida and the Bahamas. When similar types of coral are found in the ancient limestones, they provide a good estimate of the marine environment that must have existed when they were alive.

The last paragraph rounds off the passage by summarizing the main points covered in the preceding three paragraphs.

> All these factors—depth, temperature, currents, and salinity—that are revealed by fossils are important, for each detail tends to sharpen and clarify the picture of ancient geography.[2]

The four paragraphs illustrate that you will develop paragraphs coherently when you keep your mind on the central theme. If you do so, the words needed to provide proper transition will come naturally. More often than not, your transitions will be repetitions of key words and phrases, supported by such simple expressions as *also, another, of these four, because of this development, so, but,* and *however.* When you wander away from your central theme, no amount of artificial transition will wrench your writing back into coherence.

Lists and Tables

One of the simplest things you can do to ease the reader's chore is to break down complex statements into lists. Visualize the printed page. When it appears as an unbroken mass of print, it intimidates readers and makes it harder for them to pick out key ideas. Get important ideas out into the open where they stand out. Lists help to clarify introductions and summaries. You may list by (1) starting each separate point on a new line, leaving plenty of white space around it, or (2) using numbers within a line, as we have done here. Examine the following summary from a student paper, first as it might have been written and then as it actually was:

> The exploding wire is a simple-to-perform yet very complex scientific phenomenon. The course of any explosion depends not only on the material and shape of the wire but also on the electrical parameters of the circuit. In an explosion the current builds up and the wire explodes, current flows during the dwell period, and "post-dwell conduction" begins with the reignition caused by impact ionization. These phases may be run together by varying the circuit parameters.

Now, the same summary as a list:

> The exploding wire is a simple-to-perform yet very complex scientific phenomenon. The course of any explosion depends not only on the materials and shape but also on the electrical parameters of the circuit.
>
> An explosion consists primarily of three phases:
>
> 1. The current builds up and the wire explodes.
> 2. Current flows during the dwell period.
> 3. "Postdwell conduction" begins with the reignition caused by impact ionization.
>
> These phases may be run together by varying the circuit parameters.

The first version is clear, but the second version is clearer, and readers can now file the process in their minds as "three phases." They will remember it longer.

Some writers avoid using lists even when they should use them, so we hesitate to suggest any restrictions on the practice. Obviously, there are some subjective limits. Lists break up ideas into easy-to-read, easy-to-understand bits, but too many can make your page look like a laundry list. Also, some journal editors object to lists in which each item starts on a separate line. Such lists take space, and space costs money. Use lists when they clarify your presentation, but use them prudently.

Tables perform a function similar to lists. You can use them to present a good deal of information—particularly statistical information—in a way easy for the reader to follow and understand. We discuss tables and their functions in Chapter 11, "Graphical Elements."

Clear Sentence Structure

The basic English sentence structure follows two patterns, *subject-verb-object* (SVO) and *subject-verb-complement* (SVC):

Americans(S) love(V) ice cream(O).

She(S) planned(V) carefully(C).

Around such simple sentences as "Americans love ice cream," the writer can hang a complex structure of words, phrases, and clauses that modify and extend the basic idea. In this case, the writer actually wrote "Americans love ice cream, but ice cream is made from whole milk and cream and therefore contains a considerable amount of saturated fat and dietary cholesterol."

In this section on clear sentence structure, we discuss how to extend your sentences without losing clarity. We discuss sentence length, sentence order, sentence complexity and density, active verbs, active and passive voice, and first-person point of view.

Sentence Length

Many authorities have seen sentence length as an indicator of how difficult a sentence is. More recent research has found that although sentence length and word length may be indicators, they are not the primary causes of difficulty in reading sentences. Rather, the true causes may be the use of difficult sentence structures and words unfamiliar to the reader. This position is summed up well in this statement:

> A sentence with 60, 100, or 150 words needs to be shortened; but a sentence with 20 words is not necessarily more understandable than a sentence with 25 words. The incredibly long sentences that are sometimes found in technical, bureaucratic, and legal writing are also sentences that have abstract nouns as subjects, buried actions, unclear focus, and intrusive phrases. These are the problems that must be fixed, whether the sentence has 200 words or 10.
>
> Similarly, short words are not always easier words. The important point is not that the words be short, but that your readers know the words you are using. [3]

In general we agree with such advice. Sentence density and complexity cause readers more grief than does sentence length alone. Nevertheless, it's probably worth keeping in mind that most professional writers average only slightly more than 20 words per sentence. Their sentences may range from short to fairly long, but, for the most part, they avoid sentences like this one from a bank in Houston, Texas:

> You must strike out the language above certifying that you are not subject to backup withholding due to notified payee underreporting if you have been notified that you are subject to backup withholding due to notified payee underreporting, and you have not received a notice from the Internal Revenue Service advising you that backup holding has terminated.

Sentence Order

What is the best way to order a sentence? Is a great deal of variety in sentence structure the mark of a good writer? One writing teacher, Francis Christensen of the University of Southern California, looked for the answers to those two questions. He examined large samples from 20 successful writers, including John O'Hara, John Steinbeck, William Faulkner, Ernest Hemingway, Rachel Carson, and Gilbert Highet. In his samples, he included 10 fiction writers and 10 nonfiction writers.[4]

What Christensen discovered seems to disprove any theory that good writing requires extensive sentence variety. The writers whose work was examined depended mostly on basic sentence patterns. They wrote 75.5% of their sentences in plain **subject-verb-object (SVO)** or **subject-verb-complement (SVC)** order, as in these two samples:

> Doppler radar increases capability greatly over conventional radar. (SVO)
>
> Doppler radar can be tuned more rapidly than conventional radar. (SVC)

Another 23% of the time, the professionals began sentences with short **adverbial openers:**

> Like any radar system, Doppler does have problems associated with it.

These adverbial openers are most often simple prepositional phrases or single words such as *however, therefore, nevertheless,* and other conjunctive adverbs. Generally, they provide the reader with a transition between thoughts. Following the opening, the writer usually continues with a basic SVO or SVC sentence.

These basic sentence types—*SVO(C)* or *adverbial + SVO(C)*—were used 98.5% of the time by the professional writers in Christensen's sample. What did the writers do with the remaining 1.5% of their sentences? For 1.2%, they opened the sentence with **verbal clauses** based on participles and infinitives such as "*Breaking* ground for the new church," or "*To see* the new pattern more clearly." The verbal opener was again followed most often with an SVO or SVC sentence, as in this example:

> Looking at it this way, we see the radar set as basically a sophisticated stopwatch that sends out a high-energy electromagnetic pulse and measures the time it takes for part of that energy to be reflected back to the antenna.

Like the adverbial opener, the verbal opener serves most of the time as a transition.

The remaining 0.3% of the sentences (about 1 sentence in 300) were **inverted constructions,** in which the subject is delayed until after the verb, as in this sentence:

No less important to the radar operator are the problems caused by certain inherent characteristics of radar sets.

What can we conclude from Christensen's study? Simply this: professional writers are interested in getting their content across, not in tricky word order. They convey their thoughts in clear sentences not clouded by extra words. You should do the same.

Sentence Complexity and Density

Research indicates that sentences that are too complex in structure or too dense with content are difficult for many readers to understand.[5] Basing our observations on this research, we wish to discuss four particular problem areas: openers in front of the subject, too many words between the subject and the verb, noun strings, and multiple negatives.

Openers in Front of the Subject As Christensen's research indicates, professional writers place an adverbial or verbal opener before their subjects about 25% of the time. When these openers are held to a reasonable length, they create no problems for readers. The problems occur when the writer stretches such openers beyond a reasonable length. What is *reasonable* is somewhat open to question and depends to an extent on the reading ability of the reader. However, most would agree that the 27 words and 5 commas before the subject in the following sentence make the sentence difficult to read:

Opening phrase too dense

Because of their ready adaptability, ease of machining, and aesthetic qualities that make them suitable for use in landscape structures such as decks, fences, steps, and retaining walls, preservative-treated timbers are becoming increasingly popular for use in landscape construction.

The ideas contained in this sentence become more accessible when spread over two sentences:

Puts central idea before supporting evidence

Preservative-treated timbers are becoming increasingly popular for use in landscape construction. Their ready adaptability, ease of machining, and aesthetic qualities make them highly suited for use in structures such as decks, fences, steps, and retaining walls.

The second version has the additional advantage of putting the central idea in the sequence before the supporting information.

The conditional sentence is a particularly difficult type of sentence in which the subject is too long delayed. You can recognize the conditional by its *if* beginning:

Subject too long delayed

If heat (20°–35° C or 68°–95° F optimum), moisture (20%+ moisture content in wood), oxygen, and food (cellulose and wood sugars) are present, spores will germinate and grow.

To clarify such a sentence, move the subject to the front and the conditions to the rear. Consider the use of a list when you have more than two conditions:

List helps to clarify

Spores will germinate and grow when the following elements are present:

- Heat (20°–35° C or 68°–95° F optimum)
- Moisture content (20%+ moisture content in wood)
- Oxygen
- Food (cellulose and wood sugars)

Words Between Subject and Verb In the following sentence, too many words between the subject and the verb cause difficulty:

Subject and verb too widely separated

Creosote, a brownish-black oil composed of hundreds of organic compounds, usually made by distilling coal tar, but sometimes made from wood or petroleum, has been used extensively in treating poles, piles, cross-ties, and timbers.

The sentence is much easier to read when it is broken into three sentences and first things are put first:

Revised

Creosote has been used extensively in treating poles, piles, cross-ties, and timbers. It is a brownish-black oil composed of hundreds of organic compounds. Creosote is usually made by distilling coal tar, but it can also be made from wood or petroleum.

You might break down the original sentence into only two sentences if you felt your audience could handle denser sentences:

Revised

Creosote, a brownish-black oil composed of hundreds of organic compounds, has been used extensively in treating poles, piles, cross-ties, and timbers. It is usually made by distilling coal tar, but it can also be made from wood or petroleum.

Noun Strings Noun strings are another way in which writers sometimes complicate and compress their sentences beyond tolerable limits. A noun string is a sequence of nouns that modifies another noun; for example, in the phrase *multichannel microwave radiometer,* the nouns *multichannel* and *microwave* modify *radiometer.* Sometimes the string may also include an adjective, as in *special multichannel microwave radiometer.*

Nothing is grammatically wrong with the use of nouns as modifiers. Such use is an old and perfectly respectable custom in English. Expressions such as *firefighter* and *creamery butter,* in which the modifiers are nouns, go virtually unnoticed. The problem occurs when writers string many nouns together in one sequence or use many noun strings in a passage, as shown in this paragraph:

Seven noun strings in one paragraph is excessive

> We must understand who the initiators of water-oriented greenway efforts are before we can understand the basis for community environment decision making processes. State government planning agencies and commissions and designated water quality planning and management agencies have initiated such efforts. They have implemented water resource planning and management studies and have aided volunteer group greenway initiators by providing technical and coordinative assistance.[6]

In many such strings, the reader has great difficulty in sorting out the relationships among the words. In *volunteer group greenway initiators*, does *volunteer* modify *group* or *initiators*? The reader has no way of knowing.

The solution to untangling difficult noun strings is to include relationship clues such as prepositions, relative pronouns, commas, apostrophes, and hyphens. For instance, a hyphen in *volunteer-group* indicates that *volunteer* modifies *group*. The strung-out passage just quoted was much improved by the inclusion of such clues:

Relationship clues help to clarify noun strings

> We must understand who the initiators of efforts to promote water-oriented greenways are before we can understand the process by which a community makes decisions about environmental issues. Planning agencies and commissions of the state government and agencies that have been designated to plan and manage water quality have initiated such efforts. They have implemented studies on planning and managing water resources and have aided volunteer groups that initiate efforts to promote greenways by providing them with technical advice and assistance in coordinating their activities.[7]

The use of noun strings in technical English will no doubt continue. They do have their uses, and technical people are very fond of them. But perhaps it would not be too much to hope that writers would hold their strings to three words or fewer and not use more than one per paragraph.

Multiple Negatives Writers introduce excessive complexity into their sentences by using multiple negatives. By *multiple negative*, we do not mean the grammatical error of the *double negative*, as in "He does *not* have *none* of them." We are talking about perfectly correct constructions that include two or more negative expressions, such as these:

Negative statements

- We will not go unless the sun is shining.
- We will not pay except when the damages exceed $50.
- The lever will not function until the power is turned on.

The positive versions of all of these statements are clearer than the negative versions:

Positive statements

- We will go only if the sun is shining.
- We will pay only when the damages exceed $50.
- The lever functions only when the power is turned on.

Research shows that readers have difficulty sorting out passages that contain multiple negatives. If you doubt the research, try your hand at interpreting this government regulation (underlining for negatives is ours):

Excessive use of negatives

> §928.310 Papaya Regulation 10. Order. (a) No handler shall ship any container of papayas (except immature papayas handled pursuant to §928.152 of this part):
>
> (1) During the period January 1 through April 15, 1980, to any destination within the production area unless said papayas grade at least Hawaii No. 1, except that allowable tolerances for defects may total 10 percent. Provided, that not more than 5 percent shall be for serious damage, not more than 1 percent for immature fruit, not more than one percent for decay: Provided further, that such papayas shall individually weigh not less than 11 ounces each.[8]

Active Verbs

The verb determines the structure of an English sentence. Many sentences in technical writing falter because the finite verb does not comment upon the subject, state a relationship about the subject, or relate an action that the subject performs. Look at the following sentence:

Action in a noun

> Sighting of the coast was accomplished by the pilot at 7 A.M.

English verbs can easily be changed into nouns, but sometimes, as we have just seen, the change can lead to a faulty sentence. The writer has put the true action into the subject and subordinated the pilot and the coast as objects of prepositions. The sentence should read:

Action in a verb

> At 7 A.M. the pilot saw the coast.

The poor writer can ingeniously bury the action of a sentence almost anywhere. With the common verbs *make, give, get, have,* and *use,* the writer can bury the action late in the sentence in an object:

Action in an object

> The manager has the task of ensuring safe conditions on the assembly line.

or

> The speaker did not give an adequate explanation of the technique.

Properly revised, the sentences put the action where it belongs, in the verb:

Action in a verb

> The manager must ensure safe conditions on the assembly line.
>
> The speaker did not adequately explain the technique.

The poor writer can even bury the action in an adjective:

Action in an adjective

> A new discovery produces an excited reaction in a scientist.

Revised:

Action in a verb

> A new discovery excites a scientist.

When writing, and particularly when rewriting, you should always ask yourself: "Where's the action?" If the action does not lie in the verb, rewrite the sentence to put it there, as in this sample:

Action in nouns

> Music therapy is the scientific application of music to accomplish the restoration, maintenance, and improvement of mental health.

This sentence provides an excellent example of how verbs are frequently turned into nouns by the use of the suffixes *-ion, -ance* (or *-ence*), and *-ment.* If you have sentences full of such suffixes, you may not be writing as actively as you could be. Rewritten to put active ideas into verb forms, the sentence reads this way:

Action in verbs

> Music therapy applies music scientifically to restore, maintain, and improve mental health.

The rewritten sentence defines "music therapy" in one-third less language than the first sentence, without any loss of meaning or content.

Active and Passive Voice

We discuss active and passive voice sentences in Chapter 7, but let us quickly explain the concept here. In an active voice sentence, the subject performs the action and the object receives the action, as in "The heart pumps the blood." In a passive voice sentence, the subject *receives* the action, as in "The blood is pumped." If you want to include the doer of the action, you must add this information in a prepositional phrase, as in "The blood is pumped *by the heart.*" We urge you to use the active voice more than the passive. As the *CBE Style Manual* published by the Council of Biology Editors points out, "The active is the natural voice in which people usually speak or write, and its use is less likely to lead to wordiness and ambiguity."[9]

However, you should not ignore the passive altogether. The passive voice is often useful. You can use the passive voice to emphasize the object receiving the action. The passive voice in "Influenza may be caused by any of several viruses" emphasizes *influenza.* The active voice in "Any of several viruses may cause influenza" emphasizes the *viruses.*

Often the agent of action is of no particular importance. When such is the case, the passive voice is appropriate because it allows you to drop the agent altogether:

Appropriate passive

Edward Jenner's work on vaccination was published in 1796.

Be aware, however, that inappropriate use of the passive voice can cause you to omit the agent when knowledge of the agent may be vital. Such is often the case in giving instructions:

Poor passive

All doors to this building will be locked by 6 P.M.

This sentence may not produce locked doors until it is rewritten in the active voice:

Active voice

The night manager will lock all doors to this building by 6 P.M.

Also, the passive voice can lead to dangling participles, as in this sentence:

Passive with dangling modifier

While conducting these experiments, the chickens were seen to panic every time a hawk flew over.

Chickens conducting experiments? Not really. The active voice straightens out the matter:

Active voice

While conducting these experiments, we saw that the chickens panicked every time a hawk flew over.

(See also "Dangling Modifier," in Part V, "Handbook.")

Although the passive voice has its uses, too much of it produces lifeless and wordy writing. Therefore, use it only when it is clearly appropriate.

First-Person Point of View

Once, reports and scientific articles were typically written in the third person—"This investigator has discovered"—rather than first person—"I discovered." The *CBE Style Manual* labels this practice the "passive of modesty" and urges writers to avoid it.[10] Many other style manuals for scientific journals now recommend the first person and advise against the use of the third person on the grounds that it is wordy and confusing. We agree with this advice.

The judicious use of *I* or *we* in a technical report is entirely appropriate. Incidentally, such usage will seldom lead to a report full of *I's* and *we's*. After all, there are many agents in a technical report other than the writer. In describing an agricultural experiment, for example, researchers will report how *the sun shone, photosynthesis occurred, rain fell, plants drew nutrients from the soil,* and *combines harvested.* Only occasionally will researchers need to report their own actions. But when they must, they should be able to avoid such roundabout expressions as "It was observed by this experimenter." Use "*I* observed" instead. Use "We observed" when there are two or more experimenters.

A Caution about Following Rules

We must caution you before we leave this section on clear sentence structure. We are not urging upon you an oversimplified primer style, one often satirized by such sentences as "Jane hit the ball" and "See Dick catch the ball." Mature styles have a degree of complexity to them. Good writers, as Christensen's research shows, do put information before the subject. Nothing is wrong with putting information between the subject and verb of a sentence. You will find many such sentences in this book. However, you should be aware that research shows that sentences that are too long, too complex, or too dense cause many readers difficulty. Despite increasingly good research into its nature, writing is a craft and not a science. Be guided by the research available, but do not be simplistic in applying it.

Specific Words

The semanticists' abstraction ladder is composed of rungs that ascend from very specific words such as *table* to abstractions such as *furniture, wealth,* and *factor.* The human ability to move up and down this ladder enabled us to develop language, on which all human progress depends. Because we can think in abstract terms, we can call a moving company and tell it to move our furniture. Without abstraction, we would have to bring the movers into our house and point to each object we wanted moved. Like many helpful writing techniques, however, abstraction is a device you should use carefully.

Stay at an appropriate level on the abstraction ladder. Do not say "inclement weather" when you mean "rain." Do not say "overwhelming support" when you mean "62% of the workers supported the plan." Do not settle for "suitable transportation" when you mean "a bus that seats 32 people."

Writing that uses too many abstractions is lazy writing. It relieves writers of the need to observe, to research, and to think. They can speak casually of "factors," and neither they nor their readers really know what they are talking about. Here is an example of such lazy writing.

The writer was setting standards for choosing a desalination plant to be used at Air Force bases.

Too abstract

- The cost must not be prohibitive.
- The quantity of water must be sufficient to supply a military establishment.
- The quality of the water must be high.

The writer here thinks he has said something. He has said little. He has listed slovenly abstractions when, with a little thought and research, he could have listed specific details. He should have said:

Use of specific detail

- The cost should not exceed $3 per thousand gallons.
- To supply an average base with a population of 5,000, the plant should purify 750,000 gallons of water a day (AFM 88-10 sets the standard of 150 gallons a day per person).
- The desalinated water should not exceed the national health standard for potable water of 500 parts per million of dissolved solids.

Abstractions are needed for generalizing, but they cannot replace specific words and necessary details. Words mean different things to different people. The higher you go on the abstraction ladder, the truer this is. The abstract words *prohibitive, sufficient,* and *high* could be interpreted in as many different ways as the writer had readers. No one can misinterpret the specific details given in the rewritten sentences.

Abstractions can also burden sentences in another way. Some writers are so used to thinking abstractly that they begin a sentence with an abstraction and *then* follow it with the specific word, usually in a prepositional phrase. They write,

Poor

The problem of producing fresh water became troublesome at overseas bases.

Instead of

Revised

Producing fresh water became a problem at overseas bases.

Or

Poor

The circumstance of the manager's disapproval caused the project to be dropped.

Instead of

Revised

The manager's disapproval caused the project to be dropped.

We do not mean to say you should never use high abstractions. A good writer moves freely up and down the abstraction ladder. But when you use words from high on the ladder, use them properly—for generalizing and as a shorthand way of referring to specific details you have already given.

Pomposity

State your meaning as simply and clearly as you can. Do not let the mistaken notion that writing should be more elegant than speech make you sound pompous. Writing *is* different from speech. Writing is more concise, more compressed, and often better organized than speech. But elegance is not a prerequisite for good writing.

A sign at a gas station reads, "No gas will be dispensed while smoking." Would the attendants in that service station speak that way? Of course not. They would say, "Please put out that cigarette" or "No smoking, please." But the sign had to be elegant, and the writer sounds pompous, and illiterate as well.

If you apply what we have already told you about clear sentence structure, you will go a long way toward tearing down the fence of artificiality between you and the reader. We want to touch on just three more points: empty words, elegant variation, and pompous vocabulary.

Empty Words

The easiest way to turn simple, clear prose into elegant nonsense is to throw in empty words, such as these phrases that begin with the impersonal *it:* "It is evident," "It is clear that," or, most miserable of all, "It is interesting to note that." When something is evident, clear, or interesting, readers will discover this for themselves. If something is not evident, clear, or interesting, rewrite it to make it so. When you must use such qualifying phrases, at least shorten them to "evidently," "clearly," and "note that." Avoid constructions like "It was noted by Jones." Simply say, "Jones noted."

Many empty words are jargon phrases writers throw in by sheer habit. You see them often in business correspondence. A partial list follows:

to the extent that
with reference to
in connection with
relative to
with regard to
with respect to
is already stated
in view of
inasmuch as
with your permission
hence
as a matter of fact

We could go on, but so could you. When such weeds crop up in your writing, pull them out.

Another way to produce empty words is to use an abstract word in tandem with a specific word. This produces such combinations as

20 in number *for* 20
wires of thin size *for* thin wires
red in color *for* red

When you have expressed something specifically, do not throw in the abstract term for the same word.

Elegant Variation

Elegant variation will also make your writing sound pompous.[11] **Elegant variation** occurs when a writer substitutes one word for another because of an imagined need to avoid repetition. This substitution can lead to two problems: The substituted word may be a pompous one, or the variation may mislead the reader into thinking that some shift in meaning is intended. Both problems are evident in the following example:

Elegant variation

> Insect damage to evergreens varies with the condition of the plant, the pest species, and the hexapod population level.

Confusion reigns. The writer has avoided repetition, but the reader may think that the words *insect, pest,* and *hexapod* refer to three different things. Also, *hexapod,* though a perfectly good word, sounds a bit pompous in this context. The writer should have written,

Revised

> Insect damage to evergreens varies with the condition of the plant, the insect species, and the insect population level.

Remember also that intelligent repetition provides good transition. Repeating key words reminds the reader that you are still dealing with your central theme (see pages 87–88).

Pompous Vocabulary

Generally speaking, the vocabulary you think in will serve in your writing. Jaw-breaking thesaurus words and words high on the abstraction ladder will not convince readers that you are intellectually superior. Such words will merely convince readers that your writing is hard to read. We are not telling you here that you must forego your hard-won educated vocabulary. If you are writing for readers who understand words such as *extant* or *prototype,* then use them. But use them only if they are appropriate to your discussion. Don't use them to impress people.

Nor are we talking about the specialized words of your professional field. At times these are necessary. Just remember to define them if you feel your reader will not know them. What we are talking about is the desire some writers seem to have to use pompous vocabulary to impress their readers.

The following list is a sampling of heavy words and phrases along with their simpler substitutes.

accordingly: so
acquire: get
activate: begin
along the lines of: like
assist: help
compensation: pay
consequently: so
due to the fact that: because
facilitate: ease, simplify
for the purpose of: for
in accordance with: by, under
in connection with: about
initiate: begin
in order to: to
in the event that: if
in the interests of: for
in this case: here
make application to: apply
nevertheless: but, however
prior to: before
subsequent to: later, after
utilize: use

You would be wise to avoid the word-wasting phrases on this list and other phrases like them. You really don't need to avoid the single words shown, such as *acquire* and *assist.* All are perfectly good words. But to avoid sounding pompous, don't string large clumps of such words together. Be generous in your writing with the simpler substitutes we have listed. If you don't, you are more likely to depress your readers than to impress them. Don't be like the pompous writers who seek to bury you under the many-syllable words they use to express one-syllable ideas, as in this example from the U.S. Department of Transportation.

Pompous writing

> The purpose of this PPM [Policy and Procedure Memorandum] is to ensure, to the maximum extent practicable, that highway locations and designs reflect and are consistent with Federal, State and local goals and objectives. The rules, policies, and procedures established by this PPM are intended to afford full opportunity for effective public participation in the consideration of highway location and design proposals by highway departments before submission to the Federal Highway Administration for approval. They provide a medium for free and open discussion and are designed to encourage early and amicable resolution of controversial issues that may arise.

We urge you to read as much good writing—both fiction and nonfiction—as time permits. Stop occasionally as you do and study the author's choice of words. You will find most authors to be lovers of the short word. Numerous passages in Shakespeare are composed almost entirely of one-syllable words. The same holds true for the King James Bible. Good writers do not want to impress you with their vocabularies. They want to get their ideas from their heads to yours by the shortest, simplest route.

Good Style in Action

A final example will summarize much that we have said. Insurance policies were verbal bogs for so long that most buyers of insurance gave up on finding one clearly written. However, the St. Paul Fire and Marine Insurance Company decided that it was both possible and desirable to simplify the wording of its policies. The company revised one of its policies, eliminating empty words and using only words familiar to the average reader. In the revision, the company's writers avoided excessive sentence complexity and used predominantly the active voice and active verbs. They broke long paragraphs into shorter ones. The insurance company became *we* and the insured *you*. Definitions were included where needed rather than segregated in a glossary.

The resulting policy is wonderfully clear. Compare a paragraph of the old with the new.[12]

Old:

Cancellation

Passive voice

Average sentence length: 29 words

Empty words

Unfamiliar words

All one paragraph

A 66-word sentence

This policy may be canceled by the Named Insured by surrender thereof to the Company or any of its authorized agents, or by mailing to the Company written notice stating when thereafter such cancellation shall be effective. This policy may be canceled by the Company by mailing to the Named Insured at the address shown in this Policy written notice stating when, not less than thirty (30) days thereafter, such cancellation shall be effective. The mailing of notice as aforesaid shall be sufficient notice and the effective date of cancellation stated in the notice shall become the end of the policy period. Delivery of such written notice either by the Named Insured or by the Company shall be equivalent to mailing. If the Named Insured cancels, earned premium shall be computed in accordance with the customary short rate table and procedure. If the Company cancels, earned premium shall be computed pro rata. Premium adjustment may be made at the time cancellation is effected or as soon as practicable thereafter. The check of the Company or its representative, mailed or delivered, shall be sufficient tender of any refund due the Named Insured. If this contract insures more than one Named Insured, cancellation may be effected by the first of such Named Insureds for the account of all the Named Insureds; notice of cancellation by the Company to such first Named Insured shall be deemed notice to all Insureds and payment of any unearned premium to such first Named Insured shall be for the account of all interests therein.

New:

Active voice

Average sentence length: 15 words

Clear, specific language

Short paragraphs

Can This Policy Be Canceled?

> Yes it can. Both by you and by us. If you want to cancel the policy, hand or send your cancellation notice to us or our authorized agent. Or mail us a written notice with the date when you want the policy canceled. We'll send you a check for the unearned premium, figured by the short rate table—that is, pro rata minus a service charge.
>
> If we decide to cancel the policy, we'll mail or deliver to you a cancellation notice effective after at least 30 days. As soon as we can, we'll send you a check for the unearned premium, figured pro rata.

Examples that substitute specific, familiar words for the high abstractions of the original policy are used freely. For instance:

> You miss a stop sign and crash into a motorcycle. Its 28-year-old married driver is paralyzed from the waist down and will spend the rest of his life in a wheelchair. A jury says you have to pay him $1,300,000. Your standard insurance liability limit is $300,000 for each person. We'll pay the balance of $1 million.

Or:

> We'll defend any suit for damages against you or anyone else insured even if it's groundless or fraudulent. And we'll investigate, negotiate and settle on your behalf any claim or suit if that seems to us proper and wise.
>
> You own a two-family house and rent the second floor apartment to the Miller family. The Millers don't pay the rent and you finally have to evict them. Out of sheer spite, they sue you for wrongful eviction. You're clearly in the right, but the defense of the suit costs $750. Under this policy we defend you and win the case in court. The whole business doesn't cost you a penny.

Incidentally, there is no fine print in the policy. It is set entirely in 10-point type, a type larger than that used in most newspapers and magazines. Headings and even different-colored print are used freely to draw attention to transitions and important information. Most states now require insurance policies sold within their borders to meet "plain language" requirements. We can hope, therefore, that the impossible-to-read insurance policy is a thing of the past.

You can clean up your own writing by following the principles discussed in this chapter and demonstrated in the revised insurance policy. Also, if you exercise care, your own manner of speaking can be a good guide in writing. You should not necessarily write as you talk. In speech, you may be too casual, even slangy. But the sound of your own voice can still be a good guide. When you write something, read it over; even read it aloud. If you have written something you know you would not speak because of its artificiality, rewrite it in a comfortable style. Rewrite so that you can hear the sound of your own voice in it.

Planning and Revision Checklists

You will find the planning and revision checklists following Chapter 2, "Composing," Chapter 4, "Writing for Your Readers," and inside the front cover valuable in planning and revising any presentation of technical information. The following questions specifically apply to style. They summarize the key points in this chapter and provide a checklist for revising.

PLANNING

You can revise for good style, but you can't plan for it. Good style comes when you are aware of the need to avoid the things that cause bad style: ponderous paragraphs, overly dense sentences, excessive use of passive voice, pomposity, and the like.

Good style comes when you write to express your thoughts clearly, not to impress your readers. Good style comes when you have revised enough writing that the principles involved are ingrained in your thought process.

REVISION

- Do you have a style checker in your word processing software? If so, use it, but exercise the cautions we advocate in Chapter 2, "Composing."
- Are the central thoughts in your paragraphs clearly stated? Do the details in your paragraphs relate to the central thought?
- Have you broken up your paragraphs sufficiently to avoid long, intimidating blocks of print?
- Have you guided your reader through your paragraphs with the repetition of key words and with transition statements?
- Have you used lists or tables when such use would help the reader?
- Are your sentences of reasonable length? Have you avoided sentences of 60 to 100 words? Does your average sentence length match that of professional writers—about 20 words?
- Professional writers begin about 75% of their sentences with the subject of the sentence. How does your percentage of subject openers compare to that figure? If your average differs markedly, do you have a good reason for the difference?
- When you use sentence openers before the subject, do they provide good transitions for your readers?
- Have you limited sentence openers before the subject to a reasonable length?
- Have you avoided large blocks of words between your subject and your verb?
- Have you used noun strings to modify other nouns? If so, are you sure your readers will be able to sort out the relationships involved?
- Have you avoided the use of multiple negatives?
- Are your action ideas expressed in active verbs? Have you avoided burying them in nouns and adjectives?
- Have you used active voice and passive voice appropriately? Are there passive voice sentences you should revise to active voice?
- Have you used abstract words when more specific words would be clearer for your readers? Do your abstractions leave unintended interpretations open to the

reader? When needed, have you backed up your abstractions with specific detail?
- Have you avoided empty jargon phrases?
- Have you chosen your words to express your thoughts clearly for your intended reader? Have you avoided pompous words and phrases?

Exercises

1. You should now be able to rewrite the example sentence on page 85 in clear, forceful prose. Here it is again; try it:

While determination of specific space needs and access cannot be accomplished until after a programmatic configuration is developed, it is apparent that physical space is excessive and that all appropriate means should be pursued to ensure that the entire physical plant is utilized as fully as feasible.

2. Here is the pompous paragraph from page 102. Revise it into good prose:

The purpose of this PPM [Policy and Procedure Memorandum] is to ensure, to the maximum extent practicable, that highway locations and designs reflect and are consistent with Federal, State and local goals and objectives. The rules, policies, and procedures established by this PPM are intended to afford full opportunity for effective public participation in the consideration of highway location and design proposals by highway departments before submission to the Federal Highway Administration for approval. They provide a medium for free and open discussion and are designed to encourage early and amicable resolution of controversial issues that may arise.

3. Following are some expressions that the Council of Biology Editors believes should be rewritten.[13] Using the principles you have learned in this chapter, rewrite them:
   - an innumerable number of tiny veins
   - as far as our own observations are concerned, they show
   - ascertain the location of
   - at the present moment
   - at this point in time
   - bright green in color
   - by means of
   - (we) conducted inoculation experiments on
   - due to the fact that
   - during the time that
   - fewer in number
   - for the purpose of examining
   - for the reason that
   - from the standpoint of

- goes under the name of
- if conditions are such that
- in all cases
- in order to
- in the course of
- in the event that
- in the near future
- in the vicinity of
- in view of the fact that
- it is often the case that
- it is possible that the cause is
- it is this that
- it would thus appear that
- large numbers of
- lenticular in character
- masses are of large size
- necessitates the inclusion of
- of such hardness that
- on the basis of
- oval in shape, oval shaped
- plants exhibited good growth
- prior to (in time)
- serves the function of being
- subsequent to
- the fish in question
- the tests have not as yet
- the treatment having been performed
- there can be little doubt that
- throughout the entire area
- throughout the whole of this experiment
- two equal halves
- If we interpret the deposition of chemical signals as initiation of courtship, then initiation of courtship by females is probably the usual case in mammals.
- A direct correlation between serum vitamin B_{12} concentration and mean nerve conduction velocity was seen.
- It is possible that the pattern of herb distribution now found in the Chilean site is a reflection of past disturbances.
- Following termination of exposure to pigeons and resolution of the pulmonary infiltrates, there was a substantial increase in lung volume, some improvement in diffusing capacity, and partial resolution of the hypoxemia.

4. Turn the following sentence into a paragraph of several sentences. See if listing might be a help. Make the central idea of the passage its first sentence.

If, on the date of opening of bid or evaluation of proposals, the average market price of domestic wool of usable grades is not more than 10 percent above the average of the prices of representative types and grades of domestic wools in the wool category which includes the wool required by the specifications (see (f) below), which prices reflect the current incentive price as established by the Secretary of Agriculture, and if reasonable bids or proposals have been received for the advertised quantity offering 100 percent domestic wools, the contract will be awarded for domestically produced articles using 100 percent domestic wools and the procedure set forth in (e) and (f) below will be disregarded.

5. Lest you think all bad writing is American, here are two British samples quoted in a British magazine devoted to ridding Great Britain of gobbledygook.[14] Try your hand with them.

- The garden should be rendered commensurate with the visual amenities of the neighborhood.
- Should there be any intensification of the activities executed to accomplish your present hobby the matter would have to be reappraised.

6. What follows is a description of how liposome technology may lead to better medical treatment. The description is intended for an educated lay audience, an audience probably much like you. Analyze the description using the principles of this chapter: paragraph development, lists and tables, clear sentence structure, the use of specific words, and the avoidance of pomposity. You should consider such elements of style as transitions; paragraph length; sentence length, order, density, and complexity; and active and passive voice. Using your analysis, decide whether this description succeeds or not. Write a memo to your teacher stating and justifying your decision. (Use the memo format from Chapter 12, "Correspondence.")

Can microscopic artificial membranes help doctors treat cancer, angina, and viral infections more effectively, and lead to better vaccines, bronchodilators, eye drops, and sunscreens? The researchers who are developing liposome technology hope so. A liposome is a tiny sphere of fatty molecules surrounding a watery interior. Because they are made of the same material as cell surface membranes, liposomes stick to cells and are not toxic. These characteristics make them attractive candidates for drug delivery vehicles.

In 1980, two groups of researchers used liposomes filled with a common antibiotic to cure mice having a severe, but localized, infection. The infected cells were of a kind that is specialized to take up foreign bodies, and so they readily engulfed the liposomes. However, getting other kinds of cells to take up drug-filled liposomes has proven to be more difficult. A number of groups of researchers are experimenting with antibody-tagged liposomes filled with an anticancer drug. The liposomes are guided to the diseased tissue by the antibodies, which seek out cancerous cells but spare healthy ones. This selectivity allows smaller amounts of a drug to be used with greater effect, an important advantage considering the serious toxicity of many anticancer drugs.

Other research teams are developing liposome-drug compounds that would be injected into muscle to release growth hormone or anticancer agents over a period of weeks. Scientists also hope to use liposomes to improve the safety and effectiveness of vaccines, including an influenza vaccine. As the cost of both natural lipids (extracted from egg yolk and soybeans) and artificial lipids declines, the future may bring many other liposome-containing medical products as well as nonmedical items, such as cosmetics.[15]

7. Write a short report, or revise one you did earlier, using the stylistic principles of this chapter.

PART II
TECHNIQUES

In Part II, we build on the basic concepts of Part I. The three chapters in Part II describe and demonstrate the techniques you will need to inform, describe, define, and argue. With these techniques you can arrange, draft, and revise much of the writing you will do. As in Part I, we emphasize the need to consider audience and purpose no matter what technique you are using. With the techniques you master in this part, you will have many of the skills you will need for writing documents such as instructions, proposals, and feasibility reports, the subjects of Part IV, "Applications."

CHAPTER 6 INFORMING

Chronological Arrangement

Topical Arrangement

Exemplification

Analogy

Classification and Division

Our purpose in this and the next two chapters is to show you how writing techniques are tools you can choose to fulfill the various purposes you have when you write. In this chapter, we discuss the techniques that are useful when your primary purpose is to inform: chronological arrangement, topical arrangement, exemplification, analogy, and classification and division. You can use any of these techniques as an overall arranging principle for an entire report; for example, you could arrange an entire report in topical order. You will also use these techniques as subordinate methods of development within a larger framework. For example, within a paper arranged topically, you might have paragraphs or small sections based upon chronology, exemplification, classification, and so forth. The two uses of informing techniques are mutually supportive. Also, you have to incorporate many of these informing techniques into your overall strategy when your purpose is to define, describe, or argue—the subjects of Chapters 7 and 8.

Chronological Arrangement

When you need to relate a series of events for your reader, arranging the events chronologically—that is, by time—is a natural way to proceed. In your chronological narrative, keep your readers informed as to where they are in the sequence of events. In the example that follows, we have printed in boldface type the phrases the authors use to orient their readers. In this excerpt, the authors describe the history of the Hawaiian Volcano Observatory. Such a chronological narrative is frequently used in technical writing to provide a historical overview that will aid the reader in understanding some explanation or argument that is to come. The passage that follows serves such a purpose:[1]

> **Before the 20th century,** most scientific studies of volcanoes were conducted during short-lived expeditions, generally undertaken as a response to major eruptions. Thomas A. Jaggar, Jr., a geologist at the Massachusetts Institute of Technology (MIT), was not satisfied with that approach. He recognized that, to understand volcanoes fully one must study them continuously before, during, and after eruptions. Jaggar's views were profoundly affected **by a memorable visit in 1902** to the Island of Martinique (West Indies). He went as a member of the scientific expedition sent to study the catastrophic eruption of Mont Pelée that year, which devastated the city of St. Pierre and killed about 30,000 people.
>
> **In 1911,** spurred by a stimulating lecture delivered by Jaggar, a group of Hawaiian residents founded the Hawaiian Volcano Research Association (HVRA). The logo of the HVRA included the motto Ne plus haustae aut obrutae urbes (No more shall the cities be destroyed), reflecting Jaggar's memory of Mont Pelée's destructive force and his optimistic belief that better understanding of volcanoes could reduce the hazard to life and property from eruptions.
>
> **In 1912,** with support from the HVRA and the Whitney Fund of MIT, Jaggar established the Hawaiian Volcano Observatory (HVO) to study the activity of Mauna Loa and Kilauea Volcanoes on a permanent, scientific basis. "Volcanology" emerged as a modern science with the founding of the HVO, which **between 1912 and 1948** was managed by the HVRA, the U.S. Weather Bureau, the U.S. Geological Survey, and the National Park Service.
>
> **Since 1948,** it has been operated by the USGS. **During the past 75 years** of research, HVO scientists have developed and refined most of the surveillance techniques now commonly employed by volcano observatories worldwide.

For the most part, chronological narratives are related in a businesslike way. Sometimes, however, a writer may choose to use a narrative more dramatically—perhaps in brochures or advertisements for a lay audience—to simplify complex ideas and provide human interest. A writer may also use a dramatic narrative as an introduction to catch reader interest. The narrative that follows is the introduction to an article that describes and evaluates the effectiveness of humanitarian airdrops in Bosnia.

> Exactly six seconds after the navigator yelled, "Green light!" ten giant bundles of food and other supplies began gently sliding down the ramp of the C-130 and falling down into the darkness, starting a two-minute descent by parachute through the clouds blanketing eastern Bosnia.
>
> The crew members and passengers took two gulps of pure, bone dry oxygen through yellow rubber masks. By the end of the second gulp, the airplane had disgorged all ten bundles—nine containing meals, ready-to-eat (MREs), one crammed with medical supplies.

> The previously packed cargo bay, bathed in the glow of red lights, seemed positively cavernous. The airplane went into "feet dry" condition as crew and passengers donned flak jackets and helmets. Electric power was reduced. The missile warning system was switched on.
>
> Welcome to the world of precision airdropping, 1990s style. Part of a humanitarian operation dubbed "Provide Promise," the drop was a rare event. The last time USAF's computerized Adverse Weather Air Delivery System (AWADS) got such a workout was in 1968. Then, the Air Force was dropping supplies to the besieged US Marine garrison at Khe Sanh, South Vietnam.
>
> Outside, in twenty-two-knot winds, the 1,520-pound bundles swayed gently as they fell at a speed of sixty miles per hour. The elaborately wrapped bundles contained a variety of supplies: food pouches of chicken and rice, spaghetti and meatballs, Danish and British biscuits, Turkish anchovies, and other edible goods. There were vials of penicillin.
>
> The plan called for the bundles to drift no more than one-half mile from the computer-designated release point to which the Air Force C-130 had flown. The drop zone on the ground measured about 1,500 yards by 1,000 yards.[2]

Chronological narratives don't have to be restricted to past events. They can be used to forecast future events, as in this segment where the writer projects the future flight plan of the two Voyager spacecraft:

> After Neptune, the Voyager spacecraft planetary encounters will be over. Only one planet in the solar system remains unvisited—Pluto—and neither of the Voyagers can change its course to visit that planet.
>
> But the Voyager missions will continue as the two spacecraft hurtle onward through space, one above the ecliptic and one below, searching for the edge of the heliosphere—the heliopause, which is the outer boundary of the sun's energy influence. Crossing the heliopause, perhaps early in the next century, they will enter true interstellar space. These spacecraft may give us the first direct measurements of the environment outside our solar system, including interstellar magnetic fields and charged particles.[3]

Because the arrangement of your material follows the sequence of the events you relate, arrangement is not difficult when you use chronological order. Choosing the level of detail you need may be a problem. Obviously, the narrator of the history of the Hawaiian Volcano Observatory could have used more or less detail than was used. As in most kinds of writing, purpose and audience are your best guides in these matters. If your purpose is to give a broad overview for a lay or executive audience, you will limit the amount of detail. On the other hand, if you have an expert audience who will wish to analyze carefully the sequence you describe, you will need to provide considerable detail.

A major application of chronology in technical writing is describing process, that is, describing a sequence of events that progresses from a beginning to an end and results in a change or a product. Process descriptions are written in one of two ways:

- *For the interested observer*—to provide an understanding of the process.
- *For the doer*—to provide instruction for performing the process.

We cover the first type of process description in Chapter 7, "Defining and Describing," and the second in Chapter 16, "Instructions."

Topical Arrangement

Technical writing projects often begin with a topic, such as "Christmas tree farming." One way to deal with such topics is to look for subtopics under the major topic. These should serve as umbrella statements beneath which you can gather smaller sub-subtopics and related facts. In the case of the Christmas tree topic, the subtopics might very well be "production" and "marketing." "Production" could be broken down further into "planting," "maintaining," and "harvesting." "Marketing" could be broken down into "retail," "wholesale," and "cut-your-own." With some thought, you can break most topics down into umbrella-sized subtopics.

While you are settling upon the topic and subtopics for your paper, you should also be aware of the need to limit the topic. Students, in particular, often hesitate to limit their topic sufficiently. They fear, perhaps, that if they limit their material too severely they will not be able to write essays of a sufficient length. The truth of the matter is the reverse. You will find it easier to write a coherent, full essay of any length if you *limit the scope of your topic.* With your scope limited and your purpose and audience clearly defined, you can fill your paper with specific facts and examples. When your scope is broad and your purpose and audience vague, you must deal in abstract generalizations.

Suppose you wish to deal with the subject of robotics in about a thousand words. If you keep the purpose simply as "explaining robotics," what can you say in a thousand words? Probably just a few generalizations about the theory, history, and applications of robotics.

But suppose you define your purpose and audience in relation to your subject matter. Perhaps your audience is a group of executives who may have an interest in the use of robotics in their industries. You know that their major interest in robotics will be in application, not in history or theory. You limit yourself, therefore, to application. You refine your purpose. You decide you want to give them some idea of the scope of robotics. To do so, you reach into your knowledge of robotics and choose several applications that can serve as illustrative examples.

You decide to focus on the use of robotics in the following three tasks:

- installing windows in automobiles as the automobiles pass on an assembly line

- arc welding in an airplane plant
- mounting chips in a computer factory

Using these three applications as your topics, you can illustrate a wide range of current robotics practice. Within a few minutes, you can limit your topic to manageable size and make a good start on arrangement.

Figure 6–1 reproduces two pages from a pamphlet on the radioactive gas radon. The pamphlet uses a topical arrangement, each topic marked for the reader by a question such as "What is radon?" and

What is radon?

Radon is a radioactive gas which occurs in nature. You cannot see it, smell it, or taste it.

Where does radon come from?

Radon comes from the natural breakdown (radioactive decay) of uranium. Radon can be found in high concentrations in soils and rocks containing uranium, granite, shale, phosphate, and pitchblende. Radon may also be found in soils contaminated with certain types of industrial wastes, such as the byproducts from uranium or phosphate mining.

In outdoor air, radon is diluted to such low concentrations that it is usually nothing to worry about. However, once inside an enclosed space (such as a home) radon can accumulate. Indoor levels depend both on a building's construction and the concentration of radon in the underlying soil.

How does radon affect me?

The only known health effect associated with exposure to elevated levels of radon is an increased risk of developing lung cancer. Not everyone exposed to elevated levels of radon will develop lung cancer, and the time between exposure and the onset of the disease may be many years.

Scientists estimate that from about 5,000 to about 20,000 lung cancer deaths a year in the United States may be attributed to radon. (The American Cancer Society expects that about 130,000 people will die of lung cancer in 1986. The Surgeon General attributes around 85 percent of all lung cancer deaths to smoking.)

Your risk of developing lung cancer from exposure to radon depends upon the concentration of radon and the length of time you are exposed. Exposure to a slightly elevated radon level for a long time may present a greater risk of developing lung cancer than exposure to a significantly elevated level for a short time. In general, your risk increases as the level of radon and the length of exposure increase.

How certain are scientists of the risks?

With exposure to radon, as with other pollutants, there is some uncertainty about the amount of health risk. Radon risk estimates are based on scientific studies of miners exposed to varying levels of radon in their work underground. Consequently, scientists are considerably more certain of the risk estimates for radon than they are of those risk estimates which rely solely on studies of animals.

To account for the uncertainty in the risk estimates for radon, scientists generally express the risks associated with exposure to a particular level as a *range* of numbers. (The risk estimates given in this booklet are based on the advice of EPA's Science Advisory Board, an independent group of scientists established to advise EPA on various scientific matters.)

Despite some uncertainty in the risk estimates for radon, it is widely believed that **the greater your exposure to radon, the greater your risk of developing lung cancer**.

How does radon cause lung cancer?

Radon, itself, naturally breaks down and forms radioactive decay products. As you breathe, the radon decay products can become trapped in your lungs. As these decay products break down further, they release small bursts of energy which can damage lung tissue and lead to lung cancer.

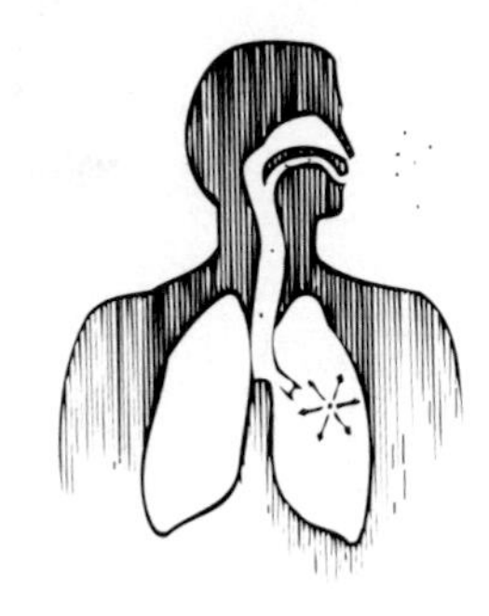

Figure 6–1 A Topical Arrangement

Source: U.S. Environmental Protection Agency, *A Citizen's Guide to Radon* (Washington, DC: GPO, 1986) 1–2.

"Where does radon come from?" The use of questions shows that the pamphlet's writer understands that headings phrased as questions are an effective device to increase reader understanding.

Exemplification

Technical writing sometimes consists largely of a series of generalizations supported by examples. The writer makes statements, such as this one about earthquakes:[4]

> The actual movement of ground in an earthquake is seldom the direct cause of death or injury.

Having made such a generalization, the writer must now support it:

> Most casualties result from falling objects and debris because the shocks can shake, damage, or demolish buildings and other structures. Earthquakes may also trigger landslides and generate huge ocean waves (seismic sea waves), each of which can cause great property damage and considerable loss of life. Earthquake-related injuries are commonly caused by: (1) partial building collapse, such as toppling chimneys, falling brick from wall facings and roof parapets, collapsing walls, falling ceiling plaster, light fixtures, and pictures; (2) flying glass from broken windows (this danger may be greater from windows in high-rise structures); (3) overturned bookcases, fixtures, and other furniture and appliances; (4) fires from broken chimneys, broken gaslines, and similar causes (this danger may be aggravated by a lack of water caused by broken mains); (5) fallen powerlines; and (6) drastic human actions resulting from panic.

There are two common ways to use examples. One way is to give one or more extended, well-developed examples. We have used this method in showing you the paragraph about earthquake damage. The other way is to give a series of short examples that you do not develop in detail, as in the following paragraph:

> I use the term culture to refer to the "system of knowledge" that is shared by a large group of people. The "borders" between countries usually, but not always, coincide with political boundaries between countries. To illustrate, we can speak of the culture of the United States, the Japanese culture, and the Mexican culture. In some countries, however, there is more than one culture. Consider Canada as an example. There is the Anglophone (i.e., English speaking) culture derived from England and there is the Francophone culture derived from France.[5]

Like almost everything else in writing, the use of examples calls for judgment on your part. Too few examples and your writing will lack interest and credibility. Too many examples and your key generalizations will be lost in excessive detail.

Analogy

Analogies are comparisons: they compare the unfamiliar to the familiar in order to make the unfamiliar more understandable for the reader. You should use short, simple analogies, particularly when you are writing for lay people. For example, many people have difficulty in understanding the immense power released by nuclear reactions. A completely technical explanation of $E = mc^2$ probably would not help them very much. But suppose you tell them that if one pound of matter—a package of butter, for instance—could be converted directly to energy in a nuclear reaction, it would produce enough electrical power to supply the entire United States for thirty-five hours (that is, more than eleven billion kilowatt hours). Such a statement reduces $E = mc^2$ to an understandable idea.

Scientists recognize the need for analogy when called upon to explain difficult concepts. A scientist working with microelectronic integrated circuits (microchips), when called upon to explain how small the circuits are, said, "You grope for analogies. If you wanted to draw a map of the entire United States that showed every city block and town square, it would obviously be a *very* big map. But with the feature sizes we're working with to create microcircuits right now, I could draw that entire map on a sheet of paper not much larger than a postage stamp."[6]

A writer who wished to express the immense age of the universe relative to humankind put it this way:

> Some 12 to 20 billion years ago, astronomers think a "primeval atom" exploded with a big bang sending the entire universe flying out at incredible speeds. Eventually matter cooled and condensed into galaxies and stars. Eons after life began to develop on Earth, humans appeared. If all events in the history of the universe until now were squeezed into 24 hours, Earth wouldn't form until late afternoon. Humans would have existed for only two seconds.[7]

Analogies serve particularly well in definitions or descriptions. If, after describing a diode, you tell a lay audience that it is similar to a water faucet in that you can use it to control the flow of electrons or shut them off completely, you make the concept more understandable.

Besides being practical, analogies can liven up your writing. Here is a writer having fun with some far-fetched analogies that help you to grasp the enormousness of the quantities he is discussing:

> If all the Coca-Cola ever produced were dumped over Niagara Falls in place of water, the falls would flow at a normal rate for 16 hours and 49 minutes ... Two ships the size of the Queen Elizabeth could be floated in the ocean of Hawaiian Punch Americans consume annually ... There are 30,000 peanut butter sandwiches in an acre of peanuts, and 540 peanuts in a 12-ounce jar of peanut butter.[8]

Analogies are sometimes fairly extended. Here is a famous extended analogy by Sir James Jeans that explains why the sky is blue:

> Imagine that we stand on an ordinary seaside pier, and watch the waves rolling in and striking against the iron columns of the pier. Large waves pay very little attention to the columns—they divide right and left and reunite after passing each column, much as a regiment of soldiers would if a tree stood in their road: it is almost as though the columns had not been there. But the short waves and ripples find the columns of the pier a much more formidable obstacle. When the short waves impinge on the columns, they are reflected back and spread as new ripples in all directions. To use the technical term they are "scattered." The obstacle provided by the iron columns hardly affects the long waves at all, but scatters the short ripples.
>
> We have been watching a sort of working model of the way in which sunlight struggles through the earth's atmosphere. Between us on earth and outer space the atmosphere interposes innumerable obstacles in the form of molecules of air, tiny droplets of water, and small particles of dust. These are represented by the columns of the pier.
>
> The waves of the sea represent the sunlight. We know that sunlight is a blend of many colors—as we can prove for ourselves by passing it through a prism, or even through a jug of water, or as Nature demonstrates to us when she passes it through raindrops of a summer shower and produces a rainbow. We also know that light consists of waves, and that the different colors of light are produced by waves of different lengths, red light by long waves, and blue light by short waves. The mixture of waves which constitutes sunlight has to struggle through the obstacles it meets in the atmosphere just as the mixture of waves at the seaside has to struggle past the columns of the pier. And these obstacles treat the light waves much as the columns of the pier treat the seawaves. The long waves which constitute red light are hardly affected, but the short waves which constitute blue light are scattered in all directions.
>
> Thus the different constituents of sunlight are treated in different ways as they struggle through the earth's atmosphere. A wave of blue light may be scattered by a dust particle, and turned out of its course. After a time a second dust particle again turns it out of its course, and so on, until finally it enters our eyes by a path as zigzag as that of a flash of lightning. Consequently the blue waves of sunlight enter our eyes from all directions. And that is why the sky looks blue.[9]

Sir James has used analogy in a very imaginative way. By comparing light waves (an unfamiliar concept) to ocean waves (a concept familiar to his English readers), he has made it easy for his readers to grasp his meaning. Such extended analogies are most useful when you are writing for lay people.

Throughout your writing, use analogy freely. It is one of your best bridges to the uninformed reader.

Classification and Division

Classification and division, like chronological and topical arrangements, are useful devices for bringing order to any complex body of material. You may understand classification and division more readily if we explain them in terms of the *abstraction ladder*. We borrow this device from semanticists—people who make a scientific study of words. We will construct our ladder by beginning with a very abstract word on top and working down the ladder to end with a specific term:

- **Factor:** Almost anything can be a factor. You could be a factor in someone's decision. So could wealth, climate, and geography, to name but a few significant concepts. Without more specific references, we cannot determine what is specifically meant at this rung of the ladder.
- **Wealth:** Now we have moved down a rung on the ladder. We have added specific information. Wealth could be money. It also could be stocks and bonds, land, furniture, or any of the other material objects that people value highly.
- **Furniture:** We now have become much more specific. We can mean beds, tables, chairs, desks, lamps, and so on.
- **Table:** Now we are zeroing in on the object. However, **table** still refers to a huge class of objects: coffee tables, kitchen tables, dining room tables, library tables, end tables, and so on.
- **Kitchen table:** Now we know a good deal more. We know the function of the object. People eat at kitchen tables. Cooks mix cakes on them. We can even generalize somewhat about their appearance. Kitchen tables are usually plain objects, compared to end tables, for example. Many of them are made of wood and have four legs.
- **John Smith's kitchen table:** Now we know precisely what we are talking about. We can describe the size, weight, shape, and color of this table. We know the material of which it is made. We know that John Smith's family eats breakfast at this table but not dinner. In the evenings John Smith's daughter does her homework there.

We must keep one important distinction in mind: even "John Smith's kitchen table" is not the table itself. As soon as we use a word for an object, the abstraction process begins. Beneath the word is the table *we see* and beneath that is the table *itself,* consisting of paint, wood, and hardware that consist of molecules that consist of atoms and space.

In classification, you move *up* the abstraction ladder, seeking higher abstractions under which to group many separate items. In division, you move *down* the abstraction ladder, breaking down higher abstractions into the separate items contained within them. We will illustrate classification first.

Suppose for the moment that you are a dietitian. You are given a long list of foods found in a typical American home and asked to comment on the value of each. You are to give such information as calorie count, carbohydrate count, mineral content, fat content, vitamin content, and so forth. The list is as follows: onions, apples, steak, string beans, oranges, cheese, lamb chops, milk, corn flakes, lemons, bread, butter, hamburger, cupcakes, and carrots.

If you try to comment on each item in turn as it appears on the list, you will write a chaotic essay. You will repeat yourself far too often. Many of the things you will say about milk will be the same things you say about cheese. To avoid this repetition and chaos you need to classify the list, to move up the abstraction ladder, seeking groups that look like the following:

Food

Vegetables	**Fruit**
Onions	Apples
String beans	Oranges
Carrots	Lemons
Meat	**Cereal**
Steak	Corn flakes
Lamb chops	Bread
Hamburger	Cupcakes
Dairy	
Milk	
Cheese	
Butter	

By following this procedure, you can use the similarities and dissimilarities of the different foods to aid your organization rather than have them disrupt it.

In division, you move down the abstraction ladder. Suppose your problem is now the reverse of the foregoing. You are a dietitian and someone asks you to list examples of foods that a healthy diet should contain. In this case, you start with the abstraction *food.* You decide to divide this abstraction into smaller divisions such as vegetables, fruit,

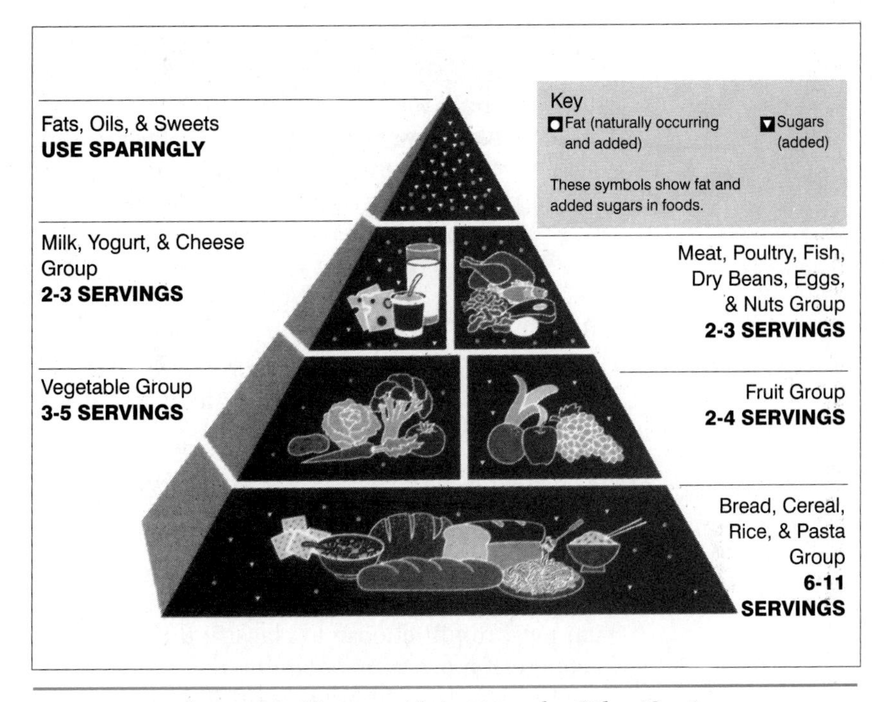

Figure 6–2 The Food Guide Pyramid: An Example of Classification
Source: U.S. Department of Agriculture, *Making Healthy Food Choices* (Washington, DC: GPO, 1993) 2.

meat, cereal, and dairy. You then subdivide these into typical examples such as cheese, milk, and butter for dairy. Obviously, the outline you could construct here might look precisely like the one already shown. But in classification we arrived at the outline from the bottom up; in division, from the top down. Figure 6-2 shows the Food Guide Pyramid, an excellent example of classification and division in action.

Very definite rules apply in using classification and division.

- **Keep all headings equal.** In the preceding example, you would not have headings of "Meat," "Dairy," "Fruit," "Cereal," and "Green Vegetables." "Green Vegetables" would not take in a whole class of food as the other headings do. Under the heading of "Vegetables," however, you could have subheadings of "Green Vegetables" and "Yellow Vegetables."
- **Apply one rule of classification or division at a time.** In the preceding example, the classification is done by food types. You would not in the same classification include headings *equal* to those of the food types for such subjects as "Mineral Content" or "Vitamin Content." You could, however, include such subheadings under the food types.

- **Make each division or classification large enough to include a significant number of items.** In the preceding example, you could have many equal major headings such as "Green Vegetables," "Yellow Vegetables," "Beef Products," "Lamb Products," "Cheese Products," and so forth. In doing so, however, you would have overclassified or overdivided your subject. Some of the classifications would have included only one item.
- **Avoid overlapping classifications and divisions as much as possible.** In the preceding example, if you had chosen a classification that included "Fruits" and "Desserts," you would have created a problem for yourself. The listed fruits would have to go in both categories. You cannot always avoid overlap, but keep it to a minimum.

As long as writers observe the rules, they are free to classify and divide their material in any way that suits their purposes. An accountant who wishes to analyze the money flow for construction of a state's highways might choose to classify them by source of funding: federal, state, county, city. An engineer, concerned with construction techniques, might choose to classify the same group of roads by surface: concrete, macadam, asphalt, gravel.

In a brochure about controlling termites, the authors chose to classify them according to their habitat:

> Based on ecological considerations, three types of termites occur in the United States: (1) drywood, (2) dampwood, and (3) subterranean. Drywood termites build their nests in sound dry wood above ground. Dampwood species initially locate their nests in moist, decaying wood but can later extend tunnels into drier parts of wood. Subterranean termites are more dependent on an external moisture source, and they typically dwell in the soil and work through it to reach wood above ground.[10]

The authors chose to classify termites in this way because, as they put it, "This information provides a foundation for control methods based on the habits and behavior of the termites." Obviously, termites can be classified in many ways. Classifying by habitat seemed the best way to serve the brochure's purpose and the needs of the audience.

Planning and Revision Checklists

You will find the planning and revision checklists following Chapter 2, "Composing," Chapter 4, "Writing for Your Readers," and inside the front cover valuable in planning and revising any presentation of technical information. The following questions specifically apply to the techniques discussed in this chapter. They summarize the key points of the chapter and provide a checklist for planning and revising.

CHRONOLOGICAL ARRANGEMENT: PLANNING

- Do you have a reason to narrate a series of events? Historical overview? Background information? Drama and human interest for a lay audience? Forecast of future events?
- What are the key events in the series?
- In what order do the key events occur?
- Do you know or can you find out an accurate timing of the events?
- How much detail does your audience need or want?

CHRONOLOGICAL ARRANGEMENT: REVISION

- Is your sequence of events in proper order?
- Are all your time references accurate?
- Have you provided sufficient guidance within your narrative so that your readers always know where they are in the sequence?
- Is your level of detail appropriate to your purpose and the needs of your audience?

TOPICAL ARRANGEMENT: PLANNING

- What is you major topic?
- What is your purpose?
- What is your audience's interest in your topic? How do their interest and purpose relate to your purpose?
- Given your purpose and your audience's purpose, how can you limit your topic?
- What subtopics would be appropriate to your purpose and your audience's purpose?
- Can you divide your subtopics further?

TOPICAL ARRANGEMENT: REVISION

- Do your topics and subtopics meet your purpose and your audience's purpose and interests?
- Did you limit your subject sufficiently so that you could provide specific facts and examples?
- Do you have headings?
- Do your headings accurately reflect how your readers will approach your subject matter?
- Are your headings phrased as questions? If not, would it help your readers if they were?

EXEMPLIFICATION: PLANNING

- Do your generalizations need the support of examples?
- Do you have or can you get examples that will lend interest and credibility to your document?

EXEMPLIFICATION: REVISION

- Have you left any generalizations unsupported? If so, have you missed a chance to interest and convince your readers?
- Have you provided sufficient examples to lend interest and credibility to your material?

ANALOGY: PLANNING

- What is your audience's level of understanding of your subject matter?
- Would the use of analogy provide your readers with a better understanding of your subject matter?
- Are there things familiar to your readers that you can compare to the unfamiliar concept, such as water pressure to voltage?

ANALOGY: REVISION

- Have you provided analogies wherever they will help reader understanding?
- Do your analogies really work? Are the things compared truly comparable?

CLASSIFICATION AND DIVISION: PLANNING

- Where is your subject matter on the abstraction ladder?
  Are you moving up the ladder, seeking higher abstractions under which you can group your subject matter (classification)?
  Are you moving down the ladder, breaking your abstractions down into more specific items (division)?
- What is your purpose in discussing your subject matter?
- What is your audience's purpose and relationship to your subject matter?
- What classification or division will best meet your purpose and your audience's needs?

CLASSIFICATION AND DIVISION: REVISION

- Are all the parts of your classification equal?
- Have you applied one rule of classification and division at a time?
- Is each classification or division large enough to include a significant number of items?
- Have you avoided overlapping classifications and divisions?
- Does your classification or division meet your purpose and your audience's needs?

Exercises

1. Write a memo to an executive. The purpose of the memo is to inform the executive about the subject matter of the memo. Base the arrangement of the memo on one of the techniques described in this chapter. Accompany your memo with a short explanation of why you chose the arrangement technique you did. Your explanation must show how your purpose and your reader's purpose and interests led to your choice. For instruction on the format of memos, see Chapter 12, "Correspondence."

2. Write a chronological narrative of several paragraphs that is intended to serve as either a historical overview or a dramatic introduction for a larger report. Choose as a subject for your narrative some significant event in your professional field. Accompany your narrative with a description of your audience and an explanation of how their purpose and yours led to the level of detail you use in your narrative.
3. Write an extended analogy of several paragraphs that will make some complicated concept in your discipline comprehensible to a fourth grade student.
4. Reproduced in this exercise is an excerpt from a longer article about the Earth's crust.[11] The article is an informational piece written primarily for the "employees and stockholders" of a large corporation. Therefore, the audience is a mixed one that will include lay people, technicians, executives, and perhaps even experts. The writer's task is to explain things so that lay people will find the article both interesting and understandable, but at the same time include facts and inferences that the other readers may find useful. The excerpt uses some of the techniques described in this chapter. Working in a group, discuss where, how, and why the author has used these techniques. Reach a judgment as to how well he has used them. In your discussion, use concepts from Chapter 4, "Writing for Your Readers," as well as the concepts in this chapter. When the group has finished its discussion, each member of the group should write a report that presents his or her own views on the questions raised in this exercise.

Typically, the Earth's crust is 12 miles thick, though it varies from three to 60 miles in depth depending on location. The distance from the surface to the very center of the globe is about 3,950 miles.

If the Earth were a beachball exactly 12 inches in diameter, the entire, quasi-solid layer of rock and dirt that we call the crust would literally be less than 2/100 of an inch thick in most areas and less than one-tenth of an inch at the very most.

The crust is the only part of the Earth we know much about, and the only part with which we've had any practical, albeit limited, experience. How limited? Take an apple and prick its skin as delicately as you possibly can with a sharp needle. If the Earth were that apple—forget the beachball—and you had a very gentle touch, you'd have made a hole roughly as deep as the deepest well ever drilled into the Earth's crust.

Seismographic Service Corporation (SSC), a subsidiary of Raytheon based in Tulsa, Okla., knows all about such wells. SSC tickles and probes, thumps, and sounds the Earth's crust with a kind of rock radar to help oil and gas drillers decide where to punch those impudent holes.

"It's much like the problems that doctors have imaging the human body," explains Dale Stone, head of SSC's geophysical research department. "Especially since seismologists use the same kind of ultrasonic energy.

"The anomaly a doctor might be looking for is cancer. Our anomaly is oil.

And for us, the equivalent of the differing densities of bone, flesh, and muscle are things like sandstone, shale, and limestone."

Not all that long ago—the debate was still going on in the 1960s—geologists were doing well to have figured out that the continents all came from one mother mass and somehow moved monumentally hither and grandly yon across the Earth's surface—the theory of continental drift. Then earth scientists discovered that it wasn't neat, tear-along-the-dotted-line continents that were drifting, but entire "plates" of crustal matter, huge slabs of Earth's cracked eggshell upon which nations and oceans ride like so many puddles and piles of dirt.

Plate tectonics—one of science's great breakthroughs—was born and became the mechanism by which the Earth "moves." The theory of plate tectonics holds that the Earth's entire crust is made up of somewhere between a dozen and 20 slabs that fit together like a badly matched jigsaw puzzle with some pieces a bit too large for each other. And not only that, but some of those pieces are growing, new matter being squeezed from deep within the Earth like sheets of hot pasta.

So one plate pushes against and even overlaps another here and there, and the plates are constantly making room for each other. The result? The edges chafe and grind and create earthquakes. One plate rams another and eventually mountains wrinkle upward when the rammed plate folds like a throw rug kicked into a corner. Plates crack from the strain and fault lines form—more earthquake potential—and magma from the bowels of the planet leaks up through rifts, ridges, and hot spots to create volcanoes and geothermal springs. We live upon a seething machine that turns "solid as a rock" into a bad joke.

Nothing quite so grand concerns Seismograph Service, which looks for oil and gas, not geological enlightenment. Still, as Carol Rorschach, SSC's communications manager, puts it, "Were it not for naturally occurring phenomena such as the movements of plates, presumably the subsurface would be more nearly homogeneous, and there would be no oil left—no place where the hydrocarbons would have been trapped."

Fortunately for the energy industry, oil and gas deposits usually gather conveniently near the surface in a variety of spots after being formed by the rare series of coincidences that turn "rotten dinosaurs," as Dale Stone puts it, into hydrocarbons. "What we look for is porous rock," Stone explains. "You can't put oil into hard rock. You need something like sand or limestone that has become porous, dead underground river channels, reefs from ancient oceans. We also look for formations such as salt domes that trap the hydrocarbons, or fault planes that can be lateral traps."

SSC "looks" by twitching the Earth's surface with short pulses of energy—with explosives, truckmounted hydraulic rams that massage the earth, air guns that blast underwater bubbles, and a variety of other tools—and recording the vectors of energy reflected from whatever is underground.

"It's like throwing a rock in a pond," says Stone. "Ripples of energy radiate outward, and wherever that energy strikes a lack of continuity, whether it's a lily pad in a pond or a layer of limestone underground, some of it scatters back toward the focal point."

It would be nice if seismograms were neat snapshots of the rust under our feet, for the equipment used lacks little in range or penetration.

"We can look down 12 miles or more," Stone says, "and image as much of the Earth's crust as you'd be interested in seeing."

But seismograms aren't photographs, for energy waves ricochet and refract, rock formations "migrate" and reappear in the wrong place, the crust under Oklahoma reacts differently than does the ground beneath Brazil.

SSC was one of several companies that pioneered the use of computer processing to clean up and amplify seismic data, beginning in the mid-1960s, but it was the only company to process that data on site.

"By 1970, everybody was processing seismic by computer," Stone explains, "but the only computers programmed to do it were in Dallas, Houston, and Tulsa. If you recorded the data in Borneo, it had to be shipped all the way back here to then stand in line waiting for computer time."

So in 1970, the subsidiary designed, assembled, and encoded an entire computer package powerful, durable, and transportable enough—by trailer, off-road truck, or helicopter sling—to do the job in the jungle or desert, on offshore platform or North Slope vastness.

"Some of the data processing techniques we've learned have been applied to earthquake research," says Stone.

"We're particularly interested in folds and fault patterns and pressures, because they're natural in oil traps, and that's also where earthquakes occur."

Safely warehoused in packing crates at the subsidiary's Tulsa headquarters are some of the world's more detailed records of the Earth's crust—files full of not only digitized data from the computer-processed era of seismology, but also long manila envelopes that shelter laboriously made magnetic analog recordings and even primitive paper traces stretching all the way back to the 1930s.

It's not too far-fetched to wonder, in fact, if some of the motivating secrets of the machine we call Earth aren't already in our hands, awaiting interpretation, in a file cabinet in Tulsa.

CHAPTER 7 DEFINING AND DESCRIBING

As a writer about technical subjects, you will have a constant need to define terms and to describe places, mechanisms, and processes. Most often, your definitions and descriptions will be part of a larger effort; for instance, you often need to define the terms used in an argument. There will also be occasions when your major objective will be to define a term or to describe some mechanism or process. The techniques of defining and describing are closely related, and we take up both in this chapter.

Definition

Everyone with a trade or a profession has a specialized vocabulary to suit that occupation. Plumbers know the difference between a *globe valve* and a *gate valve.* Electrical engineers talk easily about *gamma rays* and *microelectronics.* Statisticians understand the mysteries of *chi-square tests* and *one-way analyses of variance.* In fact, learning a new vocabulary is a major part of learning any trade or profession. Unfortunately, as you grow accustomed to using your specialized vocabulary, you may forget that others don't share your knowledge—your language may be incomprehensible to them. The first principle in understanding definition is to realize that you will have to do it frequently. You should define any term you think is not in your reader's normal vocabulary. The less expert your audience is, the more you will have to define. Sometimes, when you use a new specialized term or an old term used in a new way, you will even need to define for your fellow specialists.

Definitions range in length from a single word to long essays or even books. Sometimes, but not usually, a **synonym** inserted into your sentence will do, as in this example:

> The oil sump, that is, the oil reservoir, is located in the lower portion of the engine crankcase.

Synonym definition serves only when a common interchangeable word exists for the technical term you wish to use.

Sentence Definition

Most often you will want to use at least a one-sentence definition containing the elements of a **logical definition:**

> term = genus or class + differentia

Although you may not have heard of the elements of a logical definition, you have been giving and hearing definitions cast in the logical pattern most of your life. In the logical definition, you state that something is a member of some genus or class and then specify the differences that distinguish this thing from other members of the class.

TERM	=	GENUS OR CLASS	+ DIFFERENTIA
An ohmmeter	is	an indicating instrument	that directly measures the resistance of an electrical circuit.
A legume	is	a fruit	formed from a single carpel, splitting along the dorsal and the ventral sutures, and usually containing a row of seeds borne on the inner side of the ventral suture.

The second of these two definitions, particularly, points out a pitfall you must avoid. This definition of a legume would satisfy only someone who was already fairly expert in botany. Lay people would be no further ahead than before, because terms such as *carpel, ventral suture,* and so forth are not familiar to them. When writing for nonexperts, you may wish to settle for a definition less precise but more understandable:

> A legume is a fruit formed of an easily split pod that contains a row of seeds, such as a pea pod.

Here you have stayed with plain language and given an easily recognized example. Both of these definitions of a legume are good. The one you would choose depends on your audience.

Sometimes you may wish to begin a definition by telling what something is *not*, as in the following definition from *Chamber's Technical Dictionary:*

> metaplasm. Any substance within the body of a cell which is not protoplasm; especially food material, as yolk or fat, within an ovum.

You should, of course, avoid circular definition. "A botanist is a student of botany" will not take the reader very far. However, sometimes you may appropriately repeat on both sides of a definition common words you are sure your reader understands. In the following *Chamber's* definition it would be pointless to drag in some synonym for a word as readily understandable as *test.*

> Gmelin's test. A test for the presence of bile pigments; based upon the formation of various colored oxidation products on treatment with concentrated nitric acid.

Extended Definition

To make sure you are understood, you will often want to extend a definition beyond a single sentence. The most common techniques for extending a definition are description, example, and analogy. However, any of the arrangement techniques—chronology, topical order, classification, and division—may be used. The following definition, again from *Chamber's,* goes beyond the logical definition to give a description:

> anemometer. An instrument for measuring the velocity of the wind. A common type consists of four hemispherical cups carried at the ends of four radial arms pivoted so as to be capable of rotation in a horizontal plane, the speed of rotation being indicated on a dial calibrated to read wind velocity directly.

In our lay definition of legume an example was given: "such as a pea pod." Often analogy is valuable:

> A voltmeter is an indicating instrument for measuring electrical potential. It may be compared to a pressure gauge used in a pipe to measure water pressure.

1 Understanding GL:M

This chapter introduces the central concepts of McCormack & Dodge's general ledger system, General Ledger Millennium (GL:M). It defines the purpose of GL:M, describes its functions and outstanding features, and shows how they can help meet business needs.

A sentence definition

GL:M: A Definition

GL:M is a general ledger software package that helps you perform accounting and reporting functions with accuracy and flexibility. It automates accounting processes, to make maintaining the ledger as easy as possible.

Extension by example

The general ledger stands at the center of an integrated financial information system. Data flows into it from all related subsystems, such as accounts receivable, accounts payable, purchase order, fixed assets, and human resources.

Figure 1-1 shows how GL:M draws data from all subsystems into a centralized base of financial information.

Graphical illustration of information flow

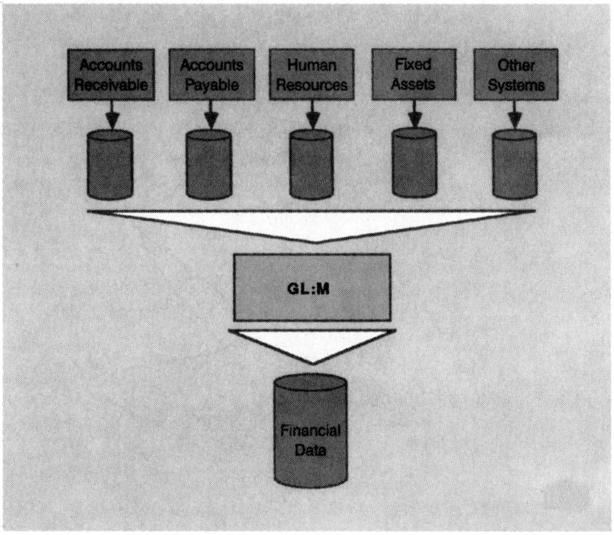

Figure 1-1. A Financial Information System

Figure 7–1 Graphic and Extended Definition

Source: GL:M General Ledger: Millennium—Overview (Atlanta, GA: Dun and Bradstreet Software Services, Inc., 1987) 1-1. Copyright © 1987 by Dun and Bradstreet Software Services, Inc. Reprinted with permission.

Look for opportunities to enhance extended definitions with graphics. Figure 7–1 demonstrates how a graphic can clarify relationships for a reader.

The following definition of a hurricane is a good example of an extended definition intended for an intelligent lay audience. In it, the writer makes extensive use of both process and mechanism descriptions. Notice, also, that the writer begins by defining other terms needed in understanding hurricanes:

Defines related terms

> A **hurricane** is defined as a rotating wind system that whirls counterclockwise in the northern hemisphere, forms over tropical water, and has sustained wind speeds of at least 74 miles/hour (119 km/hr). This whirling mass of energy is formed when circumstances involving heat and pressure nourish and nudge the winds over a large area of ocean to wrap themselves around an atmospheric low. **Tropical cyclone** is the term for all wind circulations rotating around an atmospheric low over tropical waters. A **tropical storm** is defined as a cyclone with winds from 39 to 73 mph, and a **tropical depression** is a cyclone with winds less than 39 mph.

Describes process

> It is presently thought that many tropical cyclones originate over Africa in the region just south of the Sahara. They start as an instability in a narrow east-to-west jet stream that forms in that area between June and September as a result of the great temperature contrast between the hot desert and the cooler, more humid region to the south. Studies show that the disturbances generated over Africa have long lifetimes, and many of them cross the Atlantic. In the 20th century an average of 10 tropical cyclones each year whirl out across the Atlantic; six of these become hurricanes. The hurricane season is set as being June 1 through November 30. An "early" hurricane occurs in the 3 months before the season, and a "late" hurricane takes place in the three months after the season.

Describes mechanism

> Hurricanes are well-organized. The 10-mile-thick inner spinning ring of towering clouds and rapid upper motion is defined as the hurricane's eyewall; it is here that the condensation and rainfall are intense and winds are most violent. Harbored within the eyewall is the calm eye of the hurricane—usually 10–20 miles across—protected from the inflowing winds and often free of clouds. Here, surface pressure drops to a minimum, and winds subside to less than 15 mph. Out beyond the eyewall, the hurricane forms into characteristic spiral rain bands, which are alternate bands of rain-filled clouds. In the typical hurricane, the entire spiral storm system is at least 1,000 miles across, with hurricane-force winds of 100 miles in diameter and gale-force winds of 400 miles in diameter. A typical hurricane liberates about 100 billion kilowatts of heat from the condensation of moisture, but only about 3% of the thermal energy is transferred into mechanical energy in the form of wind. Sustained wind speeds up to 200 mph have been measured, but winds of about 130 mph are more typical. It is estimated that an average hurricane produces 200 billion tons of water a day as rain.[1]

As this writer has done, extend your definition as far as is needed to ensure the level of reader understanding desired.

Placement of Definitions

You have several options for placement of definitions within your reports: (1) within the text itself, (2) in footnotes or endnotes, (3) in a glossary at the beginning or end of the paper, and (4) in an appendix. Which method you use depends upon the audience and the length of the definition.

Within the Text If the definition is short—a sentence or two—or if you feel most of your audience needs the definition, place it in the text with the word defined. Most often, the definition is placed after the word defined, as in this example:

> Besides direct electric and magnetic induction, another source of power-frequency exposure is contact currents. Contact currents are the currents that flow into the body when physical contact is made between the body and a conducting object carrying an induced voltage. Examples of contact current situations include contacts with vehicles parked under transmission lines and contacts with the metal parts of appliances, such as the handle of a refrigerator.[2]

Sometimes, the definition is slipped in smoothly before the word is used—a technique that helps break down the reader's resistance to the unfamiliar term. The following definition is a good example of the technique:

> Rocks that consist of chemical particles and rock fragments deposited by water, ice, or wind are sedimentary. Sedimentary rocks include deposits of gravel, sand, silt, clay, and the hardened equivalents of these—conglomerate, sandstone, siltstone, and shale, as well as limestone and deposits of gypsum and salt.[3]

When you are using key terms that must be understood before the reader can grasp your subject, define them in your introduction.

In Footnotes and Endnotes If your definition is longer than a sentence or two, and your audience is a mixed one—part expert and part lay—you may want to put your definition in a footnote at the bottom of the page or an endnote at the end of your report. A lengthy definition placed in the text could distract the expert who does not need it. In a footnote it is easily accessible to the lay person and out of the expert's way, as demonstrated in Figure 7-2. In an endnote it is out of the expert's way, but it is not easily accessible to the lay person.

In a Glossary If you have many short definitions to give and if you have reason to believe that most members of your audience will not read your report straight through, place your definitions in a glossary. (See Figure 7-3 and Figure 10-10, page 248.) Glossaries do have a disadvantage. Your readers may be annoyed by the need to flip around in your paper to find the definition they need. When you use a glossary, be sure to draw your readers' attention to it, both in the table of contents and early in the discussion.

In an Appendix If you need one or more lengthy extended definitions (say, more than 200 words each) for some but not all members of your audience, place them in an appendix. (See Chapter 10, pages 262–263.) At the point in your text where readers may need them, be sure to tell readers where they are.

There are two types of electric and magnetic fields, those that travel or propagate long distances from their source (also called electromagnetic waves) and those that are confined to the immediate vicinity of their source. At distances that are close to a source compared to a wavelength,[1] fields are primarily of the confined type. Confined fields decrease in intensity much more rapidly with distance from their source than do propagating fields. Propagating fields dominate, therefore, at distances that are far from the source compared to a wavelength. The fields to which most people are exposed from radio broadcast antennas are examples of propagating fields since these sources are generally much more than one wavelength (1-100 meters) removed from inhabited areas. The power-frequency fields that people encounter are of the non-propagating type because power lines and appliances are much closer to people than one 60 Hz wavelength (several thousand kilometers). Only a very minuscule portion of the energy in power lines goes into propagating fields. Because the power-frequency fields of public health concern are not of the propagating type, it is technically inappropriate to refer to them as "radiation."

[1] A wavelength is the distance that a propagating field travels during one oscillatory cycle. For fields in air this distance is c/f where c is the velocity of light and f is the frequency of the oscillating field.

Figure 7–2 Definition in a Footnote

Source: U.S. Congress, Office of Technology Assessment, *Biological Effects of Power Frequency Electric and Magnetic Fields—Background Paper* (Washington, DC: GPO, 1989) 6.

Description

In technical writing, you will chiefly have to describe three things: places, mechanisms, and processes. After explaining the use of visual language in description, we deal with the three.

Visual Language

The following brief descriptive passage shows how a combination of analogy and a few words indicating shape can help you accurately visualize a DNA molecule:

> DNA is a deceptively simple molecule, consisting of a series of subunits, called bases, linked together to form a double helix that can be visualized as an immensely long, corkscrew-shaped ladder. Each rung in the ladder is made up of two bases fitted together, and the ends of the rung are attached to chains of sugar-phosphates that are like the upright rails of a ladder.[4]

We visualize things in essentially five ways—by shape, size, color, texture, and position—and have a wide range of terms to describe all five.

Terms printed in boldface

Grammatically parallel sentence fragments used for definitions (see "Parallelism" in Part V, "Handbook")

When needed, definitions are extended in complete sentences

GLOSSARY

Alternator—a device which supplies alternating current.

Anemometer—a device for measuring wind speed.

Crossflow turbine—a drum-shaped water turbine with blades fixed radially along the outer edge. The device is installed perpendicularly to the direction of stream flow.

DC to AC inverter—a device which converts electrical current from direct to alternating.

FERC—The Federal Energy Regulatory Commission, established by Congress to regulate non-federal hydroelectric projects.

Flow—the quantity of water, usually measured in gallons or cubic feet, flowing past a point in a given time.

Generator—a device that converts mechanical energy into electrical energy—a large number of conductors mounted on an armature that rotates in a magnetic field.

Head—the vertical height in feet from the headwater (with a dam) or where the water enters the intake (no dam) to where the water leaves the turbine.

Induction generator—an alternating current generator whose construction is identical to that of an AC motor.

Intake structure—a structure that diverts the water into the penstock; a small dam.

Isolation transformer—a device used to isolate the utility grid system from an earth-grounded electric generating system.

Net energy billing—an electric metering system in which the meter turns backwards when electricity flows from the generating system to the utility lines and forward when the utility is supplying electricity to the residence.

Pelton wheel—a water turbine in which the pressure of the water supply is converted to velocity by a few stationary nozzles, and the water jets then impinge on buckets mounted on the rim of the wheel.

Penstock—the pipe that carries pressurized water from the intake structure to the turbine.

Photovoltaic array—several photovoltaic modules connected together, usually mounted in a frame.

Photovoltaic module—several solar cells connected together on a flat surface.

PURPA—Public Utility Regulatory Act, a federal regulation requiring utilities to buy back power generated by small producers.

Solar easements—a written agreement with a person's neighbors that protects his/her access to the sun through the prohibition of any structures that might block their access.

Synchronous inverter—a device that links the output from a wind generator to the power line and the domestic circuit. The varying voltage and frequency generated by the windmill is instantly converted to exactly the same type of electricity distributed by a utility's power grid.

Figure 7–3 Definitions in a Glossary
Source: U.S. Department of Energy, *Homemade Electricity: An Introduction to Small-Scale Wind, Hydro, and Photovoltaic Systems* (Washington, DC: GPO, n.d.) 57.

In addition, comparison of the unfamiliar to the familiar through analogy is a powerful visualization tool.

Shape You can describe the shape of things with terms such as *cubical, cylindrical, circular, convex, concave, square, trapezoidal,* or *rectangular.* You can use simple analogies such as *C-shaped, L-shaped, Y-shaped, cigar-shaped, corkscrew,* or *star-shaped.* You can describe things as *threadlike* or *pencillike* or as *sawtoothed* or *football-shaped.*

Size You can give physical dimensions for size, but you can also compare objects to coins, paper clips, books, and football fields.

Color You can use familiar colors such as red and yellow and also, with some care, such descriptive terms as *pastel, luminous, dark, drab,* and *brilliant.*

Texture You have many words and comparisons at your disposal for texture, such as *pebbly, embossed, pitted, coarse, fleshy, honeycombed, glazed, sandpaperlike, mirrorlike,* and *waxen.*

Position You have *opposite, parallel, corresponding, identical, front, behind, above, below, right, left, north, south,* and so forth to indicate position.

Analogy The use of analogy will aid your audience in visualizing the thing described. In the following example, a simple comparison to a balloon (which we have set in boldface) helps the reader visualize the swelling of a volcano before eruption:

> As magma enters the shallow summit reservoir, the volcano undergoes swelling or inflation **(a process similar to the stretching of a balloon being filled with air)**. This swelling in turn causes changes in the shape of the volcano's surface. During inflation, the slope or tilt of the volcano increases, and reference points (benchmarks) on the volcano are uplifted relative to a stable point and move further apart from one another. For Hawaiian volcanoes, pre-eruption inflation generally is slow and gradual, lasting for weeks to years. However, once eruption begins, the shrinking or deflation typically occurs rapidly as pressure on the magma reservoir is relieved—**a process not unlike deflating a balloon.** During deflation, changes in tilt and in vertical horizontal distances between benchmarks are opposite to those during inflation.[5]

Place Description

Writers of technical material must often describe places. Place descriptions appear in research reports when the location of the research has some bearing on the research. Highway engineers must describe the locations of projected highways. Surveyors have to describe the boundaries of parcels of property. Police officers have to describe the scene of an accident. In place description, you must pay particular attention to your point of view and your selection of detail.

Point of View When describing a place, you must be aware of your point of view—that is, where you are positioned to view the place being described. If you, in your mind's eye, position yourself before you begin to write, you'll avoid confusing and annoying your readers with an inconsistent jumping about in your description. You can shift your point of view, but be aware when you are making the shift. Have a good reason for doing it, and don't do it too often. Most often the reason will be to allow yourself to give an overview and then proceed to a close-up of selected details. Such is the case in the following passage, in which an archaeologist through imagination and scientific knowledge reconstructs for a lay audience a Roman outpost in second-century Britain:

Aerial point of view

> A new fort, built of stone, rose on the eastern part of a level plateau. The enclosure was oblong; at the north and south ends the fort's massive walls, broken by central gates, ran 85 meters; on the east and west sides the length of the walls is just under 150 meters.
>
> The overgrown ruins of the last wood fort at Vindolanda, abandoned 40 years earlier, stood just beyond the west wall of the new fort. Here the garrison engineers must have dumped many carloads of clay in the process of covering up the earlier structures and preparing a level site for the stone foundations of what we call Vicus I. On both sides of the main road leading from the west gate were erected a series of long barrackslike buildings; the masonry was dressed stone bound with lime mortar. The largest barracks was nearly 40 meters long, and all of them seem to have been about five meters wide. We deduce that the structures were one story high and probably served as quarters for married soldiers. They were divided into single rooms, one to a family, by partitions spaced about seven meters part. There was probably storage space under the roof, and a veranda outside each room would have provided cooking space. The four barracks we have located so far could have housed 64 families. The British army in India offered almost identical housing to the families of its Indian soldiers. Smaller buildings, possibly housing for the families of noncommissioned officers, stood nearby.[6]

Shift to eye-level point of view

In the passage just cited, the author begins with an aerial point of view. He looks down upon the fort and describes it in general outline—its walls, roads, and buildings. That job done, he returns to ground level, and his point of view is now that of a well-informed guide walking you through the barracks. Notice the masonry, he says. Here are the rooms. One family lived in each. See how the rooms were partitioned off and where the storage space and cooking areas were. The writer uses the descriptive language of size, shape, and position. Physical description and function are gracefully combined.

Selection of Detail Remember that in any description you are selecting details. You would find it impossible to describe everything. Very often, the details chosen support a thesis, perhaps an inference or chain of inferences. Notice how the selected details in the following passage about Jupiter's ring support the inferences (which we have set in boldface):

Voyager 2 took pictures of Jupiter's ring on the inbound leg, but more interesting were the pictures it took while behind Jupiter, looking back at the ring. Where Voyager 1's pictures were faint, the ring now stood out sharp and bright in the newest photos, telling scientists instantly that **the ring's particles scattered sunlight forward more efficiently than they scattered it backward, and therefore were tiny, dust-like motes.** (Large particles backscatter more efficiently.)

While the dust particles of the ring appeared to extend inward toward Jupiter, probably all the way to the cloud tops, the ring had a hard outer edge, as if cut from cardboard.

Close examination of Voyager photos after the encounter revealed two tiny satellites, orbiting near the outer edge of the ring and herding the particles in a tight boundary. **The source of the ring's dust probably lies within the bright portion of the ring itself. The dust may be due to micrometeorites striking large bodies in the ring.**[7]

Mechanism Description

The physical description of some mechanism is perhaps the most common kind of technical description. It is a commonplace procedure with little mystery attached to it. For example, we see a friend having little success struggling to unplug a clogged sink drain by using a plunger.

"Look," we say to her, "you ought to buy yourself a plumber's snake. It would unclog that drain in a couple of minutes."

"Really? What's a plumber's snake?"

"It's a tool for unplugging drains. Mostly, it's a flexible, springlike, steel cable about five feet long with the diameter of a pencil. The cable has a football-shaped boring head on its working end. The head is about two inches long and at its widest point is twice the diameter of the cable. The whole thing looks a bit like a snake, hence its name."

"That so? Anything to it besides the cable and head?"

"Uh-huh, there's a crank. It's a hollow steel tube in the shape of an opened-up Z. It's about ten inches long, so you can get both hands on it. You slip the crank over the cable. With it, you can rotate the cable after you've inserted it in the drain, so that the head operates something like a drill to bore through the clog."

"Sounds like a handy gadget. I'll have to get one."

In this brief passage, we have used most of the techniques of good technical description. Our purpose has been to make you *see* the object and understand its function. We have done the following:

- Described the overall appearance of the plumber's snake and named the material with which it is made (steel).
- Divided the mechanism into its component parts—cable, boring head, and crank.

- Described the appearance of the parts, given their functions, and showed how they work together.
- Pointed out an important implication, a *so-what,* of one of the descriptive facts: "It's about ten inches long, *so you can get both hands on it.*"
- Given you only information important in this description. For example, because it is of no consequence in this description, we have not told you the color of the mechanism.
- Used figurative language—*springlike* and *football-shaped*—and comparisons to familiar objects such as pencils, snakes, and drills to clarify and shorten the description.

In our discussion of mechanism description, we tell you how to plan one and provide several examples of typical mechanism descriptions.

Planning Despite certain elements that most mechanism descriptions have in common, there is no formula for writing them. You must use your judgment, weighing such matters as purpose and audience. As you plan your mechanism description, you'll need to answer questions like the ones we provide next. The answers to these questions will largely determine how you arrange your description and the details and so-whats you provide your readers.

What is the purpose of the description? To impart general knowledge? To sell the mechanism? To transfer technology?

Why will the intended reader read the description? To understand the mechanism? To consider buying the mechanism? To use the mechanism? What is the reader's level of experience and knowledge in regard to the mechanism? Would the reader understand the technical terminology involved, or would he or she need terms defined? Would the reader understand the implications of the facts presented about the mechanism or would you have to furnish the so-whats?

Consider these questions about the mechanism.

- What is the purpose and function of the mechanism?
- How can the mechanism be divided?
- What are the purpose and function of the parts?
- How do the parts work together?
- How can the parts be divided? Is it necessary to do so?
- What are the purpose and function of the subparts?
- How do the parts and subparts work together?
- Which of the following are important for understanding the mechanism and its parts and subparts: construction, materials, appearance, size, shape, color, texture, position?
- Are there any so-whats you need to express explicitly for the reader?

- Would the use of graphics—photographs and drawings, for example—aid the reader?
- Are there any analogies that would clarify the description for the reader?

When you can answer these questions, you will be better able to select the details concerning the mechanism and to arrange them in a way that will fulfill your purpose. Because purpose, audience, and mechanism vary from situation to situation, the arrangement and content of mechanism descriptions also vary. But they are all likely to include statements about function, some sort of division of the mechanism into its component parts, physical descriptions in words or graphics or some combination of the two, analogies, and so-whats.

Examples of Mechanism Descriptions In Figure 7-4, our annotations draw your attention to the various features of a mechanism description as exemplified in the descriptive overview of the Hubble Space Telescope. Figure 7-5 is one of several graphics that accompany the telescope's description. As demonstrated in Figure 7-5, such graphics are frequently annotated and are "cut away" to show the interior of the mechanism described.

Professionals often describe mechanisms for reasons of technology transfer. *Technology transfer* is a rather fancy term for a common but important part of technical and scientific life. Technical people exchange concepts through their technical journals. Sometimes the transfer is from one discipline to another. Very often the concept described concerns a mechanism. Figure 7-6 reproduces a mechanism description as it appeared in a journal devoted to technology transfer. Like most such descriptions, it uses a graphic to reduce the need for extensive visual description in words. So-whats are prominently featured. In fact, the head designed to attract the reader's attention is primarily a so-what.

As is done in most mechanism descriptions, the authors divide the mechanism into its parts—a base and a cover. They give details of construction, not forgetting the so-whats.

The purpose of the description is primarily to make a technical reader aware of the mechanism. Most people would need additional information to construct the mechanism for themselves. Therefore, the description provides a way for the reader to get further information.

Some mechanism descriptions may be sales motivated, such as catalog descriptions. In sales the implications of your facts—the so-whats—take on special importance, as in the next example:

Our English Steel Scraper Mat Is Guaranteed for Life!

This mat is assembled from strips of heavy-gauge steel which are woven and twisted together. It just can't come apart. The entire Scraper Mat has been dipped in a

Description by size and analogy

The Hubble Space Telescope is just over 13 meters (43 feet) long and 4 meters (14 feet) in diameter, about the size of a bus or tanker truck. Upright, it is a five-story tower; carried inside the Space Shuttle for the trip to orbit, it fills the payload bay.

Division of telescope into its component parts

The Hubble Space Telescope is made up of three major elements: the Optical Telescope Assembly, the focal plane scientific instruments, and the Support Systems Module, which is divided into four sections, stacked together like canisters:

Subdivision of Support System Module

Functions of key parts of Support System Module

Aperture Door and Light Shield: protecting the scientific instruments from light of the sun, Earth, and moon and also from contamination

Forward Shell: enclosing the Optical Telescope Assembly mirrors

Equipment Section: girdling the telescope to supply power, communications, pointing and control, and other necessary resources

Aft Shroud: covering the five focal plane instruments and the three fine guidance sensors.

Miscellaneous details

Solar energy arrays and communications antennas are attached to the exterior shell. Doors allow astronauts to remove instruments and components from the equipment bays. Handrails and sockets for portable foot restraints attached to the external surface aid the astronauts in performing maintenance and repair tasks.

Space Telescope Vital Statistics

Length:	*13.1 m (43.5 ft)*
Diameter:	*4.27 m (14.0 ft)*
Weight:	*11,000 kg (25,500 lb)*
Focal Ratio:	*f/24*
Primary Mirror	
Diameter:	*2.4 m (94.5 in)*
Weight:	*826 kg (1,825 lb)*
Reflecting Surface:	*Ultra-low expansion glass covered by aluminum with magnesium-fluoride coating*
Secondary Mirror	
Diameter:	*0.3 m (12 in)*
Weight:	*12.3 kg (27.4 lb)*
Reflecting Surface:	*Ultra-low expansion glass covered by aluminum with magnesium-fluoride coating*
Systems	
	Optical Telescope Assembly
	Support Systems Module
	Focal Plane Science Instruments
	Wide Field/Planetary Camera
	Faint Object Camera
	Faint Object Spectrograph
	Goddard High Resolution Spectrograph
	High Speed Photometer
	Fine Guidance Sensors (for astrometry)
Data Rate:	*Up to 1 mbps*

Use of list for ease of reference

Figure 7–4 Mechanism Description of Hubble Space Telescope
Source: National Aeronautics and Space Administration, *Exploring the Universe with the Hubble Space Telescope* (Washington, DC: GPO, n.d.) 57.

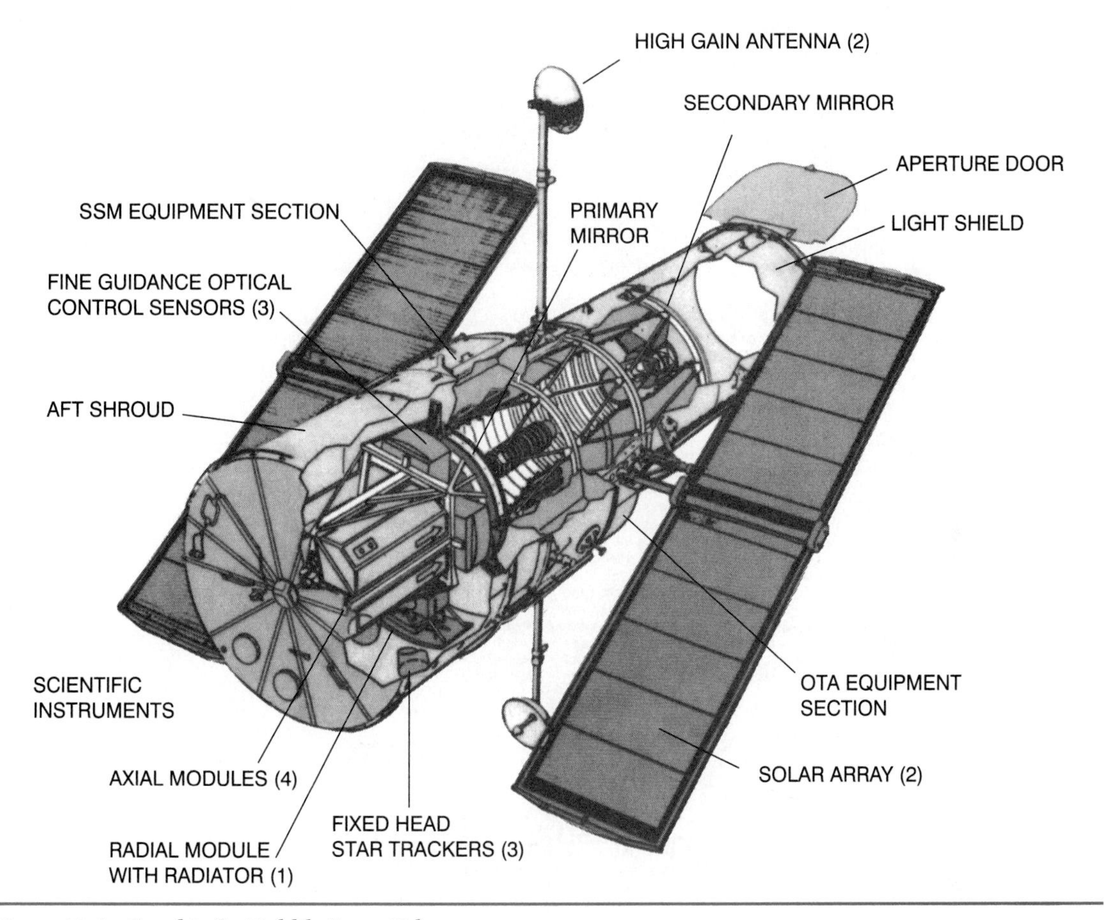

Figure 7–5 Graphic for Hubble Space Telescope
Source: National Aeronautics and Space Administration, *Exploring the Universe with the Hubble Space Telescope* (Washington, DC: GPO, n.d.) 58.

> galvanizing solution which totally protects it from rusting out. Clay, mud, dirt, or snow scraped off your shoes or boots fall through the steel strips. Ordinary door mats simply move the grime from one part of your shoes to another; Scraper Mat actually removes it altogether.[8]

Rewritten without the so-whats, the description loses most of its force:

> This mat is assembled from strips of heavy-gauge steel which are woven and twisted together. The entire Scraper Mat has been dipped in a galvanizing solution.

Catalog descriptions illustrate how important purpose and audience analysis are in mechanism description. Lay audiences, such as those who read most catalogs, are primarily interested in the function of a mechanism and the so-whats concerning it. On the other hand, the

Protective Package for a Gamma-Ray Detector

Enclosure resists contamination, voltage breakdown, and vibration.

NASA's Jet Propulsion Laboratory, Pasadena, California

Function of mechanism

A package for a germanium gamma-ray detector protects the semiconductor crystal from contamination, allows it to operate at high voltages, and isolates it from shock and vibration. The package seals the detector from its surroundings, whether in the atmosphere or in the vacuum of space.

Division into parts

The main parts of the package, a base and a cover, are made of aluminum. As shown in the figure, the cover is sealed to the base by a soft aluminum ring. Since no solder is used for the seal, there is no danger of contamination of the germanium by flux outgassing inside the package.

Construction details with so-whats

A high-voltage power connection and a low-voltage signal connection are soldered to the cover. Nitrogen is introduced into the sealed package through the evacuation port. The nitrogen gas inhibits high-voltage arcs in the package cavity, prevents oxidation of the crystal, and provides back pressure to minimize the release of gases from the package metal. A small amount of helium in the nitrogen charge allows leakage from the package to be measured with a sensitive helium leak detector.

Construction details with so-whats

A clamping ring surrounds the jaw fingers and ceramic jaws and clamps them to the germanium crystal. The device is thus supported without damage to sensitive crystal surfaces, even during high acceleration.

This work was done by Marshall Fong, Charles Lucas, Albert Metzger, Donald M. Moore, Robert Oliver, and Walter Petrick of Caltech for **NASA's Jet Propulsion Laboratory**. *For further information, Circle 28 on the TSP Request Card.*
NPO-16019

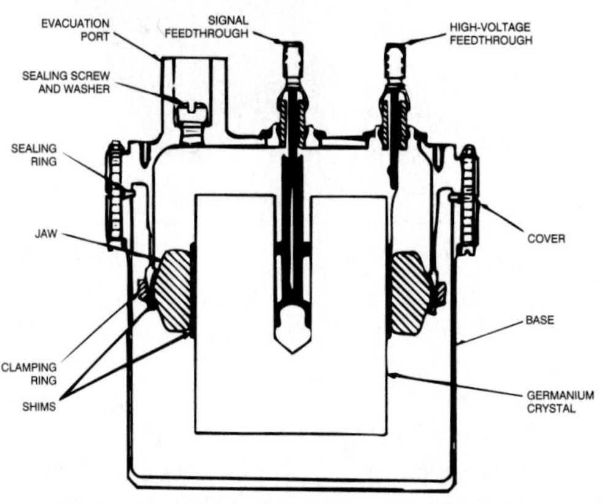

This **Housing for a Germanium Detector**, part of a gamma-ray spectrometer, holds the germanium crystal securely while protecting it from contamination. Although designed to hold nitrogen, the package can also be evacuated if necessary.

Figure 7–6 Description of a Mechanism
Source: NASA Tech Briefs (Spring 1985): 72.

more expert an audience is, the more they will be interested in the mechanism itself, as well as its functions and the so-whats.

Many of the principles found in mechanism description have non-mechanical applications. For example, we don't usually think of skin as

Skin

Function

Division into parts

The skin is the protective covering of the body. The epidermis, the outer layer of skin, contains no blood vessels or nerves. The layer of skin below the epidermis is the dermis. The dermis contains the blood vessels, nerves, and specialized structures such as sweat glands, which help to regulate body temperature, and hair follicles. The fat and soft tissue layer below the dermis is called subcutaneous (Figure 1-3). The skin is one of the most important organs of the body. The loss of a large part of the skin will result in death unless it can be replaced.

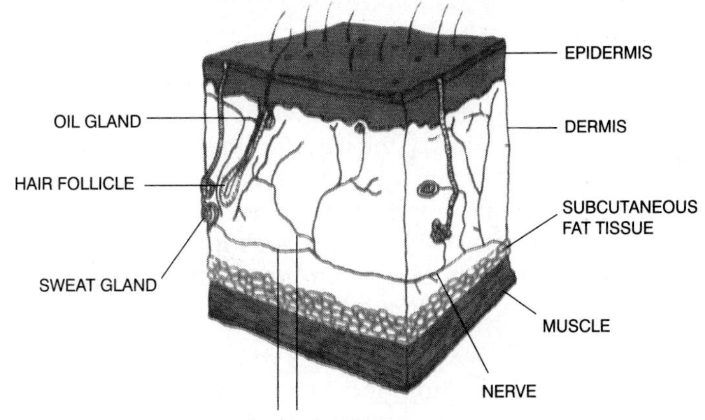

Figure 1-3 - Skin.

Function

The protective functions of the skin are many. Skin is watertight and keeps internal fluids in while keeping germs out. A system of nerves in the skin carries information to the brain. These nerves transmit information about pain, external pressure, heat, cold, and the relative position of various parts of the body.

Skin provides information to the first aider concerning the victim's condition. For example, pale, sweaty skin may indicate shock.

Figure 7–7 Skin Described as a Mechanism
Source: U.S. Department of Labor, *First Aid Book* (Washington, DC: GPO, 1991) 9.

a mechanism, but the same rational principles we have been discussing are found in Figure 7–7 in the passage about skin. The subject is divided, objective physical details are described, function is discussed, so-whats are given, and a graphic is provided.

Process Description

Process description is probably the chief use of chronological order in technical writing. By **process** we mean a sequence of events that progresses from a beginning to an end and results in a change or a product. The process may be humanly controlled, such as the manufacture of an automobile. It may be natural—the metamorphosis of a caterpillar to a butterfly, for example.

Process descriptions are written in one of two ways:

- *For the doer*—to provide instructions for performing the process
- *For the interested observer*—to provide an understanding of the process

A cake recipe is a good example of instructions for performing a process. You are told when to add the milk to the flour, when to reserve the whites of the eggs for later use. You are instructed to grease the pan *before* you pour the batter in, and so forth. Writing good instructions is a major application of technical writing, and we have devoted all of Chapter 16 to it. In this chapter we explain only the second type—providing an understanding of the process. We discuss planning and sentence structure and provide example process descriptions.

Planning Insofar as you have grasped the proper order of events, the organization of a process description will not present any particular problem. You simply follow the order of events as they occur. However, you must determine the amount of detail you need to include. A situational analysis that raises questions about purpose, reader, and the process itself will usually give you a good idea of the level of detail needed. The questions you should ask for a process description resemble those used for a mechanism description.

What is the purpose of the description? To describe a process you have followed? To describe in a general way how some process works? To describe in a detailed, highly technical way how some process works?

Why will the reader read the process description? To learn about the process in a general way? To make a decision, perhaps about using the process? To understand how the process might affect his or her work? To remain aware of activities in the discipline? To understand how an experiment was conducted?

What is the reader's level of experience and knowledge in regard to the process? Will the reader need some technical terms defined? Will the reader need some of the so-whats explained? How high is the reader's interest in the details of the process?

What is the purpose of the process?

Who or what does the process?

What are the major steps of the process? Do the steps divide into subprocesses, some of which may be going on at the same time?

Are there graphics and analogies that would help the reader?

Sentence Structure The choice of sentence structure is important in process description. You have to choose between present and past tense and decide which voice and mood to use. Many process descriptions are written in the present tense (we have set the verbs in boldface):

Present tense

> Blood from the body **enters** the upper chamber, the atrium, on the right side of the heart and **flows** from there into the lower chamber, the ventricle. The ventricle **pumps** the blood under low pressure into the lungs, where it **releases** carbon dioxide and **picks up** oxygen.

The rationale for the use of present tense is that the process is an ongoing and continuing process—the heart's pumping of blood. Therefore, it is described as going on as you are observing it.

However, processes that occurred in the past and are completed are described in the past tense. A major use of past tense in process description is found in empirical research reports where the procedures the researcher followed are described in past tense (verbs in boldface):

Past tense

> During the excavation delay we **accomplished** two tasks. First, we **installed** a temporary intake structure and **tested** the system's efficiency. Second, we **designed, built,** and **installed** a new turbine and generator.

In writing instructions, you will commonly use the active voice and imperative mood:

Imperative mood

> **Clean** the threads on the new section of pipe. **Add** pipe thread compound to the outside threads.

In following your instructions, it is the reader who is the doer. With its implied *you*, the imperative voice directly addresses the reader. But in the process description for understanding, where the reader is not the doer, the use of imperative mood would be inappropriate and even misleading.

In writing a process for understanding, therefore, you will ordinarily use the indicative mood in both active and passive voice:

Active voice

> The size of the cover opening **controls** the rate of evaporation.

Passive voice

> The rate of evaporation **is controlled by** the size of the cover opening.

In active voice, the subject does the action. In passive voice, the subject receives the action. The passive voice emphasizes the receiver of the action while deemphasizing or removing completely the doer of the action. (Incidentally, as the preceding examples illustrate, neither doer nor receiver has to be a human being or even an animate object.) When the doer is unimportant or not known, you should choose passive voice. Conversely, when the doer is known and important, you should choose the active voice. Because the active is often the simpler, more direct statement of an idea, choose passive voice only when it is clearly indicated. We have more to say on this subject on pages 96–97.

Examples of Process Description As with all technical writing, you can write process descriptions for varied audiences. In the following excerpt, the writer describes how a star such as our sun derives energy from fusion, until eventually it depletes its sources and destroys itself. The intended audience is an educated lay person. The writer uses present tense throughout and predominantly the active voice.[9]

Overview of process

> In their hot, high-pressure and high-density interiors, stars produce energy through the fusion of low-mass atomic nuclei to high-mass nuclei. In normal stars like the Sun, hydrogen nuclei are joined together to make helium, in a process that liberates large amounts of energy.

Description of process

> A star like the Sun can persist in its normal state, deriving energy from the fusion of hydrogen to helium, for some 10 billion years. Upon the inevitable depletion of its internal, hydrogen-based energy source, a star proceeds through more advanced evolutionary stages in which it converts successively more massive nuclear species into yet higher mass nuclei, to satisfy its needs for energy and prevent collapse under the influence of its strong self-gravity. After converting hydrogen to helium, it proceeds to convert the helium to carbon and oxygen, then to silicon-like nuclei, and so on, until, in the more massive stars, the nuclear fusion products approach the mass of iron nuclei. Beyond this point, no further energy can be extracted by building nuclei of increasing mass. Atomic nuclei with masses near that of iron are the most stable of nuclei; conversion of these nuclei to other species, through either nuclear fusion or nuclear fission, requires not the extraction of energy but the injection of energy.
>
> Having depleted all its nuclear energy sources, a star begins to cool and can no longer resist the pull of its own gravity. In the more massive stars, we believe that this process leads to a sudden catastrophic collapse. The gravitational collapse of the star's interior is thought to release a large amount of energy which, flowing from the star, blows the star's outer layers away into space, to disperse and mix with the interstellar matter. At the same time, the exploding material in the ensuing supernova explosion is compressed and heated to the point that fast nuclear reactions occur, resulting in a buildup of very massive atomic nuclei, which are dispersed with the star's outer layers into the preexisting interstellar matter.

In writing process descriptions to provide understanding, you'll find that extensive detail is not always necessary or even desirable. As in mechanism description, the amount of detail given relates to the technical knowledge and interest of your readers. When *Time* magazine publishes an article about open-heart surgery, its readers do not expect

Marine Consultants
42 Oceanside Avenue
East Hampton, NY 11515
(516) 286-3563

Fax (516) 286-2249

17 July 1993

Mr. Avery Brandisi
Chief Executive Officer
Maritime Transport Inc.
864 Third Avenue
New York, NY 10022

Dear Mr. Brandisi:

We have as you requested examined the condition of the pilings that support the Maritime Transport East River pier. The pilings show only minor shipworm damage. However, because of the high probability that your pilings will suffer further shipworm damage if not protected, we recommend that you take action as soon as possible.

Shipworms are mollusks that bore into submerged wood and do extensive damage to it. The shipworms, so-called because when fully grown they resemble worms, begin life as small organisms looking for a place to lodge. When a shipworm comes to rest on wood, it changes into a worm-like animal with a pair of shells on its head. Using these shells, the shipworm bores its way into the wood. As the head bores in, the end of the worm-like body remains at the entrance hole. The shipworm lives on the wood borings and the organisms in the sea water it passes through itself.

Although shipworms may reach lengths up to four feet, the entrance hole remains the same size. Thus, the wood can be completely honeycombed with shipworms and except for small entrance holes look perfectly sound. This is what can happen to your pilings if they are not properly treated.

Ironically, your firm and many other waterside businesses in New York harbor are victims of the progress made in cleansing the water in the harbor. When the harbor was polluted, it would not support most marine animals, including shipworms. In the now relatively clear water of the harbor, shipworms have returned and are causing major damage that will take millions of dollars to repair.

Either sheathing or chemical treatment would offer your pilings a good measure of protection. If you like, we can help you choose the best treatment and recommend reputable contractors to do the work.

Sincerely,

Mary Chen

Mary Chen
Consulting Entomologist

Figure 7–8 Executive Process Description

complete details on how such an operation is performed. Rather, they expect their curiosity to be satisfied in a general way. The author of the description of the sun's fusion process used a level of detail he felt would satisfy his lay audience.

Figure 7–8 shows a process description written for an executive audience. It deals with shipworms, marine organisms that attack wood

Overview and function

Use of system

One paragraph per major step

Figure 7–9 Description of a Process

Source: NASA Tech Briefs (Spring 1984): 398–399.

Automatic Coal-Mining System

Coal cutting and removal would be done with minimal hazard to people.

NASA's Jet Propulsion Laboratory, Pasadena, California

A proposed automatic coal-mining system would cut coal, grind it, mix it with water to make a slurry and transport the slurry to the surface. The system would include closed-circuit television monitoring, laser guidance, optical obstacle avoidance, proximity sensing, and other features for automatic control.

The system is intended for longwall mining in which the coal seam is divided into several blocks at least 600 feet (183 m) wide by corridors or "entries," which allow the movement of equipment and materials. Moving along an entry an extracting machine cuts coal from the face of a block (see figure) to a depth of a few feet (about a meter). An extractor has two cutting heads, one at each end. The extractor moves on crawler tracks that are reversed at the end of each pass so that coal can be cut in the opposite direction. Thus the extractor does not have to retreat in "deadhead" fashion to begin the next cut, and its productivity is increased.

The cut coal falls on loading ramps, where gathering arms move it to central screws. The screws crush the coal and feed it to the transport tube.

Water is sprayed on the extractor drums as they cut to keep down explosive dust and to promote visibility. More water is added in the screws to convert the coal powder into a low-viscosity slurry.

A laser beam guides the extractor along the face, ensuring a straight cut. Connected to the slurry-transport subsystem, the laser and its reflector are moved forward automatically with each pass of the extractor. Any deviation from the path set by the beam represents a change in the direction of travel of the extractor; and an onboard controller changes the speed of one crawler track relative to the other thereby adjusting the direction of travel.

A sonic or optical proximity sensor detects the approach of an entry and stops the extractor to prevent a collision with the far wall of the entry. The proximity sensor also guides the reversing maneuver and the start of a new cut.

The transport tube is made in repeating sections, with each section mounted on an individual skid to allow lateral motion. As the extractor moves along, it pulls each section toward it and connects with that section in turn.

The roof supports move forward in alternation, "walking" by being lifted off the floor by hydraulic pistons, then propelled by horizontal pistons. Each alternate support unit supports itself on adjacent neighbors and propels itself by pushing on its neighbors. The units that are not moving are holding up the roof. The movement of the roof supports may be controlled remotely by human operators watching via television, or they may be made to move automatically on the basis of the movement of the slurry-transport tube.

This work was done by Earl R. Collins, Jr., of Caltech for **NASA's Jet Propulsion Laboratory**. *For further information, Circle 114 on the TSP Request Card.*
NPO-15861

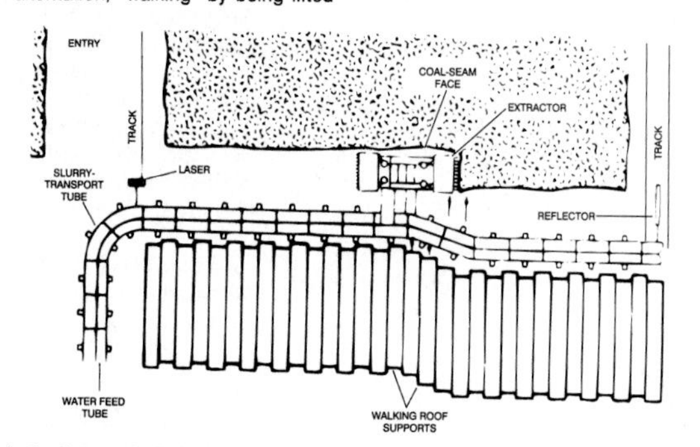

In the **Automatic Coal Mine**, the cutting, transport, and roof-support movement are all done by automatic machinery. The exposure of people to hazardous conditions would be reduced to inspection tours, maintenance, repair, and possibly entry mining.

immersed in salt water. As with the description written for a lay audience, the writer chooses her level of detail carefully. As you read it, notice that no attempt is made to give full information about the shipworm, as an entomologist might desire. We don't learn, for example, how the shipworm reproduces, nor do we even learn very clearly what it looks like. For the intended readers of the letter, such information is not needed. They need to know what is presented—the process by which the shipworm lodges on the wood and how it bores into it. They need to know that the damage done by shipworms is largely invisible from the exterior of the wood. Finally, they need to know that they must take action and what that action has to be.

Process description, like mechanism description, figures prominently in technology transfer. In their journals, scientists and technologists tell each other about useful processes just as they tell each other about useful mechanisms. Figure 7–9 describes an automatic coal-mining system. The description is particularly well done. It begins with an introduction that provides an overview of the process and the mechanisms involved and makes clear the type of coal mining for which this process is intended. In paragraphing, the author is guided by the steps of the process, one paragraph for each major step. He provides a graphic that makes the entire process clear at a glance.

The writer includes a sufficient level of detail so that a technical reader can understand the process and its advantages. To actually design the mechanisms needed and follow the process, readers would need much more information. Notice the use of present tense verbs throughout the description.

Empirical research reports are written for experts. The section in the report that describes the *methods* used by the researcher is a process description. Because the research has been completed by the time it is reported, method sections are written in past tense, usually in a mixture of active and passive voice. Writers of such reports include enough detail about the procedure so that readers equally expert as themselves could duplicate the research. Because of that requirement, the level of detail in an expert report is normally much higher than in a lay or executive report. The writer, having an expert audience, uses technical language freely. The following excerpt is the method section from a report of research that tested whether students taught to revise globally would revise better than those who were not. (To revise "globally" means to deal with such things as purpose, organization, and audience in contrast with dealing only with such surface things as spelling and grammar.)

> The study was conducted in two writing classes during the ninth week of a sixteen-week semester. Half of the students in each class were randomly assigned to the treatment group; the other half served as the control group. The mean SAT verbal score for the control group (532.2) was slightly higher than that for the treatment group (514.4); however, a t-test revealed that the difference between these means was not significant.
>
> After giving brief instructions about the nature of the study, an experimenter (not the instructor for the course) asked the students in the control group to go with another experimenter to a nearby room to complete the experiment. The experimenters read the same brief instructions to both the treatment and control groups. These task instructions (see Appendix B) informed students that they would have 30 minutes to revise a short text about the operation of a water treatment plant so that it could be used as a handout for high school students who tour the plant. The instructions specifically cued students to revise so that the handout would be "clear, organized, easy to read, and free of errors." The instructions also directed students to mark deletions, additions, changes, and movements of text in standard ways such that a typist could easily retype their revised texts.
>
> After reading the instructions and asking for questions, the experimenters reminded the students that they had 30 minutes to complete their revisions and instructed

> them to begin. The students were informed when they had 15 minutes and 5 minutes remaining. For each of the treatment and control groups, the procedure was completed within the 50-minute class period.
>
> Procedures for the two groups were identical except that an experimenter presented eight additional minutes of instruction to the treatment group. The eight minutes taken for the special instruction of the experimental groups was approximately equal to the time it took to change rooms for the control groups.
>
> The purpose of this instruction was to cue students to revise globally by illustrating how an expert writer and a novice writer revised a similar text. The experimenter illustrated differences between the revision activities of the expert and the novice writers using overhead transparencies. First, he explained the differences in basic approach and procedure—the expert writer read through the entire text to identify major problems and then focused on improving the whole text. In contrast, the novice writer began making changes immediately and proceeded to search through the text for local errors.
>
> After this overview, the experimenter illustrated differences in the amount and types of changes that the two writers made using transparencies of the two writers' actual revisions of the sample text. The transparencies illustrated that the expert writer not only made more revisions but made different kinds of revisions. The effect was rather dramatic; while the novice writer limited himself to eliminating spelling, wordiness, and grammar errors, the expert writer also addressed global issues, adding an initial purpose statement, selecting and deleting information for the specified audience, reorganizing the text, and providing explicit cues to the new overall organization.[10]

Given the premises upon which this book is based, we are pleased to report that the students taught to revise globally revised more successfully than those who were not. For more information about writing method sections in empirical research reports, see the methods discussed in Chapter 15.

As with place and mechanism descriptions, there are no easy formulas to follow in writing process descriptions. You must exercise a good deal of judgment in the matter. As in all writing, you must decide what your audience needs to know to satisfy its purpose and yours. However the checklists that follow should provide guidance to aid you in exercising your judgment (and revising globally).

Planning and Revision Checklists

You will find the planning and revision checklists following Chapter 2, "Composing," Chapter 4, "Writing for Your Readers," and inside the front cover valuable in planning and revising any presentation of technical information. The following questions specifically apply to defining and describing. They summarize the key points in this chapter and provide a checklist for planning and revising.

DEFINING

Planning

- Do your readers share the vocabulary you are using in your report? If not, make a list of the words you need to define.
- Do any of the words on your list have readily available synonyms known to your readers?
- Which words will require sentence definitions? Which words are so important to your purpose that they need extended definitions?
- How will you extend your definition: Description? Example? Analogy? Chronology? Topical order? Classification? Division? Graphics? Are there words within your definition that you need to define?
- Does everyone in your audience need your definitions? How long are your definitions? How many definitions do you have?
- Where can you best put your definitions? Within the text? In footnotes? In a glossary? In an appendix?

DEFINING

Revision

- In your sentence definitions, have you put your term into its class accurately? Have you specified enough differences so that your readers can distinguish your term from other terms in the same class?
- Will your readers understand all the terms you have used in your definitions?
- Have you avoided circular definitions?
- Have you used analogy and graphics to help your readers? If not, should you?
- Does the placement of your definitions suit the needs of your audience and the nature of the definitions?

PLACE DESCRIPTION

Planning

- What is your purpose in describing this place?
- Who are your readers? What is their purpose for reading your description?
- What will be your point of view? From what vantage point can you best view the place to be described? Will you need more than one point of view?
- Are there analogies to places familiar to the readers that will help them visualize and understand this place? Would a graphic help the readers?

PLACE DESCRIPTION

Revision

- Does your description satisfy your purpose and your readers' purpose?
- Is your point of view consistent? Do you have good reasons for any shifts in your point of view?
- Have you used well the visual language of shape, size, color, texture, and position?
- Are your inferences and implications clearly stated? Do the details you have selected support your inferences and implications?

MECHANISM DESCRIPTION

Planning

- What is the purpose of the description?
- Why will the intended reader read the description?
- What is the purpose and function of the mechanism?
- How can the mechanism be divided?
- What are the purpose and function of the parts?
- How do the parts work together?
- How can the parts be divided? Is it necessary to do so?
- What are the purpose and function of the subparts?
- Which of the following are important for understanding the mechanism and its parts and subparts? Construction? Materials? Appearance? Size? Shape? Color? Texture? Position?
- Are there any so-whats you need to express explicitly for the reader?
- Would graphics aid the reader?
- Are there analogies that would clarify the description for the reader?

MECHANISM DESCRIPTION

Revision

- Does your description fulfill your purpose?
- Does the level of detail in your description suit the needs and interests of your readers?
- Have you made the function of the mechanism clear?
- Have you divided the mechanism sufficiently?
- Do your descriptive language and analogies clarify the description?
- Have you clearly stated your so-whats?
- Have you provided enough graphic support? Are your graphics sufficiently annotated?

PROCESS DESCRIPTION

Planning

- What is the purpose of the description?
- Why will the reader read the description?
- What is the reader's level of experience and knowledge regarding the process?
- What is the purpose of the process?
- Who or what does the process?
- What are the major steps of the process?
- Are there graphics and analogies that would help the reader?

PROCESS DESCRIPTION

Revision

- Does your description fulfill your purpose?
- Does your description suit the needs and interests of your readers?
- Have you chosen the correct tense, past or present?
- Have you chosen active or passive voice appropriately?
- Are the major steps of the process clear?
- Have you provided enough graphic support? Are your graphics sufficiently annotated?

Exercises

1. Reproduced in this exercise is an extended definition of the term *microelectronics.* The definition is an excerpt from an article on microelectronic integrated circuits. The article is an informational piece written primarily for the employees and stockholders of a large corporation. Therefore, the audience is a mixed one that will include lay people, technicians, and executives, and perhaps even experts. The writer's task is to explain things so that lay people will find the article interesting and understandable, but at the same time to include facts and inferences that the other readers may find useful. Working in a group, discuss where, how, and why the author has used the techniques of definition. When the group has finished its discussion, each member of the group should write an analysis of the excerpt that shows what techniques the author has used and where, how, and why he has used them. In your analysis, use concepts from Chapter 4, "Writing for Your Readers," Chapter 6, "Informing," and this chapter.

What then *is* microelectronics? It's the design and fabrication of increasingly compact, awesomely complex electrical circuits on tiny squares of crystalline material such as silicon or gallium arsenide, and the dimensions of those circuits are literally microscopic. The silicon chips that Raytheon currently makes are typically 0.3 inch on a side, and the most advanced contain more than 100,000 "features," or individual components, *each* of them equivalent to something like one of those glowing vacuum tubes that helped snatch Jack Benny's voice out of the ether. Each feature can be etched upon the silicon in sizes as small as 1.25 microns in width, soon to be reduced to .8 micron and ultimately half a micron. (A micron is one *millionth* of a meter.)

"You grope for analogies," admits Dr. Bradford Becken of Raytheon's Submarine Signal Division, a dedicated user of microelectronics in its sonar systems. "If you wanted to draw a map of the entire United States that showed every city block and town square, it would obviously be a *very* big map. But with the feature sizes we're working with to create microcircuits right now, I could draw that entire map on a sheet of paper not much larger than a postage stamp."

Yet this tremendous sophistication, this incomprehensible tininess—"wires" etched in silicon so thinly that they are approaching a size too small for an electron to squeeze through—is all brought to market by techniques that approach the ultimate in mass production.

"You are able to apply manufacturing processes that are very repeatable, very cost-effective," explains Tom Shaw, manager of Raytheon's Missile System Division Laboratory in Bedford, Massachusetts—another heavy user of microcircuits in its missile guidance and control systems. "These repeatable processes make practical what most people would rightly assume is a highly complex operation."

It can cost $750,000 to design a single complex chip of the sort created and manufactured in Raytheon's Microelectronics Center, but the manufacturing process, while complex, is not that expensive. Some very complex devices can be sold for less than $5 apiece. Which is why grocery clerks wear watches that rival old Harrison's chronometer in accuracy, children play games on computers that exceed ENIAC in power, coffeemakers contain more computing power than a university lab of the 1950s, new generation airliners can automatically fly every phase of flight from takeoff to rollout, and pocket calculators with more power than those costing several hundred dollars just a decade ago have become giveaways.

The vacuum tube was turned into an antique by a device that at the time, in the 1950s, seemed the end of the line in electronic simplicity: the transistor. It was tough, didn't heat up, and took little space and less power. But transistors and their tiny teammates—diodes and capacitors—still had to be laboriously soldered onto circuit boards. This was fine for pocket radios and hobby kits, but a typical early-1960s computer had 100,000 diodes and 25,000 transistors. Solder that many connections and you end up with a price tag with even more digits.

When it was discovered that transistors and other semiconductors could simultaneously be made by the dozens on a wafer of pure silicon, which was then cut apart and wired to connectors and ultimately to circuit boards, it wasn't long before somebody realized how silly it was to cut them apart in the first place: Why not make the entire circuitry—the whole "device"—right on the wafer? And thus was created the "integrated" circuit, soon to be miniaturized itself and made by the dozens on each silicon wafer.[11]

2. Write an extended definition of some term in your academic discipline. Use a graphic if it will aid the reader. In a paragraph separate from your definition, explain for your instructor your purpose and audience.
3. Describe some place that could be important to you in your professional field—a river valley, a power plant, a nurses' station, a business office, a laboratory. Have a definite purpose and audience for your description, such as supporting some inference about the place for an audience of your fellow professionals. Or you might emphasize some particular aspect of the place for a lay audience. Use at least two points of view, one that allows you to give an overview of the place and one that allows you to examine it in detail. In a paragraph separate from your description, explain for your instructor your purpose and audience.
4. Choose three common household tools such as a can opener, vegetable scraper, pressure cooker, screwdriver, carpenter's level, or saw. For each tool, write a one-paragraph description that could serve as a catalog description for a particular brand of the product, such as a Black & Decker can opener or Stanley saw.

5. Figure 7-6 is a mechanism description; Figure 7-9 is a process description. Both are intended to transfer technology from one technical person to another. Choose some new mechanism or process in your field that will lend itself to such a description. Using the appropriate figure as a model, write your description. Include at least one graphic as a part of your description. In a paragraph separate from your description, explain your purpose and audience.
6. Write two versions of a process description intended to provide an understanding of a process. The first version is for a lay audience whose interest will be chiefly curiosity. The second version is for an expert or executive audience that has a professional need for the knowledge. The process might be humanly controlled, such as buying and selling stocks, writing computer programs, fighting forest fires, giving cobalt treatments, or creating legislation. It could be the manufacture of some product—paint, plywood, aspirin, digital watches, maple syrup, fertilizer, extruded plastic. Or you might choose to write about a natural process—thunderstorm development, capillary action, digestion, tree growth, electron flow, hiccuping, the rising of bread dough. In a separate paragraph accompanying each version, explain for your instructor how your situational analysis guided your strategy.
7. Write a description of a mechanism that has moving parts, such as an internal combustion engine. You may choose either a manufactured mechanism—such as a farm implement, electric motor, or seismograph—or a natural mechanism, such as a human organ, an insect, or a geyser. Consider your readers to have a professional interest in the mechanism. They use it or work with it in some way. In a paragraph separate from your description, explain for your instructor your purpose and audience.

CHAPTER 8 ARGUING

Whenever you are exercising your professional judgment and expressing an opinion, you will need the tools and techniques of argument. In a business setting, you would argue for your recommendations and your decisions. As a scientist, you would argue in the discussion section of a research report to support the conclusions you have reached. Argument is indispensable for the technical person.

You must present your argument in a persuasive way. The use of *induction, deduction,* and *comparison* is necessary in argument. *Toulmin logic,* named for its creator, Steven Toulmin, is a good technique for discovering an argument and presenting it. We cover all these points in this chapter.

Persuasive Argument

In argument, you deal with opinions that lie somewhere on a continuum between verifiable fact and pure subjectivity. Verifiable fact does not require argument. If someone says a room is 35 feet long and you disagree, you don't need an argument, you need a tape measure. Pure subjectivity cannot be argued. If someone hates the taste of spinach, you will not convince him or her otherwise with argument. The opinions dealt with in argument may be called propositions, premises, claims, conclusions, theses, or hypotheses, but under any name they remain opinions. Your purpose in **argument** is to convince your audience of the probability that the opinions you are advancing are correct.

Typically, an argument supports one major opinion, often called the **major proposition.** In turn, the major proposition is supported by a series of minor propositions. **Minor propositions,** like major propositions, are opinions, but generally they are nearer on the continuum to verifiable fact. Finally, the minor propositions are supported by verifiable facts and frequently by statements from recognized authorities.

To understand how you might construct such an argument, imagine for the moment that you are the waste management expert in an environmental consulting firm. Land developers constructing a new housing subdivision called Hawk Estates have turned to your firm for advice. Hawk Estates, like many new subdivisions, is being built close to a city, Colorful Springs, but not in a city. The problem at issue is whether Hawk Estates should build its own sewage disposal plant or tap into the sewage system of Colorful Springs. (The developers have already ruled out individual septic tanks because Hawk Estates is built on nonabsorbent clay soil.) Colorful Springs will allow the tap-in. You have investigated the situation, thought about it a good deal, and have decided that the tap-in is the most desirable alternative. The land developers are not convinced. It is their money, so you must write a report to convince them.

Major Proposition

In developing your argument, it helps to use a chart such as the one illustrated in Figure 8-1. The chart enables you to clearly separate and organize your major proposition, minor propositions, and evidence. First you must state your major proposition: "Hawk Estates should tap in to the sewage system of the city of Colorful Springs."

Minor Propositions and Evidence

Now you must support your major proposition. Your first and most relevant minor proposition is that Colorful Springs' sewage system can handle Hawk Estates' waste. Questions of cost, convenience, and so

MAJOR PROPOSITION	MINOR PROPOSITION	EVIDENCE
Hawk Estates should tap into sewage system of Colorful Springs.	Colorful Springs can handle Hawk Estates sewage.	• Estimate of waste from Hawk Estates. • City engineer's statement that Colorful Springs can handle estimated waste.
	Overall cost to Hawk Estates taxpayers only slightly higher if tapped into Colorful Springs.	• Initial cost of plant vs. cost of tap-in. • Yearly fee charged by Colorful Springs vs. operating cost of sewage lagoon. • Cost per individual tax payer.
	Proposed plant, a sewage lagoon, will be a nuisance to home owners.	• Well maintained lagoons Okay. • Lagoons hard to maintain, often smell bad, experts say. • Lagoon has to be located upwind of development.

Figure 8–1 Argument Arrangement Chart

forth would be irrelevant if Colorful Springs could not furnish adequate support. As you did with your major proposition, you begin this section of the report with your minor proposition. To support this proposition, you give the estimated amount of waste that will be produced by Hawk Estates, followed by a statement from the Colorful Springs city engineer that the city system can handle this amount of waste.

The minor proposition that states "the overall cost to Hawk Estates taxpayers will be only slightly more if they are tapped into the city rather than having their own plant" is a difficult one. It's actually a rebuttal of your argument, but you must deal with it for several reasons.

First and foremost, you must be ethical and honest with the developers. Second, it would be poor strategy not to be. If they find that you have withheld information from them, they will doubt your credibility. You decide to put this proposition second in your argument. In that way, you can begin and end with your strongest propositions, a wise strategy. To support this proposition you would list the initial cost of the plant versus the cost of the tap-in. You would further state the yearly fee charged by the city versus the yearly cost of running the plant. You might anticipate the opposing argument that the plant will save the homeowners money. You could break down the cost per individual homeowner, perhaps showing that the tap-in would cost an average homeowner only an additional 10 dollars a year, a fairly nominal amount.

Your final minor proposition is that the proposed plant, a sewage lagoon, will represent a nuisance to the homeowners of Hawk Estates. Because the cost for the tap-in is admittedly higher, your argument will probably swing on this minor proposition. State freely that well-maintained sewage lagoons do not smell particularly, but then point out that authorities state that sewage lagoons are difficult to maintain. Furthermore, if not maintained to the highest standards, sewage lagoons emit an unpleasant odor. To clinch your argument, you show that the only piece of land in Hawk Estates large enough to handle a sewage lagoon is upwind of the majority of houses during prevailing winds. With the tap-in, all wastes are carried away from Hawk Estates and represent no odor or unsightliness.

Organization

When you draft your argument, you can follow the organization shown on the chart, adding details as needed to make a persuasive case. Although your major proposition is actually the recommendation that your argument leads to, you present it first, so that your audience will know where you are heading. In executive reports, which this one is, major conclusions and recommendations are often presented first, as we point out in Chapter 4, "Writing for Your Readers."

When you sum up your argument, draw attention once again to your key points. Point out that in cost and the ability to handle the produced wastes, the proposed plant and the tap-in are essentially equal. However, you point out, the plant will probably become an undesirable nuisance to Hawk Estates. Therefore, you recommend that the builder choose the tap-in.

Throughout any argument you appeal to reason. In most technical writing situations, an appeal to emotion will make your case immediately suspect. Never use sarcasm in an argument. You never know whose toes you are stepping on or how you will be understood. Support your case with simply stated, verifiable facts and statements from recognized authorities. In our example, a statement from potential buyers that "sewage lagoons smell bad" would not be adequate. A statement to the same effect from a recognized engineering authority would be acceptable and valid.

Induction and Deduction

Much of your thought, whether you are casually chatting with friends or are on your most logical and formal behavior, consists of induction and deduction. In this section we cover induction and deduction and discuss some of the fallacies you'll want to avoid in using them.

Induction

Induction is a movement from particular facts to general conclusions. It's a method of discovering and testing the inferences you can draw from your information. Induction is the chief way we have of establishing causality, that A caused B. The inductive process consists of looking at a set of facts, making an educated guess to explain the facts, and investigating to see whether the guess fits the facts. The educated guess is called a *hypothesis.* No matter how well constructed your hypothesis is, remember it is only a guess. Be ready to discard it in an instant if it doesn't fit your facts.

For example, shortly after they attend a church picnic, 40 people out of the 100 who attended fall ill. You look at the facts and form a loose hypothesis: Something they ate at the picnic caused the illness. Investigating further, you discover that there were two lines at the picnic serving table. All the people who became ill went through the left-hand line. You refine your hypothesis to the effect that the illness had something to do with the left-hand line. You conjecture that perhaps the food handlers on the left-hand line did not follow the proper sanitary precautions.

However, upon checking with the food handlers from the left-hand line, you find that they all swear that they were models of cleanliness. Furthermore, you discover that 10 people who went through the left-hand line did not become ill. Stymied for the moment, you drop the hypothesis about the unsanitary food handlers, but you still have good reason to suspect the left-hand line.

You form another left-hand line hypothesis. Some of the people in the left-hand line must have eaten something that those in the right-hand line did not. You discover that the left-hand line served Mrs. Smith's potato salad. The right-hand line served Mrs. Olson's potato salad. Furthermore, everyone who ate Mrs. Smith's potato salad became ill. Everyone who ate Mrs. Olson's potato salad or no potato salad at all did not become ill.

Your hypothesis has probably been upgraded in your mind at this point to a *conclusion:* Mrs. Smith's potato salad caused the illnesses. Unless you could obtain some samples of Mrs. Smith's salad for testing, your conclusion would have to remain circumstantial. Because of the large number of people involved, however, you feel quite secure in your conclusion. It is possible that all 40 people came down with an infection not related to the salad, but it seems unlikely.

The whole process of gathering evidence, making hypotheses, and testing hypotheses against the evidence is the scientific method at work. Looking for **similarities** and **differences** are major tools in testing hypotheses. Examining similarities and differences in the population has led medical authorities, including the Surgeon General of the United States, to declare that cigarette smoking is hazardous to your health. They looked at the population and saw a difference: There are those who smoke and those who don't. Within these two groups, they looked for similarities. Smokers had in common a high incidence of respiratory problems, including emphysema and lung cancer. Nonsmokers had in common a low incidence of such problems. The higher incidence of such problems in the smoking group when compared to the nonsmoking group was a significant difference.

Whenever you argue inductively to support a hypothesis, you must accept the possibility that new evidence may prove you wrong. Nevertheless, far more judgments and decisions are made on inductive arguments than on direct evidence. Well-constructed inductive arguments are powerful. Yes, the Surgeon General could possibly be proven wrong, but more and more people are not betting their lives on it.

The actual practice of induction can be rather messy, as you chase down leads, retreat from blind alleys, and try out your hypotheses in a trial and error way. If for some reason you wish to show your thought processes to your readers, you might reveal some of the messiness to them when you present your argument. But, more often, after you have reached your conclusions, you will probably wish to argue for them in a straightforward way, as we demonstrated with the Hawk Estates example. In the following example, the author argues inductively for the proposition that life does not exist on Mars. The argument is supported by a series of conclusions based upon evidence gathered by the Viking spacecraft.

Major proposition

The primary objective of Viking was to determine whether life exists on Mars. The evidence provided by Viking indicates clearly that it does not.

Three of Viking's scientific instruments were capable of detecting life on Mars:

Capability of cameras

Conclusion

- The lander cameras could have photographed living creatures large enough to be seen with the human eye and could have detected growth changes in organisms such as lichens. The cameras found nothing that could be interpreted as living.

Paragraph explaining function and capability of GCMS and presenting conclusion about its findings

- The gas chromatograph/mass spectrometer (GCMS) could have found organic molecules in the soil. Organic compounds combine carbon, hydrogen, nitrogen, and oxygen and are present in all living matter on Earth. The GCMS searched for heavy organic molecules, those that contain complex combinations of carbon and oxygen and are either precursors of life or its remains. To the surprise of almost every Viking scientist, the GCMS, which easily finds organic matter in the most barren Earth soils, found no trace of any in the Martian samples.

Function of biology instrument

- The Viking biology instrument was the primary life-detection instrument. A one-cubic-foot box, crammed with the most sophisticated scientific hardware ever built, it contained three tiny instruments that searched the Martian soil for evidence of metabolic processes like those used by bacteria, green plants, and animals on Earth.

Capability of biology instruments

Appeal to authority

The three biology instruments worked flawlessly. All showed unusual activity in the Martian soil, activity that mimicked life. But biologists needed time to understand the strange chemistry of the soil. Today, according to most scientists who worked on the data, it is clear that the chemical reactions were not caused by living things.

Evidence for presence of oxidants

Conclusion

Furthermore, the immediate release of oxygen when the soil contacted water vapor in the instrument, and the lack of organic compounds in the soil, indicate that oxidants are present in the soil and in the atmosphere. Oxidants—such as peroxides and superoxides—are oxygen-bearing compounds that break down organic matter and living tissue. Therefore, even if organic compounds were present on Mars, they would be quickly destroyed.

Conclusion

Analysis of the atmosphere and soil of Mars indicated that all the elements essential to life on Earth—carbon, nitrogen, hydrogen, oxygen, and phosphorus—are present on Mars. Liquid water is also considered an absolute requirement for life. Viking found ample evidence of water in two of its three phases—vapor and ice—and evidence for large amounts of permafrost. But it is impossible for water to exist in its liquid state on Mars.

Restatement of major proposition

Possibility for future research

The conditions now known to exist on Mars and just beneath the surface of Mars do not allow carbon-based organisms to exist and function. The biologists add that the case for life sometime in Mars' distant past is still open.[1]

This passage shows well the characteristics of an inductive argument. The argument begins with the major proposition. The major proposition is then supported by a series of conclusions that are in turn supported by evidence. Remember, whatever the terminology used—conclusion, proposition, thesis, and so forth—generalizations based upon particulars are opinions, nothing more and nothing less. The frequent statements concerning the capabilities of the equipment are intended to strengthen the argument by showing that high-quality equipment was used. The references to the scientists who worked on the Viking project are appeals to authority.

Deduction

Deductive reasoning is another way to deal with evidence. While in inductive reasoning you move from the particular to the general, in

deductive reasoning you move from the general to the particular. You start with some general principle, apply it to a fact, and draw a conclusion concerning the fact. Although you will seldom use the form of a syllogism in writing, we can best illustrate deductive reasoning with a **syllogism:**

1. All professional golfers are good athletes.
2. Judy is a professional golfer.
3. Therefore, Judy is a good athlete.

Most often, the general principle itself has been arrived at through inductive reasoning. For example, from long observation of lead, scientists have concluded that it melts at 327.4° C. They have arrived at this principle inductively from many observations of lead. Once they have inductively established a principle, scientists can use it deductively, as in the following syllogism:

1. Lead melts at 327.4° C.
2. The substance in container A is lead.
3. Therefore, the substance in container A will melt at 327.4° C.

In expressing deductive reasoning, we seldom use the form of the syllogism. Rather, we present the relationship in abbreviated form. We may say, for instance, "Because Judy was a professional golfer, I knew she was a good athlete." Or "The substance is lead. It will melt at 327.4° C."

Although induction is the more common organizing technique in argument, deduction is sometimes used, as in this example:

> Methanogens, like other primitive bacteria, are anaerobic: They live only in areas protected from oxygen. This makes sense, since there was virtually no oxygen in the atmosphere when bacteria first evolved. But once bacteria developed chlorophyll a, the pigment of green plants, they began to use carbon dioxide and water for photosynthesis and produced oxygen as a waste product. When massive colonies of these photosynthetic bacteria developed, they pumped large amounts of oxygen into the atmosphere. Oxygen is a powerful reactive gas, and most early bacteria were not equipped to survive with it. Some bacterial species that were adapted to the new gas, including the oxygen producers themselves, continued to thrive. Others presumably evolved special metabolisms to protect them from oxygen, found anaerobic environments, or disappeared.[2]

Presented formally, the syllogism in this paragraph would go something like this:

1. Methanogens cannot live in oxygen.
2. Oxygen was introduced into the methanogens' environment.
3. Therefore, methanogens either evolved special metabolisms to protect them from oxygen, found anaerobic environments, or disappeared.

Logical Fallacies

Many traps exist in induction and deduction for the unwary writer. When you fall into one of these traps, you have committed what logicians call a **fallacy.** Avoid a rush to either conclusion or judgment. Take your time. Don't draw inferences from insufficient evidence. Don't assume that just because one event follows another, the first caused the second—a fallacy that logicians call *post hoc, ergo propter hoc* (that is, *after this, therefore, because of this*). You need other evidence in addition to the time factor to establish a causal relationship.

For example, in the sixteenth century, tobacco smoking was introduced into Europe. Since that time, the average European's life span has increased severalfold. It would be a fine example of a *post hoc* fallacy to infer that smoking caused the increased life span, which in fact probably stems from improvements in housing, sanitation, nutrition, and medical care.

Another common error is applying a syllogism backwards. The following syllogism is valid:

1. All dogs are mammals.
2. Jock is a dog.
3. Therefore, Jock is a mammal.

But if you reverse statements (2) and (3) you have an invalid syllogism:

1. All dogs are mammals.
2. Jock is a mammal.
3. Therefore, Jock is a dog.

Jock could be a cat, a whale, a Scotsman, or any other mammal. You can often find flaws in your own reasoning or that of others if you break the thought process down into the three parts of a syllogism.

Comparison

In business and technical situations, you frequently have to choose between two or more **alternatives.** When such is the case, the method of investigating the alternatives will usually involve comparing the alternatives one to another. (Contrast is implied in comparison.) To be meaningful, the comparisons should be made by using standards or **criteria.** Perhaps you have bought a car recently. When you did, you had your choice of many alternatives. In reaching your decisions, you undoubtedly compared cars using criteria such as price, comfort, appearance, gas mileage, and so forth. Perhaps you even went so far as to rank the criteria in order of importance, for example, giving price the highest priority and appearance the lowest. The more conscientiously you applied your criteria, the more successful your final choice may have been.

After you bought your car, no one asked you to make a report to justify your decision. However, in business it's common practice for someone to be given the task of choosing among alternatives. The task involves a report that makes and justifies the decisions or recommendations made. When such is the case, a comparison arrangement is a good choice. You can arrange comparison arguments by **alternatives** or by **criteria.**

Alternatives

Assume you work for a health organization and that you are comparing two alternative contact lenses: daily wear and extended wear. Your criteria are cost, ease of use, and risk of infection. After the necessary explanations of the lenses and the criteria, you might organize your material this way:

- Daily wear
    - Cost
    - Ease of use
    - Risk of infection
- Extended wear
    - Cost
    - Ease of use
    - Risk of infection

In this arrangement, you take one alternative at a time and run each through the criteria. This arrangement has the advantage of giving you the whole picture for each alternative as you discuss it. The emphasis is on the alternatives.

Criteria

In another possible arrangement, you discuss each criterion in terms of each alternative:

- Cost
    - Daily wear
    - Extended wear
- Ease of use
    - Daily wear
    - Extended wear
- Risk of infection
    - Daily wear
    - Extended wear

Memorandum

Date: September 22, 1994

To: Professor Marie Chavez
Department of Fisheries and Wildlife

From: David M. Zellar DZ

Subject: Choice of Camcorders

At your request I compared camcorders to see which one might be most suitable for use on departmental field trips. I narrowed the field to three camcorders: the Canon A1 Digital, the Panasonic PV 660, and the RCA Pro 850. The criteria used for the comparison, listed in priority order, are cost, color accuracy, clarity, low-light capability, zoom ratio, and weight.*

The Canon has the best performance characteristics overall but has a high price tag ($1850). The Panasonic meets all the criteria acceptably. It is the heaviest of the three (7 3/4 lbs), but its weight isn't excessive for field trips. Its cost ($1530) is high. The RCA has the lowest price ($980) but its performance characteristics are the poorest of the three. Based upon performance characteristics and cost, I recommend the Panasonic. However, if its cost is too high, the RCA would be marginally acceptable.

Findings

The following table summarizes my findings.

Criteria	Camcorders		
	Canon A1 Digital	Panasonic PV 660	RCA Pro 850
Cost ($)	1850	1530	980
Color accuracy	excellent	good	excellent
Clarity	good	good	fair
Low light capability	excellent	excellent	fair
Zoom ratio	10:1	12:1	6:1
Weight (lb)	4	7 3/4	2 1/4

*Based upon "Guide to the Gear," Consumer Reports 56 (1991): 178–79.

Figure 8–2 Comparison Report

The arrangement by criteria has the advantage of sharper comparison. It also has an advantage for readers who read selectively. Not every reader will have equal interest in all parts of a report. For example, an executive reading this report might be most interested in cost, a consumer in ease of use, an optometrist in risk of infection.

Recommendation reports based upon such comparisons are frequently cast in the form of a memorandum (see Chapter 12).

Figure 8–2 shows a student report based upon this comparison arrangement. Notice that because the report is for an executive, the writer states his conclusions and recommendations before presenting his data.

Toulmin Logic

When you construct an argument, it can be difficult to see the flaws in it. When you expose the same argument to your friends, even in casual conversation, they will be more objective and can often spot the flaws you have overlooked. **Toulmin logic** provides a way of checking your own arguments for those flaws.[3]

Discovering Flaws in Argument

Because Toulmin logic is a way of raising questions readers may ask, its use will make your arguments more reader oriented. Toulmin logic comprises five major components:

1. Claim — The major proposition or conclusion of the argument
2. Grounds — The evidence upon which the claim rests—facts, experimental research data, statements from authorities, and so forth
3. Warrant — Evidence that justifies the grounds and makes them relevant to the claim
4. Backing — Further evidence for accepting the warrant
5. Rebuttal — Counterarguments; exceptions to the claim, warrant, or backing; or reasons for not accepting them

We'll illustrate how Toulmin logic works, first with a simple example and then with one more realistically complex. In this first sample we indicate the kinds of questions readers might ask that would lead to the next consideration.

Claim:	We can't go on our picnic tomorrow.
Reader:	How come?
Grounds:	The weather will be too nasty. The National Weather Service predicts rain and low temperatures for tomorrow.
Reader:	Can we trust the forecast?
Warrant:	The Weather Service forecasts are accurate approximately 80% of the time.
Reader:	How come the forecasts are so accurate?
Backing:	Today's weather forecasting is based upon extensive observations, the application of scientifically sound principles, and the use of modern technology, such as computers.
Reader:	What about the 20% of the time the forecasts are wrong?
Rebuttal:	Of course, the forecasts are wrong about 20% of the time.

Claim, grounds, warrant, and backing, in this case, indicate that we have a strong argument, but the rebuttal demonstrates that it is not strong enough to say with absolute conviction that we can't go on the picnic. The situation calls for a **qualifier.** The claim would be better phrased as "We *probably* can't go on our picnic tomorrow." However, if a decision had to be made based upon this argument, calling off the picnic would seem to be justified.

Even this simple example shows that arguments are rather complex chains of reasoning in which you have to argue not only for your claim but for the grounds upon which the claim is based. Toulmin logic helps you to construct the chain.

For a more complex example, let's consider the greenhouse effect hypothesis.[4]

Claim

The accumulation of gases, particularly carbon dioxide (CO_2), emitted from the burning of fossil fuels will trap heat in the atmosphere, which will cause global warming, resulting in droughts, floods, and food shortages.

Grounds

In the past 100 years, CO_2 concentration in the atmosphere has risen from 270 parts per million (ppm) to 350 ppm. That this rise has been caused by the increased burning of fossil fuels seems indisputable. Researchers from Ohio State University report that central Asia has warmed by 1 to 3° C in the past century. Various computer models predict that global temperatures will rise by as much as 4° C in the next 50 years.

Warrant

Reputable scientists agree with this hypothesis. Mathematicians from AT&T Bell Labs report that "There is a 99.9% chance that the warming and the CO_2 rise are causally related. Climatologist James Hansen of NASA's Goddard Institute for Space Studies says that temperature data of the last 100 years show a worldwide rise of .4° C. He has further stated in congressional testimony that the greenhouse effect has begun and will worsen.

Warrant

Worldwide, environmentalists have called for stabilization of CO_2 by 2000. Fred Krupp, Executive Director of the Environmental Defense Fund, has said that "We now realize that the burning of fossil fuels—especially in the United States and other industrial nations—produces far more carbon dioxide and other greenhouse gases than our global ecosystems can absorb." In June of 1992, the nations of the world at the Earth Summit conference in Rio de Janeiro signed a treaty that states a dangerous global warming has already started.

Up to this point, the argument for the greenhouse hypothesis and its effects seems to be going well. But, if you dig further, you will find rebuttals.

Rebuttal

A study of ocean temperatures by scientists from the Massachusetts Institute of Technology shows no rise in ocean temperatures in the past 100 years.

Rebuttal

Scientists at the National Oceanographic and Atmosphere Institute reviewed U.S. climatic records and concluded, "There is no statistically significant evidence of an overall change in annual temperature or change in annual precipitation for the contiguous U.S. 1895–1987." A scientist at the National Center for Atmospheric

Rebuttal

Research in Boulder, Colorado, says that "it's possible that everything in the last 30 years of temperature records is no more than noise," that is, little numerous fluctuations that mean nothing.

Rebuttal

Michael Schlesinger, a respected climatologist from the University of Illinois, says that Hansen's "statements have given people the feeling that the greenhouse effect has been detected with certitude. Our current understanding does not support that. Confidence in its detection is now down near zero." Greenpeace, an organization much concerned with environmental safety, surveyed climatologists and discovered that 47% of them believe a "runaway" greenhouse effect is unlikely.

And so on. Digging for evidence on the greenhouse effect shows a sharp division with reputable scientists coming down on both sides of the question. The claim has to be qualified, perhaps something like this: "Some scientific studies show a correlation between the rise of CO_2 in the atmosphere and global warming, but the evidence and methodology of such studies have not convinced all scientists of their validity." Applying Toulmin logic has resulted in a weaker claim, but it is a claim that can be supported with the existing evidence.

Arranging Argument for Readers

Toulmin logic can help you arrange your argument as well as discover it. Though you would not want to follow Toulmin logic in a mechanical way, thinking in terms of claim, grounds, warrant, backing, rebuttal, and qualifier can help you to be sure you have covered everything that needs to be covered. Obviously, claim and grounds must always be presented. In most business situations, as we have pointed out, the claim is likely to be presented first, particularly in executive reports. However, in a situation where the readers might be hostile to the claim, it might be preferable to reverse the order. If the grounds are strong enough, they may sway the readers to your side before they even see the claim. On the other hand, if a hostile audience sees the claim first they may not pay enough attention to the grounds to be convinced.

Rebuttals should always be considered and, if serious, should be included in your presentation. You have an ethical responsibility to be honest with your readers. Furthermore, if your readers think of rebuttals that you do not deal with, it will damage your credibility. If you can counter the rebuttals successfully, perhaps by attacking their warrant or backing, your claim can stand. If you cannot counter them, you will have to qualify your claim.

How deeply you go into warrants and backing depends upon your readers. If your readers are not likely to realize what your warrant is (for example, "Respected scientists agree with this hypothesis"), then you had better include the warrant. If your readers will be likely to disagree with your warrant or discount its validity, then you had better include the backing. All in all, Toulmin logic can be a considerable help in discovering and arranging an argument. It is also extremely useful in analyzing the soundness of other people's arguments.

Planning and Revision Checklists

You will find the planning and revision checklists following Chapter 2, "Composing," Chapter 4, "Writing for Your Readers," and inside the front cover valuable in planning and revising any pre-

sentation of technical information. The following questions specifically apply to argument. They summarize the key points in this chapter and provide a checklist for planning and revising.

PLANNING

- What is your claim, that is, the major proposition or conclusion of your argument?
- What are your grounds? What is the evidence upon which your claim rests—facts, experimental research data, statements from authorities, and so forth?
- Do you need a warrant that justifies your grounds and makes them relevant?
- Do you need further backing for your grounds and warrant?
- Are there rebuttals—counterarguments, exceptions to the claim, warrant, or backing, or reasons for not accepting them? Can you rebut the rebuttals? If not, should you qualify your claim? Will you present your argument unethically if you do not state the rebuttals and deal with them honestly?
- Are you choosing among alternatives? If so, what are they?
- What are the criteria for evaluating the alternatives?
- Is your audience likely to be neutral, friendly, or hostile to your claim? If your audience is hostile, should you consider putting your claim last rather than first?

REVISION

- Is your claim clearly stated?
- Do you have sufficient grounds to support your claim?
- If needed, have you provided a warrant and backing for your grounds?
- Does any of your evidence cast doubt upon your claim? Have you considered all serious rebuttals?
- Have you dealt responsibly and ethically with any rebuttals?
- Have you remained fair and objective in your argument?
- Have you presented evidence for causality beyond the fact that one event follows another?
- If you have used deductive reasoning, can you state your argument in a syllogism? Does the syllogism demonstrate that you have reasoned in a valid way?
- Is your argument arranged so that it can be read selectively by readers with different interests?

Exercises

1. We have reprinted here excerpts from a report printed in the *FDA Consumer,* a publication of the U.S. Food and Drug Administration.[5] The report argues, primarily for a lay audience, for the acceptance of food irradiation. Working in a group, analyze the argument's arrangement and presentation. Use this chapter's Planning and Revision Checklists to guide your analysis. Pay particular attention to how the argument's parts work together to defend its major proposition that food irradiation is safe and effective. Judge whether the argument is effective or ineffective. Write an analysis for your teacher that presents and supports your judgment.

A measure FDA announced in the *Federal Register* this year may go unused because of consumer apprehension. On May 2, 1990, FDA issued a rule defining the use of irradiation as a safe and effective means to control a major source of food-borne illness—*Salmonella* and other food-borne bacteria in raw chicken, turkey, and other poultry. However, FDA has received written objections that it must evaluate before the rule can go into effect.

Experts believe that up to 60 percent of poultry sold in the United States is contaminated with *Salmonella*, according to Joseph Madden, Ph.D., acting director of FDA's division of microbiology. Madden adds that studies suggest that all chicken may be contaminated with the *Campylobacter* organism.

People often become ill after eating contaminated poultry. Symptoms may range from a simple stomachache to incapacitating stomach and intestinal disorders, occasionally resulting in death.

As equipment used to irradiate food is regulated as a food additive, the FDA rule is the first step in permitting irradiation of poultry. However, although the U.S. Department of Agriculture will soon propose a companion rule finalizing guidelines for commercial irradiation of poultry, industry groups cite consumer apprehension as a drawback to implementing the procedure. And reaction to FDA's new rule has elicited more questions than answers.

A Scary Word

Irradiating food to prevent illness from food-borne bacteria is not a new concept. Research on the technology began in earnest shortly after World War II, when the U.S. Army began a series of experiments irradiating fresh foods for troops in the field. Since 1963, FDA has passed rules permitting irradiation to curb insects in foods and microorganisms in spices, control parasite contamination in pork, and retard spoilage in fruits and vegetables.

But, to many people, the word irradiation means danger. It is associated with atomic bomb explosions and nuclear reactor accidents such as those at Chernobyl and Three Mile Island. The idea of irradiating food signals a kind of "gamma alarm," according to one British broadcaster. (Gamma rays are forms of energy emitted from some radioactive materials.)

But when it comes to food irradiation, the only danger is the bacteria that contaminate the food. The process damages their genetic material, so the organisms can no longer survive or multiply.

Irradiation does not make food radioactive and, therefore, does not increase human exposure to radiation. The specified exposure times and energy levels of radiation sources approved for foods are inadequate to induce radioactivity in the products, according to FDA's Laura Tarantino, Ph.D., an expert on food irradiation. The process involves exposing food to a source of radiation, such as to the gamma rays from radioactive cobalt or cesium or to x-rays. However, no radioactive material is ever added to the product. Manufacturers use the same technique to sterilize many disposable medical devices.

Tarantino notes that in testing the safety of the process, scientists used much higher levels of radiation than those approved for use in poultry. But even at these elevated levels, researchers found no toxic or cancer-causing effects in animals consuming irradiated poultry.

Beyond the Gamma Alarm

Market tests show that once consumers learn about irradiation, they will buy irradiated food. For example, Christine Bruhn, Ph.D., of the University of California's Center for Consumer Research in Davis, Calif., reports that irradiated papayas outsold the nonirradiated product by more than 10 to 1 when in-store information is available. And, Danny Terry, Ph.D., a consumer researcher at Central Missouri State University in Warrensburg, Mo., says that a recent market test he conducted with irradiated strawberries showed that consumers who received written information about irradiation along with the fruit were slightly more interested in buying irradiated products in the future.

Nevertheless, concern about the process remains strong. Since 1989, three states (Maine, New York, and New Jersey) have either banned or issued a moratorium on the sale of irradiated foods. According to a U.S. General Accounting Office report prepared in May 1990 at the request of Rep. Douglas Bosco (D-Calif), "officials of these states told us that their states took the actions in response to public concern rather than as a result of scientific evidence questioning the safety of food irradiation."

"Something quite aside from food safety appears to lie at the root of the entire controversy, which may explain why it continues to flourish in the face of all safety assurances," says Carolyn Lochhead in the August 1989 issue of *Food Technology* magazine. "Many opponents charge that the Food and Drug Administration, the World Health Organization, and the nuclear power industry are conspiring to promote the technique as a way to dispose of nuclear waste."

Lochhead discusses concern that one source of radioactive material for food irradiation, cesium 137, is recovered from spent fuel rods in nuclear power plants. The conspiracy charge promotes unwarranted fear among consumers, says Lochhead.

"For economic, as well as other, reasons," says Department of Energy official Barbara Thomas, "the U.S. commercial nuclear power industry does not attempt to recover material, such as cesium 137, from spent fuel."

According to the DOE, commercial irradiators in the United States choose their irradiation source (whether the gamma-emitting radioactive materials cesium 137 or cobalt 60, or accelerators that can produce electrons, x-rays or both) based on practical requirements, such as cost. The product to be irradiated also influences the choice. Many foods require low energy levels to kill harmful organisms, while medical supplies may need higher doses for sterilization.

However, the fallout from a falsely characterized cesium recovery plan has charged the legislative atmosphere. George Giddings, Ph.D., a consultant food scientist and expert in food irradiation matters, sees it as the "single most inciting issue in the food irradiation area." Giddings suggests that legislators are wary of supporting food irradiation measures some critics say are linked to increased nuclear activity, including the production of nuclear weapons.

A 1982 congressional amendment bars using spent commercial fuel for military purposes. The Department of Energy has no interest in changing this law.

Michael Colby, director of Food and Water, Inc., one of the more vocal groups lobbying against food irradiation, says the new poultry regulation will lead to nuclear hazards, including "the continued generation of radioactive wastes for which a secure isolation technology has yet to be developed." Colby submitted the comment during a 30-day objection period following publication of the final rule. In the case of food additives, FDA evaluates objections in order to determine whether any changes in the final rule are appropriate. Based on FDA's findings, those raising the objection may be entitled to a hearing before the commissioner.

FDA inspections of all irradiation plants conducted from 1986 to 1989 showed no violations of the food irradiation regulations.

Giddings contends that groups such as Food and Water play on the public's fear of nuclear energy and misrepresent the safety questions surrounding food irradiation. They frame it as a "populist" issue to legislators and pressure them to introduce legislation banning food irradiation.

Consumer Uncertainty

Other consumer groups have taken more moderate positions. The Center for Science in the Public Interest, for instance, says that "at a minimum, irradiated foods should be labeled" so that consumers know what they're buying.

Since 1966, FDA has required that irradiated foods be labeled as such. In 1986, a mandatory logo was added to this labeling requirement. The international logo, first used in the Netherlands, consists of a solid circle, representing an energy source, above two petals, which represent the food. Five breaks in the outer circle depict rays from the energy source.

Consumer surveys show mixed reactions. According to an article in the October 1989 issue of *Food Technology* magazine, which reviewed surveys conducted by various academic and consumer research groups, consumers are more concerned about chemical sprays and pesticide residues, preservatives, and food-borne illnesses than about food irradiation. A Louis Harris poll, conducted from 1984 through 1986, however, found that 76 percent of Americans consider irradiated food a hazard.

"Consumer acceptance of irradiation as a treatment for foods is showing only minimal positive change, at best," said Fred Shank, Ph.D., director of FDA's Center for Food Safety and Applied Nutrition, in a symposium on food irradiation at the 1990 annual meeting of the Institute of Food Technology. Shank said that the greatest concern about the process is its perceived association with radioactivity and nuclear power.

Another concern, raised often in comments to FDA when it proposed the use of radiation to kill microorganisms in spices and insects in fresh foods, is that irradiation may produce substances not known to be present in nonirradiated foods.

These substances, described by scientists as "radiolytic products" sound more threatening than they actually are, says George Pauli, Ph.D., an FDA food irradiation expert and policy maker. For instance, Pauli says, when we heat food it often creates new substances that produce new tastes and smells. These substances could be called thermolytic products—an intimidating word for a harmless change.

In 1979, FDA established the Bureau of Foods Irradiated Food Committee (BFIFC) to review safety assessments of irradiated food. Experiments have shown that very few of these radiolytic products are unique to irradiated foods. In fact, the BFIFC estimated that approximately 90 percent of the substances identified as radiolytic products are found in foods that have not been irradiated—including raw, heated, and stored foods. Moreover, many of these substances are not well known because foods usually have not been studied at the minute (parts per million) levels scrutinized by chemists who analyzed the irradiated foods...

The Future of Food Irradiation

The World Health Organization believes irradiation can substantially reduce food poisoning. According to a 35-year WHO study, there has been a constant increase in the incidence of food-borne diseases, as well as the emergence of "new" disease-causing organisms, such as *Campylobacter* and *Listeria.*

Food irradiation would be another weapon in the arsenal against food-borne illness. FDA and WHO, however, emphasize that irradiation is not a substitute for careful handling, storage, and cooking of food. Irradiated poultry can become recontaminated, for instance, if placed next to contaminated, nonirradiated poultry, or left unrefrigerated so that the remaining organisms can grow.

To date, 35 countries have issued unconditional or provisional clearances allowing irradiation of commercial foods. Of the more than 140 industrial gamma irradiators in over 40 countries, 29 are used part-time to irradiate food items and conduct food-related research. (They are used mostly for sterilizing disposable medical supplies.) A 1989 Library of Congress report prepared for Congress estimates that by the early 1990s, 55 facilities worldwide will be used for food irradiation and related food irradiation research.

However, as Tanya Roberts of USDA's Economic Research Service stresses, the future of irradiation depends upon consumer acceptance—based largely on proof that the process can produce safer foods at a lower cost. Roberts estimates that the cost of medical treatment and lost productivity for five food-borne diseases—trichinosis, toxoplasmosis, salmonellosis, campylobacteriosis, and beef tapeworm—totals more than $1 billion annually.

The last chapter in the story of food irradiation still remains to be written. Will the fear of nuclear energy prevent this technology from being used to its fullest potential? Or will education win acceptance for a procedure that can lower the incidence of food-borne illness? Only consumers can supply the answers.

2. For this exercise, you may use the material provided for Exercise 1. "George Watts, president of the National Broiler Council says 'The U.S. poultry industry has always been a consumer-driven business, demonstrated by the variety of new products developed over the years to meet the American public's demand.' He says that should consumers desire irradiated food products, 'the industry will respond.'"[6] Imagine that you work for Mr. Watts. Write him a memo that argues pro or con for the following proposition: For the National Broiler Council to lead the American consumer to accept food irradiation in poultry would be both good business and a responsible and ethical thing to do.
3. Write a memo to an executive that recommends the purchase of some product or service the executive needs to conduct his or her business. Your memo should establish criteria and justify the choice of the product or service you recommend against other possible choices. See Chapter 12 for information on memo format.
4. Your new boss on your first job knows how important it is for the organization to stay aware of trends that may affect the organization. He or she asks you to explore such a trend. The possibilities are limitless, but you may wish to explore some trend in your own field. For example, are you in computer science? Then you might be interested in the latest trends in artificial intelligence. Are you in forestry? Are there trends in the uses and kinds of wood products?

   Develop a claim about the trend; for example:

If trend A continues, surely B will result.

Trend A will have great significance for X industry.

   Support your claim with a well-developed argument that demonstrates your ability to use induction, deduction, and Toulmin logic. Write your argument as a memorandum to your boss. See Chapter 12 for information on memo format.
5. You are a member of a consulting firm. Your firm has been called in to help a professional organization deal with a question of major importance to the members of the organization. For example, nurses have an interest in whether nurses should be allowed to prescribe medication and therapy. You will probably be most successful in this exercise if

you deal with organizations and questions relevant to your major. Investigate the question and prepare a short report for the executive board of the organization. Your report should support some claim; for example, nurses should be allowed to prescribe medication and therapy. Use Toulmin logic in discovering and presenting your argument. That is, be aware of the need to provide grounds, warrants, backings, and qualifiers. Anticipate rebuttals and deal with them ethically and responsibly. Use Part III, "Document Design," to help you format your report.

PART III
DOCUMENT DESIGN

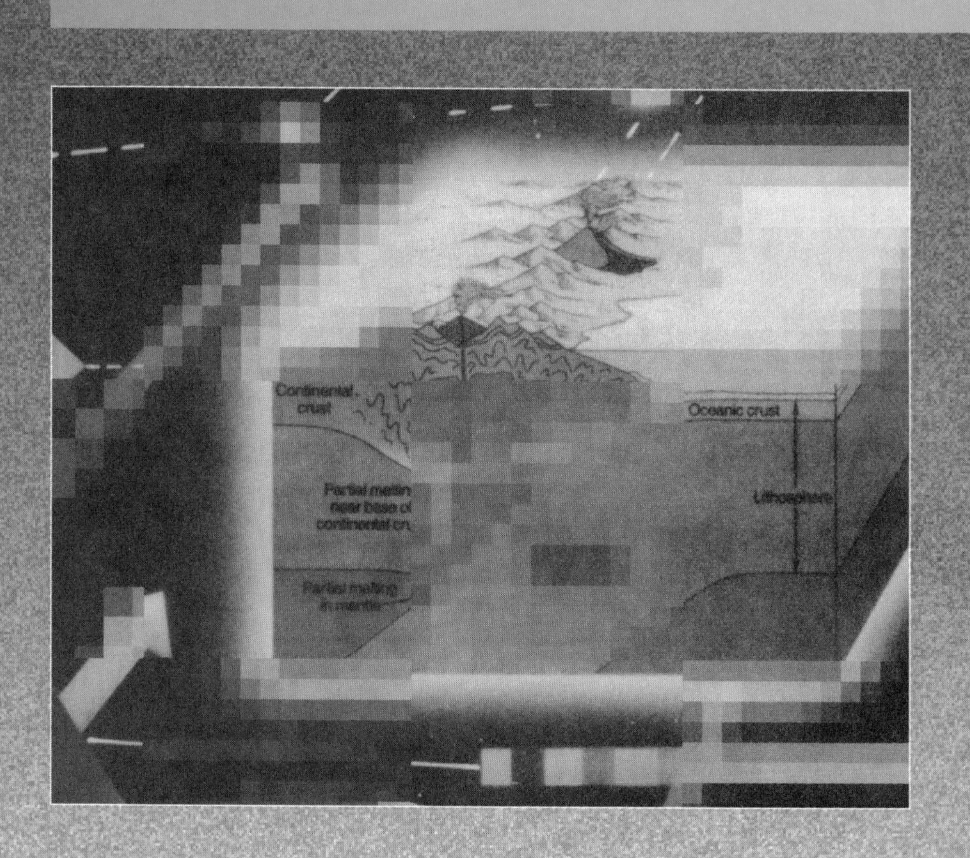

Part III deals with document design and graphics. Good design—creating a format that helps readers to find information and to read selectively—is vitally important in technical writing. Chapter 9, "Document Design," deals primarily with the format and appearance of the document. Chapter 10, "Design Elements of Reports," tells you how to construct all the elements that full reports need, such as covers, tables of contents, introductions, discussions, summaries, and notes. Chapter 11, "Graphical Elements," tells you how to use tables, graphs, drawings, and photographs to inform your reader about concepts, processes, trends, relationships, and objects. Technical writing is marked by an extensive use of graphics.

CHAPTER 9

DOCUMENT DESIGN

As the previous chapters have shown, effective writing requires a number of composing strategies based on your audience's needs. But writing is more than the generation of words that will achieve the writer's purpose with the audience, more than development of correct sentences arranged in logical paragraphs. Research has shown that effective writing is also visually effective.[1] Within the past decade, an increasing number of documents have been drafted or produced in final form on computers. With the increased capabilities of word processing software, writers can produce documents with a variety of type sizes, type styles, and artistic effects. They can arrange words on a page in a variety of ways.

Because so much of the writing we see in magazines, trade journals, and advertisements is artistically effective, we have become accustomed to writing that is visually appealing. To understand this point, look at Figure 9–1 and then at Figure 9–2. Which one would you rather read? Which one is easier for you to read? Why? The purpose of this chapter is to show you how to write visually and how to use word processing software to make your writing easy for your reader to understand.

Understanding the Basics of Document Design

Recall from Chapter 2 that format and document design are elements of the third, fifth, and sixth steps in the composing process. These four principles will help you plan your document's visual design:

- Know what decisions you can make.
- Choose a design that fits your document.
- Reveal your arrangement to your reader.
- Keep your format consistent.

Once you have considered these four principles, you can begin to plan for graphics, which will be discussed in Chapter 11.

Know What Decisions You Can Make

Many times you are limited by the software that is available. Software varies in capability and cost. Learn what your software can do; even with the most basic word processing package you can produce a document that is visually effective. In Figure 9-2, the writer has used the grouping of related information, numbered lists of steps, and boldfaced headings to redesign the way information is presented in Figure 9-1.

Many companies have a standard format for reports, letters, or proposals. Many journals have standard formats that all manuscripts must follow. Find out what format requirements are already in place and follow them. If you think that the format you are being asked to use won't work well for your readers and your document, find out who makes decisions on format and present a case for the changes you want.

Choose a Design That Fits Your Document

Provide no more complexity in the format than the situation requires. You'll impress readers most by providing just the information they need in a way that makes it easy for them to find and understand. Do not use a table of contents or a glossary for reports under five pages. Add appendix material only if it is necessary and will be useful to your reader.

Subject line does not state the purpose of the memo

Paragraph 1 states only the background

Here's part of the news

The bulk of the paragraph tells readers what to do in narrative form

Here's the rest of the news

TO: All Level 4 Technicians DATE: June 2, 1993

FROM: Allen Cranford

SUBJECT: Time Sheets

During the recent meeting of our installation group, we came to the conclusion that time reporting procedures are needed and that changes need to be initiated. I have discussed the reporting problems that we are currently experiencing with Todd Marley, who has agreed to increase our overtime capability if we can document our needs by more accurately reflecting the increasing amount of time we are spending on routine installation maintenance. While a certain amount of additional time on routine maintenance is expected of everyone in the installation group, we can make a good case for increased budget requirements if we can document, during the next three months, all repairs that exceed scheduled maintenance by at least 15%.

I want to have a meeting to go over the first draft submitted by the forms design people July 2, 1993. Beginning August 15, 1993, a packet of newly designed time sheets will be available at the equipment check-out desk. Please study these time sheets and be ready to begin using them by September 1, 1993. The forms will include the equipment, cable, and miscellaneous hardware that you use on each job. A check list will be provided with room for you to note additional materials used, also the start-stop time for each maintenance job, any unusual circumstances you encounter. The time sheet will likely have a box for typical problems you encounter, such as TM-ASK, BK-ASK, BKK-ASKK, RF-ASK, and GM-ASK. There will also be space for others and a line for you to explain. The form will also contain space for any recommendations you wish to make about equipment malfunction. Please complete this space, as the information may be useful in the next budget planning period.

If you have suggestions for the time sheet, please contact me at 2856 by Tuesday, June 15. I need your input to be sure our forms design department can provide the time sheet we need. If we all work on documenting what we do, the time required, and the equipment and facilities we use, we will have grounds for requests for budget overruns for the first half of 1994.

This memo is not visually compelling. It does not say "Read me." Important information is not concentrated in the opening segments. Instructions in paragraph 2 are difficult to locate. Because paragraph 1 does not tell readers important information—meeting date and required action date—the memo will likely not be read.

Figure 9–1 Memo with Ineffective Visual Presentation

Purpose, action, date, and problem announced first.

Boldface headings help readers to group information.

Numbered list helps readers see what they must do.

TO: All Level 4 Technicians DATE: June 2, 1993

FROM: Allen Cranford, Supervisor
Plant Engineering

SUBJECT: Time sheets to be developed by July 15, 1993

ACTION REQUIRED: Your suggestions by June 15, 1993

Here's the Current Situation

Tom Marley has agreed to increase our overtime capability if we can document our needs by more accurately reflecting the increasing amount of time we are spending on routine installation maintenance. To develop a reporting document, I need your suggestions by June 15, 1993.

You are welcome to attend an open meeting to go over the draft of the design:

July 2, 1993, Room 216A

During the recent meeting of our installation group, we concluded that time-reporting procedures are needed and that changes need to be initiated. While a certain amount of added time on routine maintenance is expected of everyone in the installation group, we can make a good case for increased budget requirements if we can document, during the next three months, all repairs that exceed scheduled maintenance by at least 15%.

What You Need to Do

We'll use this procedure to develop the time sheets:

1. Pick up your packet of newly designed time sheets at the equipment check-out desk. These will be available beginning August 15, 1993.

2. Please study these time sheets and be ready to begin using them by September 1, 1993.

The form will initially include the following items in this order:

1. equipment
2. cable
3. miscellaneous hardware used on each job
4. space to list additional materials used

NOTE: What else do you think should be included?

Figure 9–2 Revision of Figure 9-1 to Improve Visual Accessibility and Reader Interest

June 2, 1993

All Level 4 Technicians
Page 2

The form will also provide space for you to note

5. the start-stop time for each maintenance job
6. any unusual circumstances you encounter
7. a box for typical problems, such as TM-ASK, BK-ASK, BKK-ASK, RF-ASK, and GM-ASK
8. other problems
9. any recommendations about equipment malfunction

NOTE: Please think about what other information needs to be included, as it may be useful in the next budget-planning period.

Your Input Is Vital

Last paragraph reinforces importance of reader response.

If you have suggestions for the time sheet, please contact me at ext. 2856 by Tuesday, June 15. I need your input to be sure our forms design department can provide the time sheet we need. If we all work on documenting what we do, the time required, and the equipment and facilities we use, we will have grounds for requests for budget overruns for the first half of 1994.

Figure 9-2 (continued).

Note the difference in the subject lines of Figure 9–1 and Figure 9–2. Figure 9–2, with its use of a descriptive subject line and action required in boldface type lets readers know immediately the subject of the memo, the action that is required, and the date by which the action is required. Listing and highlighting, which we will discuss in this chapter, help the reader follow the progression of ideas and see important information. The strategies for effective design you will see in this chapter are designed to help readers find their way through any technical document.

Reveal Your Arrangement to Your Reader

Research on how people read and process information shows that readers have to see the arrangement of material before they can make sense of it.[2] Headings and perhaps subheadings reveal the organization of your document so that your reader can see what you are writing before trying to absorb your message on the sentence level. Seeing the text is the first step in the reading process. Arrangement of material is the crucial difference between Figure 9–1 and Figure 9–2. The subject line and the grouping of information in Figure 9–2 help you see the content of the memorandum immediately.

Keep Your Format Consistent

Consistency in format is essential to easy reading. To see what inconsistent format can do, compare Figure 9–3, which illustrates major violations of principles for good formatting, with Figure 9–4, which incorporates the following seven suggestions:

- Design the page for easy reading.
- Leave ample margins.
- Use a medium line length.
- Set the spacing for easy reading.
- Use a ragged right margin.
- Use white space to group information.
- Avoid placing too much information on a page.

Long line and uninterrupted flow of text obscures the new procedures to be followed for regular orders, large orders, and pick-up

TO: All Department Heads

SUBJECT: New Copy Procedures

A recent study of our copy room request procedures indicates that we are not fulfilling copy requests as efficiently as possible. A number of problems surfaced in the survey. First, many requests, particularly large orders, are submitted before the copy center opens. Others are submitted after the copy room closes. As a result, the copy center has an enormous backlog of copy orders to fill before it can begin copy orders submitted after 8:30 A.M., when the center officially opens. This backlog may throw the center two or three hours behind schedule. All copy requests throughout the day then require over two hours to complete. By 2:00 P.M., any copy requests submitted may not be filled that day. P.M. If large orders arrive unexpectedly even a routine copy request may take two days to complete.

To remedy the situation, we will change to the following copy request procedure beginning Monday, February 2. The copy center will close at 3:00 every afternoon. Two work-study employees will work at the center from 3:00 until 5:00 to complete all orders by 4:00. If you submit copy requests by 3:00, the center will have them ready by 4:00. In short, all requests will be filled the day they are submitted. However, do not leave copy requests after 3:00, as these will not be processed until the following day. However, we guarantee that if you leave your request for copies with us between 8:30 and 2:00, you will have them that day.

Request for copies of large orders—over 100 copies of one item, single/multiple copies of any document over 25 pages, or front/back photocopying of one item up to 50 copies—will require that a notice be given the copy center one day in advance. That way, the center can prepare for your copy request and be sure to have it ready for you. Copies of the request form are attached. Please complete one of these and send it to Lynda Haynes at the copy center so that she can schedule all big jobs. If you submit a big copy request without having completed the form, your request will be completed after other requests are complete.

Allow plenty of time for routine jobs—at least two hours and three if possible. Beginning February 2, give all copy requests to the receptionists at the copy center. Be sure you attach complete instructions. Give your name, your phone number, and your office number. State the number of copies required, any special instructions. Specify staples or clips, color paper, and collation on multi-page copies.

Pick-up procedures will also change February 2. All copy jobs, after they are complete, will be placed in each department's mail box. No copies will be left outside the copy center after closing time. No copies will be left with the receptionist. Large orders that will not fit mail boxes will be delivered to your office.

If you have questions about this new procedure, please contact Lynda Haynes at 2257.

Figure 9–3 Memo That Violates Format Guidelines

Headings separate procedures. Instructions are numbered and spaced for easy reading. Time is highlighted.

MEMORANDUM

TO: All Department Heads DATE: January 27, 1990

FROM: Lynda Haynes

SUBJECT: **Change in Copy Procedures**

EFFECTIVE DATE: MONDAY, FEBRUARY 2, 1990

In order that the center can handle orders more quickly and efficiently, please ask everyone in your department to observe the following procedures:

Routine Copy Requests:

1. Attach complete instructions to your copy request. Please include:
   a. Name, phone number, and office number;
   b. Number of copies required;
   c. Choice of staples or clips and collation instructions;
   d. Color of paper or other special instructions.

2. Give all copy requests to the copy center receptionists.

3. Allow 2–3 hours for your order to be filled.

4. Order requests left between 8:30 a.m. and 3:00 p.m. will be processed by 4:00 p.m. the same day.

5. The copy center will close at 3:00 p.m. Orders left after that time will be processed the next day.

Large Order Procedures:

1. Large orders are defined as
   a. Over 100 copies of one item;
   b. Single/multiple copies of any document over 25 pages;
   c. Front/back photocopying of one item up to 50 copies.

2. Complete one of the attached request forms.

3. Send completed request form to Lynda Haynes at the copy center one day in advance to avoid a delay in processing.

Distribution:

Copy orders will be placed in each department's mailbox. Orders too large for your mailbox will be delivered to your office.

Questions? Contact Lynda Haynes at ext. 2257.

Figure 9–4 Revision of Figure 9-3

Design the Page for Easy Reading Examine the content you are going to present. After you organize it, choose a layout that achieves the following goals:

- The page looks inviting.
- The page looks open—has space inside the text as well as in the margins.
- The design helps readers find information quickly.
- The design is appropriate for the type of document.

Leave Ample Margins Margins are the white space around the text. White space, used carefully, makes the page look inviting. Too little space, as in Figure 9–3, makes the page look dense and uninviting. If you are putting your work in a binder, be sure to leave room for the binding; don't punch holes through the text. Think about whether a reader will want to punch holes in a copy later or put the work into a binder. Figure 9–5 shows the following guidelines for an 8 ½- by 11-inch page:

top margins	1 inch
bottom margins	1 ½ inch
left margin	1 inch, if not being bound; 2 inches, if being bound
right margin	1 inch

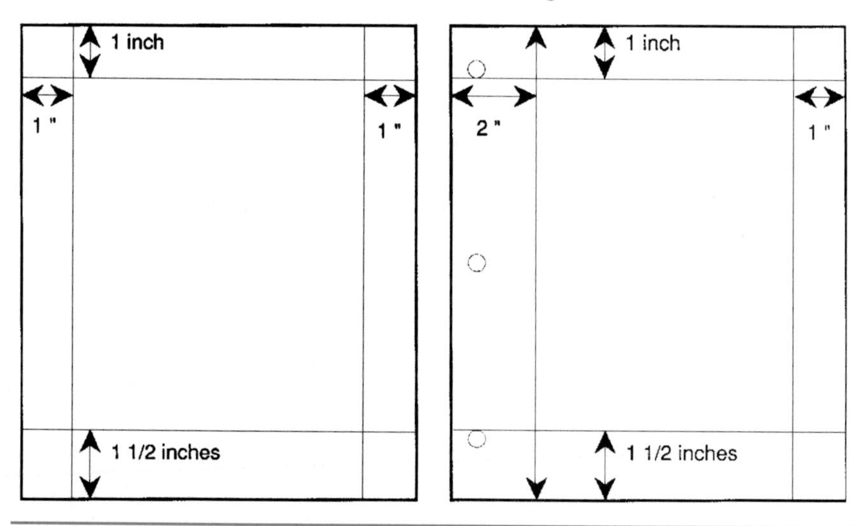

Figure 9–5 Page Layouts Showing Margins for 8 1/2- by 11-Inch Paper

Short lines reduce readability of instructions. Lack of spacing creates dense text.

Attachment A is a CRAS 24 report designed for the claims organization containing the required information to identify all potential claims. This report will be meaningless if the recommended procedure for tacking wet cables, damaged cables, and major cable failure is not followed as described below:

Cut cables (Attachments A, C, D)
The tracking of cut cables will require the
use of the special study and remarks field.
The special study field is six characters in
length, and when the ticket is input into
CRAS, C U T _ _ should be entered in the
special study field. If the cut cable was
located, this information is obtained from
your log of Request to Locate Telephone Plant
(GK6391) or equivalent. C U T L O C should be
entered in the special study field. This section
should be used to enter the claim number and,
if known, the name of the person or company
responsible for the damage and any other
pertinent information. The entry would be
C N 1 2 3 - 4 5 6 7 - 8 9 0.
Wet Cables (Attachments B, C, D)
When it is determined that a cable was wet,
enter W E T _ _ _ in the special field.
Major Failure (Attachments G, D)
If the cable failure has the potential of
affecting ten or more customers, enter Y
in MAJ FAILURE.

Figure 9–6 Text with Too Short Lines

If you are going to photocopy on both the front and back of the page, leave space for the binding in the left margin of odd-numbered pages and in the right margin of even-numbered pages. Some word processing programs let you set the margins so that they alternate for right-hand (odd-numbered) pages and left-hand (even-numbered) pages. If you cannot set alternating margins on your word processing system, set both the right and left margin at about 1 ½ inches to allow for binding two-sided copies.

Use a Medium Line Length As you can see from Figure 9–3, long lines of uninterrupted type are tiring to read. Moreover, readers are likely to lose their place in moving from the right margin of a long line back to the left margin of the next line.

As Figure 9–6 shows, very short lines are also hard to follow. Readers have trouble keeping the sense of what they are reading. Too many short lines can create excessive white space.[3]

As shown in Figures 9–2 and 9–4, lines of type should be about 50–70 characters or about 10–12 words per line. The number of characters that fit in a certain amount of space depends in part on the size and style of type you are using (see pages 204–206). In a two-column format, keep each column to about 35 characters, about 5 words. Note how Figure 9–8 is much easier to read than Figure 9–7, and Figure 9–10 is much easier to read than Figure 9–9 because the page layout of Figures 9–8 and 9–10 is much less dense.

Two-column format can be particularly difficult to read if the writer is not sensitive to the problems that multiple columns can create for readers. Thus, effective use of headings, ragged right margins, and spacing are crucial, as shown in Figure 9–10.

Set the Spacing for Easy Reading When you type your reports, when should you use single-spacing? When should you use double-spacing? What other options could you consider?

It is common in the workplace to use single-spacing in documents such as letters and memos unless they are very short. When you use single-spacing, leave an extra space between paragraphs, as in Figures 9–2, 9–4, and 9–8.

Drafts are usually double-spaced to give writers and editors more room to write their corrections and notes. When you use double-spacing in drafts, indent the first line of each paragraph or add one or two lines between each paragraph. Reports are often done in double-space or space-and-a-half.

Use a Ragged Right Margin Since you first began writing by hand on ruled notebook paper and then typing, you have been accustomed to having a firm left margin in your documents. The text in Figures 9–1, 9–2, 9–4, 9–6, 9–7, and 9–8 has a firm left margin and a ragged or uneven right margin. With word processing, you can have firm margins on both left and right margins, as shown in Figure 9–9. The pages in this text are also even on both the right and left margins. The technique of making all the lines end at exactly the same distance from the margin is called *justification.*

Be careful if you decide to justify type on the right margin, as some word processing programs create unsightly gaps that make the text very difficult to read, as illustrated in Figure 9–3. Research shows that regular spacing is part of the reading process and that readers move more easily from one word to the next when the space between words is approximately the same.[4] This is one of the reasons Figure 9–10 is more appealing than Figure 9–9. But Figure 9–10 also unpacks the text and reveals content by clearly phrased boldface headings.

Even if your computer can microjustify—put extra space evenly across the line so that you can't tell where it is—think about the purposes and audiences for your work. Justified type gives a document a formal tone; unjustified type gives a document a more friendly, personal tone. Evidence exists that readers like unjustified text[5] and that poor readers have more difficulty reading justified text.[6] The dense, justified text in Figure 9–9 makes the cable splicing procedures difficult to follow. Figure 9–11, with its two-column unjustified text and clearly worded headings, reveals its content in a less forbidding way. The information is much easier to find. You can see the content as it is suggested by the informative headings. Whether you choose to justify your text on the right should be

-- 3 --

With the substantial growth in computing in the College of Engineering during the past decade, the issue of linking the departments through a computer network has become critical. The network must satisfy a number of criteria to meet the needs of all of the engineering departments. We first state these criteria and then discuss them individually in detail.

To adequately serve both faculty and student needs in the present environment, the network must be able to handle the number of computers currently in use. In addition, the system must be able to expand and link in additional computers as the number of computers increases over the next few years. The different types of computers that the departments presently possess must all be linkable to the network, and the types of computers that are scheduled for purchase must also be able to be connected to the network. The network should permit the transfer of files in both text and binary form in order to facilitate student access to files and collaborative exchange among faculty and research associates. The network must also have adequate bandwidth in order to handle the expected traffic. Finally, the network must permit both students and faculty to link to the existing national networks.

The page with just text looks dense and uninviting.

Readers can't tell at a glance what the text is about.

Each department currently has both computer laboratories for students and computers that are associated with faculty research projects. The various departments possess different numbers of computers. The Aeronautical Engineering Department at present has 27 computers, while Civil Engineering has 12. The Electrical Engineering Department has the most in the College with 46. Mechanical has 22, and Nuclear Engineering, the smallest department in the College, presently has 7. This means that the entire College presently has 114 computers which will need to be networked.

In order to meet their different needs, each department has focused on the purchasing of computers with differing strengths. The computers provide for faculty and advanced students to program in a variety of languages including Pascal, C, and Fortran.

Figure 9–7 An Example of Poor Formatting

With the substantial growth in computing in the College of Engineering during the past decade, the issue of linking the departments through a computer network has become critical. The network must satisfy a number of criteria to meet the needs of all of the engineering departments. We first list these criteria and then discuss them individually in detail.

Large headings make the topics and structure obvious.

What must the network do?

To serve both faculty and students, the network must be able to

A bulleted list makes the points more memorable.

Each item in the list becomes the heading for a subsection.

- handle the number of computers currently in use
- link different types of computers
- expand as the number of computers increases
- link to the national networks
- transfer and store both text and binary files

The network must also have adequate bandwidth in order to handle the expected traffic.

The subsection headings are also bold but smaller than the main section heading.

Handling the number of computers currently in use

The shorter line length makes the text easier to head and makes the headings stand out.

Each department has both computer laboratories for students and computers that are associated with faculty research projects. The following table shows the number of computers in each department at the end of the last fiscal year.

Department	Computers
Aeronautical Engineering	27
Civil Engineering	12
Electrical Engineering	46
Mechanical Engineering	22
Nuclear Engineering	7
Total	114

The numbers are much clearer in a table.

Linking different types of computers

In order to meet their different needs, each department has focused its purchasing on machines with different strengths. The computers

The footer on every page reminds readers of the overall topic.

A Proposal to Install a Computer Network for the College of Engineering **page 3**

Figure 9–8 The Same Page Reformatted

Bell System Practices
AT&TCo Standard

Section 680-120-013
Issue 1, July 1977

TIE CABLE PAIR ADMINISTRATION

1. GENERAL

1.01 This section covers the administration of Form E-6536, Tie Cable Record in a manual assignment office that administers tie pairs terminating on conventional frames. The procedures described should be used as a guide to provide effective control of the tie cable and all interrelated records.

1.02 When this section is reissued, the reason for reissue will be listed in this paragraph.

1.03 A tie cable generally is used to provide a facility path between all types of frames within a central office (CO) building or between CO buildings that are adjacent. The physical layout of the frames and equipment within each building will determine the quantity of tie cables used.

1.04 In larger CO buildings, many tie cables may terminate on a tie pair distributing frame (TPDF) to facilitate interconnection between various types of equipment and feeder cable pairs.

1.05 Following are reasons for standardizing the administration of tie cable pairs:

- Continuous increase in service orders which require miscellaneous equipment
- Continuous increase in service orders which require interconnection of cable pairs for special services and associated equipment
- Mechanized recordkeeping systems which require accurate data input.

2. DESCRIPTION AND USE

2.01 Form E-6536, Tie Cable Record (Fig. 1) will be used when applicable to administer tie cable pairs terminating on main distributing frames (MDFs). The E-6536 form is numbered 1 through 50 on the front side and 51 through 00 on the reverse side. Appropriate hundreds are to be added as required. Form E-6536 has been designed for use with a conventional frame or the COSMIC frame (Section 680-830-012, COSMIC Frame Manual Assignment Procedures), as well as for modular-type frames (Section 680-535-009, Short Jumper Assignment for Modular Main Distributing Frames).

2.02 To maintain Tie Cable Record integrity and efficient tie pair administration, a random sample comparison of the various tie pair connections versus the assignment office Tie Cable Records should be made at approximately 6-month intervals.

2.03 It is recommended that a specific portion of the Tie Cable Record associated with tie cables terminating on the conventional frame be committed by the assignment office to the design group associated with special service order circuit design and makeup to assure interdepartmental operating efficiency. This will allow tie pair used to connect special circuit equipment to be posted on the circuit layout card. This will also aid CO personnel in installation and repair work and ensure that these tie pairs are removed on the conventional frames and assignment records when a disconnect order is completed. The assignment office Tie Cable Record will have the notation "record maintained by design group." If this recommended procedure is not established, the design group will have to call the assignment office for the tie pair assignments associated with special service orders in order to have the necessary tie pair information for entry on the circuit layout card.

2.04 The following procedures are used to assign a tie cable on a conventional MDF:

(1) Assign a cable and pair.

Figure 9–9 Two-Column Instructions, Left- and Right-Justified

TIE CABLE PAIR ADMINISTRATION

BSP: 680-120-013
DATE: March 1990
SUBJECT: Tie Cable Pair Administration - manual records

The Purpose of This Practice

This practice ensures that the manual assignment office has established guidelines for administering records for tie cable pair administration. The following are reasons for standardizing the administration of tie pairs:

— Continuous increase in service orders which require miscellaneous equipment assigned.
— Continuous increase in service orders which require interconnection of cable pairs for special services and associated equipment.
— Mechanized recordkeeping systems which require accurate data input.

Definitions of Abbreviated Terms

CO — Central Office
TPDF — Tie Pair Distributing Frame
MDF — Main Distributing Frame
ECCR — Exchange Customer Cable Record
DPAC — Dedicated Plant Assignment Card

References

Section 680-830-012—COSMIC Frame Manual Assignment Procedures
Section 680-535-009—Short Jumper Assignment for Modular MDF

Forms to Be Used

Form E6536, Tie Cable Record will be used to administer records. The form is numbered 1 through 50 on the front side and 51 through 00 on the reverse side. Appropriate hundreds are to be added as required. The form has been designed for use with COSMIC and modular-type frames. Form E6536 may be ordered from Procurement in packages of 25 sheets per package.

Scope

A tie cable generally is used to provide a facility path between all types of frames within a CO building or between CO buildings that are adjacent. The physical layout of the frames and equipment within each building will determine the quantity of tie cables used. In larger CO buildings, many tie cables may terminate on a TPDF to allow for interconnection between various types of equipment and feeder cable pairs.

Figure 9–10 Revision of Figure 9–9

determined by your word processing program. Can it microjustify? If not, leave your right margin ragged. If your word processing program can microjustify, look at a justified version and a ragged version of your text. Then decide which looks best and serves the needs of your audience.

Use White Space to Group Information White space around text and headings can help you group your ideas so that your reader can see

680-120-012
Page 2 of 2
March 1990

Responsibilities

Tie cable record administration of non-designed services will be maintained on Form E6536 by a clerk in the assignment office. A specific portion of the record will be committed to the design group associated with special service circuits. A designated clerical employee in the design group will maintain the record of tie pairs used to connect special circuit equipment to be posted on the circuit layout card. This employee will compare the various pair connections versus the manual tie cable records at approximately six-month intervals. The second level manager of the assignment office and the special services group will be responsible for ensuring the procedure is followed for their respective work groups. This will maintain tie cable record integrity and efficient tie pair administration.

Procedures to Assign Tie Cable Pairs

(1) Assign a cable and pair.

(2) Choose available line equipment from the list of available line equipment.

(3) If interconnection of cable pairs is required or miscellaneous CO equipment is needed, assign the necessary tie pairs.

(4) Enter the information on the service order.

(5) Post the ECCR and/or DPAC.

(6) Post the E6536.

Procedures for Unusual Circumstances

In those instances where the need for a tie cable pair is not recognized at the time the order is assigned, the following procedures apply:

(1) CO will call the assignment office.

(2) Assignment will provide the CO with the required tie pair information.

(3) Assignment will post the E6536, ECCR, and/or DPAC.

(4) If only a tie pair is needed, CO must call the assignment office to verify the tie pair assigned and assure removal of the circuit information posted on the E6536.

Figure 9-10 (continued).

the arrangement. As we have noted, research in document design has shown that readers must see the organization of text before they can process it.[7] Lists allow you to group related ideas and draw your readers' attention to them. Separate sections and headings with ample white space. Figure 9-10 and Figure 9-12, which is a redesigned version of Figure 9-13, show the value of white space and listing for producing a readable page. Figures 9-14 and 9-15 show segments of an instructional brochure on immunizations. Notice how type size and white space enhance the visual effectiveness of the information in 9-14 as compared to 9-15.

The large bold type and the dark line make the title stand out.

U.S. Department of Transportation
National Highway Traffic Safety Administration

Office of Public Affairs
Washington, D.C. 20590
(202) 426-9550

Testing How Well New Cars Perform In Crashes

April 1983

The title tells readers what the document is about.

The headings are easy to see because they are in bold type, larger than the text type, and separated from the text by white space.

What is the New Car Assessment Program?

The Department of Transportation has an experimental program in which it tests cars to see how well they perform during a crash. NHTSA (the National Highway Traffic Safety Administration) is the agency within the Department of Transportation that conducts these tests. NHTSA publishes this fact sheet to give consumers information that can help them to compare the relative safety of cars they may be planning to buy.

Every new car sold in the United States must meet minimum safety standards that are set by the Federal government. NHTSA's crash test program goes beyond these minimum standards. It tests cars at 35 mph, a much stricter test than the 30 mph test that is required by the standards. A 35 mph crash is about 35 percent more violent than a 30 mph crash.

If you are interested in only one question, you can find that one easily.

The graphs that come with this fact sheet show how different makes and models compare in the way they perform during these high speed crashes. In interpreting these test results, it is important to remember two points:

The text of each section is easy to read because the lines are short.

(1) **Drivers and passengers should always wear safety belts.** The human-like dummies used in the crashes are wearing safety belts. These tests measure how cars perform, not how people perform. The crashes are intended to illustrate potential injuries to people who are **properly seated and belted** in the car. Fifty percent of the deaths from road accidents could be avoided if drivers and passengers wore their safety belts.

(2) **Large cars usually offer more protection in a crash than small cars.** These test results are only useful for comparing the performance of cars in the **same size class**.

What do the graphs show?

The graphs show how badly a person's head could be injured in a head-on collision between two identical cars, if both were going at 35 mph.

In general, the lower the score on the head injury criteria, the **less** likely drivers and front-seat passengers will be to be seriously injured or killed in a frontal crash at 35 mph. If the dummy in a car model scores substantially higher than 1,000 on the head injury criteria, human drivers and front-seat passengers in that model will be **more** likely to suffer a serious head injury or be killed in a frontal crash at 35 mph.

How does NHTSA select the cars it will test?

NHTSA tests about 25 cars every year. The graphs show the test results for a variety of cars. Because NHTSA buys its test cars off the lot, just as you would, the tests cannot begin until after a new model year has started. Test results on new cars generally begin to be available in the late winter of each year.

NHTSA chooses the cars it will test to give useful information to as many consumers as possible. For example, a popular model is more likely to be chosen, because information on that model would be of interest to many consumers. For the same reason, very expensive cars are not tested as often.

What happens when model years change?

If a car that NHTSA has already tested remains essentially the same for the new model year, it will probably not be tested again. But a car that undergoes substantial changes in the new model year will be likely to be retested. In using any crash test data, you should always check the model year that was tested and whether NHTSA believes that the test results should be used to evaluate other model years.

Figure 9–11 Two Equal Columns Is a Good Format for This Fact Sheet (Note: Reduced 59% from Original)

Source: Testing How Well New Cars Perform in Crashes (Washington, DC: National Highway Traffic Safety Administration, U.S. Department of Transportation, April 1983).

DATE: March 15, 1990

SUBJECT: Preparation of 4-1-90 Payroll Change Lists (PCLs) for Merit Increases

ACTION REQ: Due 4-2-90

REF: Joint Practice No. 74—Payroll Change Reports

Please refer to the letter from the Vice-President–Personnel dated 1-16-90. Use the following guidelines to complete the attached blanket payroll change lists.

Payroll Activity Reflected on Payroll Change Reports

All payroll change activity effective April 1, 1990 which was entered into the payroll system before March 18, 1990, is reflected. These will be used for authorizing 4-1-90 merit salary increases for management employees:

Procedures for Completing Blanket Payroll Changes

1. Make all entries in RED ink.

2. Note the change amount and proposed rate of pay:

    1. Enter the correct amount.
    2. Enter XXXX in these columns if no change is recommended.
    3. Refer questions concerning calculation of new salary rates to your segment personnel coordinator.

3. Line off employee names and prepare individual PCRs (Payroll Change Records) if

    * Pre-printed information is incorrect.
    * Employee has other payroll activity effective 4-1-90.
    * Employee is on disability or leave of absence.

4. Do not add names to the lists.

5. Return completed PCLs to the payroll office by 10:00 A.M., on 4-2-90.

Payroll will forward a copy to the VP-Personnel for review.

Questions regarding the Payroll Change Lists may be referred to me at 713 893-7146.

Gayle Johnson
Assistant Manager—Payrolls

Figure 9–12 Memo after Revision

Management Merit Increase
Payroll Change List
Effective April 1, 1990

Attached are blanket Payroll Change Lists (PCLs), effective April 1, 1990, to be used to authorize Merit Salary Increases for management employees. These changes are covered in the letter dated January 16, 1990, file number 450.0701 from the Vice-President–Personnel.

Three copies of the PCL are attached. Reflected on the PCL is all payroll change activity which has been entered to the Payroll System prior to March 18, 1990 and effective through April 1, 1990. The performance code changes effective March 1, 1990 are reflected on the PCL. The Change Amount and Proposed Rate of Pay have been left blank and should be completed by your organization. Enter "XXXX" in these columns if no money amount is to be entered. All entries should be made in red ink.

If the preprinted information is incorrect, line the employee off the list and submit an individual Payroll Change Report (PCR). If an employee has other payroll change activity effective April 1, 1990, the entry should be lined off the PCL and a PCR submitted showing all proposed information applicable for that date. Names may not be added to the list.

Any employee on disability or leave of absence may not receive a change to their basic rate of pay until he/she returns to an active management assignment. Please refer to Joint Practice 74 for special procedures to be followed.

The authorized original and first copy of the PCL should be forwarded so as to be received in the Houston Payroll Office no later than 10:00 A.M., April 2, 1990. A copy of this list will be forwarded to the Vice-President–Personnel for review.

Any questions concerning the calculation of the new salary rates on the Management Salary Plan should be referred to your segment personnel representative.

Questions regarding the PCL may be referred to me at 713 893-7146.

Gayle Johnson
Assistant Manager–Payrolls

Figure 9–13 Memo before Revision

As shown in Figure 9–12,

- Keep items in lists parallel.
- Indent the list to set it off from surrounding text.
- Use bullets when the order of items is not important.
- Number items when the arrangement and number of items is important.

THINGS THAT YOU SHOULD KNOW ABOUT MEASLES, MUMPS AND RUBELLA

WHAT IS MEASLES ? Measles is the most serious of the common childhood diseases. Usually it causes a rash, high fever, cough, runny nose, and watery eyes lasting 1 to 2 weeks. Sometimes it is more serious. It causes an ear infection or pneumonia in nearly 1 out of 10 children who get it. Approximately 1 child out of every 1,000 who get measles has an inflammation of the brain (encephalitis). this can lead to convulsions, deafness, or mental retardation. About 2 children in every 10,000 who get measles die from it. Measles can also cause a pregnant woman to have a miscarriage or give birth to a premature baby. Wide use of measles vaccine has nearly eliminated measles from the United States. However if children are not vaccinated they have a high risk of getting measles, either now or later in life.

WHAT IS MUMPS ? Mumps is a common disease of children. Usually it causes fever, headache, and inflammation of the salivary glands, which causes the cheeks to swell. Sometimes it is more serious. It causes a mild inflammation of the coverings of the brain and the spinal cord (meningitis) in about 1 child in every 10 who get it. More rarely, it can cause inflammation of the brain (encephalitis) which usually goes away without leaving permanent damage. Mumps can also cause deafness. About 1 out of every 4 adolescent or adult men who get mumps develops painful inflammation and swelling of the testicles. While this condition usually goes away, on rare occasions it may cause sterility. Because of wide use of mumps vaccine, the number of cases of mumps is much lower. However if children are not vaccinated they have a high risk of getting mumps.

WHAT IS RUBELLA ? Rubella is also called German measles. It is a common disease of children and may also affect adults. Usually it is very mild and causes a slight fever, rash and swelling of glands in the neck. The sickness lasts about 3 days. Sometimes especially in adult women, there may be swelling and aching of the joints for a week or two. Very rarely rubella can cause inflammation of the brain (encephalitis) or cause a temporary bleeding disorder (purpura).

The most serious problem with rubella is that if a pregnant woman gets this disease, there is a good chance that she may have a miscarriage or that the baby will be born crippled, blind, or with other defects. Wide use of rubella vaccine has caused the number of cases of rubella to be much lower than in the past. However, if children are not immunized, they have a high risk of getting rubella and possibly exposing a pregnant woman to the disease. If an unimmunized woman later becomes pregnant and catches rubella, she may have a defective baby. Many women of childbearing age do not recall if they had rubella as a child. A simple blood test can show whether a person is immune to rubella or is not protected against the disease.

Figure 9–14 Information in an Appropriate Type Size

Figure 9–12 is a routine memo that is designed to be read quickly. Note how the listing arrangement immediately reveals the actions the reader should take. Figure 9–13 shows this memo before it was reformatted.

Planning for Document Design

Word processing enables writers to use a wide variety of visual aids: tables, line graphs, bar graphs, pie graphs, drawings, icons, flowcharts, and clip art. You need to consider which ones you will use (if any) as you plan your document, as well as where you will use them and for what purpose. These visual aids will be discussed further in Chapter 11.

THINGS THAT YOU SHOULD KNOW ABOUT MEASLES, MUMPS AND RUBELLA

WHAT IS MEASLES ? Measles is the most serious of the common childhood diseases. Usually it causes a rash, high fever, cough, runny nose, and watery eyes lasting 1 to 2 weeks. Sometimes it is more serious. It causes an ear infection or pneumonia in nearly 1 out of 10 children who get it. Approximately 1 child out of every 1,000 who get measles has an inflammation of the brain (encephalitis). This can lead to convulsions, deafness, or mental retardation. About 2 children in every 10,000 who get measles die from it. Measles can also cause a pregnant woman to have a miscarriage or give birth to a premature baby. Wide use of measles vaccine has nearly eliminated measles from the United States. However if children are not vaccinated they have a high risk of getting measles, either now or later in life.

WHAT IS MUMPS ? Mumps is a common disease of children. Usually it causes fever, headache, and inflammation of the salivary glands, which causes the cheeks to swell. Sometimes it is more serious. It causes a mild inflammation of the coverings of the brain and the spinal cord (meningitis) in about 1 child in every 10 who get it. More rarely, it can cause inflammation of the brain (encephalitis) which usually goes away without leaving permanent damage. Mumps can also cause deafness. About 1 out of every 4 adolescent or adult men who get mumps develops painful inflammation and swelling of the testicles. While this condition usually goes away, on rare occasions it may cause sterility. Because of wide use of mumps vaccine, the number of cases of mumps is much lower. However if children are not vaccinated they have a high risk of getting mumps.

WHAT IS RUBELLA ? Rubella is also called German measles. It is a common disease of children and may also affect adults. Usually it is very mild and causes a slight fever, rash and swelling of glands in the neck. The sickness lasts about 3 days. Sometimes especially in adult women, there may be swelling and aching of the joints for a week or two. Very rarely rubella can cause inflammation of the brain (encephalitis) or cause a temporary bleeding disorder (purpura).

The most serious problem with rubella is that if a pregnant woman gets this disease, there is a good chance that she may have a miscarriage or that the baby will be born crippled, blind, or with other defects. Wide use of rubella vaccine has caused the number of cases of rubella to be much lower than in the past. However, if children are not immunized, they have a high risk of getting rubella and possibly exposing a pregnant woman to the disease. If an unimmunized woman later becomes pregnant and catches rubella, she may have a defective baby. Many women of childbearing age do not recall if they had rubella as a child. A simple blood test can show whether a person is immune to rubella or is not protected against the disease.

Figure 9–15 Information in an Inappropriate Type Size

In planning your design you also need to consider choosing a readable type, designing useful headings, and using other devices to reveal content and organization.

Choosing Readable Type

Many printers and typewriters allow you to change the type size and style. These characteristics may be predetermined by companies that have firm style guidelines for their official reports and letters, but you may still have choices for documents you write as part of your job.

To choose a type that will be easily readable,

- Choose a type that is legible.
- Choose a typeface that is appropriate for the document.
- Use highlighting effectively.
- Use mixed case letters.
- Use color carefully.
- Use special typefaces sparingly.

Choose a Legible Type Many word processing programs offer a wide choice of point sizes, as you can see in Figure 9–16.

In general, use a 9- to 12-point type for regular text.[8] Use larger sizes for titles and headings. Type that is smaller than about 8 points is difficult to read. The two basic type groups are *serif,* which has extenders on the letters, and *sans-serif,* which does not use extenders. Figure 9–17 shows you the difference. Figure 9–9 is typed in 4-point sans-serif type, and Figure 9–15 is in 8-point sans-serif type.

This is 8-point type.
This is 10-point type.
This is 12-point type.
This is 14-point type.
This is 18-point type.
This is 24-point type.
This is 36-point.

Figure 9–16 Type Comes in Different Sizes

Choose an Appropriate Typeface On many computers, you can choose not only the size but also the style of type, called the *typeface* or *font*. There are hundreds of type fonts, both serif and sans-serif. Research has shown that serif type is generally more readable than sans-serif type; whatever type you choose, be sure to examine it for readability.

Serifs draw the readers' eyes across the page, so most books and other long documents are printed with serif type. Sans-serif type works

Serif	Sans-Serif
Times	Helvetica
Palatino	Geneva
New York	Chicago
Schoolbook	Helvetica Narrow
Courier	Monaco

Figure 9–17 Some Traditional and New Typefaces (Fonts) on the Macintosh Computer

Figure 9–18 Unusual Typefaces Should Be Used with Caution and Only in Appropriate Situations
Source: Fonts from Arts & Letters, Computer Support Corporation, Dallas, TX 75244.

well in brochures and in visual aids such as graphs and tables. Sans-serif also gives documents a contemporary look. Figures 9–17 and 9–18 show examples of some traditional typefaces and some of the newer ones that are available on computers.

Use Highlighting Effectively To highlight material, you can use boxes, underlining, boldface, italics, and perhaps color, as many of the pages in this text do. You can even adjust the type size of selected words, phrases, and headings.

Lines above and below the sentence, for example, for a warning.

---

boldface *italics* SMALL CAPITALS

Large, bold headings stand out

Boldface works well for making a **word**, short phrase, or short sentence **stand out**.

Italics work well for emphasizing single words, such as *special*.

Italics are correct for book titles, such as *Reporting Technical Information*.

On a typewriter, use underlining for book titles, such as <u>Reporting Technical Information</u>.

Because of the increasing ways to highlight material, several cautions are in order:[9]

1. Don't overdo it. Highlighting calls attention to special features. If you use too many different kinds of highlighting or use one kind too often (as in Figure 9–19), you dilute the effect of highlighting.
2. Be consistent. Highlighting helps readers find their way through your text. Once you choose a feature to highlight a particular kind of information, use that feature only for that information. If you use italics for specially defined words, use italics throughout. If you decide to box all warnings, be sure that all warnings are boxed.
3. Match the highlighting to the importance of the information.
4. Don't use any highlighting techniques for more than a short sentence at a time. Whole paragraphs underlined or in boldface or in italics are difficult to read. Figure 9–19 is a memo that uses excessive underlining in an attempt to emphasize major points. The result, however, is that these points are obscured rather than revealed. Figure 9–20 shows an effective use of highlighting to display the same information shown in Figure 9–19. Notice how the judicious use of white space and headings improves readability.

Use Lowercase and Mixed Case Letters Don't use all capitals for text. A sentence in all capitals takes about 13% more time to read than a sentence typed in the regular uppercase and lowercase letters that we expect.[10] Research shows that boldface works better for headings than all caps.[11]

Difficult to read

CAPITAL LETTERS GIVE US NO CLUES TO DISTINGUISH ONE LETTER FROM ANOTHER. THEREFORE, THE LETTERS BLUR INTO EACH OTHER VERY QUICKLY, AND WE WANT TO STOP READING. LOWERCASE LETTERS GIVE US CLUES TO THE SHAPES OF THE WORDS, AND WE USE THOSE SHAPES AS WE READ.

Easier to read

Capital letters give us no clues to distinguish one letter from another. Therefore, the letters blur into each other very quickly, and we want to stop reading. Lowercase letters give us clues to the shapes of the words, and we use those shapes as we read.

Use Color Carefully Excessive use of color can detract from the appearance of the page. Avoid using color for text: black on white paper provides the best contrast. High contrast between paper and ink is necessary for easy reading. Colored paper may be a better choice if it enhances the effectiveness of your document. For example, some organizations use light gray, cream, or light blue papers for major reports and proposals. The covers may be black, brown, or dark blue. If you choose to use color, choose a light shade to keep the contrast between ink and paper high. This text uses varying shades of blue to highlight divisions, but the text itself is basically black on white paper.

TO: David Stewart DATE: December 2, 1993

FROM: Kathy Hillman

SUBJECT: Short Course Request from Ocean Drilling

Because I will be away on a three-week teaching assignment, I would appreciate your handling the following request, which came in just as I was preparing to leave today.

Randy Allen, Director of the offshore drilling research team would like a short-course in writing offshore safety inspection reports. He would like the short-course taught from 2–4 P.M. Monday–Friday afternoons, beginning week after next. The class must be scheduled then, as the team leaves the following week for their next research cruise.

The drilling research team spends two weeks each month on cruise. After they return, they have one week to complete their reports before briefing begins for the next research expedition. Because of their rigid schedule, they cannot attend our regularly scheduled writing classes.

Allen says that the cultural and educational backgrounds of the team are varied. Five of the ten regular researchers are native Europeans who attended only European universities. Of the remaining five, two have American degrees, and three attended school in Canadian universities. As a result of their varied educational backgrounds, their reports lack uniform handling of English and organization. All the researchers have expressed interest in having a short review of standard English usage so that their reports to management will be more uniform.

Sarah Kelley says she can develop a class for the drilling team. We have materials on reports, style, and standard usage in the files. She can work with Ocean Drilling to determine the best report structure and develop a plan. These items can be easily collected and placed in binders. We also have summary sheets on each topic that will be good reference aids when the researchers write their reports following their cruise.

Sarah will contact you Monday morning. If her teaching the class meets with your approval, please give Randy Allen a call, at extension 721, before noon. He has a staff meeting scheduled at 1:30 and would like to announce the short course then. In fact, if the course cannot be scheduled this month, it cannot be taught for seven months because of off-season cruise schedules. Allen wants this course before the team begins a series of four reports during the off-season.

Please arrange a time for Sarah to meet with Allen so they can go over several previous reports. Sarah wants to be sure that what she covers in the course is what they need.

If you need to talk to me about this request, I will be staying at the Hyatt in New Orleans.

Figure 9–19 Memorandum with Overuse of Underlining and Lack of Headings

Use Special Typefaces Sparingly Some computers offer unusual typefaces, as you can see in Figure 9–18. These typefaces are not generally appropriate for technical reports, and they are not as readable as the traditional typefaces shown in Figure 9–17.

TO: David Stewart DATE: December 2, 1993

FROM: Kathy Hillman

SUBJECT: Request from Offshore Drilling Team for a Special Short Course

ACTION REQUIRED: By Monday, noon, April 2

Because I will be away on a three-week teaching assignment, I would appreciate your handling the following request, which came in just as I was preparing to leave today.

What Does Offshore Drilling Want?

Randy Allen, Director of the offshore drilling research team, would like a short course in offshore safety inspection reports. He would like the short course taught from 2–4 P.M. Monday–Friday afternoons, beginning week after next. The class must be scheduled then, as the team leaves the following week for their next research cruise.

Why Do They Want a Specialized Short Course?

The drilling research team spends two weeks each month on cruise. After they return, they have one week to complete their reports before briefing begins for the next research expedition. Because of their rigid schedule, they cannot attend our regularly scheduled writing classes.

Allen says that the cultural and educational backgrounds of the team are varied. Five of the ten regular researchers are native Europeans who attended only European universities. Of the remaining five, two have American degrees, and three attended school in Canadian universities. As a result of their varied educational backgrounds, their reports lack uniform handling of English and report organization. Allen says that all the researchers have expressed interest in having a short review of standard English usage so that their reports to management will be more uniform.

How Can We Handle Their Request?

Sarah Kelley says she can develop a class for the drilling team. We have materials on reports, style, and standard usage in the files. She can work with Ocean Drilling to determine the best report structure and develop a plan. These items can be easily collected and placed in binders. We also have summary sheets on each topic that will be good reference aids when the researchers write their reports following their cruise.

What Needs to Be Done Immediately?

Sarah will contact you Monday morning. If you approve her teaching the class, please give Randy Allen a call, at extension 721, **before noon. He has a staff meeting scheduled at 1:30 and would like to announce the short course then. In fact, if the course cannot be scheduled this month, it cannot be taught for seven months because of off-season cruise schedules. Allen wants this course before the team begins a series of four reports during the off-season.**

Please arrange a time for Sarah to meet with Allen so they can go over several previous reports. Sarah wants to be sure that what she covers in the course is what they need.

Figure 9–20 Memo Designed with Questions as Headings

Designing Useful Headings

Figures 9-2, 9–4, 9–8, 9–10, and 9–20 all exemplify effective use of headings. Even short documents such as memos and letters can benefit from headings because they allow readers to locate information quickly and see the general structure of your document. In long documents, the first few levels of headings become the table of contents.

Headings are the road map to your document. They come from your outline, but you usually do not use roman numerals or letters in headings. A few guidelines will help you use headings effectively.

Figure 9–21 Table of Contents Compiled from Generic Headings

Make Headings Informative Generic headings such as *Part I* or *Section 2* give readers no clues to the content of your work. Compare the report headings in Figures 9–21 and 9–22. Note that the headings in 9–22, unlike those in 9–21, clearly indicate the content of each report section. Also note the difference between Figure 9–19 and Figure 9–20—the ways in which headings contribute to your ability to see the meaning of the text immediately.

Avoid using single nouns or noun strings as headings. Better yet, make sure your heading clearly expresses the content that follows. Figure 9–22 illustrates this point. As a rule, use a heading that is a short descriptive phrase, perhaps containing a verb. You should be able to read your headings, apart from the text, and see that they suggest the content of the document. For example, in Figure 9–23, you can read the subject line, the opening segments (which most people usually read), and the headings and get the main message. The supporting material beneath the heading expands on the information presented in the subject line and headings.

You can create headings in various ways:

- **Questions,** which announce that the answer will follow in the text beneath the question heading.

  Figure 9–22 is a table of contents derived from headings that ask questions. The text of the report answers those headings. Research shows that headings presented as questions are an effective way to introduce content.[12] See Figures 9–2 and 9–20 for effective use of questions as headings.

List of Illustrations

Foreword

Summary

Introduction

Terms Potential Buyers of Condominiums Need to Understand

> Condominiums: What are they?
> Cooperatives
> Commercial Condominium Properties
> Other Forms of Condominiums
> Ownership: Who Owns What?

The Basic Question: Should You Rent or Buy?

> Look at Your Current Needs and Your Future Needs.
> Consider Mobility. Are You Ready for a Long-Term Mortgage?
> What Are Your Equity and Tax Advantages?

What Kind of Condominium? Single or Multi-Family Housing?

> Advantages of the Single-Family Unit
> Advantages of the Multi-Family Unit

What Are Your Rights and Responsibilities?

Conclusion—Guide for Decision Making

Recommendation—Ask Questions, Think, and Look before You Buy

References—Reading Material on Condominiums

Figure 9–22 Table of Contents Compiled from Clearly Worded Headings

- **Verb phrases,** such as *Adding a graphic, Selecting the data, Selecting the type of graph to use, Labeling the axes, Adding a title,* or *Adding a legend.*
- **Short sentences.** An instructions manual for new teachers might use the following headings:

Make your attendance policy clear.
Explain your grading policy.
Announce your office hours.
Supply names of texts to be purchased.
Go over assignments and their due dates.

FROM: Division Manager—Plant Services—Atlanta

SUBJECT: **Company Policy: Interduct shall not be placed for buried lightweight cable**

Contrary to the alleged claims of some overzealous vendors, the above policy remains unchanged.

A field trial was recently conducted where the interduct was buried and cable placement was attempted several days later. The negative observations were as follows:

1. Cable lengths are reduced between costly splices.

The interduct conformed to the high and low spots in the trench. These numerous bends introduced added physical resistance against the cable sheath during cable pulling. Even with application of cable lubricant, the average length pulled was 500 feet before the 600-pound pulling tension was exceeded.

2. Maintenance liability is increased.

Due to shorter cable lengths, the number of splices increased. As the number of splices increases, maintenance liability increases.

3. Added material costs are counterproductive.

Interduct, while adding 16% to the material cost, offers little or no mechanical protection to buried fiber-optic cable. In fact, when an occupied interduct was pulled at a 90-degree angle with a backhoe bucket, the fibers were not broken only at the place of contact, as they would have been with a direct buried cable. Instead, due to the stress being distributed along the interduct, the fibers shattered and cracked up to 100 feet in each direction.

4. Increased labor costs are unnecessary.

In about the same amount of time required to place interduct, the fiber-optic cable could be placed. The added 28% of labor hours expended to pull cable after the interduct is placed cannot be justified.

Please ensure that this policy is conveyed to and understood by your construction managers. If you have questions about the reasons that the policy remains unchanged, please call me at 404 597-8677.

Corporate Engineering Manager

Figure 9–23 Memo with Effective Headings That Reveal Content

- **Key words**

A proposal might use the following headings:

Project Summary
Project Description
Rationale and Significance

Plan of Work
Facilities and Equipment
Personnel
Budget

Like list items, headings must be parallel.

Design Headings to Show Levels of Information Headings do more than outline your document. They also help readers find specific parts quickly, and they show the relationships among the parts. To help readers, each level of heading has to be easily distinguished from the others. To show levels of headings, you have several options, which are illustrated in Figure 9–24. Your choices depend on the kind of word processing program you have available.

You can vary the size of the type:

Very Large Type

Large Type

Regular Size Type

You can vary the weight or style:

boldface

italic

underline

color

You can vary the placement on the page:

Centered Heading

At the Left Margin

Indented and above the Text
Text, text, text...

Indented and run-in. Text, text, text...

You can vary the capitalization scheme:

ALL CAPITALS FOR ALL WORDS
Initial Capitals for Important Words
Sentence style, first word capitalized

Level 1: 18-point boldface, centered initial capitals

Controlling Soil-Borne Pathogens in Tree Nurseries

Level 2: 14-point boldface, at the left margin, initial capitals

Types of Soil-Borne Pathogens and Their Effects on Trees

Simply stated, the effects of soil-borne pathogens..............................
....

Level 3: 11-point boldface, at the left margin, first word capitalized

The soil-borne fungi

At one time it was thought that the soil-borne fungi..........................
....

Level 4: 11-point (same as the text), boldface, run-in, first word capitalized

Basiodiomycetes. The basiodiomycetes are a class of fungi whose species
....

Phycomycetes. The class phycomycetes is a very diversified type of fungi. It is the..

The plant parasitic nematodes

Nematodes are small, nonsegmented..
....

Treatments and Controls for Soil-Borne Pathogens

....
....

Figure 9–24 Four levels of heading that you can create on a word processor. This format might work well in a report. You might use a different format for a software manual.

You can vary the spacing of the letters:

REGULAR SPACING
EXPANDED CAPITALS

You can use features such as lines above or below headings:

Main Heading

Second level heading

For an example of lines used above and below headings, see the student feasibility study, Figure 15-4 in Chapter 15.

Basically, these examples illustrate the following principles: First-level headings should be larger than second-level headings; second-level should be larger than third-level. Distinguish all headings in a uniform way. That is, use the same size type for all first-level headings throughout your report, the same size type for all second-level headings, and so on. When readers see a specific size of type, they should know that they are reading a specific level of heading. Figure 9-8 illustrates this concept, as does the student feasibility report (Figure 15-4) in Chapter 15. Through consistent use of type style and size for headings you can use design to reveal the organization of the text.

Limit Number of Heading Levels to Four or Less Student papers and technical documents rarely need more than four levels of headings. Readers may get lost and confused in a document that goes deeper than four levels of headings.

Keep your readers' expectations in mind. We all expect bigger headings to be more important. First-level headings should be larger than second-level and so on down to the level that is the same size as the text. Headings should never be smaller than the text. Most people also see all-caps headings as being more important than initial-caps or sentence-style headings. If you use all capitals, use it for the highest level of heading. You may want to again examine the student feasibility report (Figure 15-4, in Chapter 15) to see how the writer has varied type size to show rank of headings.

Now examine Figure 15-4, the student feasibility study in Chapter 15. If you use a rule (line) as a feature of page layout, consider putting the rule above the heading to emphasize that the heading goes with the text below it. This report illustrates this technique.

Keep the Heading with the Section That It Covers Don't leave a heading at the bottom of a page when the text appears on the next page. Make sure you have at least two lines of the first paragraph on the page with the heading. Or, you may want to have the heading and the complete text on a page so that the text is not broken. You make these decisions when you repage your document. If you are typing a report, as you approach the bottom of a page, stop and estimate what material you will need to type next and how much space it will require. The formal reports—Figures 15-2 and 15-4—in Chapter 15 illustrate effective ways of breaking pages.

Using Other Devices to Reveal Content and Organization

In addition to clearly worded, visually accessible headings, several other devices are useful in helping reveal your content and organization.

Number the Pages You have been numbering the pages of essays, research papers, and book reports for years, but you may not have considered the importance of page numbering as a document design tool. Page numbers help readers keep track of where they are and provide easy reference points for talking about a document. Always number the pages of your drafts and final documents.

Brief manuscripts and reports that have little prefatory material almost always use arabic numerals, like those used to number the pages of this book. The commonly accepted convention is to center the page number below the text near the bottom of the page or to put it in the upper right-hand corner. Always leave at least one double space between the text and the page number. Put the page number in the same place on each page. If you decide to place page numbers at the bottom of the page, these often have a hyphen on each side, like this:

- 17 -

As reports grow longer and more complicated, the page-numbering system also may need to be more complex. If you have a preface or other material that comes before the main part of the report, it is customary to use small roman numerals (i, ii, iii) for that material and then to change to arabic numerals for the body of the report. The beginning pages of this text use roman numerals, and the actual chapters use arabic numerals. The introduction may be part of the prefatory material or the main body. The title page doesn't show the number but is counted as the first page. The following page is number 2 or ii.

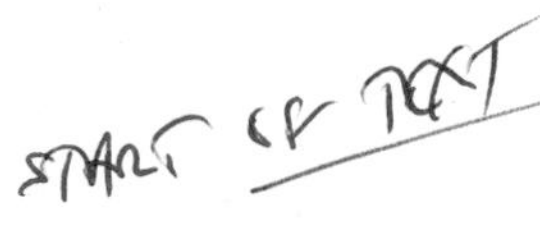

When numbering the pages, you have to know whether your document is going to be single-sided or double-sided. If both sides of the paper will have printing on them, you may have to number some otherwise blank pages in your word processing files. New chapters usually start on a right-hand page, that is, on the front side of a page that is photocopied on both sides. The right-hand page always has an odd number. If the last page of your first chapter is page 9, for example, and your document will be photocopied double-sided, you have to include an otherwise blank page 10 so that the first page of your second chapter will be a right-hand page when the document is bound.

In reports, the body is usually paginated continuously from page 1 to the last page. For the appendixes, you may continue the same series of numbers or you may change to a letter-plus-number system. In that system, Appendix A is numbered A-1, A-2, etc.; Appendix B is numbered B-1, B-2, etc. If your report is part of a series or if your company has a standard report format, you will need to make your page numbering match that of the series or standard.

Numbering appendixes with the letter-plus-number system has several advantages.

- It separates the appendixes from the body. Readers can tell how long the body of the report is as well as how long each appendix is.
- It clearly shows that a page is part of the appendixes and which appendix it belongs to. It makes pages in the appendixes easier to locate.
- It allows the appendixes to be printed separately from the body of the report. Sometimes, the appendixes are ready before the body of the report has been completed.
- It allows changes to either the appendixes or the body without disturbing the other parts.

Include Headers or Footers In multipage documents, it helps readers if you give them information about the document as well as the page number at the top or bottom of each page. If the information is at the top of the page, it is a header. If it is at the bottom of the page, it is a footer. Figure 15-4, at the end of Chapter 15, uses footers. Many word processing programs allow you to set up headers and footers so that they appear automatically in the same place on each page. This text uses headers printed in dark blue to help you quickly find chapters as you look through the pages.

A typical header for a report might look like this:

Jane Fernstein
Feasibility Study
June 1994

A typical header for a letter might look like this:

Dr. Jieru Chen -2- June 16, 1994

The header does not appear on the title page of the report or on the first page of a letter. Most word processing programs allow you to start the header on the second page. Either you put the header at the beginning of the file and then suppress it for the first page, or you print it in the file at the top of the second page.

Consider Numbers with Your Headings In many companies and agencies, the standard for organizing reports and manuals is to include a numbering system with the headings. The three most commonly used systems are

- The traditional outline system
- The century-decade-unit system (often called the Navy System)
- The multiple-decimal system

Figure 9-25 shows the three systems.

The rationale for these systems is that you can refer to a section elsewhere in the report by its number. The numbering systems, however, have several disadvantages. In all these systems, if you want to add or remove a section, you have to renumber at least part of the report. Unless your software does this for you automatically, renumbering is tedious and highly susceptible to error. Many readers find it very difficult to follow these numbering systems, especially if you need more than three levels. The multiple-decimal system is particularly difficult for most people to use. For example, if you see a heading marked 1.1.1.1 (and some government reports can be found that go to a fifth level, 1.1.1.1.1), you will likely have a difficult time remembering what the main division 1.0 actually was.

Better alternatives are available now. You can show the hierarchy of heading levels distinctively with changes in type size and placement instead of numbers. Then the numbering is redundant to the design and wording of the headings. The example student report in this book, Figure 15-4 in Chapter 15, uses type size and placement of headings.

If you are not required to use a numbering system, we suggest that you not institute one. Figure 9-10 omits the numbering system used in Figure 9-9 without damaging the clarity of the presentation. In short, clear headings are more useful than numbers. However, many government agencies and many companies, especially those that prepare documents for the government, require one of these numbering systems. Therefore, it pays to be familiar with them.

If you use a numbering system with your headings, you must also put the numbers before the entries in your table of contents.

Using Word Processors

Generating documents on a computer has become the most common and practical way of developing memoranda, letters, and reports. Because of the growing number and power of word processing programs available for every computer, you need to learn the best ways of using word processing power.

Strategies for Using Word Processors Effectively

Word processors may have many advantages over typewriters and paper-and-pencil, but they won't by themselves make you a better

Traditional outline system

TITLE
I. FIRST-LEVEL HEADING
    A. Second-Level Heading
        1. Third-level heading
        2. Third-level heading
    B. Second-Level Heading

II. FIRST-LEVEL HEADING
    A. Second-Level Heading
        1. Third-level heading
        2. Third-level heading
    B. Second-Level Heading

Century-decade-unit system

TITLE
100 FIRST-LEVEL HEADING
    110 Second-Level Heading
        111 Third-level heading
        112 Third-level heading
    120 Second-Level Heading
200 FIRST-LEVEL HEADING
    210 Second-Level Heading
        211 Third-level heading
        212 Third-level heading
    220 Second-Level Heading

Multiple-decimal system

TITLE
1 FIRST-LEVEL HEADING
    1.1 Second-Level Heading
        1.1.1 Third-level heading
        1.1.2 Third-level heading
    1.2 Second-Level Heading
2 FIRST-LEVEL HEADING
    2.1 Second-Level Heading
        2.1.1 Third-level heading
        2.1.2 Third-level heading
    2.2 Second-Level Heading

Figure 9–25 Three Types of Numbering Systems

technical writer. As the composing process model in Chapter 2 shows, thinking, planning, and organizing are still the most crucial steps, and come before you ever begin the actual writing step.[13] The word processor, like the typewriter, is only a tool. It may be a fancier tool, but like the typewriter, it only puts on paper what you type.

Learn to Type You'll never regret knowing how to type. Even if you plan to be a scientist, engineer, technical specialist, or a manager or executive, you'll find yourself using a keyboard. Even if you use a computer with a different type of device, such as a mouse, a touch screen, or a light pen, you'll still need the keyboard for typing your text.

Be Aware of Your Writing Habits Many people find that when they first start using a computer, they get bogged down in revising too early. Because making changes is so easy, they spend too much time polishing single sentences instead of plowing ahead with the draft.

Learn to Compose by Using the Composing Process You can do much of your planning while you are sitting at your computer. For example, you can make a list of main ideas you want to include, if you are writing a short document, such as a memo or letter. Or, you can make a list of main sections you think will be needed in a report. Using the insert function, you can then place ideas beneath each main idea or main section. This method allows you to add ideas wherever they seem appropriate. You can use the delete function to quickly erase ideas. Many of the phrases you use in the list can be turned into headings and subheadings. Then, you can expand each idea by developing sentences and paragraphs under each heading and subheading.

Let's take a simple example to show how this works. Assume that you have to write a memo to everyone in your university professional society telling them about a meeting they are to attend and what they need to do before the meeting.

As you plan what you will say, you can type on the screen the information you must include in the memo:

Date, Time, Location
Purpose of Meeting
What to Do Before the Meeting

Adding to this basic list of ideas, you might come up with the following:

Date, Time, Location

We will meet Monday, February 17, 1993 at 7:00 P.M. in Room 300, Research Annex.

Purpose of the Meeting

The social committee needs to complete the agenda of all summer activities so that these can be presented to the Executive Committee for approval by the March 4 meeting. A final schedule of summer activities will be voted on March 19 and then made available to the general membership by April 1 to allow everyone plenty of time to include all club activities on their summer schedules.

What to Do Before the Meeting

Determine which events you are responsible for planning.
List three possible dates.
Check the university calendar for conflicts.

Bring at least two possible dates to the meeting.
Outline activities for each event. Bring ten copies to the meeting to hand out to committee members.

Note that you can add or delete information easily. After you write your draft, you can look for ways to enhance the format to improve readability, add any information you have forgotten, and add the appropriate memo format:

DATE: February 2, 1993
TO: Beta Club Members
FROM: Ben Irvin
SUBJECT: *Important* Meeting to Establish Summer Social Agenda

Date, Time, Location

We will meet Monday, February 17, 1993, 7:00 P.M. Room 300, Research Annex.

Purpose of the Meeting

The social committee needs to complete the agenda of all summer activities so that these can be presented to the Executive Committee for approval by the March 4 meeting. A final schedule of summer activities will be voted on March 19 and then made available to the general membership by April 1 to allow everyone plenty of time to include all club activities on their summer schedules.

What to Do Before the Meeting

1. Determine which events you are responsible for planning.
2. List three possible dates.
3. Check the university calendar for conflicts. Helen Brown in 401 Bowers Hall has copies of the calendar. Her number is 7205.
4. Bring at least two possible dates to the meeting.
5. Outline activities for each event. Bring ten copies to the meeting to hand out to committee members.

If you have any problems in attending this meeting, please *let me know by Wednesday, February 12. I will need your written suggestions even if you cannot attend.*

Think About the Overall Arrangement of Your Text With most computers, you can see only a small portion of your draft at one time. A typical computer screen shows you about one-third of a page of text. Sometimes, writers who compose with a word processor focus too much on the words that are in front of them. Their paragraphs are well structured, but the overall organization of the draft suffers. If you learn to plan and then compose on the computer, you will be less likely to lose sight of the general arrangement of your text, as you will build your text from a small model to a large version.

Print Out Your Drafts Often To check on the overall organization of your draft and the effectiveness of your document design, print out what you have written often. Then you can read through the entire draft, checking for coherence and consistency. How does the draft look? Does the format reveal the content? Can you see main ideas? Is the progression of ideas logical? Are the type size and arrangement appropriate for the information?

You cannot see your mistakes just by watching the screen. Printing your drafts frequently gives you an entirely new approach to your document.

Start Assignments Early Leaving the writing to the last minute is just as disastrous with a word processor as it is with a typewriter or paper-and-pencil. The computer may make typing your assignment easier and may help you to prepare a clean, neat copy, but it doesn't shorten the composing process. Effective writing still requires you to go through all the steps in the composing process.

Advantages of Using Word Processors

If you are hesitant about using word processors to compose your document, you need to consider the many advantages they offer:

You Can Edit and Revise Easily If you change your mind about what you want to say, you can type over what you have written. You can add to your text at any point. You can delete words, sentences, even whole paragraphs or pages. One major advantage over using paper-and-pencil or a typewriter is that, no matter how many changes you make, your copy always looks clean.

You Can Move Text Around Easily With many word processing programs, you can "cut and paste" so that reorganizing is easy. You can move words, sentences, paragraphs, or larger blocks. You can see how a section reads with one organization, try it in a different order, and then go back to the first if that was better.

You Can Copy Material from One Place to Another If you can cut and paste, you can probably also copy material from one place in your paper to another. In many programs, when you copy material, you copy both the text and the format of the text. This can be a handy feature for making sure that you use the same format for all the headings at a certain level. If you have set up a way to do the heading in one place but have trouble remembering how you did it, copy the first heading to the place for the second one and then edit it.

You Can Save What You Have Typed and Work on It Again Later On a typewriter, if you stop in the middle of a page, it is difficult to put the page back into the typewriter later and get it lined up exactly as it was before. With a computer or word processor, writing and printing are two separate stages of preparing the final product.

You Can Save Different Drafts of the Same Text You can save several versions, even on the same diskette, as long as you give each draft a different name. For example, if you are not sure what approach to take in the early stages of writing a report, you might try a few different approaches. You can save them all. You might want to save the drafts of an assignment to compare the final version to an earlier version. If you are working on a collaborative writing assignment, each writer can prepare and save his or her part, and then the group can put the entire assignment together on the computer.

You Can Search and Replace Words and Phrases Another advantage that many word processing programs give you is the ability to search for specific words and phrases and to replace them one at a time or all at once. For example, if you realize that the *Smith* you have been referring to actually spells his name *Smythe,* you can use the search and replace feature to change all the occurrences of *Smith* to *Smythe.*

You can also use the search and replace feature in interesting ways. Here are two:

1. *You can find out more about your own writing style.* Do you overuse a word, such as *however, is,* or *are*? You can search for the occurrences of each of these words to see how often you use them. Or, you can tell the computer to mark (underline) each word as it occurs, print your document, and study ways to eliminate words you overuse.
2. *You can check some specific common errors.* Do you sometimes type a possessive when you mean a plural? That is, do you sometimes type *student's* when you mean *students*? A program that checks for spelling errors won't find this type of error because *student's* is an English word; it's just not the right one here. You can search for all occurrences of the apostrophe (') and make sure that you have used each one correctly.

Recognizing the Power of Document Design

In several before/after examples, we have shown how principles of document design may enhance the readability of text, if they are used correctly. We have also shown how violation of document design can obscure the meaning of text. We now provide a real example of poor document design that causes the directive to fail in its effort to communicate.

Situation 1

Figure 9–26 was sent to all employees of Mega University informing them that a payroll change would be enacted. The memo instructs employees to complete the change form, which was attached, and return it.

The problem? Few employees returned the completed change form by the deadline, and many were returned to the wrong office. The memo was reissued shortly before the deadline, because of lack of compliance, with a phrase hand-written at the top: Please comply immediately. Phone lines were jammed with faculty trying to contact the office of the Controller.

What happened? Employees saw the subject line, which did not clearly state what the memo was about (or its importance). Most employees were already having the university automatically deposit their paychecks, so these employees did not read the memo. They assumed that automatic deposit and electronic deposit were the same. Lack of effective document design buried the important information items—what to do, why, and where to return the form.

The result? A major directive from the vice president's office was ignored because faculty did not read it and were not encouraged to do so. More than 2,500 memos had to be redistributed, and the electronic deposit of paychecks could not be initiated on time for all employees.

By employing document design principles discussed in this chapter, technical writing students were able to redesign the memo. Figures 9–27 and 9–28 represent two efforts by students like you. As you study them, you can see that each is effective but different. You can also see the how document design principles can reshape a document to instantly reveal the importance of the message.

Which revision to you like best? Study the original and design your own revision.

Situation 2

Now you are ready to try your hand at a revision on your own. Assume that the memo in Figure 9–29 was sent to instructors at your school. The memo announces that they have been selected to use the experimental teaching evaluation forms that the university wants to evaluate. The main part of the memo, paragraph 2, provides instructions on how to give the evaluations. The teachers who receive this memo are not expecting to receive it and are totally unfamiliar with the new teaching evaluation, a copy of which is attached to the memo.

Reformat the memo to help them administer the evaluation correctly and let them know what they should do after they give the evaluation.

Situation 3

Revise the memorandum in Figure 9–30, which is to be sent to plant engineering managers, so that it clearly explains the company's policy for paying for continuing education courses.

FISCAL DEPARTMENT MEMORANDUM NO. 92-15

TO: Vice Presidents, Deans, Directors, and Department Heads

SUBJECT: Electronic Deposit of Paychecks

Senate Bill 3, as passed during the first called session of the 72nd Legislature, mandated that the State Comptroller's Office pay state employees through the Federal Reserve System's Automated Clearing House system for those state employees paid from funds on deposit in the State Treasury. The Mega-State University System has decided to extend this procedure to all employees, regardless of funding source.

An informational memorandum and enrollment forms are currently being distributed to all Mega University employees through the payroll clerks in their campus departments. Employees are being asked to return completed enrollment forms to the Budget and Payroll Services Office, Room 009, YMCA Building, by May 1, 1994. In order to demonstrate compliance with Senate Bill 3 for audit purposes, all budgeted employees (including graduate assistants) will be required to complete and return the enrollment forms whether they request participation in or seek exemption from the direct deposit program. Wage employees (including student workers) are also eligible and must complete an enrollment form if they choose to participate.

Direct deposit will begin for biweekly-paid employees on the June 19 pay date and for monthly-paid employees on the July 1 pay date. Beginning with these dates, our previous practice of delivering checks to local financial institutions will be discontinued. Paychecks for employees who delay return of the enrollment form or request exemption from direct deposit will be given to their campus departments for distribution.

Provision of the direct deposit service should be beneficial to employees in the conduct of their financial affairs. Please encourage your employees to complete and return their enrollment forms by the May 1 deadline. If you have questions, please call Danny Smith or Ed Jones in the Budget and Payroll Services Office at 645-1711.

____

Jane Brunari,
Assistant Vice President & Controller

Figure 9–26 Original Memo—Situation 1

April 4, 1994

Fiscal Department Memorandum No. 92-15

TO: Vice Presidents, Deans, Directors, and Department Heads

FROM: Jane Brunari, Assistant Vice President & Controller

SUBJECT: **Mandatory Enrollment Forms for All Budgeted Employees Due May 1, 1994**

New Direct Deposit Enrollment Forms Required

The payroll clerks in all departments are distributing new direct deposit enrollment forms to employees. **All university budgeted employees (including graduate assistants)** must complete an enrollment form for our new direct deposit system, even if they are currently having their checks direct deposited. Those who do not wish to have their checks direct deposited must also fill out the new forms. **Wage employees (including student workers)** must complete an enrollment form only if they wish to have their checks direct deposited. Employees must return the enrollment form to the **Budget and Payroll Services Office, Room 009, YMCA Building by May 1, 1994.** Those employees who fail to return the completed forms by the deadline will automatically have their paychecks sent to their departments for distribution.

Description of New Direct Deposit Systems

Although the university has direct deposited employee paychecks in the past, we are changing our system to comply with the mandates of Senate Bill 3. This new direct deposit system (the Federal Reserve System's Automated Clearing House system) extends direct deposit to all state employees paid with State Treasury funds.

Transition from Old System to New Direct Deposit System

We will begin direct deposit under this new system on the **June 19** pay date (for biweekly-paid employees) and the **July 1** pay date (for monthly-paid employees). To facilitate the change from the old system to the new one, please encourage employees to complete the new form, even if they are currently having their checks delivered to local financial institutions or do not wish to have their checks direct deposited.

With your cooperation, the transition to the new direct deposit system should be smooth and relatively painless. If you have any questions about this new system, please call Danny Smith or Ed Jones in the Budget and Payroll Office at **645-1711**. Thank you for helping us with this process.

Figure 9–27 Revision 1—Situation 1

April 9, 1994

FISCAL DEPARTMENT MEMORANDUM NO. 92-15

TO: Vice Presidents, Deans, Directors, and Department Heads

FROM: Jane Brunari
Assistant Vice President & Controller

SUBJECT: Payroll Deadline for All Employees: May 1, 1994

New Payroll Procedure

As of June 19, 1994, all paychecks will be distributed through the Federal Reserve System's Automated Clearing House. To facilitate our change to the new system, the Budget and Payroll Services Office must receive a completed enrollment form from every budgeted employee (including graduate assistants) by May 1.

What You and Your Employees Must Do

- Pick up forms and informational memos for the new program from payroll clerks in all campus departments.
- Return completed forms to the Budget and Payroll Services Office, YMCA Building Room 009, by May 1, 1994.

Employees who delay returning the form will not receive their paychecks on time, regardless of whether or not they choose to participate. These employees, along with those who request exemption from direct deposit, will receive their paychecks through their campus departments. Paychecks will no longer be delivered to local financial institutions.

Employees already being paid through electronic deposit must complete an enrollment form to be entered into the new system. Wage employees (including student workers) may participate in the program, but are not required to do so.

If you have any questions, please call Ed Jones or Danny Smith in the Budget and Payroll Services Office at 645-1711.

Figure 9–28 Revision 2—Situation 1

DATE: November 11, 1994

FROM: Karen Jones
Associate Director of Testing

TO: Faculty

As you know, the university has made every effort to see that teaching evaluations, which are given once a year, are as accurate a reflection as possible on the effectiveness of your teaching. We know that this goal is your desire. To help us better achieve this goal, we are launching a pilot program to test a new kind of evaluation. You were one of 50 faculty who agreed to test the new evaluation system. Because you are getting this memorandum, you are one of the faculty chosen for the trial evaluations. After you receive your scores, we will send you a response form to allow you to express your views on the evaluation. We will then set up an interview with you so that we can more fully discuss your views of the accuracy of the results and changes you think should be made.

When you receive the questionnaire, a copy of it is attached, we want you to do a number of things. Please announce that the questionnaire will be given and urge students to attend class that day. If some students are absent the day you give the questionnaire, give those students a questionnaire the next class period. For the trial questionnaire, it is imperative that every student in the trial sections complete a questionnaire. Have someone else administer the questionnaire—either a colleague or your department secretary. Be sure you are not present while the students are completing the questionnaire. Have the person who is monitoring the questionnaire collect all of them and place them back in the envelope. These should be sealed in front of the students. The person monitoring the questionnaire should return these to the testing office (104 Haggarty) immediately after the test. The tests should be left at the test desk, which is the first desk on the right after you enter the office. Give the test to the clerk in charge of the trial test evaluation. Her name is Micki Nance. She will be there from 8–11 and 1–4 every class day during the test week, which will be the first week in December (December 2–6). Sign the sheet to indicate that you have returned your trial test. You will receive your printout by the first week in February. When you receive your printout, it will include a date that tells you when we will want to talk to you further. The response card, indicating your feelings about the accuracy of the trial evaluation, should be completed and returned immediately.

If you have any questions, please call Sammy Carson at ext. 9912.

Figure 9–29 Original Memo—Situation 2

TO: Plant Engineering Managers DATE: September 1, 1994

FROM: John Bridgers
District Superintendent

SUBJECT: Company Education Policy

Here is the company's policy on education, which many of you have asked about. Please keep it for your files for reference.

Policy 44.7. Advanced Education and Training. This policy applies to all employees except technicians and maintenance personnel. In order to encourage management personnel to achieve increasing professional competence in their disciplines and to enhance advancement potential, personnel who register for credit at the undergraduate or graduate level in accredited institutions will be reimbursed for tuition costs, registration fees, required textbooks, lab equipment, and other required materials upon completion of these courses. Certification that the specific course(s) will enhance the employee's professional growth must be provided by the employee's direct supervisor and countersigned by the supervisor's superior, unless the employee's supervisor holds the rank of vice president. Successful completion, defined as a grade of B or higher, must be attained in any course before the employee can apply for compensation. Costs of travel to the institution and costs of nonrequired materials such as paper, photocopies, and clerical help will not be reimbursed nor submitted to the company clerical workers. Submission to the Training Division of all receipts for all expenses, approval of the direct supervisor that the course fills the requirements of this policy, and documentation of successful completion are required before reimbursement will be permitted by the Training Division. Supervisors may allow release time for their employees to enroll in credit courses when work schedules permit. Release time is encouraged only when scheduled meetings of extremely important courses occur during regular working hours. If possible and necessary, personnel may be required to make up working time outside normal working hours. If the credit college course can be taken outside the individual's normal working hours, no release time will normally be allowed. To receive reimbursement, personnel should submit Training Division Form 6161 to the Training Division in accordance with the instructions on that form.

Figure 9–30 Original Memo—Situation 3

Planning and Revision Checklists

The following questions are a summary of the key points in this chapter, and they provide a checklist when you are planning and revising any document for your readers.

GENERAL QUESTIONS

Planning

- Have you planned the format?
- Have you considered how people will use your document?
- Have you checked on the software and hardware that you will use to prepare both drafts and final copy? (Do you know what options are available to you?)
- Have you found out whether you are expected to follow a standard format or style sheet?

- Have you thought about all the features (headings, pictures, tables) that you will have, and have you planned a page design that works well with those features?
- Have you thought about how you will make the arrangement obvious to your readers? (What will you do to make it easy for people to read selectively in your document?)

Revision

- Is your document clean, neat, and attractive?
- Is your text easy to read?
- Will your readers be able to find a particular section easily?
- If your document is supposed to conform to a standard, does it?

QUESTIONS ABOUT SETTING UP A USEFUL FORMAT

Planning

- Have you set the margins so that there is enough white space around the page, including space for binding, if necessary?
- Have you set the line length and line spacing for easy reading?
- Have you decided how you are going to show where a new paragraph begins?
- Have you decided whether to use a justified or ragged right margin and set the software appropriately?
- Have you planned which features to surround with extra white space, such as lists, tables, graphics, and examples?

Revision

- Have you left adequate margins? Have you left room for binding?
- Are the lines about 50–70 characters long? If you are using two columns, are the lines about 35 characters long in each column?
- Is the spacing between the lines and paragraphs consistent and appropriate?
- Can the reader tell easily where sections and paragraphs begin?
- Have you left the right margin ragged? If not, look over the paper to be sure that the justification has not made overly tight lines or left rivers of white space.
- Have you used the white space to help the reader find information on the page?
- Have you put white space around examples, warnings, pictures, and other special elements?
- Have you used lists for steps in procedures, options, and conditions?

QUESTIONS ABOUT MAKING THE TEXT READABLE

Planning

- Have you selected a type size and typeface that will make the document easy to read?
- Have you planned for highlighting? Have you decided which elements need to be highlighted and what type of highlighting to use for each?
- Do you know whether you can use color? If you can, have you planned what color to use and where to use it in the document?

Revision

- Is the text type large enough to be read easily?
- Have you been consistent in using one typeface?
- Have you used uppercase and lowercase letters for the text and for most levels of headings?
- Have you used highlighting functionally? Is the highlighting consistent? Does the highlighting make important elements stand out?

QUESTIONS ABOUT MAKING INFORMATION EASY TO LOCATE

Planning

- Have you planned your headings? Have you decided how many levels of headings you will need? Have you decided on the format for each level of heading?
- Will the format make it easy for readers to tell the difference between headings and text? Will the format make it easy for readers to tell one level of heading from another?
- Have you decided where to put the page numbers and what format to use?
- Have you decided on headers or footers (information at the top or bottom of each page)?
- Have you found out whether you are expected to use a numbering system? If you are, have you found out what system to use and what parts of the document to include?

Revision

- Have you checked the headings? Are the headings informative? Unambiguous? Consistent? Parallel?
- Will readers get an overall picture of the document by reading the headings?
- Is the hierarchy of the headings obvious?
- Can readers tell at a glance what is heading and what is text?

- If readers want to find a particular section quickly, will the size and placement of the headings help them?
- Have you checked the page breaks to be sure that you do not have a heading by itself at the bottom of a page?
- Are the pages numbered?
- Are there appropriate headers and footers?
- If you are using a numbering system, is it consistent and correct?

A QUESTION BEFORE YOU PRINT

- Does the printer have a clean ribbon or cartridge? Do you have enough of the correct size, color, and weight paper?

Exercises

1. Take an earlier assignment and revise it with the design guidelines you have learned in this chapter.
2. Have you recently seen a document that you found difficult to read or follow because it was not designed according to the guidelines in this chapter? Write a brief report to the company or agency that put out the document. Give specific recommendations for changing the design. Support your recommendations.
3. Redesign the document following your recommendations.
4. When college students receive a package of forms about financial aid, they also get a page of information about the financial aid program. Here are the headings from one state's version of this information page:

   The Financial Aid Program
   Eligibility
   Deferments
   Terms and Conditions
   Loan Institutions
   Eligible Schools
   Liability for Repayment
   Application Process

   Are these headings meaningful to you? Are they in the most logical order? Plan an information page that would be more useful to you. Write a set of headings that you would like to see on the information page. (Hints for this exercise: If you would ask questions, write the headings as questions. Don't just translate the nouns in this example into other words. Think about the content and arrangement that you, as the reader, would want to see.)

5. Prepare a tentative table of contents for your final report. Show the sentence style that you will use for each level of heading. Show the typography that you will use for each level of heading. Check your table of contents for parallelism.
6. Plan the page layout for your final report. Will you use a header or a footer? Where will you put the page numbers? Will you use a numbering system? If so, which one will you use? Prepare one or more sample pages using the word processor and printer that you will be using for the report. The pages do not have to have real text on them. For example, you can type, "Level One Heading," and "This is what the text will look like." The pages should indicate the margins, line length, spacing, type choices, and so on. Show all the levels of heading that you are planning to use. Also show how you will handle graphics with the text.

CHAPTER 10

DESIGN ELEMENTS OF REPORTS

Our approach to format is descriptive, not prescriptive; that is, we describe some of the more conventional practices found in technical reporting. We realize fully, and you should too, that many colleges, companies, and journals call for practices different from the ones we describe. Therefore, we do not recommend that you follow at all times the practices in this chapter. If you are a student, however, your instructor may, in the interests of class uniformity, insist that you follow this chapter fairly closely.

Chapter 9, "Document Design," discusses strategies for achieving good design. In this chapter, we discuss the elements—the tools you can use to carry out those strategies. We divide the elements into three groups: prefatory elements, main elements, and appendixes. Finally, we provide a section on documentation.

Prefatory Elements

Prefatory elements help your readers to get into your report. The letter of transmittal or preface may be the readers' first introduction to the report. The table of contents reveals the structure of your report. In the glossary, readers will find the definitions of terms that may be strange to them. All the prefatory elements discussed in this section contribute to the success of your report.

Letter of Transmittal and Preface

We have placed the letter of transmittal and preface together because they are often quite similar in content. They usually differ in format and intended audience only. You will use the letter of transmittal when the audience is a single person or a single group. Many of your reports in college will include a letter of transmittal to your professor, usually placed just before or after your title page. When on the job, you may handle the letter differently. Often, it is mailed before the report, as a notice that the report is coming, or it may be mailed at the same time as the report but under separate cover.

Generally, you will use the preface for a more general audience when you may not know specifically who will be reading your report. The preface or letter of transmittal introduces the reader to the report. It should be fairly brief. Always include the following basic elements:

- Statement of transmittal or submittal (included in the letter of transmittal only)
- Statement of authorization or occasion for report
- Statement of subject and purpose

Additionally, you may include some of the following elements:

- Acknowledgments
- Distribution list (list of those receiving the report—used in the letter of transmittal but not in the preface)
- Features of the report that may be of special interest or significance
- List of existing or future reports on the same subject

- Background material
- Summary of the report
- Special problems (including reasons for objectives not met)
- Financial implications
- Conclusions and recommendations

How many of the secondary elements you include depends upon the structure of your report. If your report's introduction or discussion includes background information, there may be no point in including such material in the preface or letter of transmittal. See Figures 10-1 and 10-2 for a sample letter of transmittal and a sample preface.

Gatlin Hall
Weaver University
Briand, MA 02139
July 27, 1994

Dr. Ross Alm
Associate Professor of Geography
Department of Geography
Weaver University
Briand, MA 02139

Dear Dr. Alm:

Statement of transmittal

Occasion for report

I submit the accompanying report entitled "Selected Characteristics of the People of the Commonwealth of Independent States" as the final project for Geography 334, The Commonwealth of Independent States.

Statement of subject and purpose

This report explores the characteristics of the People of the European states of the CIS (Russia, Belarus, Ukraine, Moldova, Georgia, Armenia, and Azerbaijan) and the Central Asian States (Kazakhstan, Uzbekistan, Turkmenistan, Kyrgyzstan, and Tajikistan). The report considers population size, nationality, ethnic divisions, language, literacy rate, and labor force skills. I have attempted to provide a base for understanding how differences in these major characteristics will influence the relationships of these states within the Commonwealth.

Acknowledgment

I am indebted to Professor Janet Mattson who has allowed me to draw on her unpublished monograph on the Commonwealth.

Sincerely,

Anne K. Chimato

Anne K. Chimato
Geography 334

Figure 10-1 Letter of Transmittal

PREFACE

Occasion for report

Statement of subject and purpose

Feature of special interest

Another report on similar subject

Following the breakup of the USSR, the Commonwealth of Independent States (CIS) was formed from the former Soviet republics. These states have a need and a desire to coexist in harmony. How well they can do that will depend in some measure on how well these diverse states can work together without the power of a central government to hold them together. This report—part of the Central Intelligence Agency's public information program—lays a groundwork for understanding the difficulties involved by exploring the characteristics of the people of the European states of the CIS (Russia, Belarus, Ukraine, Moldova, Georgia, Armenia, and Azerbaijan) and the Central Asian States (Kazakhstan, Uzbekistan, Turkmenistan, Kyrgyzstan, and Tajikistan). The report considers population size, nationality, ethnic divisions, language, literacy rate, and labor force skills.

Of particular interest is the wide distribution of different ethnic groups within these new states. Such ethnic divisions may cause future conflict as has been seen in Bosnia and in the dispute between Armenia and Azerbaijan over the ethnically Armenian enclave of Nagorno-Karabakh in Azerbaijan.

For detailed information on the geography, government, economy, and defense force of each of these states, see The World Factbook, published yearly by the Central Intelligence Agency and available through the U.S. Government Printing Office, Superintendent of Documents, Washington, DC 20402.

Figure 10–2 Preface

If the report is to remain within an organization, the letter of transmittal will become a memorandum of transmittal. This changes nothing but the format. See Chapter 12, "Correspondence," for letter and memorandum formats.

Cover

A report's cover serves three purposes. The first two are functional and the third aesthetic and psychological.

First, covers protect pages during handling and storage. Pages ruck up, become soiled and damaged, and may eventually be lost if they are not protected by covers. Second, because they are what readers first see as they pick up a report, covers are the appropriate place to display identifying information such as the report title, the company or agency by or for which the report was prepared, and security notices if the report contains proprietary or classified information. Incidentally, students should not print this sort of information directly on the cover. Rather, they should put the information onto gummed labels readily obtainable at the college bookstore, and then fasten the labels to the cover. A student label might look like the one in Figure 10–3.

SELECTED CHARACTERISTICS OF THE PEOPLE OF THE COMMONWEALTH OF INDEPENDENT STATES

by

Anne K. Chimato

Geography 334 July 27, 1994

Figure 10–3 Student Label

Finally, covers bestow dignity, authority, and attractiveness. They bind a bundle of manuscript pages into a finished work that looks and feels like a report and has some of the characteristics of a printed and bound book.

Suitable covers need not be expensive and sometimes should not be. Students, particularly, should avoid being pretentious. All three purposes may be served by covers of plastic or light cardboard, perhaps of 30 or 40 pound substance. Students can buy such covers in a variety of sizes, colors, and finishes.

While you are formatting your report, remember that when you fasten it into its cover, about an inch of the left margin will be lost. If you want an inch of margin, you must leave two inches on your paper. Readers grow irritated when they must exert brute force to bend open the covers to see the full page of text.

Title Page

Like report covers, title pages perform several functions. They dignify the reports they preface, of course, but far more important, they provide identifying matter and help to orient the report users to their reading tasks.

To give dignity, a title page must be attractive and well designed. Symmetry, balance, and neatness are important. The most important items should be boldly printed; items of lesser importance should be subordinated. These objectives are sometimes at war with the objective of giving the report users all the data they may want to see at once. Here, we have listed in fairly random order the items that sometimes appear on title pages. A student paper, of course, would not require all or even most of these items. The first four listed are usually sufficient for simple title pages.

- Name of the company (or student) preparing the report
- Name of the company (or instructor and course) for which the report was prepared
- Title and sometimes subtitle of the report
- Date of submission or publication of the report
- Code number of the report
- Contract numbers under which the work was done
- List of contributors to the report (minor authors)
- Name and signature of the authorizing officer
- Company or agency logo and other decorative matter
- Proprietary and security notices
- Abstract
- Library identification number
- Reproduction restrictions
- Distribution list (A list of those receiving the report. If the letter of transmittal does not contain this information, the title page should.)

Of course, placing all of these items on an 8½ by 11 inch page would guarantee a cluttered appearance. Put down what you must, but no more.

Word processing now allows report writers to use different type sizes and styles on a title page to indicate what is important and what is subordinate. Use this capability discreetly. Don't turn your title page (or any other part of your report) into a jumble of different typefaces. Generally, different type sizes and the use of boldface and plain style will suffice.

In Figure 10-4 we illustrate a title page in plain style. Figure 10-5 illustrates the use of word processing.

Pay particular attention to the wording of your title. Titles should be brief but descriptive and specific. The reader should know from the title what the report is about. A title such as "Effects of Incubation Temperatures on Sexual Differentiation in Turtles: A Geographic Comparison" is illustrative. From it, you know specifically the research being reported. To see how effectively this title works, leave portions of it out, and see how quickly your understanding of what the article contains changes. For example, "Sexual Differentiation in Turtles" would suggest a much more comprehensive report than does the actual title. A title such as "Effects of Incubation Temperatures" could as well be about chickens as turtles. On the other hand, adding the words "An Investigation into" to the beginning of the title would add nothing useful. The test of whether a title is too long or too short isn't in the number of words it contains but what happens if words are deleted or added. Keep your titles as brief as possible, but make sure they do the job.

SELECTED CHARACTERISTICS OF THE PEOPLE OF
THE COMMONWEALTH OF INDEPENDENT STATES

Prepared for

Dr. Ross Alm

Associate Professor of Geography

Geography 334

The Commonwealth of Independent States

by

Anne K. Chimato

Abstract

This report explores selected characteristics of the people of the European states of the CIS (Russia, Belarus, Ukraine, Moldova, Georgia, Armenia, and Azerbaijan) and the Central Asian States (Kazakhstan, Uzbekistan, Turkmenistan, Kyrgyzstan, and Tajikistan). The report considers population size, nationality, ethnic divisions, language, literacy rate, and labor force skills. It provides a base for understanding how differences in these major characteristics will influence the relationships of these states within the Commonwealth.

July 27, 1994

Figure 10–4 Typed Title Page

Table of Contents

A table of contents (TOC) performs at least three major functions. Its most obvious function is to indicate the page on which discussion of

Final Report of the Committee on Word Processing Alternatives for the Oxford Insurance Company

Prepared for
Dennis Colcombet
Vice President for Information Services

Committee Members
Betty Robinett, Chair
Terry Collins
Ann Duin
Donald Ross

10 December 1994

Distribution List
Ann Bailly, Office Manager
Vickie Mikelonis, Computer Technician
Arthur Walzer, Purchasing Officer

Figure 10–5 Word Processed Title Page

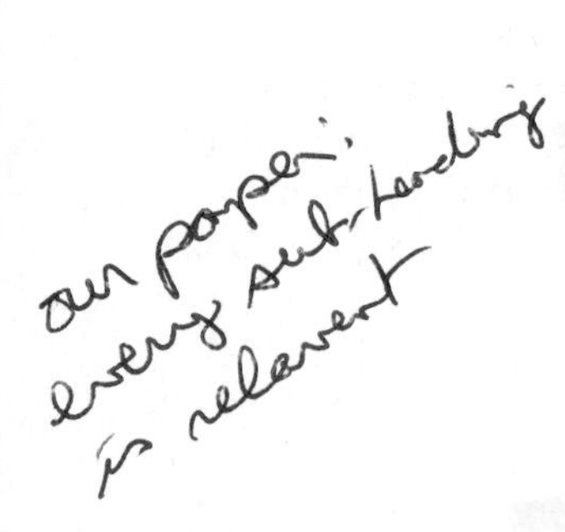

each major topic begins; that is, it serves the reader as a locating device. Less obviously, a TOC forecasts the extent and nature of the topical coverage and suggests the logic of the arrangement and the relationship of the parts. Still earlier, in the prewriting stage, provisional drafts of the TOC enable the author to "think on paper"; they act as outlines to guide the composition.

A system of numbers, letters, type styles, indentations, and other mechanical aids has to be selected so that the TOC will perform its intended functions. Figure 10–6 shows a TOC suitable for student reports. We have annotated the figure to draw your attention to a few key points. However, the annotations are suggestions only. There are many acceptable variations in TOCs. For example, the leader dots in Figure 10–6 are used to carry the reader's eyes from the end of each title to the

Make all major headings distinctive. All capitals or boldface are good choices.

Line up all numbers, arabic and roman, on right-hand digits.

Show hierarchy with indentation. If you have used a numbering system in the report, repeat it in table of contents.

In subheadings, capitalize first and last words and all principal words. Do not capitalize articles, prepositions, and coordinating conjunctions unless they are a first or last word.

CONTENTS

Page

LIST OF ILLUSTRATIONS.................................iii
INTRODUCTION.................................1
EUROPEAN STATES.................................2
    Population Size.................................4
    Nationality.................................6
    Ethnic Divisions.................................7
    Language.................................9
    Literacy Rate.................................10
    Labor Force Skills.................................11
CENTRAL ASIAN STATES.................................12
    Population Size.................................14
    Nationality.................................15
    Ethnic Divisions.................................16
    Language.................................18
    Literacy Rate.................................19
    Labor Force Skills.................................20
SUMMARY.................................21
NOTES.................................22

ii

Figure 10–6 Typed Table of Contents

page number and to tie the page together visually. However, this practice is by no means universal. Some people feel the leader dots clutter the page, and therefore do not use them. If you use a numbering system in your report (see pages 218–219), the TOC should reflect that system.

In Figure 10–7, we reproduce a professionally created TOC that shows a good use of word processing capabilities. When you design your own TOC avoid overcrowding. Seldom is there justification for going beyond three levels of headings; beyond three levels, users have almost as much trouble locating items in the TOC as they have in locating them by flipping through the report. Be sure that page numbers in your TOC match the page numbers in your final draft. Remember that the wording of the TOC entries and the headings on the text pages must be exactly the same. Every entry in the TOC must also be in the report. However, every heading in the report need not be in the TOC. That is, if you have four levels of headings in your report, you might list only the top three levels in your TOC.

Contents

v

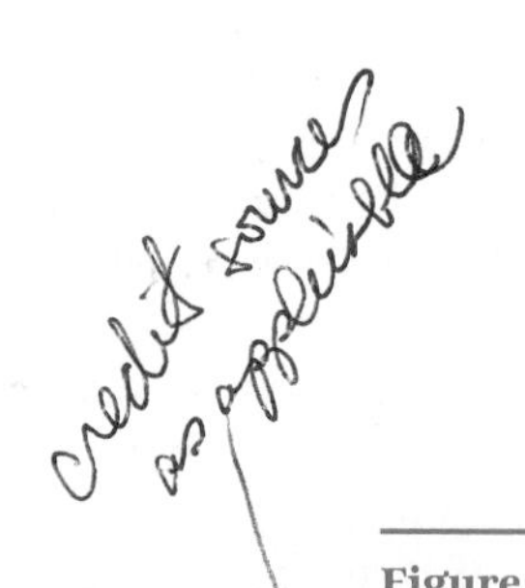

Figure 10–7 Word Processed Table of Contents
Source: U.S. Department of Health and Human Services, *The Public Health Consequences of Disasters 1989* (Atlanta: Centers for Disease Control, 1989) v.

List of Illustrations

If a report contains more than a few illustrations, say more than three or four, it is customary to list the illustrations on a separate page or on the TOC page. Illustrations are of two major types: tables and figures. A

Line up all illustration and page numbers, arabic and roman, on the right-hand digit.

In titles, capitalize first and last words and all principal words. Do not capitalize articles, prepositions, and coordinating conjunctions unless they are a first or last word.

When titles run two or more lines, indent all lines after the first.

ILLUSTRATIONS

Figure 10–8 List of Illustrations

table is any array of data, often numerical, arranged vertically in columns and horizontally in rows, together with the necessary headings and notes. **Figures** include photographs, maps, graphs, organization charts, and flow diagrams—literally any illustration that does not qualify as a table by the preceding definition. (For further details, see Chapter 11, "Graphical Elements.")

If the report contains both tables and figures, it is customary to use the page heading "Illustrations" or "Exhibits" and to list all the figures first and then all the tables. If you have all of one kind of figure in your report, you can use the appropriate term as your heading; for example, "Maps" is used as a heading in Figure 10-8. If you have various kinds of figures, use "Figures" as your heading.

Illustration titles should be as brief and yet as self-explanatory as possible. Avoid a cumbersome expression such as "A Figure Showing Characteristic Thunderstorm Recording." Say, simply, "Characteristic Thunderstorm Recording." On the other hand, do not be overly economical and write just "Characteristic" or "A Comparison." At best, such generic titles are only vaguely suggestive.

Figure 10-8 shows a simple list that should satisfy most needs. Notice in the figure that arabic numerals are used for figures and roman numerals for tables. This practice is common but by no means standard. Make sure that your page numbers are correct and that the titles listed accurately repeat the titles in the report.

Glossary and List of Symbols

Reports dealing with technical and specialized subject matter often include abbreviations, acronyms, symbols, and terms not known to the nonspecialist. Thus, a communication problem arises. Technically trained persons have an unfortunate habit of assuming that what is well known to them is well known to others. This assumption is seldom justified. Terms, symbols, and abbreviations change in meaning with time and context. In one context, ASA may stand for American Standards Association; in another context, for Army Security Agency. The letter K may stand for Kelvin or some mathematical constant. The meaning given to Greek letters may change from one report to the next, even though both were done by the same person.

Furthermore, writers seldom have complete control over who will read their reports. A report intended for an engineering audience may have to be read by managers, legal advisors, or sales representatives. It is wise to play it safe by including a **list of symbols** or a **glossary** or both. Readers who do not need these aids can easily ignore them; those who do need them will be immeasurably grateful.

The list of symbols is normally a prefatory element. A glossary may be a prefatory element or placed as an appendix at the end of the report. When you first use a symbol found in your list of symbols or a term found in the glossary, tell your reader where to find the list or the glossary. Figure 10-9 illustrates a list of symbols. Figure 10-10 illustrates a glossary.

Abstracts and Summaries

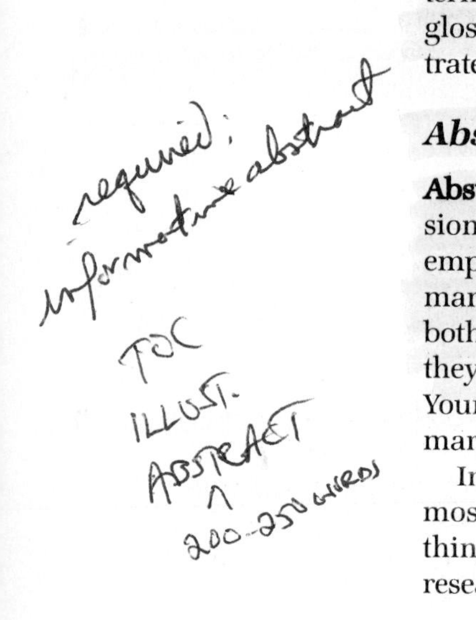

Abstracts and **summaries** are overviews of the facts, results, conclusions, and recommendations of a report. In many formats, such as empirical research reports and feasibility reports, abstracts or summaries will be placed near the front of the report. In that position, they both summarize the report and allow busy readers to decide whether they want to read further. As we pointed out in Chapter 4, "Writing for Your Readers," both executives and experts expect the abstract or summary to come early in the report.

In more discursive reports, such as magazine articles, summaries most often come at the end of the report, where they serve to draw things together for the reader. Many reports, particularly empirical research reports, have an abstract at the beginning and a summary at

SYMBOLS

A	Mass number
A.W.	Atomic weight
c	Velocity of light (2.998 x 10^{10} cm/sec)
D	H^2 atom (deuterium)
E	Energy
e-	Electron
e	Electronic charge (1.602 x 10^{-10} abs. coulomb)
ev	Electron volt
F	Free energy
(g)	Gas phase
H	Heat content
h	Planck's constant (6.624 x 10^{-27} erg/sec)
I_{sp}	Specific impulse
k	Boltzmann's constant (1.3805 x 10^{-16} erg/deg)
ln	Natural logarithm

* * * *

α	Alpha particle
γ	Gamma ray
ζ	Bond energy
μ	Micro
ρ	Density

v

Figure 10–9 List of Symbols

the end. These facts raise the question of when an overview is an "abstract" and when it is a "summary." In general, these principles hold true:

- Abstracts are placed before technical reports, such as empirical research reports, meant for technical audiences.
- Summaries are placed before business and organizational reports, such as proposals and feasibility reports. When the audience is primarily an executive audience, the summary will be known as an executive summary.

GLOSSARY

Use parallel sentence fragments for glossary definitions. (See "Parallelism" section in Handbook.)

Use complete sentences to add information to definition.

Btu	the amount of heat required to raise the temperature of one pound of water one degree Fahrenheit.
degree day	a temperature standard around which temperature variations are measured.
design temperature	the maximum reasonable temperature expected during the heating or cooling season. Design calculations are based upon this number.
heat transmission coefficient	the quantity of heat in Btu transmitted per hour through one square foot of a building surface.
infiltration	the air leaking into a building from cracks around doors and windows.
sensible heat	heat that the human body can sense.
thermal conductivity	the quantity of heat in Btu transmitted by conduction per hour through one square foot of a homogeneous material for each degree Fahrenheit difference between the surfaces of the material.
thermal resistance	the reciprocal of thermal conductivity.

vii

Figure 10–10 Glossary

- An overview placed at the end of a report will probably be called a "summary."

Because abstracts and executive summaries always appear as prefatory elements, we discuss them in this section. We discuss other types of summaries and conclusions and recommendations on pages 257–262, where we discuss how to end a report.

Abstracts Discussed here are abstract style and two kinds of abstracts: **informative** and **descriptive.**

Never use "I" statements in either kind of abstract. Report your information impersonally, as though it were written by someone else. The informative abstract in Figure 10–11 shows the style. This is not an arbitrary principle. If you publish your report, your abstract will probably be reprinted in an abstracting journal, where the use of "I" is inappropriate. Also, many companies, in the interest of good intracompany

ABSTRACT

Research objectives

Methodology

Findings

This study investigated the role of signaling in helping good readers comprehend expository text. As the existing literature on signaling, reviewed in the last issue of the Journal, pointed to deficiencies in previous studies' methodologies, one goal of this study was to refine prose research methods. Two passages were designed in one of eight signaled versions each. The design was constructed to assess the individual and combined effect of headings, previews, and logical connectives. The study also assessed the effect of passage length, familiarity, and difficulty. The results showed that signals do improve a reader's comprehension, particularly comprehension two weeks after the reading of a passage and comprehension of subordinate and superordinate inferential information. This study supports the hypothesis that signals can influence retention of text-based information, particularly with long, unfamiliar, or difficult passages.

Figure 10–11 Informative Abstract
Source: Jan H. Spyridakis, "Signaling Effect: Increased Content Retention and New Answers—Part II," *Journal of Technical Writing and Communication* 19 (1989): 395.

communication, publish the abstracts of all company research reports. The restriction on the use of "I" makes the use of passive voice common in abstracts. Because your full report contains complete documentation, you need not footnote or otherwise document the information in abstracts.

Figure 10–11 illustrates an **informative abstract.** Informative abstracts are most often intended for an expert audience; therefore, their authors can use the technical language of the field freely. Like most informative abstracts, the one in Figure 10–11 summarizes three major elements of the full report:

- The objectives of the research or the report
- The methodology used in the research
- The findings of the report, including the results, conclusions and recommendations

Most professional journals or societies publish stylebooks that include specifications about how to write an abstract. Many journals, because of high publication costs, will set an arbitrary limit of 200 words for abstracts. Because abstract writing uses many of the same techniques as summary writing, you might want to read what we say about summaries on pages 255–260.

The main purpose of the **descriptive abstract** is to help busy readers decide whether they need or want the information in the report enough to read it entirely. The descriptive abstract merely tells what the full report contains. Unlike the informative abstract, it cannot serve as a substitute for the report itself. Many reports contain descriptive abstracts, and many abstracting journals print them. The abstract in

Figure 10–4 and the following example are typical of the content and style of a descriptive abstract:

> The management of the process by which technical documents are produced usually proceeds according to one of two models, the "division of labor" model or the "integrated team" model. This article reports on a survey that suggests the prevalence of each model and that gives insights into how the choice of a management model affects the practice of technical communication and the attitudes of technical communicators.[1]

The descriptive abstract discusses the *report,* not the subject. After reading this abstract, you know that this article "gives insights into how the choice of a management model affects the practice of technical communication and the attitudes of technical communicators," but you must read the article to gain the insights. Whether a report is 10 or 1,000 pages long, a descriptive abstract can cover the material in less than 10 lines.

Executive Summary Placed at the front of a report, the **executive summary** ensures that the points of the report important to an executive audience are immediately accessible. To that end, it is written in a nontechnical language suited to an executive audience. Seldom more than one page long, double spaced, it emphasizes the material that executives need in their decision-making process. It need not summarize all the sections of the report. For example, writing for a combined audience of scientists and executives, a writer might include a theory section in a report. The executive summary might skip this section altogether or treat it very briefly.

In their decision making, executives weigh things such as markets, risks, rewards, costs, and people. If your report recommends buying new equipment, they want to be assured that you have examined all reasonable alternatives and considered cost, productivity, efficiency, profits, and staffing. If you are reporting research, executives take your methodology for granted. They care very little for the physics, chemistry, or sociology behind a development. What they want to know are your results and the implications of those results for the organization.

Figure 10–12 lists the questions that executives ask in different situations and that, therefore, an executive summary should answer. Before writing an executive summary, look at that figure and read pages 59–61, where we discuss the reading habits of executives. The annotated executive summary in Figure 10–13 illustrates the technique and the major parts of an executive summary.

Place an executive summary immediately before the introduction and label it "Summary" or "Executive Summary." In short reports and memorandum reports, the executive summary often replaces the introduction and is followed immediately by the major discussion.

Problems
What is it?
Why undertaken?
Magnitude and importance?
What is being done? By whom?
Approaches used?
Thorough and complete?
Suggested solution? Best? Consider others?
What now?
Who does it?
Time factors?

New Projects and Products
Potential?
Risks?
Scope of application?
Commercial implications?
Competition?
Importance to Company?
More work to be done? Any problems?
Required manpower, facilities, and equipment?
Relative importance to other projects or products?
Life of project or product line?
Effect on Westinghouse technical position?
Priorities required?
Proposed schedule?
Target date?

Tests and Experiments
What tested or investigated?
Why? How?
What did it show?
Better ways?
Conclusions? Recommendations?
Implications to Company?

Materials and Processes
Properties, characteristics, capabilities?
Limitations?
Use requirements and environment?
Areas and scope of application?
Cost factors?
Availability and sources?
What else will do it?
Problems in using?
Significance of application to Company?

Field Troubles and Special Design Problems
Specific equipment involved?
What trouble developed? Any trouble history?
How much involved?
Responsibility? Others? Westinghouse?
What is needed?
Special requirements and environment?
Who does it? Time factors?
Most practical solution? Recommended action?
Suggested product design changes?

Figure 10–12 What Managers Want to Know
Source: James W. Souther, "What to Report," *IEEE Transactions on Professional Communication* PC-28 (1985): 6.

Main Elements

The body of a report contains detailed information and interpretation. The body needs to be introduced and, normally, finished off with an ending of some sort that may include a summary, conclusions, recommendations, or simply a graceful exit from the report. We discuss all these elements in this section on main elements.

Introduction

A good introduction forecasts what is to follow in the rest of the report. It directs the reader's mind to the subject and purpose. It sets limits on the scope of the subject matter and reveals the plan of development of the report. Early in your paper, you also give any needed theoretical or historical background, and this is sometimes included as part of the introduction.

Subject Never begin an introduction with a superfluous statement. The writer who is doing a paper on Read Only Memory in computers and begins with the statement "The study of computers is a vital and

SUMMARY

Problem definition

The University is steadily falling behind in the faculty and student use of computers. Our existing computer labs have insufficient numbers of computers, and those we do have are badly dated. We have faculty who are capable of designing computer programs for instructional use but who are reluctant to do so because their students do not have access to computers. Too many graduates are leaving the University as computer illiterates.

Information Resources has considered three solutions to the problem:

Alternatives considered

1. Require all freshmen to buy a microcomputer at an approximate cost of $1,500 each. At an interest rate of 8%, students could repay the University for their computers in 16 quarterly payments of $108.31 each.

2. Provide those students and faculty who want them with microcomputers through the University bookstore at deep discounts. Purchasers would arrange their own financing if needed.

3. Upgrade the University computer labs by providing $1 million over the next fiscal year to provide microcomputers, printers, software, and new furniture. Student computer lab fees of $25 per quarter will pay the cost of material and employees to run the labs.

Recommendation

Effect of recommendation

We recommend both alternatives 2 and 3. We reject alternative 1 on the grounds that we are a public school and must not put educational costs out of reach for our students. Solutions 2 and 3 would make enough computers available for the immediate future to encourage their use by both students and faculty.

Figure 10–13 Executive Summary

interesting one" has wasted the readers' time and probably annoyed them as well. Announce your specific subject loud and clear as early as possible in the introduction, preferably in the very first sentence. The sentence "This paper will discuss several of the more significant applications of the exploding wire phenomenon to modern science" may not be very subtle, but it gets the job done. The reader knows what the subject is. Often, in conjunction with the statement of your subject, you will also need to define some important terms that may be unfamiliar to your readers. For example, the student who wrote the foregoing sentence followed it with these two:

> A study of the exploding wire phenomenon is a study of the body of knowledge and inquiry around the explosion of fine metal wires by a sudden and large pulse of current. The explosion is accompanied by physical manifestations in the form of a loud noise, shock waves, intense light for a short period, and high temperatures.

In three sentences the writer announced the subject and defined it. The paper is well under way.

Sometimes, particularly if you are writing for nonspecialists, you may introduce your subject with an interest-catching step. The following example introduces an article on fighting today's forest fires with a World War II bomber.

> It's a losing battle, attempting to complete a fire break before the prevailing wind blows a raging forest fire into the laps of you and your coworkers. Only 300 yards away, the fire crackles threats of destruction. Relief comes with a sudden sound overhead. Flying straight for you, just above treetop level, thunders a huge four-engine bomber.
>
> Had this been 1945, you might have been terrified by the same sight, had you been a Japanese soldier in the South Pacific. Now, however, forty-six years later, the same aircraft, a Consolidated PB4Y-2 Privateer, is on its way to dispatch a different enemy. Dropping thousands of gallons of fire retardant on the blaze that menaces the lives of the forest and your crew, the bomber effectively controls the leading edge of the fire, giving you a chance to do your job.[2]

Or you may simply extract a particularly interesting fact from the main body of your paper. For example,

> Last year, the Federal Aviation Administration attributed 16 aircraft accidents to clear air turbulence. What is known about this unseen menace that can cripple an aircraft, perhaps fatally, almost without warning?

In this example, the writer catches your interest by citing the accident rate caused by clear air turbulence and nails the subject down with a rhetorical question in the second sentence. Interest-catching introductions are used in brochures, advertisements, and magazine and newspaper articles. You will rarely see an interest-catching introduction in business reports or professional journals. If you do, it will usually be a short one.

Purpose Your statement of purpose tells the reader *why* you are writing about the subject you have announced. By so doing, you also answer the reader's question "Should I read this paper or not?" For example, a Department of Labor report on job hunting includes this sentence: "Whether you are involuntarily unemployed, changing jobs, or looking for your first job, this Guide is designed to help you negotiate the many phases of the job search process."[3] Readers who have no reason to be interested in such a discussion will know there is no purpose in reading further.

Another way to understand the purpose statement is to realize that it often deals with the *significance* of the subject. Writers who had human–computer interaction as their subject announced their purpose this way:

> Why take yet another look at the way humans and computers interact? Because an important part of human–computer interaction is the way that people feel about themselves and the computer's actions both during the interaction and after it is completed. These feelings affect their decision to buy or use a program, their attitude toward computing, and their effectiveness in future interactions. People may be happy and comfortable with the dialog, or they may not be. They may feel belittled and bewildered, or they may feel they are leading along a plodding machine that does not serve them well. Parallels can be drawn between human–computer dialogs and the dialogs that people have with each other.[4]

Scope The statement of scope further qualifies the subject. It announces how broad the treatment of the subject will be. Often it indicates the level of competence expected in the reader for whom the paper is designed. For example, a student who wrote, "In this report I explain the application of superconductivity in electric power systems in a manner suitable for college undergraduates" declared his scope as well as his purpose. He is limiting the scope to superconductivity in electric power systems and stating that his target audience is not composed of high school students or graduate physicists but of college undergraduates.

Plan of Development In a plan of development, you forecast your report's organization and content. The principle of psychological reinforcement is at work here. If you tell your readers what you are going to cover, they will be more ready to comprehend as they read along. The following, taken from the introduction to a paper on enriching flour with iron, is a good example of a plan of development:

> This study presents a basic introduction to three major areas of concern about iron enrichment: (1) which form of iron is most suitable; (2) potential health risks from overdoses of iron, such as cardiovascular disease, hemochromatosis, and the masking of certain disorders; and (3) ignorance of the definitions, extent, and causes of iron deficiency.

You need not think of the announcement of subject, purpose, scope, and plan of development as four separate steps. Often, subject and purpose or scope and plan of development can be combined. In a short paper, two or three sentences might cover all four points, as in this example:

Subject

Purpose

Scope and plan of development

> Concern has been expressed recently over the possible presence in our food supply of a class of chemicals known as "phthalates" (or phthalate esters). To help assess the true significance of this concern as to the safety and wholesomeness of food, this article discusses the reasons phthalates are used, summarizes their toxicological properties, and evaluates their use in food packaging materials in the context of their total use in the environment.[5]

Also, introductions to specialized reports may have peculiarities of their own. These will be discussed in Part IV, "Applications."

Theoretical or Historical Background When theoretical or historical background is not too lengthy, you can incorporate it into your introduction. In the following excellent introduction, the writer begins with background, makes his subject and purpose clear, and closes with a paragraph that states his scope and plan of development:

> In recent years, two questions have received a good deal of attention in the field of business and technical writing, or professional communication, as I will call it. One question is old and one new, but both are closely related—at least to professional communication teachers and researchers who are trained in the profession of literary criticism. The first question is as old as technical and business writing courses, dating back to the 1910s: To what extent, if any, should business and technical writing courses serve the pragmatic needs of business and industry, and to what extent, if any, should those courses teach the concerns of literary studies? The second question is relatively new, but it has received a great deal of attention in recent years: What is the responsibility, if any, of the instructor of these courses to teach ethics? The two questions are related in complex ways because, for some teachers in the field of professional communication, putting business and technical writing courses at the service of business and industry is viewed as ethically suspect, and there have been a number of articles recently that argue or suggest that business and technical writing courses for students majoring in science, technology, and business should ask those students to critique the ethical basis of science, business, and industry from what is essentially the perspective of literary studies, as we shall see.
>
> Let me say from the outset that I believe that all teachers—and all professions and all institutions and indeed all human beings—have a responsibility to promote ethical behavior. So, too, every profession, institution, and human being ought to engage in critical reflection at times. The question is not whether teachers, courses, disciplines, professions, and institutions should promote ethical behavior but how, when, and for what purposes they should be promoted. These are far more complex questions, and they cannot be answered without instructors considering their methods, timing, and motives for raising ethical issues. In this article, I want to point out some potential difficulties—which are essentially ethical difficulties—in literature-trained faculty teaching ethics in professional communication courses, and I will warn against a too-hasty—uncritical, if you will—pursuit of certain kinds of critical reflection as a goal of these courses.
>
> First, I will examine the historical and an institutional context of my two central questions and look at two recent answers to them. Then I will turn to an obscure chapter in the history of business and technical communication—the teaching of "engineering publicity" at Massachusetts Institute of Technology (MIT) in the early 1920s—to see the unusual answer of one institution to these questions at a crucial point in its history. Finally, I will suggest why I believe that answer is worth serious consideration by those of us involved in writing instruction and curricular planning for students who will enter business and professional communities, both outside and within academia (for we must remember that academics—even those in literary studies—are professionals as well).[6]

If the background material is extensive, however, it more properly becomes part of the body of your paper.

Discussion

The discussion will be the longest section of your report. Your purpose and your content will largely determine the form of this section. Therefore, we can prescribe no set form for it. In presenting and discussing your information, you will use one or more of the techniques described in Part II, "Techniques," or the special techniques described in Part IV, "Applications." In addition to your text, you will probably also use headings (see pages 209–216) and visual aids such as graphs, tables, and illustrations.

When thinking about your discussion, remember that almost every technical report answers a question or questions: What is the best method of desalination to create a water supply at an overseas military base? How are substances created in a cell's cytoplasm carried through a cell's membranes? What is the nature of life on the ocean floor? How does single parenthood affect the children in the family? Ask and answer the reporter's old standbys: Who? What? When? Where? Why? How? Use the always important "so-what?" to explore the implications of your information. However you approach your discussion, project yourself into the minds of your readers. What questions do they need answered to understand your discussion? What details do they need to follow your argument? You will find that you must walk a narrow line between too little detail and too much.

Too little detail is really not measured in bulk but in missing links in your chain of discussion. You must supply enough detail to lead the reader up to your level of competence. You are most likely to leave out crucial details at some basic point that, because of your familiarity with the subject, you assume to be common knowledge. If in doubt about the reader's competence at any point, take the time to define and explain.

Many reasons exist for too much detail, and almost all stem from writers' inability to edit their own work. When you realize that something is irrelevant to your discussion, discard it. It hurts, but the best writers will often throw away thousands of words, representing hours or even days of work.

You must always ask yourself questions like these: Does this information have significance, directly or indirectly, for the subject I am explaining or for the question I am answering? Does the information move the discussion forward? Does it enhance the credibility of the report? Does it support my conclusions? If you don't have a yes answer to one or more of these questions, the information has no place in the report, no matter how many hours of research it cost you.

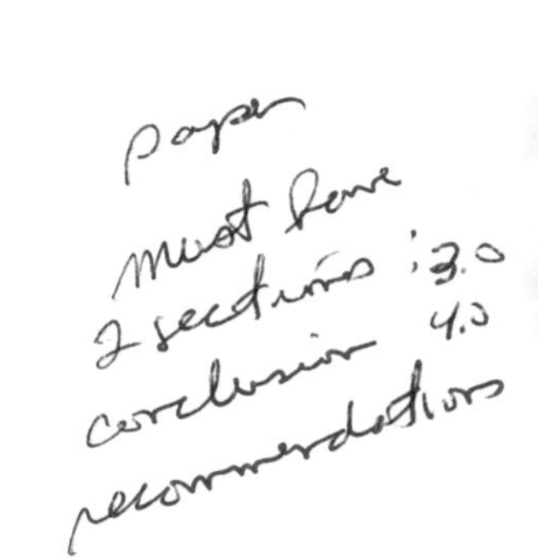

Ending

Depending upon what sort of paper you have written, your ending can be a summary, a set of conclusions, a set of recommendations, or a graceful exit from the paper. Frequently, you'll need some combination of these. We'll look at the four endings and at some of the possible combinations.

It's also possible in reports written with executives as the primary audience that the "ending" may actually be placed at the front of the report. Remember the audience analysis research we discussed in Chapter 4, "Writing for Your Readers." It indicates that executives are more interested in summaries, conclusions, and recommendations than they are in the details of a report. Thus, many writers in business and government move these elements to the front of their reports. They may be presented in separate sections labeled "Summary," "Conclusions," and "Recommendations" or combined into an executive summary. (See page 250.) In either case, the body of the report may be labeled "Discussion" or even "Annexes."

Summary Many technical papers are not argumentative. They simply present a body of information that the reader needs or will find interesting. Frequently, such papers end with summaries. In a summary, you condense for your readers what you have just told them in the discussion. Good summaries are difficult to write. At one extreme, they may lack adequate information; at the other, they may be too detailed. In the summary you must pare down to material essential to your purpose. This can be a slippery business.

Suppose your purpose is to explore the knowledge about the way the human digestive system absorbs iron from food. In your discussion you describe an experiment conducted with Venezuelan workers that followed isotopically labeled iron through their digestive systems. To enhance the credibility of the information presented, you include some details about the experiment. You report the conclusion that vegetarian diets decreased iron absorption.

How much of this should you put in your summary? Given your purpose, the location and methodology of the experiment would not be suitable material for the summary. You would simply report that in one experiment vegetarian diets have been shown to decrease iron absorption.

In general, each major point of the discussion should be covered in the summary. Sometimes you may wish to number the points for clarity. The following, from a paper of about 2,500 words, is an excellent summary:

> The exploding wire is a simple-to-perform yet very complex scientific phenomenon. The course of any explosion depends not only on the material and shape of the wire

but also on the electrical parameters of the circuit. An explosion consists primarily of three phases:

1. The current builds up and the wire explodes.
2. Current flows during the dwell period.
3. "Post-dwell conduction" begins with the reignition caused by impact ionization.

These phases may be run together by varying the circuit parameters.

The exploding wire has found many uses: it is a tool in performing other research, a source of light and heat for practical scientific application, and a source of shock waves for industrial use.

Summaries should be concise, and they should introduce no material that has not been covered in the report. Read your discussion over, noting your main generalizations and your topic sentences. Blend these together into a paragraph or two. Sometimes you will represent a sentence from the discussion with a sentence in the summary. At other times you will shorten such sentences to phrases or clauses. The last sentence in the foregoing example represents a summary of four sentences from the writer's discussion. The four sentences themselves were the topic sentences from four separate paragraphs.

If you are working with word processing, you might do well to copy the material you are summarizing and then go through it, eliminating unwanted material to make your summary. Such a technique may be easier and more accurate than retyping the material.

Conclusions Some technical papers work toward a conclusion. They ask a question, such as "Are nuclear power plants safe?," present a set of facts relevant to answering the question, and end by stating a conclusion: "Yes," "No," or sometimes, "Maybe." The entire paper aims squarely at the final conclusion. In such a paper, you argue inductively and deductively. You bring up opposing arguments and show their weak points. At the end of the paper, you must present your conclusions. Conclusions are the inferences drawn from the factual evidence of the report. They are the final link in your chain of reasoning. In simplest terms, the relationship of fact to conclusion goes something like this:

Facts	Conclusion
Car A averages 25 miles per gallon. Car B averages 40 miles per gallon.	On the basis of miles per gallon, Car B is preferable.

Because we presented a simple case, our conclusion was not difficult to arrive at. But even more complicated problems present the same relationship of fact to inference.

In working your way toward a major conclusion, you ordinarily have to work your way through a series of conclusions. In answering the question of nuclear power plant safety, you would have to answer many subquestions concerning such things as security of the radioactive materials used, adequate control of the nuclear reaction, and safe disposal of nuclear wastes. The answer to each subquestion is a conclusion. You may present these conclusions in the body of the report, but it's usually a good idea to also draw them all together at the end of the report to support the major conclusion.

Earlier, we showed you the introduction to a report that questioned whether the class of chemicals known as phthalates endangered public health. Here are the conclusions to that report:

> Based on the observations made thus far, there is no evidence of toxicity in humans due to phthalates, either from foods, beverages, or household products as ordinarily consumed or used.
>
> These observations, coupled with the limited use of phthalate-containing food packaging materials and the low rate of migration of the plasticizers from packaging material to food, support the belief that the present use of phthalates in food packaging represents no hazard to human health.[7]

All the conclusions presented are supported by evidence in the report.

In larger papers or when dealing with a controversial or complex subject, you would be wise to precede your conclusions with a summary of your facts. By doing so, you will reinforce in your reader's mind the strength and organization of your argument. In any event, make sure your conclusions are based firmly upon evidence that has been presented in your report. Few readers of professional reports will take seriously conclusions based upon empty, airy arguments. Conclusions are frequently followed by recommendations.

Recommendations A conclusion is an inference. A **recommendation** is the statement that some action should be taken or not taken. The recommendation is based upon the conclusions and is the last step in the process. You conclude that Brand X bread is cheaper per pound than Brand Y and just as nutritious and tasty. Your final conclusion, therefore, is that Brand X is a better buy. Your recommendation is "Buy Brand X."

Many reports such as feasibility reports, environmental impact statements, and research reports concerning the safety of certain foods or chemicals are decision reports that end with a recommendation. For example, we are all familiar with the government recommendations that have removed certain artificial sweeteners from the market and that have placed warnings on cigarette packages. These recommendations were all originally stated at the end of reports looking into these matters.

Recommendations are simply stated. They follow the conclusions, often in a separate section, and look something like this:

> Based upon the conclusions reached, we recommend that our company
>
> - Not increase the present level of iron enrichment in our flour.
> - Support research into methods of curtailing rancidity in flour containing wheat germ.

Frequently, you may have a major recommendation followed by additional implementing recommendations, as in the following:

Major recommendation

> We recommend that the Department of Transportation build a new bridge across the St. Croix River at a point approximately three miles north of the present bridge at Hastings.

Implementing recommendations

> - The Department's location engineers should begin an immediate investigation to decide the exact bridge location.
> - Once the location is pinpointed, the Department's right-of-way section should purchase the necessary land for the approaches to the bridge.

You need not support your recommendations when you state them. You should have already done that thoroughly in the report and in the conclusions leading up to the recommendations. It's likely, of course, that a full-scale report will contain a summary, conclusions, and recommendations.

Graceful Close A short, simple, nonargumentative paper often requires nothing more than a graceful exit. As you would not end a conversation by turning on your heel and stalking off without a "good-bye" or a "see you later" to cover your exit, you do not end a paper without some sort of close. In a short informational paper that has not reached a decision, the facts should be still clear in the readers' minds at the end, and they will not need a summary. One sentence, such as the following, which might end a short speculative paper on superconductivity, will probably suffice:

> Because superconductivity seems to have numerous uses, it cannot fail to receive increasing scientific attention in the years ahead.

Sometimes, even a long, involved paper can profit from a graceful close. In the next example, the author gracefully exits a long, scholarly paper with a reference to other work that supports his own:

> Teachers of professional communication have a unique interdisciplinary perspective and thus a unique responsibility. They can—indeed they must—daily negotiate the

> distance between "the two cultures." C. P. Snow, who coined the phrase, was both a physicist and a man of letters, and it is salutary to recall that his famous essay (perhaps more often cited than read) is not an ethical indictment of the ethical position of scientific "culture," but just the opposite. Writing to his fellow literati, he says, "the greatest enrichment the scientific culture could give us is...a moral one." Snow praises scientific "culture" for its commitment to human improvement manifested in active involvement. Snow takes to task the other culture, the "mainly literary" one, for an ethical complacency "made up of defeat, self-indulgence, and moral vanity," a complacency to which "the scientific culture is almost totally immune." And he concludes, "It is that kind of moral health of the scientists which, in the last few years, the rest of us have needed most; and of which, because the two cultures scarcely touch, we have been most deprived" (414). Both cultures have changed much in the four decades since Snow published his essay, but perhaps each culture still has much to learn from the other, even about ethics.[8]

Combination Endings We have treated summaries, conclusions, and recommendations separately. Indeed, a full-scale report leading to a recommendation will often contain in sequence separate sections labeled "Summary," "Conclusions," and "Recommendations." When such is the case, the summary will often be restricted to a condensation of the factual data offered in the body. The implications of the data will be presented in the conclusions, and the action to be taken in the recommendations.

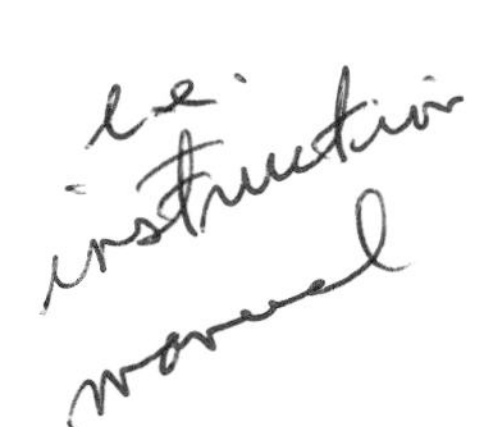

However, in many reports, the major elements of factual summary, conclusions, and recommendations may be combined. A combination of summary, conclusions, and recommendations placed at the front of a report for a technical audience will probably be labeled an abstract. It will be, in fact, what we describe on pages 248–249 as an informative abstract. The same combination located at the end of a report for any audience would probably be called a summary. A summary written specifically for an executive audience and located at the front of the report will be an executive summary. (See page 250.)

It's unfortunate that there is a slight confusion of terms when these elements are used in different ways. Don't let the confusion in terminology confuse the essence of what is involved here. In all but the simplest reports, you must draw things together for your readers. You must condense and highlight your significant data and present any conclusions and recommendations you have. Notice how this summary of a scientific research report smoothly combines all these elements:

SUMMARY

Summary

> In many turtles the hatchling's sex is determined by the incubation temperature of the egg, warm temperatures causing femaleness and cool temperatures maleness. Consequently, the population sex ratio depends upon the interaction of (i) environmental temperature, (ii) maternal choice of nest site, and (iii) embryonic control of sex determination. If environmental temperature differs between populations, then

sex ratio selection is expected to adjust either maternal behavior or embryonic temperature-sensitivity to yield nearly the same sex ratio in the different populations.

Conclusion

Recommendation

To test this hypothesis in part, we have compared sex determining temperatures among embryos of emydid turtles in the northern and southern U.S. We predicted that embryos of southern populations should develop as male at higher temperatures than those of northern populations. The data offer no support for this prediction among the many possible comparisons between northern and southern species. The data actually refute the prediction in both of the North–South intraspecific comparisons. Further study is needed, in particular, of nest temperatures in the different populations.[9]

Appendixes

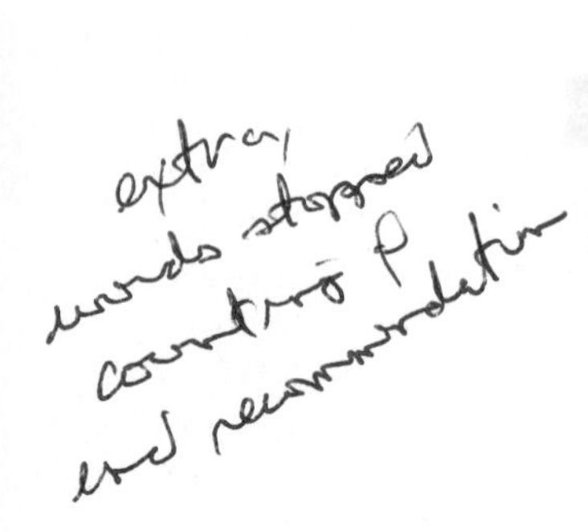

Appendixes, as the name implies, are materials appended to a report. They may be materials important as background information or needed to lend the report credibility. They will not in most cases be necessary to meet the major purpose of the report or the major needs of the audience. For example, if you are describing research for an executive audience, they will likely be more interested in your results and conclusions than in your research methodology. If your audience consists totally of executives, you might include only a bare-bones discussion of your methodology in your report.

But suppose you have a primary audience of executives and a secondary audience of experts. You could satisfy both audiences by placing a detailed discussion of your methodology in an appendix—out of the executives' way but readily accessible for the experts. Like most decisions in technical writing, what goes into the body of a report, what goes into an appendix, and what is eliminated altogether are determined by your audience and purpose.

During the final stages of arranging your report, determine whether materials such as the following should be placed in appendixes:

- Case histories
- Supporting illustrations
- Detailed data
- Transcriptions of dialogue
- Intermediate steps in mathematical computation
- Copies of letters, announcements, and leaflets mentioned in the report
- Samples, exhibits, photographs, and supplementary tables and figures
- Extended analyses
- Lists of personnel

- Suggested collateral reading
- Anything else that is not essential to the sense of the main report

Before you place anything in an appendix, consider the effect on the report. Be certain that shifting an item to an appendix does not undermine your purpose or prevent the reader from understanding major points of the report.

Documentation

Different documentation systems are in use from college to college, journal to journal, company to company. Therefore, we cannot claim a universal application for the instructions that follow. Use them barring conflicting instructions from your instructor, college, employer, or the stylebook of the journal or magazine in which you hope to publish.

Before we go into the mechanics of documentation, it might be wise to discuss why and when you need to document.

First of all, documenting fulfills your moral obligation to give credit where credit is due. It lets your readers know who was the originator of an idea or expression and where his or her work is found. Second, systematic documentation makes it easy for your readers to research your subject further.

When do you document? Established practice calls for you to give credit when you borrow the following:

- Direct quotes
- Research data and theories
- Illustrations, such as tables, graphs, photographs, and so forth

You do not need to document general information or common knowledge. For example, even if you referred to a technical dictionary to find that creatinine's more formal name is methyglycocyamidine, you would not be obligated to show the source of this information. It is general information, readily found in many sources. If, on the other hand, you include in your paper an opinion that the cosmos is laced with strands of highly concentrated mass energy called *strings*, you would need to document the source of this opinion.

In general, give credit where credit is due, but do not clutter your pages with references to information readily found in many sources. If in doubt as to whether to document or not, play it safe and document.

We explain two systems of documentation. The first involves using notes; the second, author-date parenthetical references keyed to a list of works cited. Notes—that is, footnotes and endnotes—are often used in business and academic reports. Parenthetical documentation is widely used in scientific papers and journals. Some general rules apply in both systems. We cover these rules first and then the note system and the author–date parenthetical system.

Teachers may specify one method or the other for class papers. In our explanation we follow *The MLA Handbook for Writers of Research Papers,* a guide published by the Modern Language Association.[10] We also include a brief comparison of the MLA author–date system to the author–date system of the American Psychological Association (APA).[11]

General Rules

The rules that follow apply whether you are using the note system or the author–date parenthetical system.

- You may use a short form of the publisher's name, such as *Wiley* for John Wiley & Sons. Be consistent throughout your notes or citations, however.
- When page numbers are two digits, use both digits in the second number: *22–24,* not *22–4.*
- When page numbers are three or four digits, use only the last two digits in the second number: *112–15; 1034–39.*
- When citing inclusive page numbers of an article, if the article is continued over to pages later in the publication, cite the first page of the article only, followed immediately by a plus sign: *22+.*
- When a city of publication is not well known, include an abbreviation for the state, province, or country in your note or citation. For states and provinces, use postal abbreviations, such as *NY* for New York and *BC* for British Columbia. (See Figure 12-7 for a list of state and province abbreviations.) For countries, use the abbreviations that can be found in most college dictionaries, such as *Arg.* for Argentina.
- Use three-letter abbreviations for months with more than four letters in them: thus, *Dec.* and *Apr.* but *May* and *June.*
- When information on pagination, publisher, or date is missing in your source, at the point where you would put that information, put one of these abbreviations: *n. pag.* for no pagination, *n.p.* for no place of publication or no publisher, *n.d.* for no date.
- Illustrations are documented separately from the text. As we explain in Chapter 11, "Graphical Elements," they are documented directly on the table or figure. Because illustrations have complete documentation internally, their notes do not appear with page footnotes or on a list of endnotes or works cited. The form of illustration notes, however, does follow the

form of the notes we describe here. For guidance, see Chapter 11 and the many notes we have provided on illustrations throughout this book, such as those on pages 244 and 249.

Notes

In this section we first explain the note system of documentation and then provide a series of model notes that you can use as guides in constructing your notes.

The Note System Notes may be displayed at the bottom of the page on which the documented material appears or gathered at the end of the report under the heading *Notes.* In the first method, the notes are called *footnotes* and in the second, *endnotes.* Footnotes are illustrated in Figure 10–14, endnotes in Figure 10–15. Endnotes are more common than footnotes in student and business reports and in journal articles. We use them in this book; see our "Chapter Notes" on pages 673-680.

Whether using footnotes or endnotes, number your notes in sequence through your paper. If your paper is divided into chapters, as is this book, number your notes in sequence through each chapter. Except for spacing, the note form is the same for both footnotes and endnotes. In footnotes, single-space each note and double-space between notes. In endnotes, double-space the notes themselves and between notes.

In both footnotes and endnotes, the note number is indicated in the text by a superscript number, that is, a number placed above the line of type, as you can see in the many note numbers in this book and in Figure 10–14. Place the number in the text where it is relevant and where it disrupts the text the least. Generally, you should place the note at the end of the grammatical unit—sentence or clause—that contains the material you are documenting.

Notes, whether footnotes or endnotes, appear in the order that the superscript note numbers occur in the paper. Indent the superscript number corresponding to the note number five spaces from the left margin. Begin the body of the note one space to the right of the number. If the note is longer than one line, begin subsequent lines at the left margin. Figures 10–14 and 10–15 illustrate these details.

Word processing can make placing and numbering footnotes and endnotes much easier than on a typewriter. Many programs will number the notes automatically. Write the note at the place in the text where it is relevant. If you have specified endnotes, the program saves the note with the other endnotes and puts them all after the text. If you have specified footnotes, the program puts the note on the bottom of the page, adjusting the rest of the text for it. If you move the relevant text to another place later, the note moves with it. When you add or delete a note, the program renumbers all the subsequent notes.

Current rhetorical theory indicates that this attempt, through analogy, to call on schemata for newspapers could affect readers' expectations about the writing in the newsletters, which in turn could influence the way these readers process the writing. Genre theory, for example, posits that generic patterns such as those in a newspaper, as part of our "cultural rationality,"[1] alert readers to ways of perceiving and interpreting documents.[2] In addition, theories of intertextuality, the concept that all texts contain explicit or implicit traces of other texts,[3] suggest that creating an analogy between newspapers and newsletters would affect readers' expectations, encouraging them to perceive and interpret material in a particular way.[4] We must ask, therefore, what readers' expectations about newspapers and hence, by analogy, about the newsletters, might be.

Place superscript note number after any punctuation at end of grammatical unit cited.

Use four spaces between text and notes.

Use one space between number and body of note.

Indent five spaces and repeat superscript number.

[1] C. R. Miller, "Genre as Social Action," Quarterly Journal of Speech 70 (1984): 165.

[2] Miller 159.

[3] J. E. Porter, "Intertextuality and the Discourse Community," Rhetoric Review 5 (1986): 34.

Single-space each note; double-space between notes.

[4] Porter 38.

Use no punctuation between name and page.

Figure 10–14 Footnotes on a Page
Source: Adapted with permission from Nancy Roundy Blyler, "Rhetorical Theory and Newsletter Writing," *Journal of Technical Writing and Communication* 20 (1990): 144.

Model Notes To construct your notes, consult the sample notes in Figures 10–14 and 10–15 and the examples that follow. We have annotated Figures 10–14 and 10–15 to draw your attention to certain distinctive note features, such as italics, underlining, punctuation, and spacing. (If you are typing your report, use underlining for italics. With word processing, you may be able to use italics if you wish.) We categorize the model notes under the headings of **Books, Periodicals, Other,** and **Subsequent References,** and provide an example of most of the notes you are likely to need in school or business. If you need a more complete set of notes than we have provided, consult *The MLA Handbook for Writers of Research Papers* or the stylebook of the organization or journal where you intend to publish.

Books The following examples illustrate various forms of book notes.

Notes

Use comma between author and book title.

Indent 5 spaces and use superscript number.

Underline or italicize book titles. Capitalize first and last words and all principal words between. Do not capitalize articles, prepositions, and coordinating conjunctions unless they are a first or last word.

[1] R. John Brockmann, Writing Better Computer User Documentation (New York: Wiley, 1986) 13.

Put publication information in parentheses.

[2] David G. Byrd, Paula R. Feldman, and Phyllis Fleishel, The Microcomputer and Business Writing (New York: St. Martin's, 1986) 22–24.

Put author's or authors' names in normal order.

Use specific page number or numbers of reference.

[3] Brockmann 19.

Use no punctuation between author and page.

[4] Brockmann 24.

[5] Producing Quality Technical Information (San Jose: IBM, 1986) 5–11.

Place quotation marks around article titles that appear in periodicals or anthologies. Capitalize first and last words and all principal words between. Do not capitalize articles, prepositions, and coordinating conjunctions unless they are a first or last word.

Place comma between article and journal title. Note that comma goes inside quotation marks.

[6] Beverly L. Sauer, "Sense and Sensibility in Technical Communication," Journal of Business and Technical Communication 7 (1993): 65–66.

Place date in parentheses.

[7] Bob Schulman, Eric C. W. Dunn, and George Shackelford, Quicken, version 1.5, computer software, Intuit, 1989.

[8] Sauer 69.

[9] Rebecca E. Burnett and Ann Hill Duin, "Collaboration in Technical Communication: A Research Continuum," Technical Communication Quarterly 2 (1993): 9.

Double-space notes and between notes.

Use arabic numbers for volume number, even if periodical uses roman numerals.

Figure 10–15 List of Endnotes

ONE AUTHOR

[1] Robert A. Day, *Scientific English: A Guide for Scientists and Other Professionals* (Phoenix, AZ: Oryx, 1992) 13.

TWO OR MORE AUTHORS

[2] Edward von Koenigseck, James N. Irvin, and Sharon C. Irvin, *Technical Writing for Private Industry: The A-to-Z of O & M Manuals* (Malabar, FL: Krieger, 1991) 22–24.

AN ANTHOLOGY

[3] Linda P. Driskill, June Ferrill, and Marda Nicholson Steffey, eds., *Business and Managerial Communication: New Perspectives* (New York: Harcourt Brace Jovanovich, 1992) 112–15.

Use the abbreviation *ed.* for one editor, *eds.* for two or more. Use *trans.* for one or more translators.

AN ESSAY IN AN ANTHOLOGY

[4] Lee Odell, Dixie Goswami, Anne Herrington, and Doris Quick, "Studying Writing in Non-Academic Settings," *New Essays in Technical and Scientific Communication: Research, Theory, Practice,* eds. Paul V. Anderson, R. John Brockmann, and Carolyn R. Miller (Farmingdale, NY: Baywood, 1983) 27–28.

SECOND OR SUBSEQUENT EDITION

[5] Michael H. Markel, *Technical Writing: Situations and Strategies,* 3rd ed. (New York: St. Martin's, 1992) 17–18.

ARTICLE IN REFERENCE BOOK

[6] "Petrochemical," *The New Columbia Encyclopedia,* 1975.

When a reference work is well known, you need cite only the name of the article, the name of the reference work, and the date of publication. If the work is arranged alphabetically, do not cite page numbers.

For less well-known reference works, give complete information. Cite the author of the article if known; otherwise, begin with the name of the article.

[7] "Perpetual Calendar, 1775–2076," *The New York Public Library Desk Reference,* eds. Paul Fargis and Sheree Bykofsky (New York: Webster's New World, 1989) 10–13.

A PAMPHLET

[8] *Cataract: Clouding the Lens of Sight* (San Francisco: American Academy of Ophthalmology, 1989) 1–2.

If the author of a pamphlet is known, give complete information in the usual manner.

GOVERNMENT OR CORPORATE PUBLICATION

[9] U.S. National Aeronautics and Space Administration, *Voyager at Neptune: 1989* (Washington, DC: GPO, 1989) 16.

Treat government and corporate publications much as you would any book, except that the author is often a government agency or a division within a company. *GPO* is the abbreviation for the U.S. Government Printing Office.

ANONYMOUS BOOK

[10] *Producing Quality Technical Information* (San Jose: IBM, 1986) 5–11.

When no human, government, or corporate author is listed, begin with the title of the book.

PROCEEDINGS

[11] Mary Fran Buehler, "Rules that Shape the Technical Message: Fidelity, Completeness, Preciseness," *Proceedings 31st International Technical Communication Conference* (Washington, DC: Society for Technical Communication, 1984) WE-9.

UNPUBLISHED DISSERTATION

[12] Penny Hutchinson, "Trauma in Emergency Room Surgery," diss., U of Chicago, 1994, 16.

Abbreviate *University* as *U*; a university press would be *UP*, as in *Indiana UP*.

Periodicals The following examples illustrate various forms of periodical notes.

JOURNAL WITH CONTINUOUS PAGINATION

[1] Thomas T. Barker, "Word Processors and Invention in Technical Writing," *The Technical Writing Teacher* 16 (1989): 127.

JOURNAL THAT PAGES ITS ISSUES SEPARATELY

[2] David P. Gardner, "The Future of University/Industry Research," *Perspectives in Computing* 7.1 (1987): 5.

In this note, 7 is the volume number, 1 the issue number, and 5 the page number.

An alternative to the preceding format is to treat such journals like commercial magazines that page their issues separately.

[3] David P. Gardner, "The Future of University/Industry Research," *Perspectives in Computing* Spring 1987: 5

COMMERCIAL MAGAZINES AND NEWSPAPERS

A weekly or biweekly magazine:

[4] Robert J. Samuelson, "The Health-Care Crisis Hits Home," *Newsweek* 2 Aug, 1993: 38.

Monthly or bimonthly magazine:

[5] Scott Beamer, "Why You Need a Charting Program," *MacUser* June 1990: 126.

Newspaper

[6] Louise Levathes, "A Geneticist Maps Ancient Migrations," *New York Times* 27 July 1993, national ed., sec. C: 1+.

When the masthead of the paper specifies the edition, put that information in your note. Newspapers frequently change from edition to edition on the same day.

Anonymous article

[7] "Absolute," *New Yorker* 18 June 1990: 28.

When no author is given for an article, begin with the title of the article.

Other Here, we show you model notes for computer software, information services, letters, and interviews.

COMPUTER SOFTWARE

[1] Bob Schulman, Eric C. W. Dunn, and George Shackelford, *Quicken,* version 1.5, computer software, Intuit, 1989.

INFORMATION SERVICES

[2] R. Berdan and M. Garcia, *Discourse-Sensitive Measurement of Language Development in Bilingual Children* (Los Alamitos, CA: National Center for Bilingual Research, 1982) (ERIC ED 234 636).

LETTERS

[3] John S. Harris, letter to the author, 19 July 1993.

INTERVIEWS

[4] Herman Estrin, personal interview, 16 Mar. 1993.

Subsequent References After the first complete note on an item, subsequent references need only briefly identify the item. If the page reference is the same, use only the author's last name:

[5] Duin.

If the page numbers are different, include the new page numbers:

[6] Duin 188–90.

If you have two works by the same author, include a shortened version of the title to identify the correct work:

[7] Conniff, "Eye on the Storm" 21.

Author–Date Documentation

Author-date documentation combines parenthetical references in the text with an alphabetized list of all the works cited. The system is a common method of documentation in the sciences. Shortly we show you how to make parenthetical references, but first we discuss how to construct the list of works cited.

Works Cited In a section headed "Works Cited," you will list all the works that you cite in your paper. Figure 10–16 illustrates how it is done. We have annotated Figure 10–16 to draw your attention to certain distinctive citation features, such as underlining, punctuation, and spacing.

Works Cited

Brockmann, R. John. 1986. <u>Writing Better Computer User Documentation</u>. New York: Wiley.

Burnett, Rebecca, and Ann Hill Duin. 1993. "Collaboration in Technical Communication: A Research Continuum." <u>Technical Communication Quarterly</u> 2: 5–21.

Byrd, David G., Paula R. Feldman, and Phyllis Fleishel. 1986. <u>The Microcomputer and Business Writing</u>. New York: St. Martin's.

<u>Producing Quality Technical Information</u>. 1986. San Jose: IBM.

Sauer, Beverly L. 1993. "Sense and Sensibility in Technical Communication." <u>Journal of Business and Technical Communication</u> 7: 63–83.

Schulman, Bob, Eric C. W. Dunn, and George Shackelford. 1989. <u>Quicken</u>. Version 1.5. Computer software. Intuit.

Place date after author's name.

Invert order of first author's name.

Underline or italicize periodical and book titles. Capitalize first and last words and all principal words between. Do not capitalize articles, prepositions, and coordinating conjunctions unless they are a first or last word.

For anonymous books and articles, use book or article title and put date after title.

Use normal order for all names after first.

Use quotation marks around article titles. Capitalize first and last words and all principal words between. Do not capitalize articles, prepositions, and coordinating conjunctions unless they are a first or last word.

Use arabic numbers for volume number even if periodical uses roman numerals.

Use periods between all major elements of citations.

Double-space all entries and between entries.

Use inclusive page numbers for articles in periodicals and anthologies.

Figure 10–16 List of Works Cited

The models that follow show you how to construct the citations you are likely to need in school and business. They look much like notes, but you will do some things quite differently.

- Use periods between the basic components of the citation.
- Transpose the author's name: last, first, middle name or initial. If you have two or three authors, print the subsequent authors' names in normal order. If you have four or more authors, list the first author's name followed by "et al." for "and others."

- The year of publication follows the author's name and precedes the title. When you have no author's name, place the date after the title.
- Arrange the entries in alphabetical order. You need enter each entry only once. Determine alphabetical order by the author's last name or, if no author is listed, by title (disregarding *the, a,* or *an*).
- Omit page numbers from whole-book entries. Give inclusive page numbers for articles in anthologies and periodicals.
- Do not number entries. Begin the first line at the left margin and indent the second and subsequent lines five spaces.

We have categorized the models under "Books," "Periodicals," and "Other."

Books The following examples illustrate various forms of book citations.

ONE AUTHOR

Day, Robert A. 1992. *Scientific English: A Guide for Scientists and Other Professionals.* Phoenix, AZ: Oryx.

TWO OR THREE AUTHORS

Von Koenigseck, Edward, James N. Irvin, and Sharon C. Irvin. 1991. *Technical Writing for Private Industry: The A-to-Z of O & M. Manuals* Malabar, FL: Krieger.

FOUR OR MORE AUTHORS

Wilkinson, C. W., et al. 1986. *Writing and Speaking in Business.* 9th ed. Homewood, IL: Irwin.

Use the first author's name with et al. (and others) for four or more authors.

AN ANTHOLOGY

Driskill, Linda P., June Ferrill, and Marda Nicholson Steffey, eds. 1992. *Business and Managerial Communication: New Perspectives.* New York: Harcourt Brace Jovanovich.

Use the abbreviation *ed.* for one editor, *eds.* for two or more. Use *trans.* for one or more translators.

AN ESSAY IN AN ANTHOLOGY

Odell, Lee, et al. 1983. "Studying Writing in Non-Academic Settings." *New Essays in Technical and Scientific Communication: Research, Theory, Practice.* Eds. Paul V. Anderson, R. John Brockmann, and Carolyn R. Miller. Farmingdale, NY: Baywood. 17–40.

SECOND OR SUBSEQUENT EDITION

Markel, Michael H. 1992. *Technical Writing: Situations and Strategies.* 3rd ed. New York: St. Martin's.

ARTICLE IN REFERENCE BOOK

"Petrochemical." 1975. *The New Columbia Encyclopedia.*

When a reference work is well known, you need cite only the name of the article, the name of the reference work, and the date of publication. If the work is arranged alphabetically, do not cite page numbers.

For less well-known reference works give complete information. Cite the author of the article if known; otherwise begin with the name of the article.

"Perpetual Calendar, 1775–2076." 1989. *The New York Public Library Desk Reference.* Eds. Paul Fargis and Sheree Bykofsky. New York: Webster's New World. 10–13.

A PAMPHLET

Cataract: Clouding the Lens of Sight. 1989. San Francisco: American Academy of Ophthalmology.

If the author of a pamphlet is known, give complete information in the usual manner.

GOVERNMENT OR CORPORATE PUBLICATION

U.S. National Aeronautics and Space Administration. 1989. *Voyager at Neptune: 1989.* Washington, DC: GPO.

Treat government and corporate publications much as you would any book, except that the author is often a government agency or a division within a company. *GPO* is the abbreviation for the U.S. Government Printing Office.

ANONYMOUS BOOK

Producing Quality Technical Information. 1986. San Jose: IBM.

When no human, government, or corporate author is listed, begin with the title of the book.

PROCEEDINGS

Buehler, Mary Fran. 1984. "Rules that Shape the Technical Message: Fidelity, Completeness, Preciseness." *Proceedings 31st International Technical Communication Conference.* Washington, DC: Society for Technical Communication. WE 9-12.

UNPUBLISHED DISSERTATION

Hutchinson Penny. 1994. "Trauma in Emergency Room Surgery." Diss. U of Chicago.

Abbreviate *University* as *U*; a university press would be *UP*, as in *Indiana UP.*

Periodicals The following examples illustrate various forms of periodical citations.

JOURNAL WITH CONTINUOUS PAGINATION

Barker, Thomas T. 1989. "Word Processors and Invention in Technical Writing." *The Technical Writing Teacher* 16: 126–35.

JOURNAL THAT PAGES ITS ISSUES SEPARATELY

Gardner, David P. 1987. "The Future of University/Industry Research." *Perspectives in Computing* 7.1: 4–10.

In this citation, 7 is the volume number, 1 is the issue number, and 4-10 are the inclusive pages of the article.

An alternative to the preceding is to treat such journals similarly to commercial magazines that page their issues separately.

Gardner, David P. 1987. "The Future of University/Industry Research." *Perspectives in Computing.* Spring: 4–10.

COMMERCIAL MAGAZINES AND NEWSPAPERS

A weekly or biweekly magazine:

Samuelson, Robert J. 1993. "The Health-Care Crisis Hits Home." *Newsweek.* 2 Aug.: 38.

Monthly or bimonthly magazine:

Beamer, Scott. 1990. "Why You Need a Charting Program." *MacUser* June: 126–38.

Newspaper:

Louise Levathes. 1993. "A Geneticist Maps Ancient Migrations." *New York Times* 27 July, national ed., sec. C: 1+.

When the masthead of the paper specifies the edition, put that information in your note. Newspapers frequently change from edition to edition on the same day.

Anonymous article

"Absolute." 1990. *New Yorker* 18 June: 28–29.

When no author is given for an article, begin with the title of the article.

Other Here, we show you model citations for computer software, information services, letters, interviews, and two or more entries by the same author.

COMPUTER SOFTWARE

Schulman, Bob, Eric C. W. Dunn, and George Shackelford. 1989. *Quicken.* Version 1.5. Computer software. Intuit.

INFORMATION SERVICE

Berdan, R., and M. Garcia. 1982. *Discourse-Sensitive Measurement of Language Development in Bilingual Children.* Los Alamitos, CA: National Center for Bilingual Research. ERIC ED 234 636.

LETTER

Harris, John S. Letter to the author. 19 July 1993.

INTERVIEW

Estrin, Herman. Personal interview. 16 Mar. 1993.

TWO OR MORE WORKS BY THE SAME AUTHOR

When you have two or more works by the same author or authors in your list of works cited, replace the author's name in the second citation with three unspaced hyphens and alphabetize by title.

Barker, Thomas. 1985. "Video Field Trip: Bringing the Real World into the Technical Writing Classroom." *The Technical Writing Teacher* 11: 175–79.

—. 1989. "Word Processors and Invention in Technical Writing." *The Technical Writing Teacher* 16: 126–35.

When you have two or more works in the same year by the same author or authors in your list of works cited, mark the years with lowercase letters, beginning with "a" and alphabetize by title.

Crowhurst, M. 1983a. *Persuasive Writing at Grades 5, 7, and 11: A Cognitive-Development Perspective.* Paper presented at the annual meeting of the American Educational Research Association, Montreal, Canada. ERIC ED 230 977.

—. 1983b. *Revision Strategies of Students at Three Grade Levels.* Final report. Educational Research Institute of British Columbia. ERIC ED 238 009.

PARENTHETICAL REFERENCE

When you have completed your list of works cited, refer your reader to it through parenthetical references in your text. Figure 10–17 is a page using parenthetical references. We have annotated the figure to show you how the system works in the text. As Figure 10–17 illustrates, parenthetical references are placed before the punctuation at the end of the clause or sentence that contains the material cited. Their purpose is to guide the reader to the corresponding entry in the list of works cited

Use specific page reference.

Place comma between date and page reference.

Current rhetorical theory indicates that this attempt, through analogy, to call on schemata for newspapers could affect readers' expectations about the writing in the newsletters, which in turn could influence the way these readers process the writing. Genre theory, for example, posits that generic patterns such as those in a newspaper, as part of our "cultural rationality" (Miller 1984, 165), alert readers to ways of perceiving and interpreting documents (Miller 1984, 159). In addition, theories of intertextuality, the concept that all texts contain explicit or implicit traces of other texts (Porter 1986, 34), suggest that creating an analogy between newspapers and newsletters would affect readers' expectations, encouraging them to perceive and interpret material in a particular way (Porter 1986, 38). We must ask, therefore, what readers' expectations about newspapers and hence, by analogy, about the newsletters, might be.

Place parenthetical notes before any punctuation that ends the material cited.

Figure 10–17 Parenthetical References on a Page
Source: Adapted with permission from Nancy Roundy Blyler, "Rhetorical Theory and Newsletter Writing," *Journal of Technical Writing and Communication* 20 (1990): 144.

and, when appropriate, cite the specific pages of the reference. Some model references follow:

AUTHOR AND DATE

(Asher 1992)

Refers the reader to Asher's 1992 work in the list of works cited. Use this form when you are not citing a specific page.

AUTHOR, DATE, AND PAGE

(Asher 1992, 93)

Refers the reader to page 93 of Asher's 1992 work. Use this form when you are citing a specific page or pages.

DATE AND PAGE

(1992, 97)

Use this form when you have already mentioned the author's name in the passage leading up to the parenthetical reference, for example, "As Asher's research shows ... "

PAGES ONLY

(324–27)

Use this form when you have mentioned both the author's name and the date in the passage leading up to the parenthetical reference; for example, "As Asher's research in 1992 shows ... "

GOVERNMENT OR CORPORATE AUTHOR

(U.S. National Aeronautics and Space Administration 1994)

When a government agency or corporate division is listed as the author, you may use that in your parenthetical reference. If the name is long and clumsy, you might do better to work it into the passage leading up to the parenthetical reference. Alternatively, you might provide the reader with a shortened form, such as *NASA.*

TITLE OF WORK

(Producing Quality Technical Information 1986, 14.)

Use this form when you have no author's name and have listed the work by its title. Omit anything from the reference that you have mentioned in the passage leading up to the reference.

TWO OR THREE AUTHORS

(Berdan and Garcia 1982)

Name all the authors of a work by two or three authors.

FOUR OR MORE AUTHORS

(Odell et al. 1983, 28)

Use first author's name with et al. (and others) for four or more authors.

TWO WORKS BY THE SAME AUTHOR IN DIFFERENT YEARS

(Jarrett 1992, 18)

(Jarrett 1993)

When you have listed two works by the same author but written in different years, the dates will distinguish between them.

TWO OR MORE WORKS BY THE SAME AUTHOR IN THE SAME YEAR

(Jarrett 1991a, 18)

(Jarrett 1991b)

When you list two or more works by the same author in the same year, to distinguish them, mark them with lowercase letters, both in the parenthetical reference and in the list of works cited.

MLA–APA Comparison

Becoming familiar with the MLA author–date system as described in this book will prepare you to use most author–date systems used in documenting technical reports. Another prominent author–date system is that found in the *Publication Manual of the American Psychological Association* (APA). To alert you to the kinds of variations that exist from system to system, we provide a few samples to demonstrate the differences between the MLA and APA systems.

Both systems use parenthetical references in the text, but the APA system is slightly more complex:

MLA (Asher 1992, 93)
APA (Asher, 1992, p. 93)

The APA system places a comma between name and date and puts a "p." for "page" before the page number.

Both systems key their parenthetical references to a list of works cited. MLA calls the list "Works Cited." APA uses the title "References." Note the differences between typical book entries and periodical entries:

Book Entries

MLA
Title italicized; first, last, and principal words capitalized

Brockmann, R. John. 1986. Writing Better Computer User Documentation. New York: Wiley.

APA
Date in parentheses; first word only capitalized

Brockmann, R. J. (1986). Writing better computer user documentation. New York: Wiley.

Periodical Entries

MLA
Article title in quotes; first, last, and principal words capitalized

Barker, T. T. 1989. "Word Processors and Invention in Technical Writing." The Technical Writing Teacher 16: 126–35.

APA
Date in parentheses; article title not in quotes; first word only capitalized

Barker, Thomas T. (1989). Word processors and invention in technical writing. The Technical Writing Teacher, 16, 126–35.

A full comparison of the two systems would reveal similar differences running throughout the various kinds of entries. A comparison with other systems would reveal still other differences. The moral of such comparisons is clear. You must obtain and use whatever style manual governs the publications or reports you write.

Copyright

Stringent copyright laws protect published work. When you are writing a student report that you do not intend to publish, you need not concern yourself with these laws. If you intend to publish a report, however, you should become familiar with copyright law. You must get permission from the copyright holder to use illustrations and extended quotations. Look for information on the copyright holder on the title page of a publication. You can find a good summary of copyright law in *The Chicago Manual of Style.*[12]

Planning and Revision Checklists

You will find the planning and revision checklists that follow Chapter 2, "Composing" (pages 38–39 and inside the front cover), and Chapter 4, "Writing for Your Readers" (pages 81–82), valuable in planning and revising any presentation of technical information. The following questions specifically apply to the elements of reports. They summarize the key points in this chapter and provide a checklist for planning and revising.

PLANNING

- Which does your situation call for, a letter of transmittal or a preface?
- Will you bind your report? If so, remember to leave extra left-hand margin.
- Does your situation call for any information on your title page beyond the basic items: name of author, name of person or organization receiving the report, title of report, and date of submission?
- Have you used a system of headings that must be repeated in your table of contents?
- Have you used a numbering system that must be repeated in your table of contents?
- Do you have enough illustrations to warrant a list of illustrations?
- Have you used symbols, abbreviations, acronyms, and terms that some of your readers will not know? Do you need a list of symbols or a glossary?
- Does your report require an abstract? Should you have an informative or descriptive abstract or both?
- Is your primary audience an executive one? Does the length of your report require an executive summary?
- What will be the major questions in the executives' minds as they read your report? Plan to answer these questions in the executive summary.
- Are your subject, purpose, scope, and plan of development clear enough that you can state them in your introduction?
- Do you need an interest-catching step in your introduction?
- Do you need definitions or theoretical or historical background in your introduction?
- What information do your readers really need and want in your discussion? What questions will they have? What details do they want?
- What kind of ending do you need: summary, conclusion, recommendations, graceful exit, or a combination of these?
- Do you have an executive audience? Should your "ending" come at the beginning of the report?
- Do you have material that would be better presented in an appendix rather than in the discussion? Should you leave it out altogether?
- Do you have material you need to document: direct quotes, research data and theories, illustrations?
- Does your situation call for notes or author–date documentation? If notes, will you want footnotes or endnotes? Do you have word processing capability for your notes?
- Do you plan to publish your report? Does it contain material protected by copyright law? If so, begin seeking permission to use the material as early as possible.

REVISION

- Does your letter of transmittal contain clear statements of transmittal, the occasion for the report, and subject and purpose of the report? Do you need any other elements, such as a distribution list?
- Does your preface contain clear statements of the occasion for the report and the subject and purpose of the report? Do you need any other elements, such as acknowledgments?
- Is your cover suitable for the occasion of the report? Is it labeled with all the necessary elements?
- Is your title page well designed? If you have used word processing, have you kept your design simple? Does your title page contain all the needed elements? Does your title describe your report adequately?
- Have you an effective design for your table of contents? Do the headings in your table of contents match their counterparts in the report exactly? Are the page numbers correct?
- Do you have a simple but clear numbering system for your illustrations? Do the titles in your list of illustrations match exactly the titles in the report? Are the page numbers correct?
- If needed, do you have a glossary and a list of symbols? Have you written your glossary definitions correctly?
- Are your abstracts written in the proper impersonal style? Does your informative abstract cover all the major points of your report? Have you avoided excessive detail? Does your abstract conform to the length requirements set for you?
- Does your executive summary answer the questions your executive audience will have? Have you included clear statements of your conclusions and recommendations? Have you held the length to one double-spaced page?
- Does your introduction clearly forecast your subject, purpose, scope, and plan of development? If they are needed, does the introduction contain definitions or theoretical or historical background?
- Does your discussion answer all the questions you set out to answer? Does it contain any material irrelevant to your subject and purpose?
- Does your ending draw things together for the readers? Does it condense and highlight your significant data? Does it present your conclusions and recommendations? Should the "ending" come at the front of the report?
- Does the appendix material belong in the appendix? If any material in an appendix is a key to a major point, move it to the discussion. If it seems irrelevant to the report, remove it altogether.
- Have you documented everything that needs documenting in your report? Have you documented accurately? Do your notes or citations follow the appropriate format rules?

Exercises

1. Reprinted in this exercise is the discussion section from a report titled *Smoking and Your Digestive System.*[13]

   It was written for both health care professionals and intelligent lay people. Its purpose is to examine the possible dangers of smoking to the digestive system. It reaches conclusions but does not offer recommendations. For the purpose of this exercise, pretend that you have written the report as a class assignment. Provide the following elements:

- Letter of transmittal to your writing teacher
- Title page that includes a descriptive abstract
- Introduction
- Summary
- Conclusions

How Does Smoking Contribute to Heartburn?

Heartburn is a very common disorder among Americans. Heartburn is especially common among pregnant women, with 25 percent reporting daily heartburn and more than 50 percent experiencing occasional distress.

Most people will experience heartburn if the lining of the esophagus comes into contact with too much stomach juice for a long period of time. This stomach juice consists of acid produced by the stomach, as well as bile salts and digestive enzymes that may have washed into the stomach from the intestine.

Normally, a muscular valve at the lower end of the esophagus, the lower esophageal sphincter (LES), keeps the acid solution in the stomach and out of the esophagus. Sometimes the LES is weak and allows stomach juice to reflux, or flow backward, into the esophagus.

Many people have occasional reflux episodes. Persons with heartburn usually have frequent episodes or fail to return the refluxed material to the stomach promptly. The prolonged contact of acid stomach juice with the esophageal lining injures the esophagus and produces burning pain. Smoking decreases the strength of the esophageal valve, thereby allowing more refluxed material into the esophagus.

Smoking also seems to promote the movement of bile salts from the intestine to the stomach to produce a more harmful reflux material. Finally, smoking may directly injure the esophagus, making it less able to resist further damage because of contact with refluxed material from the stomach.

Does Smoking Cause Peptic Ulcers?

An ulcer is an open sore in the lining of the stomach or duodenum, the first part of the small intestine. The exact cause of ulcers is not known. A *relationship* between smoking cigarettes and ulcers, especially duodenal ulcers, does exist. The 1989 Surgeon General's Report stated that ulcers are more likely to occur, less likely to heal, and more likely to cause death in smokers than in nonsmokers.

Why is this so? Doctors are not really sure, but smoking does seem to be one of several factors that work together to promote the formation of ulcers.

Stomach acid is important in producing ulcers. Normally, most of this acid is buffered by the food we eat. Most of the unbuffered acid that enters the duodenum is quickly neutralized by sodium bicarbonate, a naturally occurring alkali produced by the pancreas. Some studies show that smoking reduces the bicarbonate produced by the pancreas, interfering with the neutralization of acid in the duodenum. Other studies suggest that chronic cigarette smoking may increase the amount of acid secreted by the stomach. There also is some evidence suggesting that smoking increases the speed at which the stomach empties its acid contents into the small intestine. Although the evidence is inconclusive on some of these issues, all are possible explanations for the higher rate and slower healing of ulcers among smokers.

Whatever causes the link between smoking and ulcers, two points have been repeatedly demonstrated: persons who smoke are more likely to develop an ulcer, especially a duodenal ulcer, and ulcers are less likely to heal quickly among smokers in response to otherwise effective treatment. This research tracing the relationship between smoking and ulcers strongly suggests that a person with an ulcer should stop smoking.

How Does Smoking Affect the Liver?

The liver is a very important organ that has many tasks. Among other things, the liver is responsible for processing drugs, alcohol, and other toxins to remove them from the body. There is evidence that smoking alters the ability of the liver to handle these substances. In some cases, this may influence the dose of medication necessary to treat an illness. One theory, based on current evidence, also suggests that smoking can aggravate the course of liver disease caused by excessive alcohol intake.

Does Smoking Help Control Weight?

A common belief is that smoking helps to control weight. Smokers do, indeed, weigh less, on the average, than nonsmokers. And those who quit smoking *are* more likely to gain weight. Most people think this is because smokers eat less than nonsmokers.

Some researchers have found, however, that smokers actually eat *more* than nonsmokers. How can they weigh less? What happens to the extra calories? Scientists are not really sure about the answers to these questions, but they caution smokers not to think that just because they weigh less, they are healthier than if they didn't smoke. Research shows that the bodies of smokers use food less efficiently than nonsmokers. Scientists are still studying what implications this has on the long-range health of smokers.

Can the Damage to the Digestive System Be Reversed?

Some of the effects of smoking on the digestive system appear to be of short duration. For example, the effect of smoking on bicarbonate production by the pancreas does not appear to last. Within a half-hour after smoking, the production of bicarbonate returns to normal. The effects of smoking on how the liver handles drugs also disappear when a person stops smoking. While doctors suspect that most other digestive abnormalities caused by smoking would also disappear soon after stopping smoking, this question has received little study.

2. This exercise uses the report on smoking in Exercise 1. Pretend you have written that report for the executives and physicians of a health maintenance organization (HMO). The HMO, for a set annual fee, provides health care to members. As part of its service, the HMO publishes pamphlets for its members describing various medical problems and telling them ways of leading healthier lives. Write an executive summary for the report that ends with a recommendation for a new pamphlet about the digestive problems caused by smoking.
3. Compile a list of difficult and unusual terms you may have to introduce in the final report of your own selected project. Will you need a glossary? If so, prepare a draft of the glossary.
4. Prepare at least a tentative table of contents for your proposed final report. Use at least two levels of headings.
5. Decide what material, if any, you should place in the appendix of your final report. If it is placed in an appendix, who would use the material and for what purpose? Why would you not include this material somewhere in the discussion of your report or eliminate it altogether?
6. In a published report or textbook, locate as many of the elements discussed in this chapter as possible. What variations and departures do you find?
7. The following sources are given in the order in which they are cited in a report but without attention to proper format. Following the guidelines in this chapter, list them in proper order and with proper format as they would appear in (a) a list of notes and (b) a list of works cited.

   David N. Dobrin. Writing and Technique. Urbana, Illinois. National Council of Teachers of English. 1989. Page 22 cited.

   Myra Kogen, editor. Writing in the Business Professions. Urbana, Illinois. National Council of Teachers of English. 1989. Page 46 cited.

   Jeanne W. Halpern. An Electronic Odyssey. Printed on pages 157 to 189 of Writing in Nonacademic Settings. Lee Odell and Dixie Goswami, editors. New York, New York. The Guilford Press. 1985. Pages 160 to 161 cited.

   Nancy Allen, Diane Atkinson, Meg Morgan, Theresa Moore, Craig Snow. Shared-Document Collaboration: A Definition and Description. Printed on pages 70 to 90 of Iowa State Journal of Business and Technical Communication. Volume 1. Issue number 2. September, 1987. Page 72 cited. Journal pages its issues separately.

   Jo Ann C. Gutin. Good Vibrations. Printed on pages 45 to 54 of Discover. June 1993. Page 51 cited.

   David N. Dobrin. Writing and Technique. Urbana, Illinois. National Council of Teachers of English. 1989. Page 32 cited.

   U.S. National Aeronautics and Space Administration. Search for

Extra Terrestrial Intelligence. Washington, District of Columbia. U.S. Government Printing Office. No date. Page 18 cited.

Jo Ann C. Gutin. Good Vibrations. Printed on pages 45 to 54 of Discover. June 1993. Page 52 cited.

Richard C. Freed and David D. Roberts. The Nature, Classification, and Generic Structure of Proposals. Printed on pages 317 to 352 of Journal of Technical Writing and Communication. Volume 19. Issue number 4. 1989. Page 319 cited. Journal with continuous pagination.

Design and Typography. Printed on pages 561 to 584 of The Chicago Manual of Style. 13th Edition. Chicago, Illinois. The University of Chicago Press. 1982. Page 70 cited.

CHAPTER 11

GRAPHICAL ELEMENTS

In creating technical reports, you have two main ways of communicating information: verbal expression (words) and visual display (pictures). We live in a society that emphasizes visual forms of communication—movies, advertising, and television particularly—but even familiar forms of verbal communication, such as newspapers and magazines, rely on visual as much as verbal communication. Most report writers soon realize the value of effective visual presentation, as most readers, products of a visual society, quickly tire of writing that relies too heavily on verbal presentation. Document design and format, discussed in Chapter 9, are the basic tools for making verbal expression easy to see and understand. Then, by knowing how to use the many varieties of graphs, drawings, figures, and tables, you have additional tools for turning words into visual images that help readers understand numbers and concepts quickly and accurately.

As in all aspects of technical writing, you have an ethical responsibility to develop graphics that present an accurate picture of your data. You may want to refer frequently to Chapter 2 to help you remain aware of issues that pertain to ethical presentation of technical information. After discussing guidelines on choosing the best graphic to present a specific type of concept, we will shift our focus to emphasize accurate graphic development. In learning how to present technical information effectively, you will need to know a few basic rules for developing graphs, drawings, figures, and other visual aids so that these are accurate, easy to see and understand, and appropriate for the document you need to write.

Today, word processing software and graphics software offer an almost unlimited range of visual display, as the examples on the color insert illustrate. It is important to be aware of ways that graphics can be used to communicate your message, to know the basic types of visual presentation available, and to learn how to use them effectively. Most graphics programs give you choices for displaying data, but the visual form you choose will be yours. Because increasingly powerful graphics software can create artistic but highly distorted visual representation, you need to remember that only you can determine whether your graphic presents your data or concept in an effective, accurate way that enhances your reader's understanding of your message.

Because the focus of this chapter is on graphics, we will present it through graphics, rather than verbal discussion. We will show you how to use verbal presentation to support visual presentation, but our emphasis will be on showing you an array of graphics—many good ones, some of questionable quality. By the time you complete the exercises in this chapter, you should have a basic understanding of how to use common types of graphics to help your readers visualize either data or concepts. We provide guidelines and methods for the following:

- Choosing graphics
- Designing tables
- Designing line graphs
- Developing other forms of graphics
- Integrating text and visuals
- Designing graphics that are accurate and effective
- Choosing the best graphic for the information and the audience

Choosing Graphics

Why use graphics? History shows us that ancient peoples used pictures and drawings to communicate long before they developed writing. Art itself exists as a monument to the human tendency to express ideas through visual expression. In spite of the importance of writing in modern cultures, visual ways of expressing information are quicker and more effective than writing for helping readers visualize trends, processes, relationships among steps or parts of processes, and comparisons. Graphics can help you show a reader what information is the most important. Because thinking and reasoning are enhanced if we can *see* the idea or the information being presented, effective graphics can be powerful, persuasive tools for helping readers understand and remember. For example, the drawing in Figure 11-1 comes from an early sixteenth-century medical book. The drawing helps you see how broken limbs were straightened in the early 1500s, even though the process is described in some detail in several pages of text. The modern drawings of coastlines, shown in Figure 11-2, help you understand the meaning of the term *barrier island* and related coastal surfaces. The modern drawing works just as effectively as the sixteenth-century drawing to convey its concept.

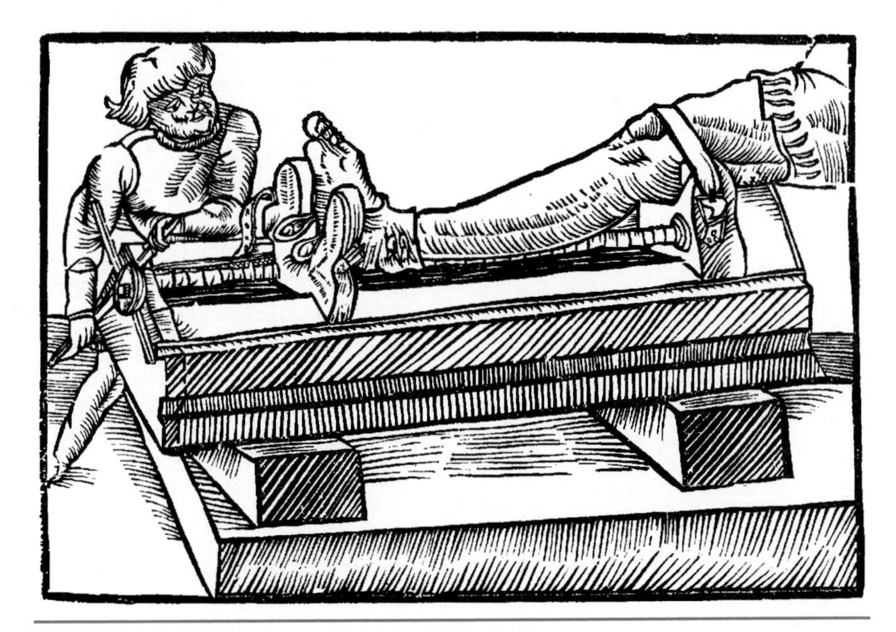

Figure 11-1 Hieronymus Von Braunschweig, *Experyence of the Warke of Surgeri,* London, 1525.

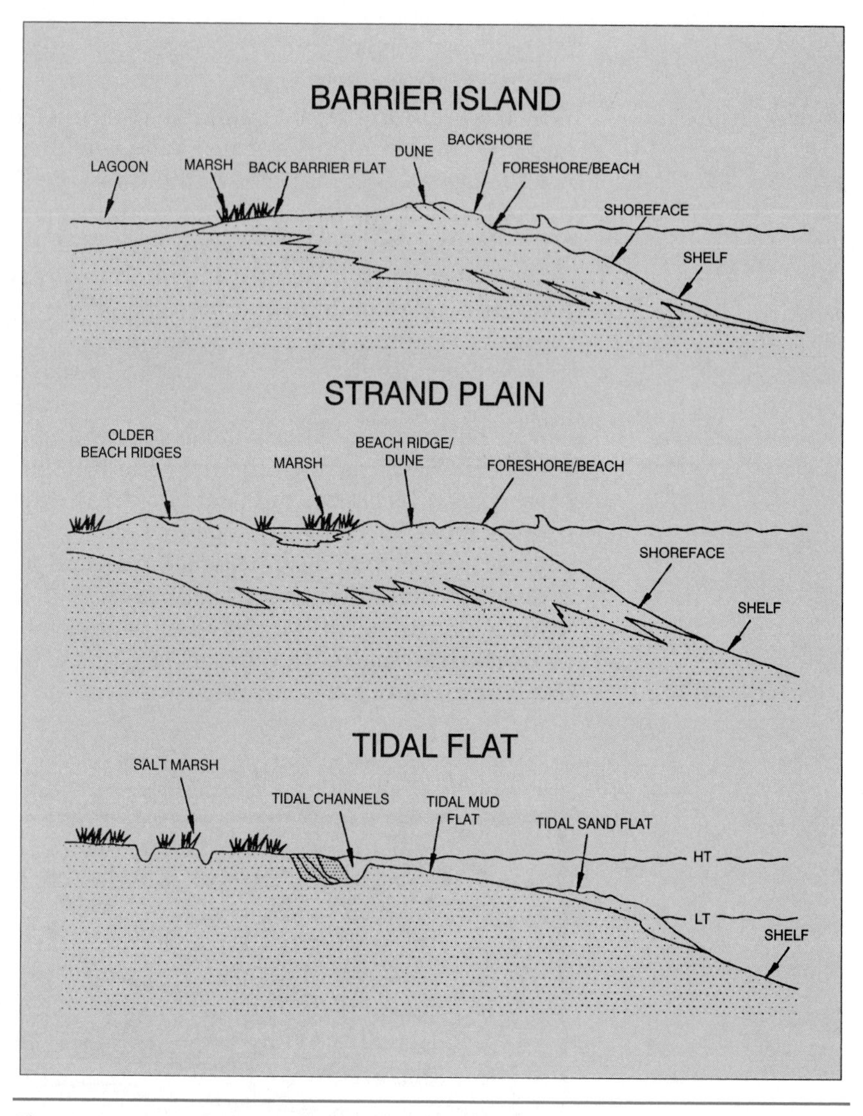

Figure 11–2 Drawing Used to Visualize Features Hidden from Normal View
Source: Joann Mossa and Edward P. Meisburger, *Geomorphic Variability in the Coastal Zone* (Washington, DC: Department of the Army, Waterways Experiment Station, Corps of Engineers, Vicksburg, MS, May 1992) 37.

In planning your report, be aware of ways in which information can best be presented in visual form. As you write and revise, continue to be aware of making your material more visually accessible. In Chapter 5, we emphasize that style should make your meaning clear to your reader; that is, you want your reader to *see* what you mean. Graphics can also help your reader visualize data, relationships, processes, and concepts. If words will achieve clarity, then graphics may not be necessary. If graphics supported by verbal discussion will aid your reader, then look for ways to visualize your point. As illustrated in Figure 11-2, visual presentation is an excellent way to make abstract information immediately obvious to readers. In short,

- To help your readers see trends, use graphs, such as bar graphs, line graphs, and surface graphs.
- To summarize information, use bar graphs, line graphs, and surface graphs.
- To help your readers see relationships within quantities of data, use tables.
- To emphasize or reinforce the meaning of data, begin with a table and then use a graph to accentuate the differences suggested by the data.

For example, Figure 11-3 shows you two graphics. The table, as its title indicates, allows readers to see how usage of birth control methods in India has changed during a 20-year period. The graph allows us to see much more clearly how birth control has increased and what methods are used. When you collect data, you will likely need to use a table to arrange it initially. From the table, many graphics programs will allow you to select a more powerful visual way of presenting the data.

- To help your reader visualize processes, use drawings and flowcharts.

Figure 11-4 visualizes the development cycle of the termite and shows the different individuals that compose the termite colony. The representation of the members of the termite caste system and the reproductive stages of the termite help readers better understand the overall purpose of the publication that contains this graphic, the recognition and control of termites, and their damage to wood.

- To improve the impact of your message, use photographs and creative graphics.

The map in Figure 11-5 shows immediately what states and geographical areas are targets for termite damage. The photograph (Figure 11-6) of termite damage in a piece of wood achieves two goals: It accentuates the problem caused by termites and provides a method of identifying termite damage.

- To help simplify concepts, use drawings, flowcharts, diagrams, algorithms, and schematics.
- To present objects, use photographs if you need detail or realism and if cost is not a factor. Use drawings if you want to show your reader only selected features of an object.

Table 2.
Couples Effectively Protected, by Method of Family Planning: Selected Years

	Percent currently protected			Percent distribution of protected couples		
Method	1970-71	1980-81	1989-90	1970-71	1980-81	1989-90
All	10.4	22.8	43.3	100	100	100
IUD	1.4	1.0	6.3	13	4	15
Sterilization	8.0	20.1	30.1	77	88	70
Other modern	1.0	1.7	6.9	10	7	16

CHILDLESS WOMEN: 1991
Percent of ever-married women
aged 45 to 49. . . . . 4.4

Source: IDFW, 1991, table E.1; and IRG, 1984, table C6.

Figure 4.

Number of Couples Effectively Protected, by Method: 1971 to 1990

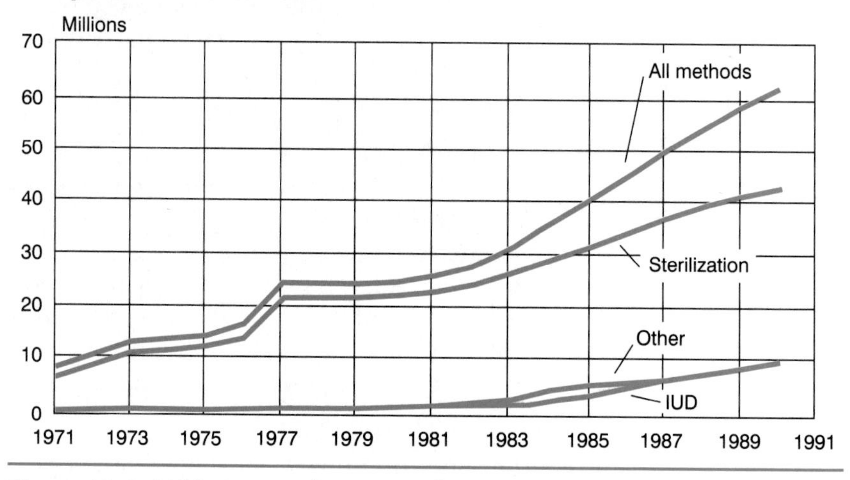

Figure 11–3 Table Converted into a Graph
Source: Population Trends: India (Washington, DC: U.S. Department of Commerce, Bureau of the Census, Center of International Research, Oct. 1992) 13.

Figure 11–7 provides a drawing that allows visual comparison of the termite and the ant, both of which may exist in proximity. The drawings highlight the main physical differences to enable recognition.

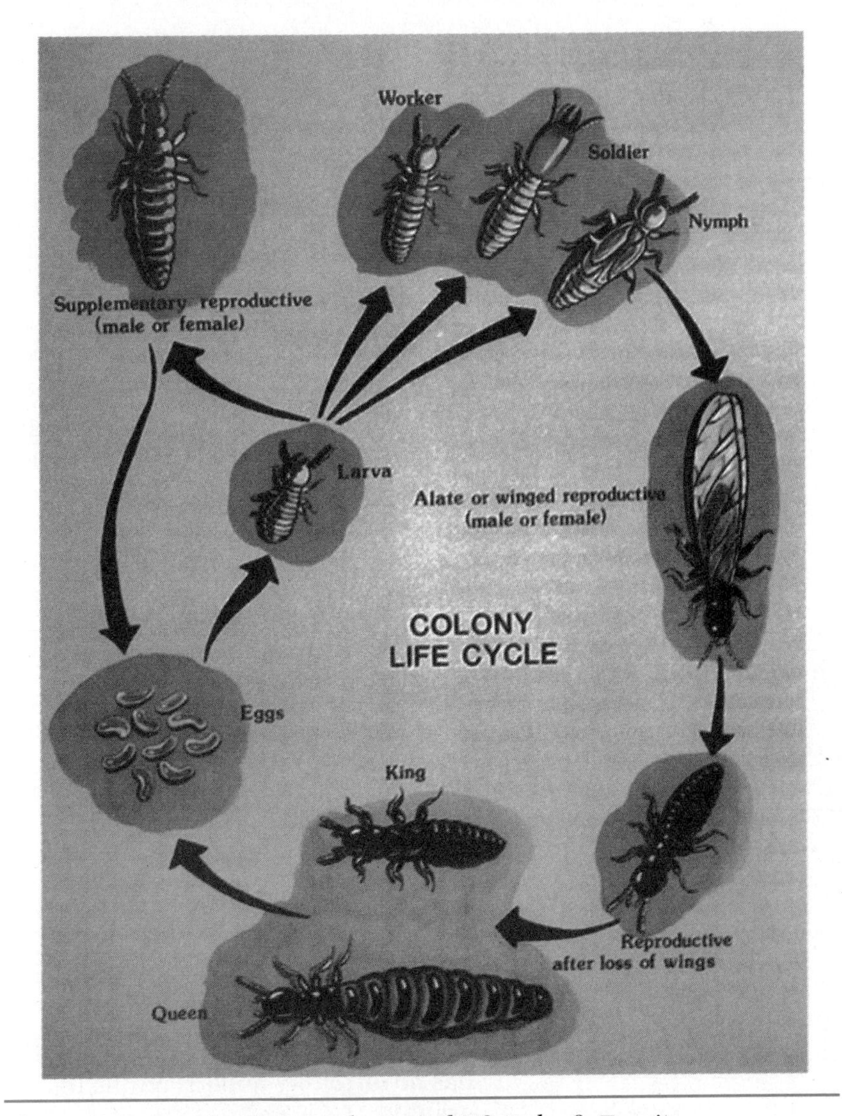

Figure 11–4 Drawing Showing the Typical Life Cycle of a Termite
Source: Forest Service, United States Department of Agriculture, *Subterranean Termites—Their Prevention and Control in Buildings* (Washington, DC: U.S. GPO, 1989) 5.

Figure 11–8, on an entirely different subject, shows the levels of education in the United States and how lower levels feed into higher levels.

As each of these graphics illustrates, you have a variety of ways to help your reader see what you mean. In choosing a graphic, always consider your purpose and your reader's profile: the kind of information you need to present, the context in which you are presenting it, and your reader's preferences. Consider graphics when you are planning

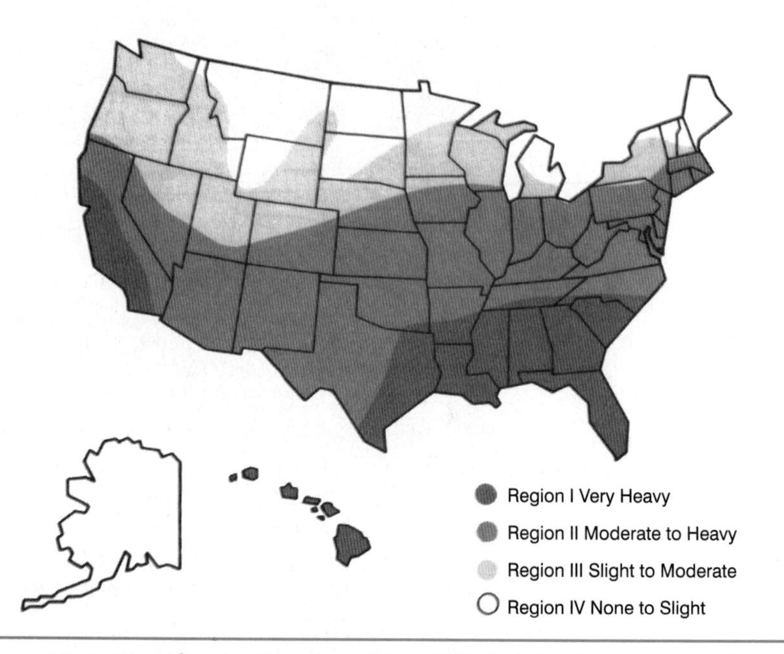

Figure 11–5 Map Showing Location of Termite Infestation Areas by Intensity
Source: Forest Service, United States Department of Agriculture, *Subterranean Termites—Their Prevention and Control in Buildings* (Washington, DC: GPO, 1989) 2.

your document, but continue to look for ways to visualize your point even as you are writing and revising.

In deciding when and how to use graphics as you plan your document, remember these five guidelines:

- Keep your graphic as simple as possible so that your reader has no difficulty understanding the message conveyed by the graphic.
- Place the graphic as close to the point it explains as possible. Announce the graphic—what it is or shows, insert the graphic, then add any verbal explanation your reader will need to fully understand the graphic.
- Know the expected length of your document. Graphics can extend the size of any report and increase reproduction costs.
- Realize that you usually have a *choice* of graphics. If a graphic you decide to use initially does not convey your point quickly and accurately, consider other ways of presenting the information visually.

Engineers design circuit boards with systems like IBM's Engineering Design System.

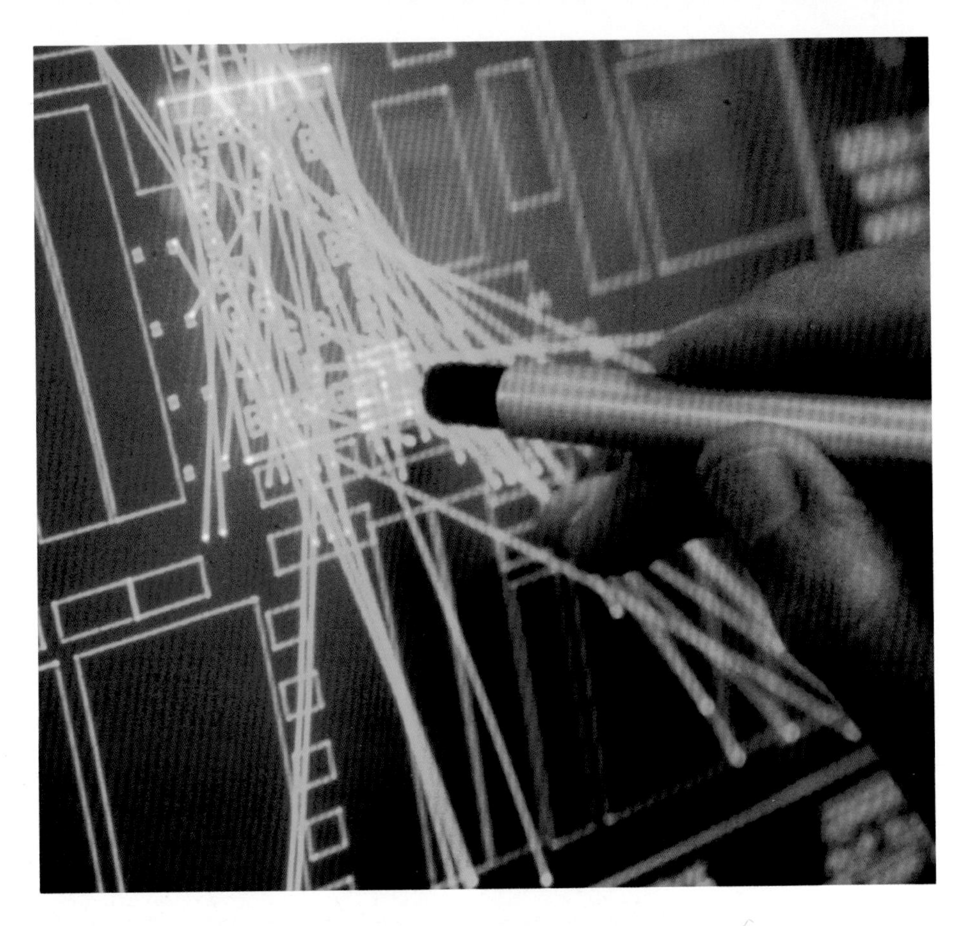

Source: *Courtesy of IBM Corporation*

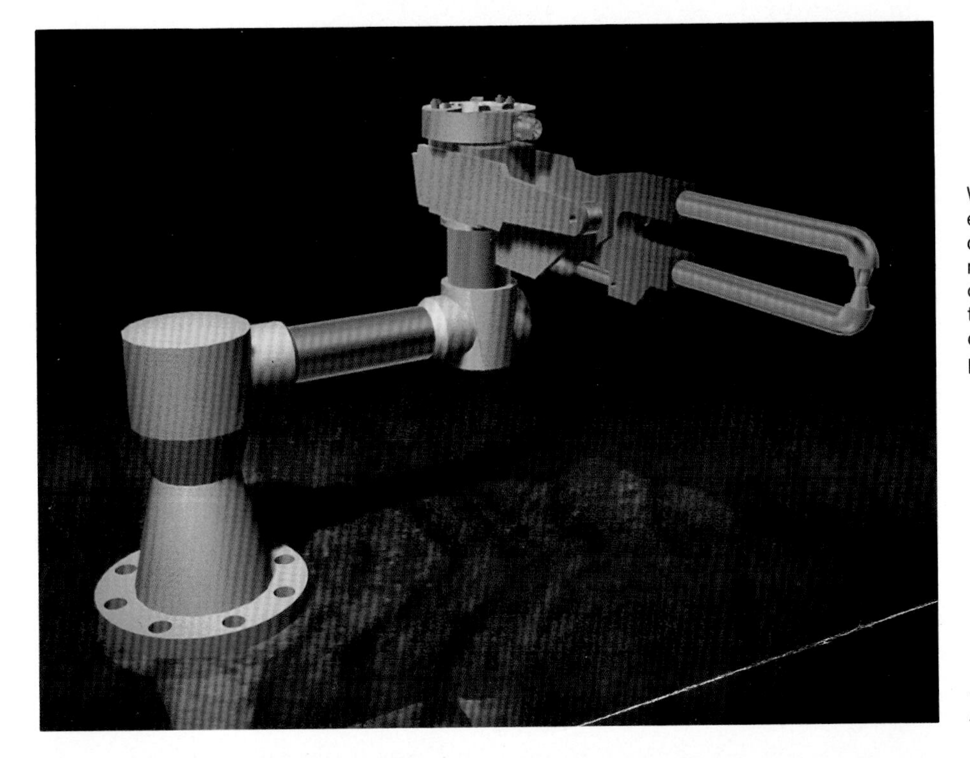

With computers, engineers can see three-dimensional images of machines that they are designing. They can rotate the image to view the design from different perspectives.

Source: *Courtesy of Autodesk, Inc.*

You can use computers to turn data into contour maps.

The three-dimensional contour map represents elevation. The underlying two-dimensional contour map represents the acquifier beneath the surface.

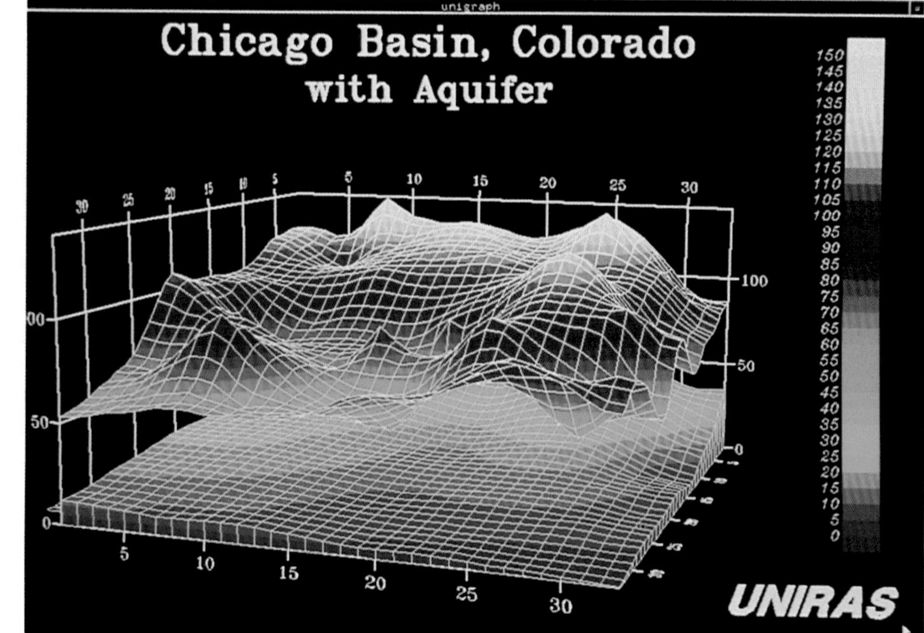

Source: *Courtesy of UNIRAS*

Computers can be used to track weather data.

Here data on Hurricane Andrew is superimposed on a map of Southern Florida and nearby islands.

Source: *Courtesy of Sunset Photo*

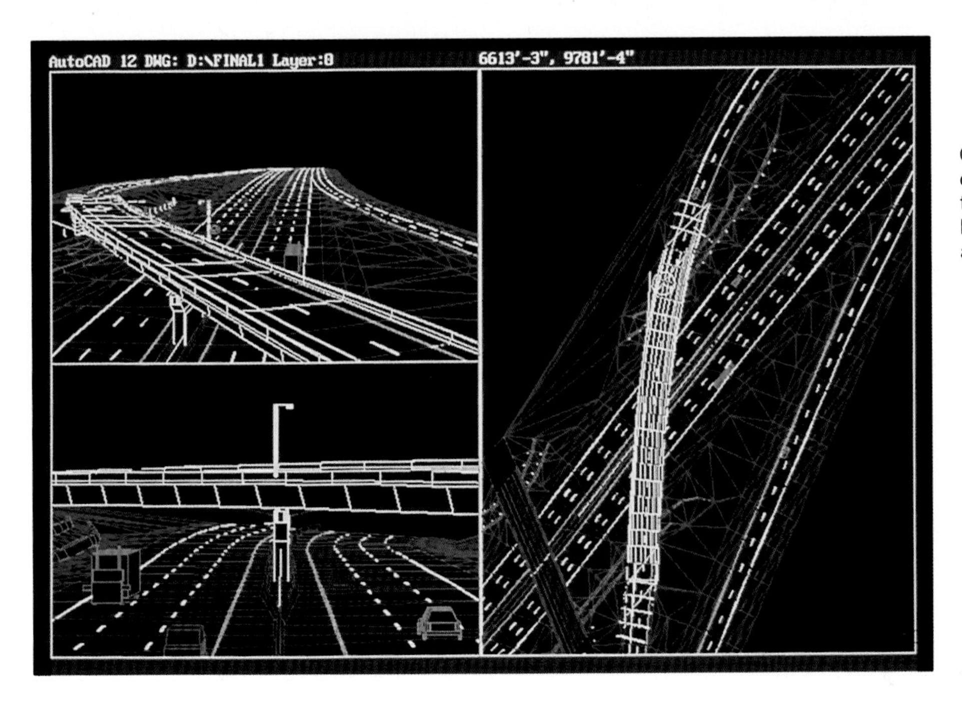

Civil engineers use computers to simulate traffic flow and plan highways with overpasses and intersections.

Source: *Courtesy of Autodesk, Inc.*

Here you see several views of a model of the Formula One racing car developed for the Penske racing team.

Source: *Courtesy of Computervision Corporation*

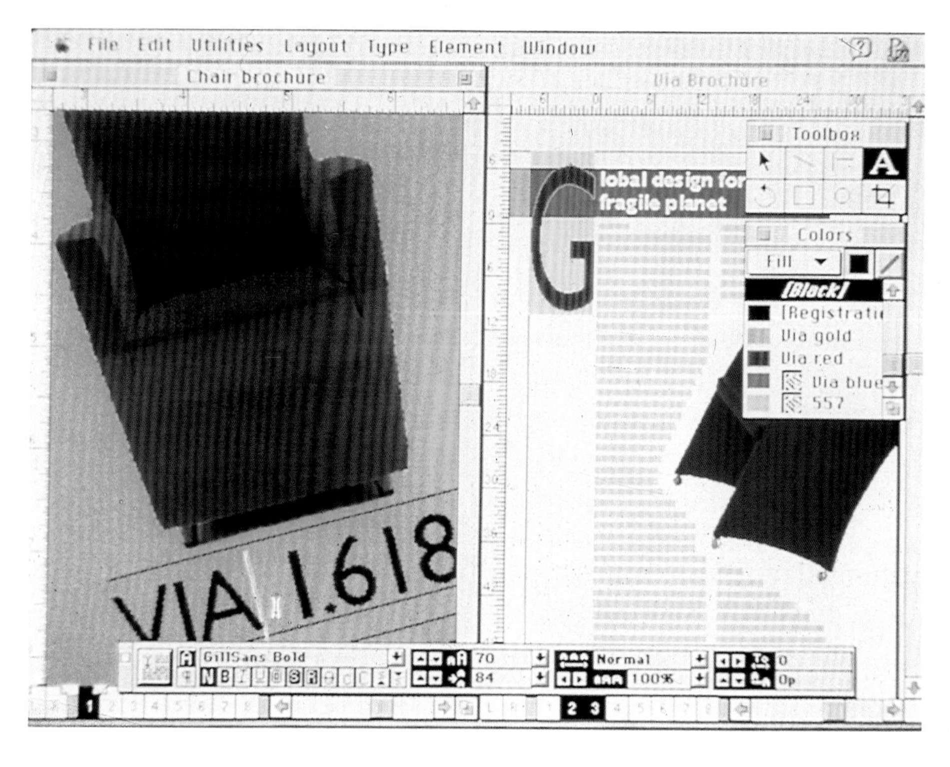

Technical writers can use several related programs to combine text and graphs.

Source: *Courtesy of Aldus Corporation*

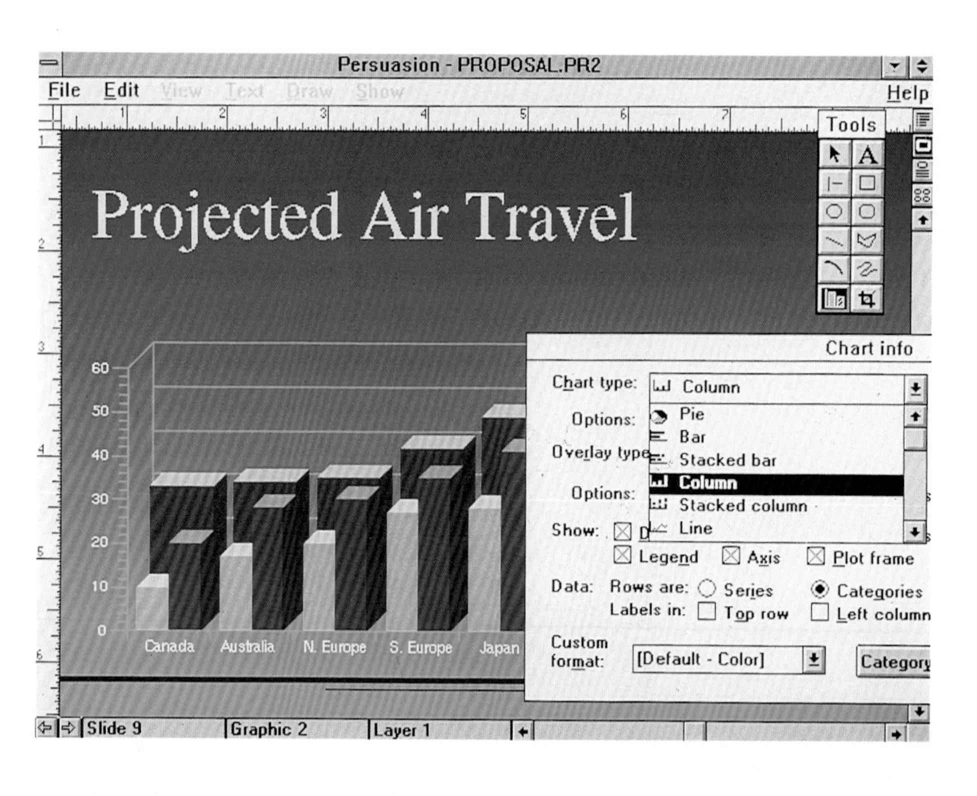

Numerical information is much more easier to grasp in a well-designed chart than in prose.

Before completing this chart, technical writers need to add

- a time period to the title
- a legend that tells what each color means
- a label to the y-axis so that readers will know what the numbers from 0 to 60 represent

This type chart works well only when all the bars in the back are larger than the bars in the front.

Source: *Courtesy of Aldus Corporation*

Technical writers can use several related programs to combine text and graphs.

Here the writer combines text from a word processor with data for a spread sheet. The writer can put the data in a table, as shown in the underlying window, or in a graph, as shown in the window at the bottom right of the screen.

Source: *Courtesy of Microsoft Corporation*

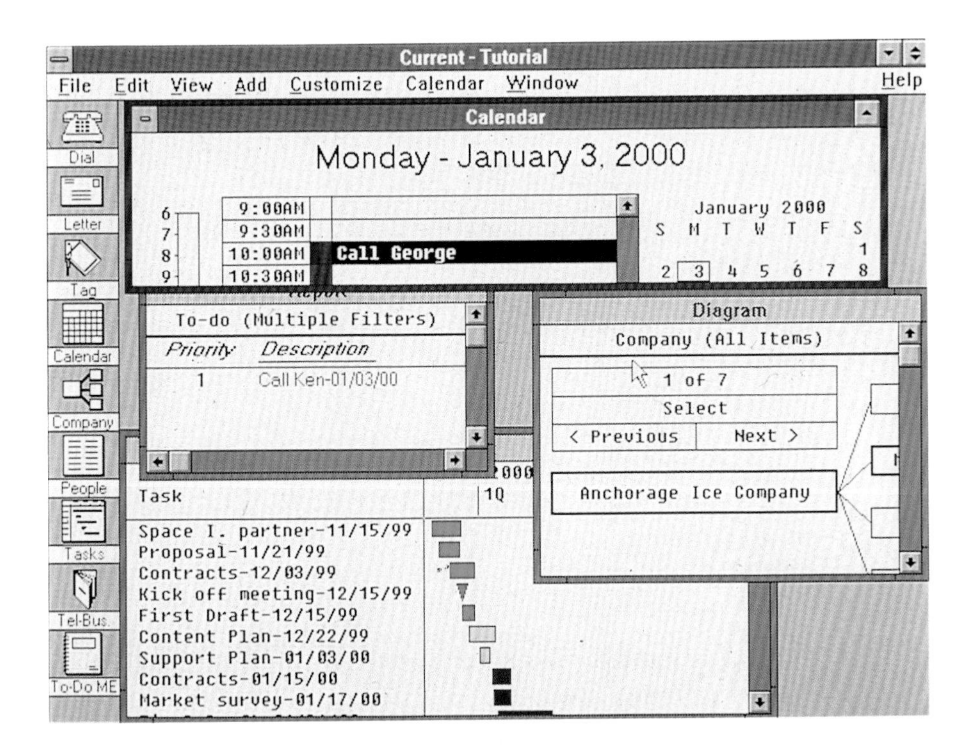

All of the people who are working on a project can keep track of plans and schedules with computer programs like these.

Technical writers have an important role in helping software designers decide on menus names and icons (small pictures like the telephone for "dial") that users will understand easily.

Source: *Courtesy of Microsoft Corporation*

On workstations like this one, you can have several views of a project open at the same time.

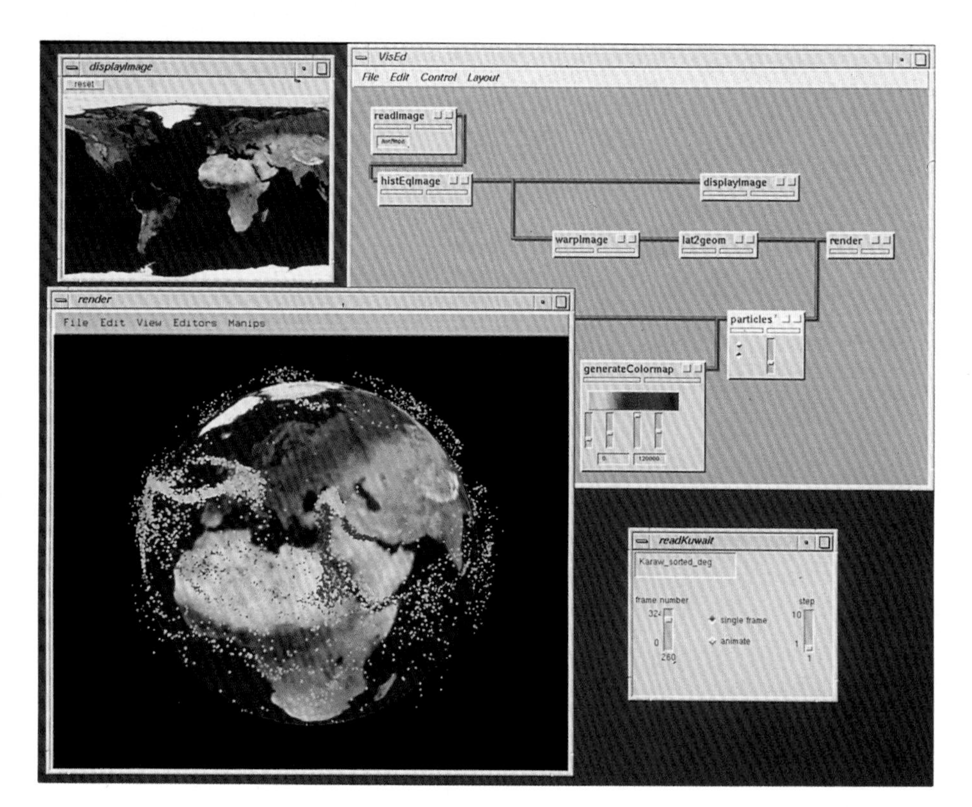

Source: *Courtesy of Silicon Graphics*

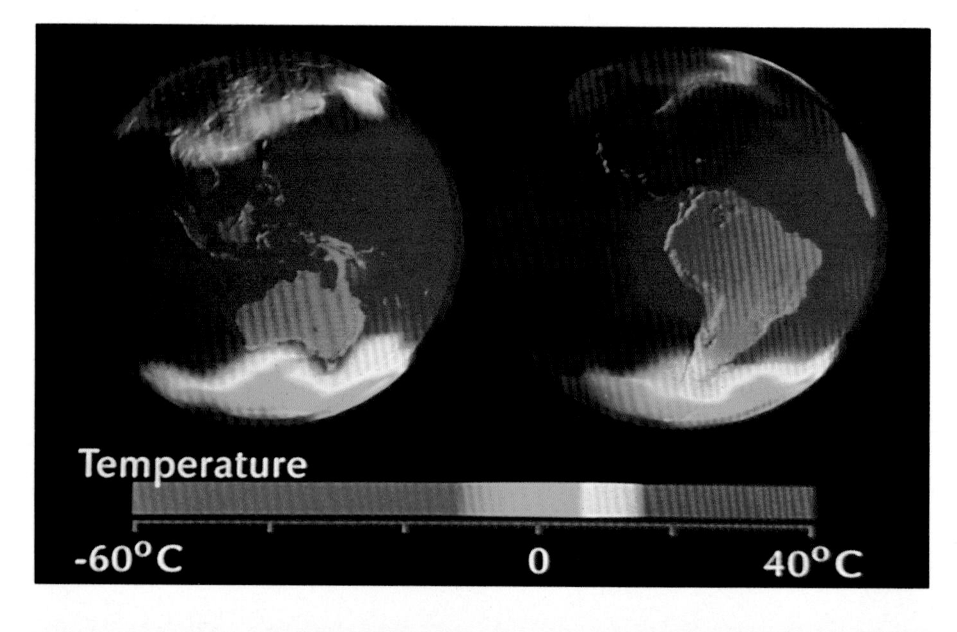

You can use color to make a point dramatically with computers.

In this picture, red signifies "hot" and shows the effects of global warming throughout the earth.

Source: *Courtesy of National Center for Atmospheric Research, photo by Thomas Bettge*

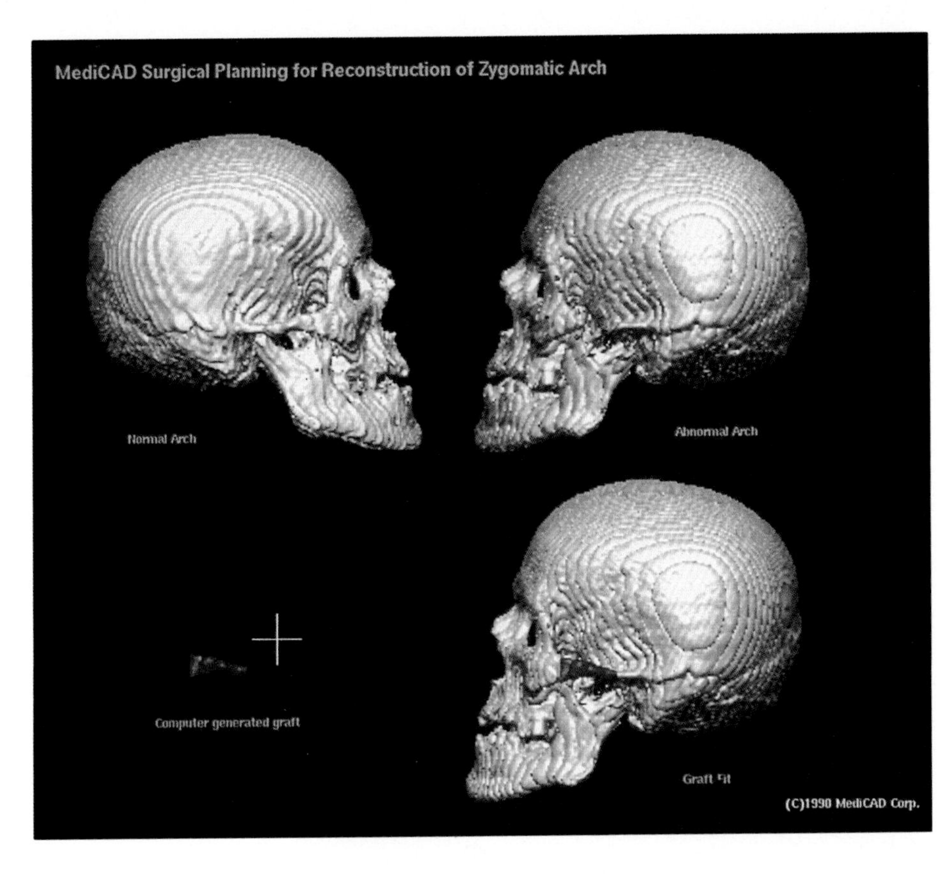

Medical doctors use computers to visualize the results as they plan reconstructive surgery.

Source: *Courtesy of Silicon Graphics*

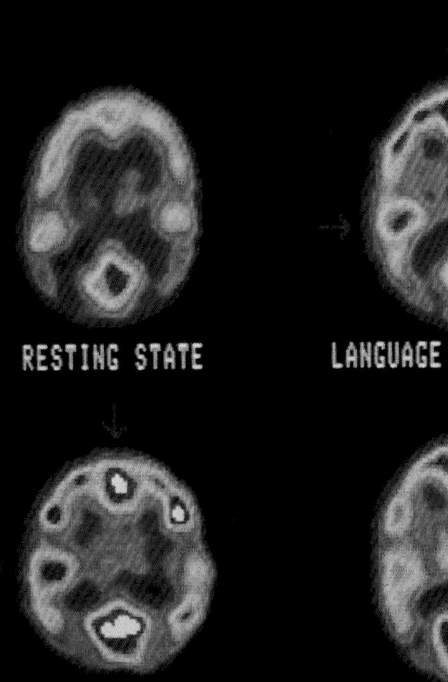
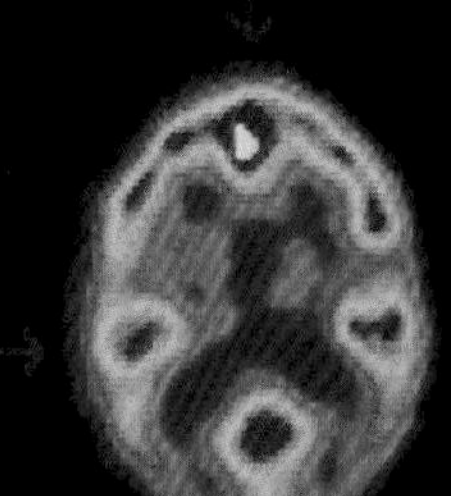
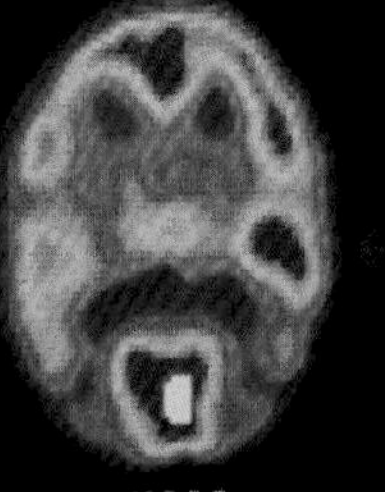

We use different parts of the brain for different activities. In this picture, researchers are using computers to display what happens when a person listens to language or music or both.

Source: *Courtesy of M.E. Phelps & J.C. Mazziotta, UCLA School of Medicine, #36–23a*

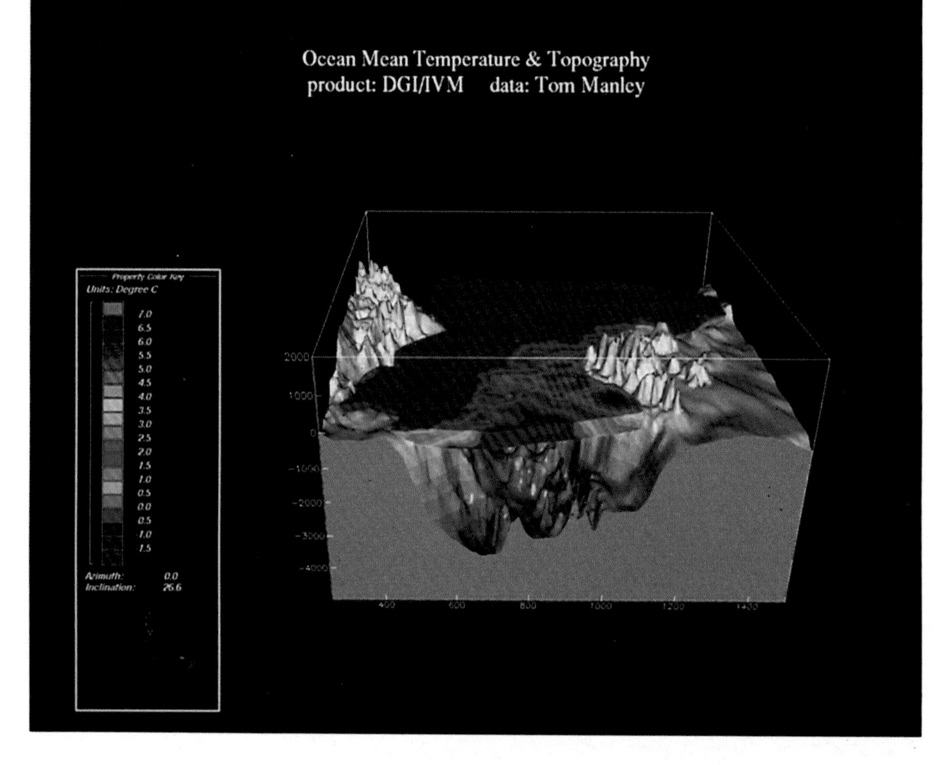

With computers, you can combine data about two features, for example, temperature and altitude, in one picture.

Source: *Courtesy of Silicon Graphics*

Civil engineers can use computers to show how cement particles in concrete react with water and different minerals.

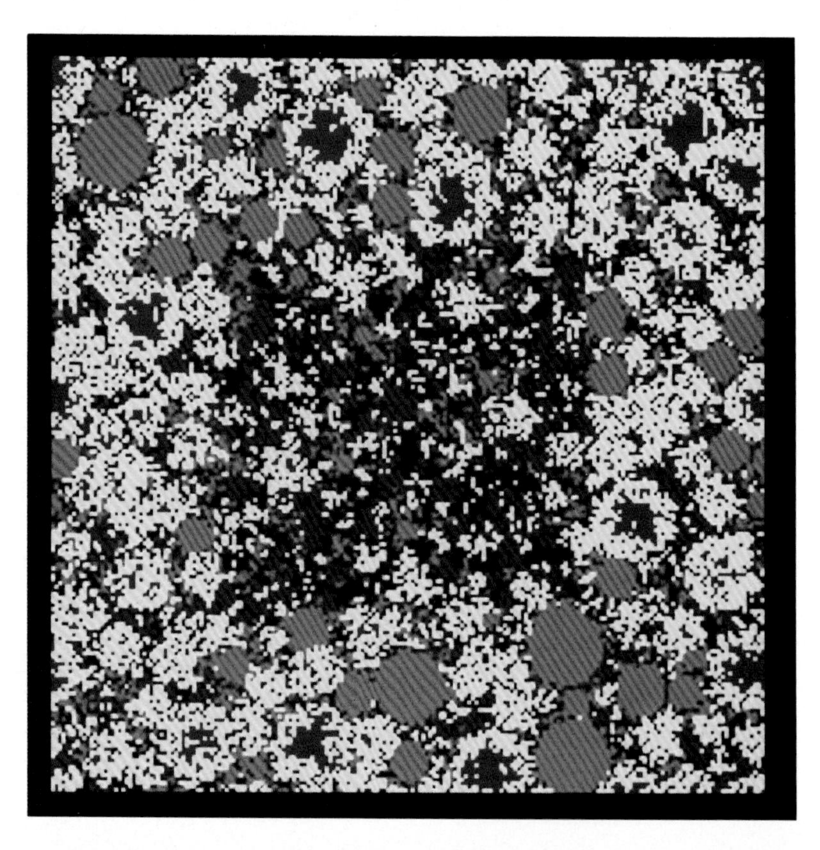

This simulation was done on a Cray-2 supercomputer by the National Institute of Standards and Technology.

Source: *Courtesy of Edward Garbocczi, National Institute of Standards and Technology*

Figure 11–6 Photograph Showing Termites and Their Damage to Wood
Source: Forest Service, United States Department of Agriculture, *Subterranean Termites—Their Prevention and Control in Buildings* (Washington, DC: GPO, 1989) 8.

- Computer graphics software makes available an increasing number of graphics. However, the effectiveness of your information in any form is a decision that you, not the computer, must make.

After you make your graphics selection, consult the following guidelines to be sure that the image your graphics program generates is as simple, clear, and effective as possible. As you plan and develop your graphics, you may want to keep the summary chart on page 325 handy, as it provides a quick review of the principles we will present here.

Designing Tables

As you examine the following guidelines for developing tables, refer to Figures 11-9 and 11-10, which illustrate each point:

- Tables need titles. If you use several tables in your report, number the tables. Place the table number and title above the table. Be sure that the title clearly indicates the content of the table.
- If you use a table from another source, give that source beneath the table.

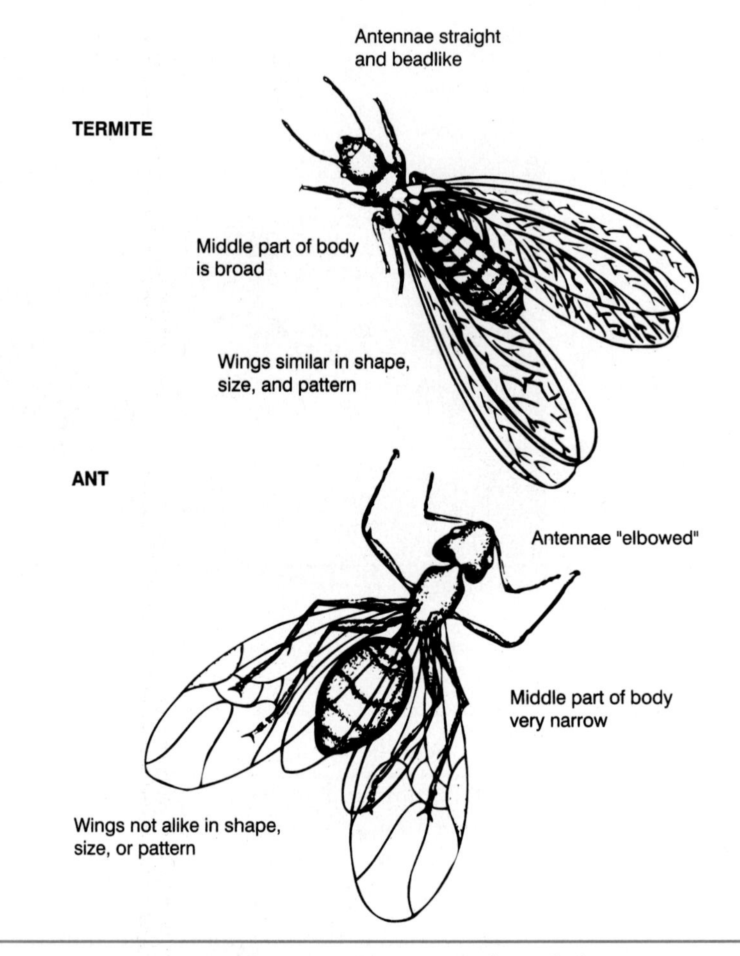

Figure 11–7 Drawing Showing Differences in Physical Characteristics of Ants and Termites

Source: Forest Service, United States Department of Agriculture, *Subterranean Termites—Their Prevention and Control in Buildings* (Washington, DC: GPO, 1989) 8.

- Every column in a table should have a heading that identifies the information beneath it.
- Headings should be brief. If headings need more explanation, include this information in a footnote beneath the table. Use lowercase letters, numbers, or symbols (e.g., *, +, or §).
- If possible, box your table.
- Keep tables as simple as possible. Do not include data not relevant to your purpose. Avoid excessive lines or data that give your table a crowded appearance.
- Always alert your readers to graphics that will appear.

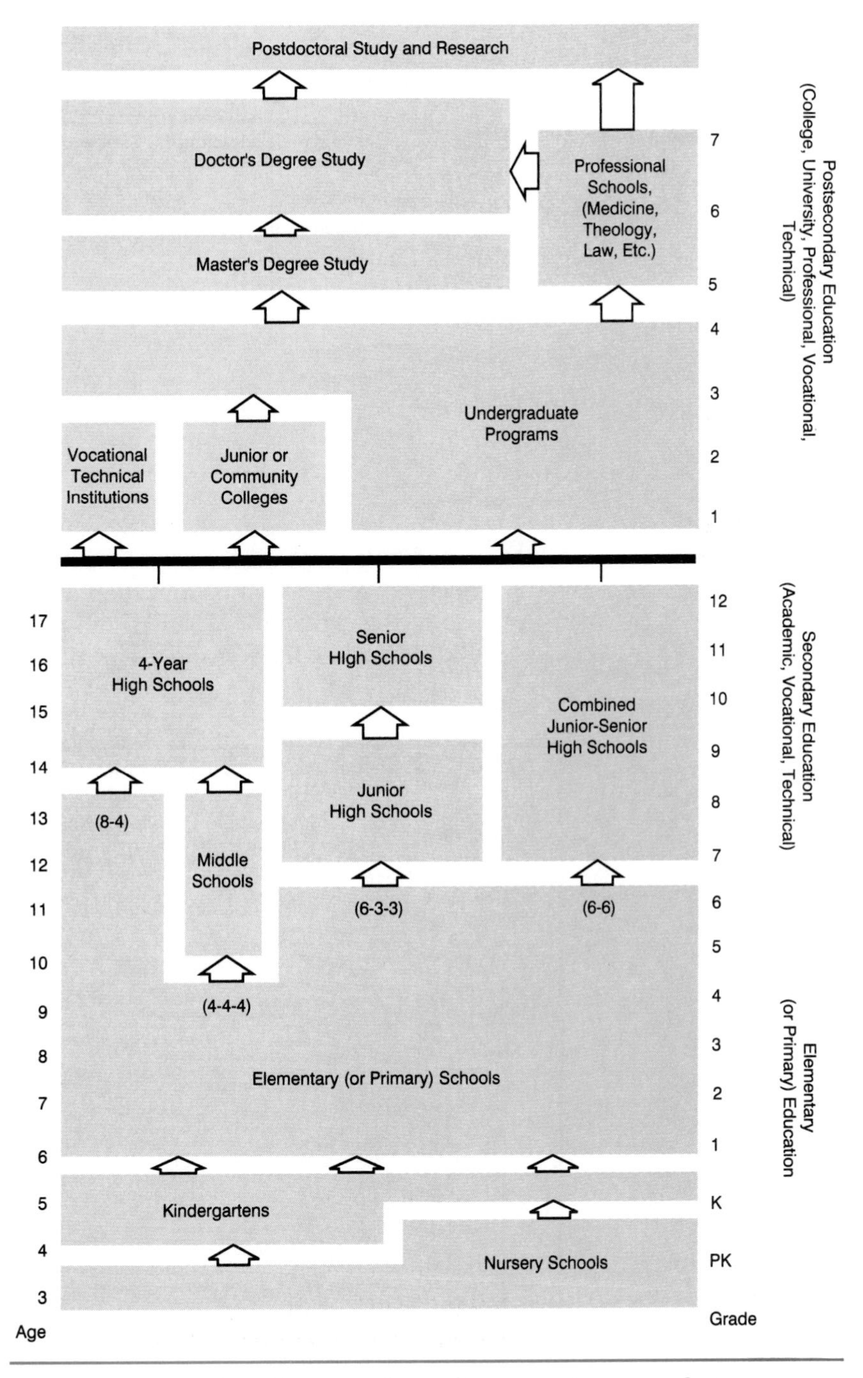

Figure 11–8 Flowchart Showing Relationship among Segments of a Process
Source: U.S. Department of Education, *Digest of Education Statistics* (Washington, DC: Office of Educational Research and Improvement, 1991) 59.

TABLE NUMBER
TITLE

Stub Heading[a]	Column Heading[b]	Column Heading	
		Subheading	Subheading
Line Heading	Individual "cells" for tabulated data		
Subheading			
Subheading			
Line Heading			

[a]Footnote
[b]Footnote

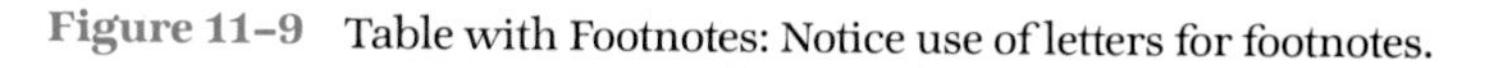

Figure 11–9 Table with Footnotes: Notice use of letters for footnotes.

- Attempt to place the table at the point in the discussion where it will help your reader. Do not lead readers through a complicated prose explanation and then refer to the graphic that simplifies the explanation. Send them to the graphic immediately, and they can cut back and forth between the prose and the

Table 2. Total annual energy inputs to various irrigation systems in equivalent gallons of diesel per acre to apply 24 inches of net irrigation.

Irrigation System[1]	Installation Energy[2]	Labor Energy	Pumping Energy[3]		Total Energy	
			Zero	500 ft	Zero	500 ft
Traditional Surface	?	20	0	396	20	416
Modern Surface	31[4]	1	16	214	48	246
Hand-move Sprinkle	17	1	58	285	76	303
Center-Pivot Sprinkle	42	1	51	249	94	292
Trickle	57	1	41	217	99	275

[1]All of the systems occupy a 160 ac (½-x ½-mile) square field with the water supply in one corner. The traditional system is surface irrigated and has an assumed efficiency of 40%.
The modern surface system has precision land leveling, gated pipe with automatic gates and a return flow system. The assumed inlet pressure is 17 psi and efficiency of 80%.
The hand-move sprinkle system utilizes low-pressure sprinklers and has an assumed system inlet pressure of 56 psi and efficiency of 70%.
The center-pivot system is a standard ¼-mile lateral fitted with low-pressure sprinklers and has an assumed inlet pressure of 56 psi and efficiency of 80%.
The trickle system has an assumed inlet pressure of 50 psi and efficiency of 90%.
[2]Taken from Batty et al. (1975) and spread out over the expected life of the system and assuming aluminum can be recycled.
[3]Pumping Energy/Acre = $\frac{\text{79.2 x inches applied x total feet of head}}{\text{Pumping efficiency x irrigation efficiency}}$
For this table the pump efficiency was assumed to be 60 percent.
[4]Land grading 11.2, aluminum pipes 20 = total installation energy.

Figure 11–10 Table with Footnotes: Notice use of numbers for footnotes.
Source: Office of Research and Development, *Design Manual: Municipal Wastewater Stabilization Ponds* (Washington, DC: U.S. Environmental Protection Agency, 1991).

graphic as necessary. For example, we introduced Figures 11-1 and 11-2 before we began using these two examples to help you visualize the verbal guidelines for designing tables.

- Integrate all tables with your verbal discussion. Introduce the table, present it, and then follow the table with appropriate analysis. Any time you refer to a table, use the table number.

Many computer programs will insert tables and other graphics into the text when you instruct them to do so. However, effective integration of the pertinent text with the graphic is your responsibility.

Designing Graphs

Information presented in tables can often be converted into a graph, as we have already shown in Figure 11-3. As you looked at Figure 11-3, your eyes were probably drawn first to the line graph rather than the table because the line graph allowed you to see the relationships among the numbers quicker than the table did. The basic kinds of graphs you can use to visualize data are bar graphs, circle graphs, divided bar graphs, and line graphs. See Figure 11-11. Guidelines for graphs are similar to those for tables. Figure 11-12 exemplifies two effectively presented graphs. Examine these two graphs and the discussion that accompanies them as you consider the following guidelines:

- Graphs need clear titles. Number graphs if you include more than one in a report.
- Introduce the graph in the text and discuss its significance. Try to place the graph as close to the discussion to which it relates as possible.
- Make the graph big enough to be legible.
- Avoid placing too much information on a graph. Computer graphics will allow you to pack a tremendous amount of

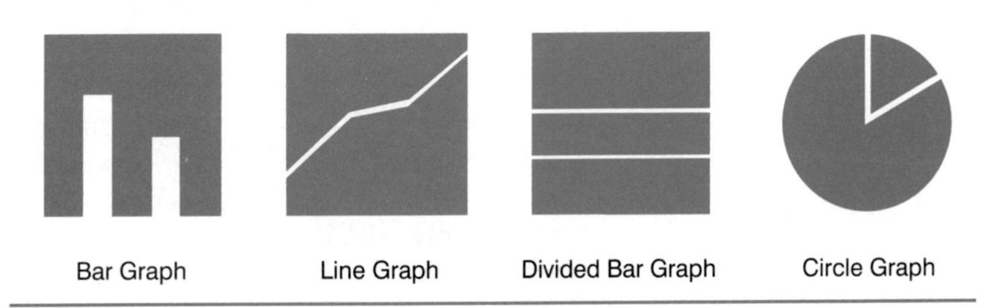

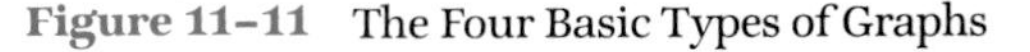
Figure 11-11 The Four Basic Types of Graphs

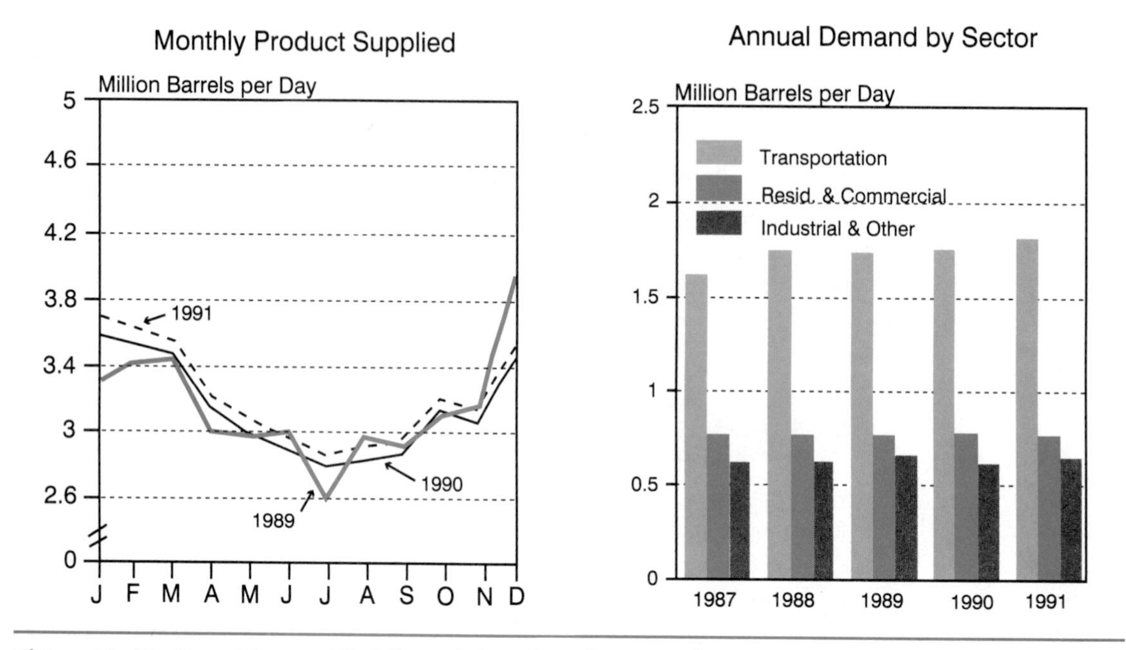

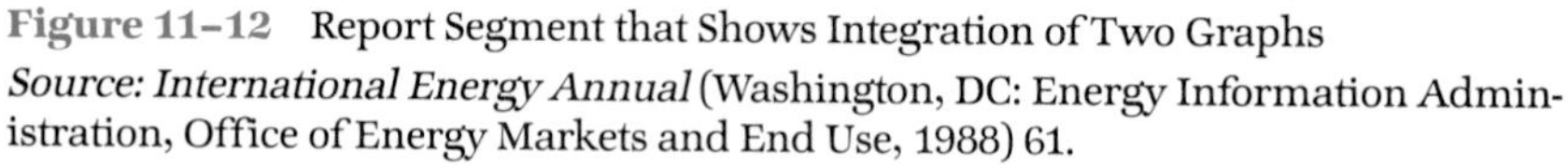
Figure 11–12 Report Segment that Shows Integration of Two Graphs
Source: International Energy Annual (Washington, DC: Energy Information Administration, Office of Energy Markets and End Use, 1988) 61.

information into a graph. Placing too much information in a graph can reduce reader comprehension by obscuring the point being made. The bar graph in Figure 11–12 includes about as much information as you can place in one bar graph.

- Be sure to label the x-axis and the y-axis on all graphs—what each measures and the units in which each is measured.
- If you borrow a graph from another source, be sure to acknowledge the source, generally by placing a source note beneath the graph. The examples we use in this chapter follow this guideline.
- Color can enhance the effect of a graphic, but excessive color can reduce comprehension and distort information. For example, examine the color graphics used in the insert section. Which colors do you find visually pleasing? Which would you change?
- Make the graph accurate. Computer graphics allow a tremendous range of special effects. However, artistic graphs are not always either effective or accurate. Three-dimensional graphics are often difficult to interpret. Understanding the relationships of the three lines in Figure 11–13 is difficult because of the 3-D grid, which can create an optical illusion. A simple two-dimensional line graph of the same information would be less artistic but easier to comprehend.

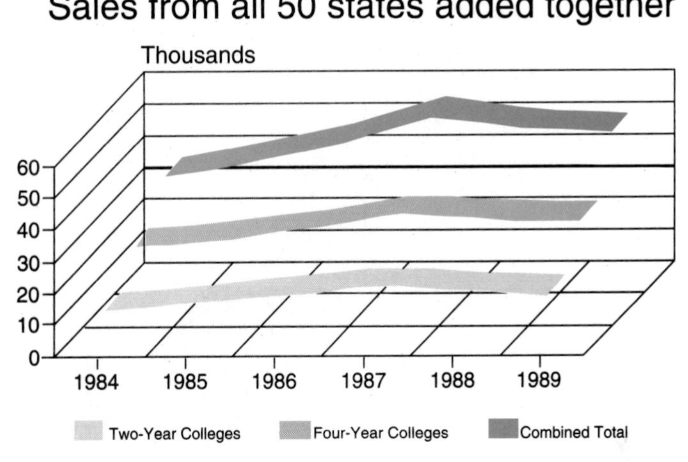

Figure 11–13 Three-Dimensional Line Graph with Shifting Depth: Three-dimensional graphs can create optical illusions.

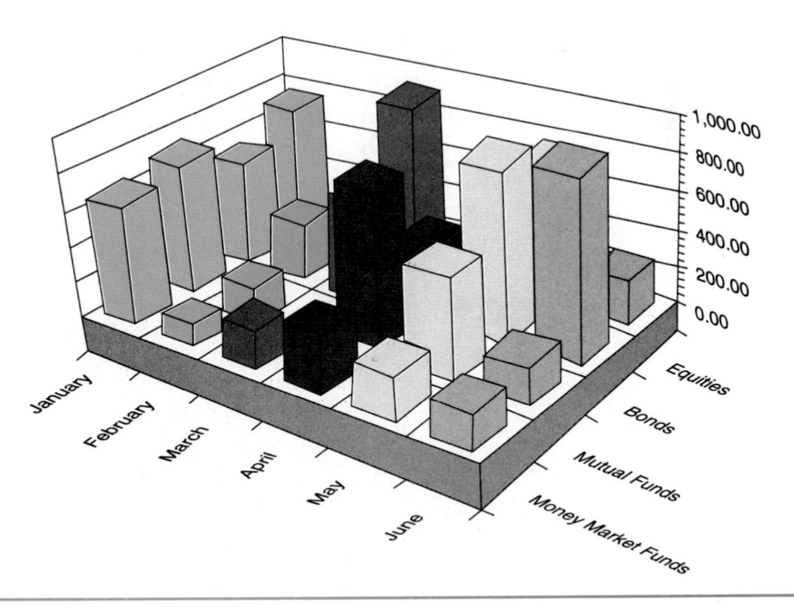

Figure 11–14 Three-Dimensional Graph: Most readers find three-dimensional graphs hard to interpret. Because of the upward slope of the plane, the relative heights of the bars are difficult to determine.

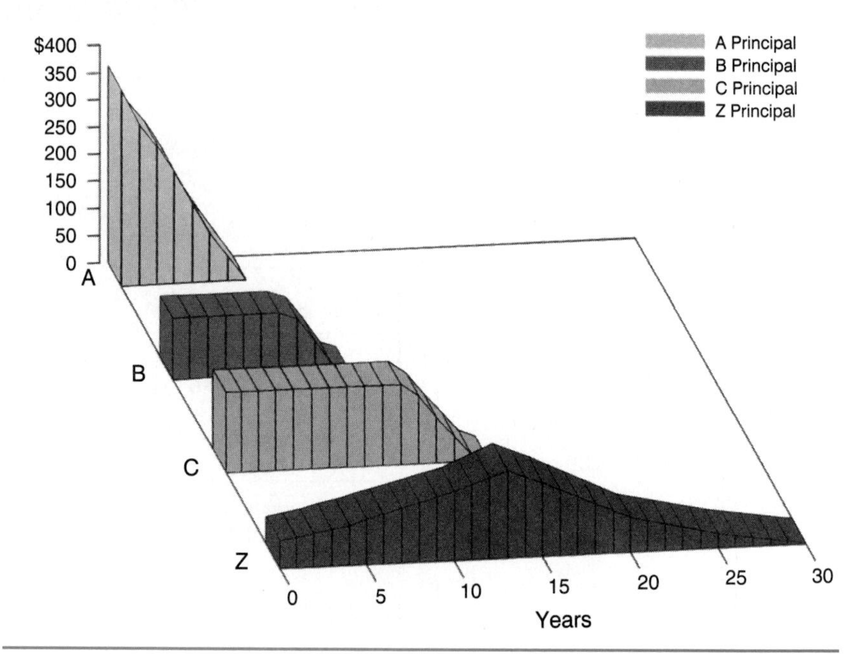

Figure 11–15 Artistic but Unclear Bar Graph: The upward slope of this graph makes the heights of the bars difficult to compare.
Source: Fannie Mae, ***Investing in REMIC Securities.*** Washington, DC, 1992.

Because of an upward slope of the graphs in Figures 11–14 and 11–15, which illustrate three-dimensional bar graphs, readers have difficulty visually comparing the relative heights of the bars. The bars located in the back of the graph appear higher than the ones in the front.

Because of the characteristics of various kinds of graphs, you will need to remember basic guidelines for each.

Bar Graphs

Bar graphs are useful in showing comparisons among quantities of information. Bar graphs may be drawn either vertically (Figure 11–16) or horizontally (Figure 11–17). Page layout—how much space you have—may often determine how you decide to draw your bar graph. Bar graphs are used frequently in material written for general audiences—newspapers, annual corporate reports, and magazines—but they are useful any time you need to show relationships among fixed quantities of data that change over time. Bar graphs are also useful in oral presentations because they quickly reveal comparisons. Examine Figures 11–16 and 11–17 as you consider the following guidelines:

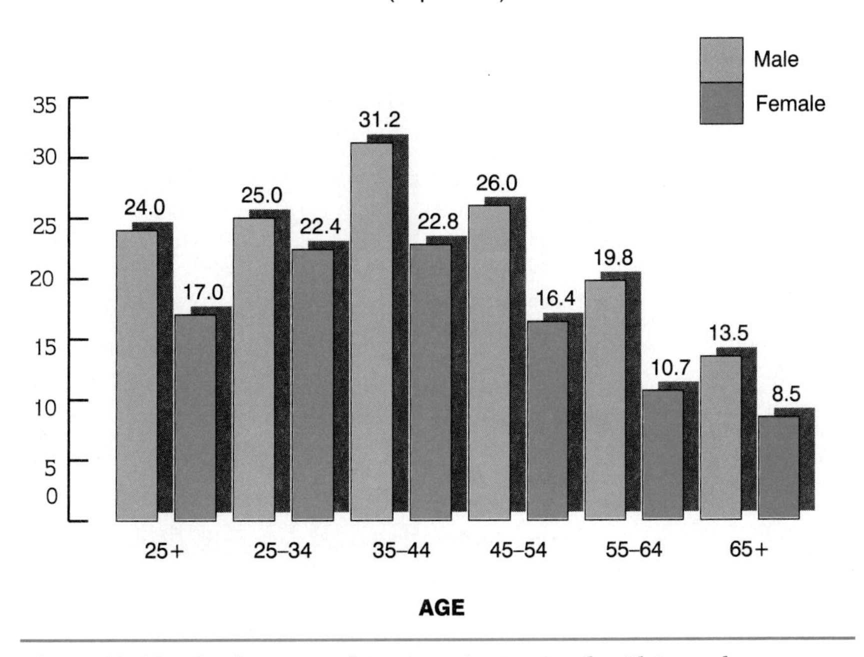

Figure 11–16 Clearly Presented Comparative Bar Graph with Legend
Source: U.S. Bureau of the Census, *Statistical Abstract of the United States: 1986*, 106th ed. (Washington, DC: GPO, 1986) xxviii.

- Be sure that you clearly label the x-axis and the y-axis on bar graphs, as shown in the examples that follow. Readers need to know what is being measured.
- If you choose three-dimensional bar graphs, watch for distortion. Again, as in Figures 11–14 and 11–15, the upward tilt of the graph makes it difficult to determine the relative heights of the bars presented as blocks.
- Try to write captions on or near the bars, as shown in Figures 11–16 and 11–17. Avoid legends (or keys) if possible because they slow reader comprehension. Legends can be particularly difficult to use with divided bar graphs, such as Figure 11–18. When bars are divided into too many divisions and when the reader must consult a legend to interpret these divisions, the result can be confusion rather than effective communication.
- For divided bar graphs with extensive divisions, try to use color or shading to distinguish divisions, as illustrated in Figure 11–19, rather than cross-hatching patterns, illustrated in Figure 11–20.

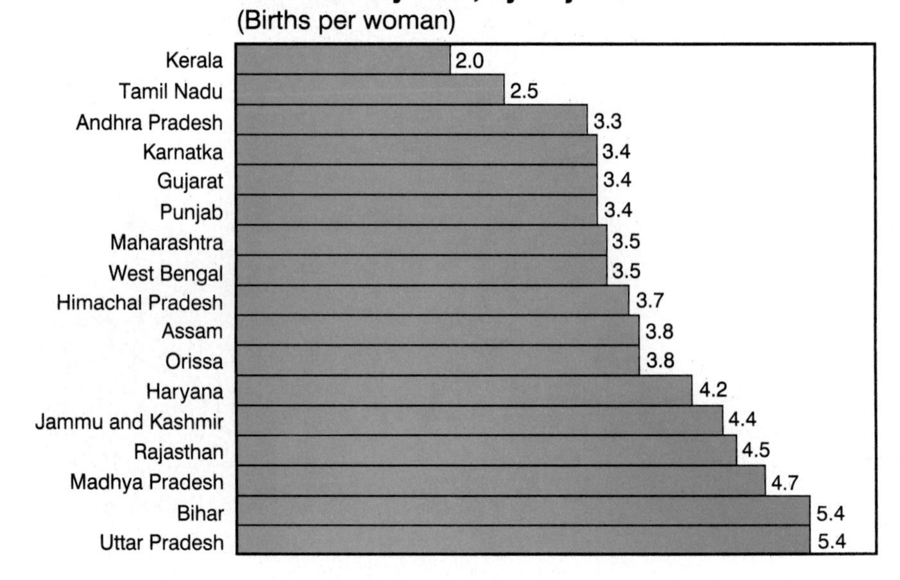

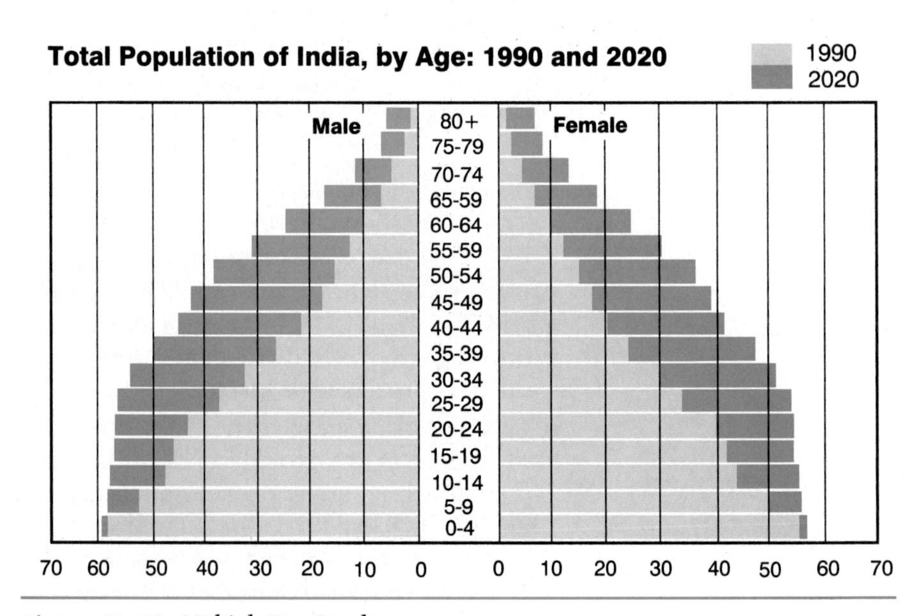

Figure 11–17 Multiple Bar Graphs

Source: Population Trends: India (Washington, DC: U.S. Department of Commerce, Bureau of the Census, Center for International Research, Oct. 1992) 12.

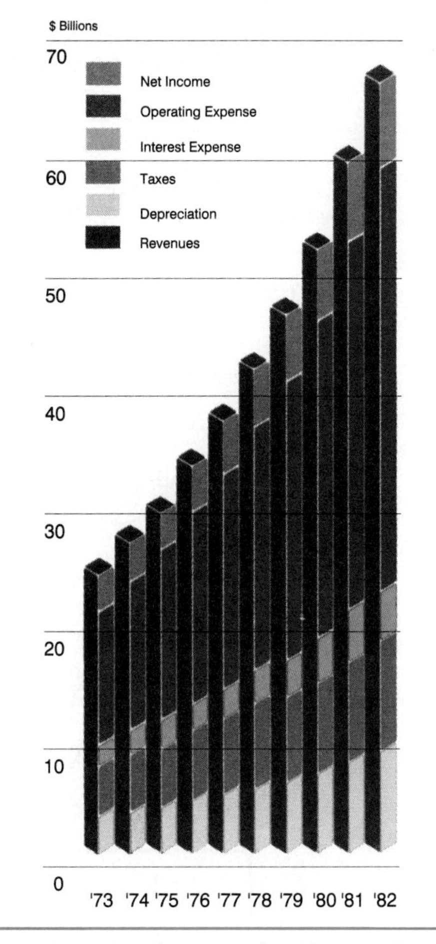

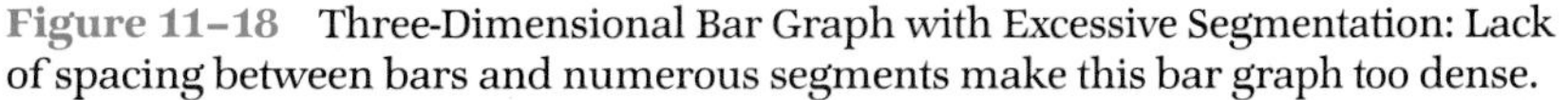

Figure 11–18 Three-Dimensional Bar Graph with Excessive Segmentation: Lack of spacing between bars and numerous segments make this bar graph too dense.

Three-dimensional bar graphs, such as those in Figures 11–14, 11–15, and 11–16 are generally not any better than standard two-dimensional ones, such as Figure 11–21.

- Do not use a bar graph if you will have to use an excessive number of bars, which can lead to visual clutter. Using three-dimensional bar graphs, such as those shown in Figure 11–18, will also reduce the number of bars you can use.
- Allow adequate space between bars. For example, Figures 11–19, 11–20, and 11–21 are more visually accessible than Figure 11–18 because spacing between bars provides easier viewing.

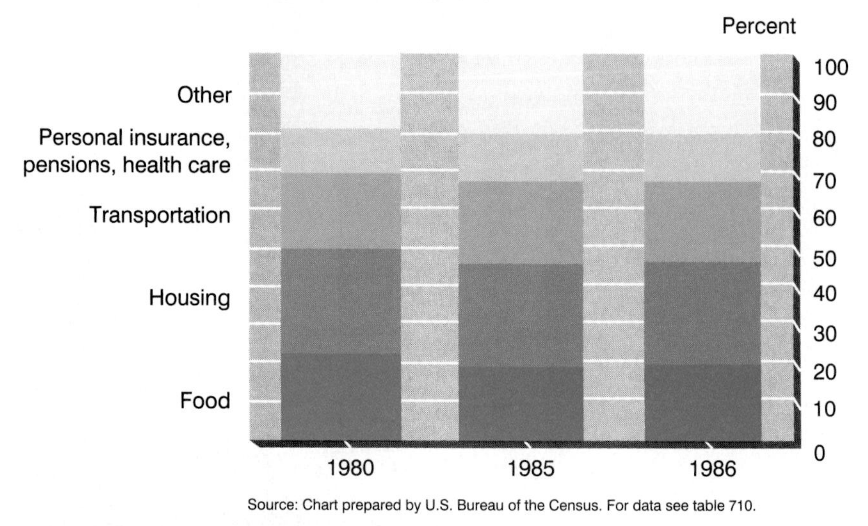

Figure 11–19 Divided Bar Graph Using Shading
Source: U.S. Bureau of the Census, *Statistical Abstract of the United States: 1989*, 109th ed. (Washington, DC: GPO, 1989) 439.

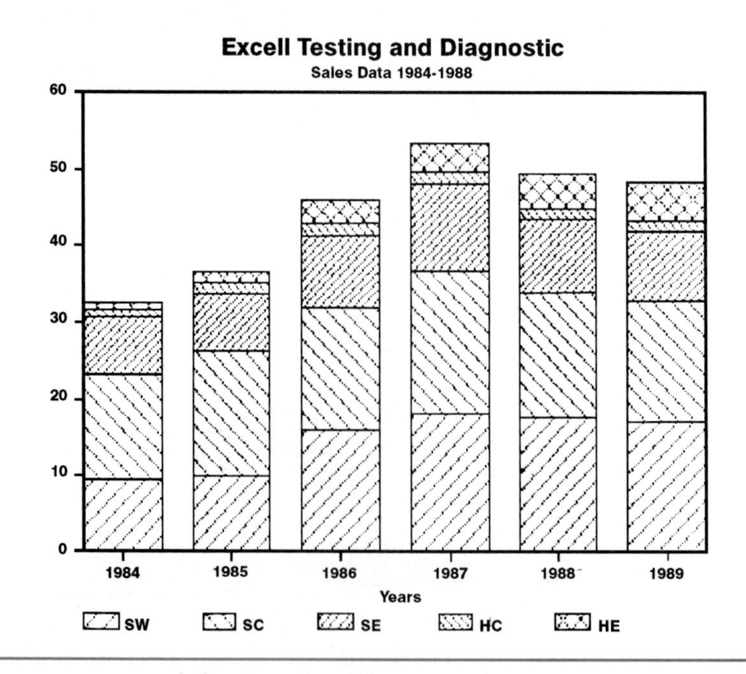

Figure 11–20 Divided Bar Graph with Cross Hatching

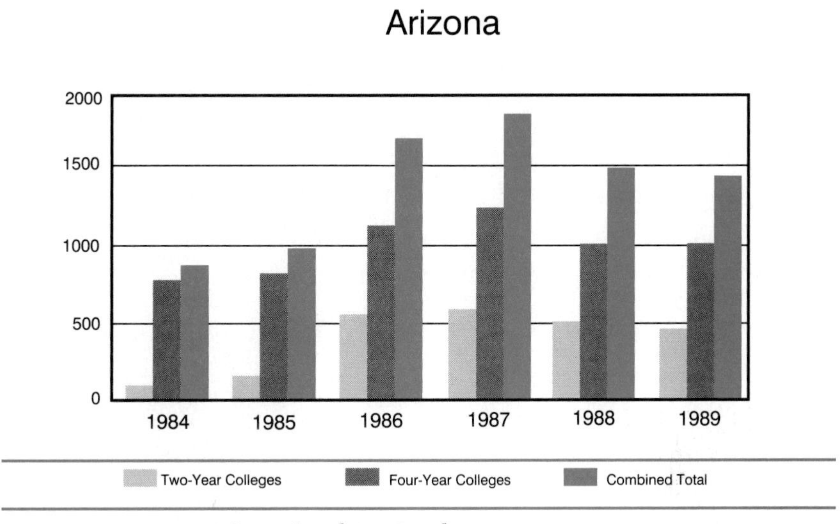

Figure 11–21 Two-Dimensional Bar Graph

Circle Graphs

Circle graphs or pie graphs are most often used to portray percentages or proportions. One problem with the pie graph is the limitation on the number of segments into which the circle can be divided before comprehension of the relative sizes is reduced. As your computer program builds circle graphs, watch for the number of segments that you will need. Sometimes shading can be used to differentiate segments, as in Figure 11–22, so that the differences between the segments are easier for the reader to see. In Figure 11–22, a different shade is used in both circle graphs to denote each tuition source: sales and services, endowment income, private sources, state government, etc. Because the percentage of revenue for these sources is presented in the same order on each graph, parallel use of colors for each source allows easy visual comparison of the percentages.

Compare Figure 11–22 with the black-and-white circle graph in Figure 11–23, which uses shades of gray and black to emphasize segments. High-quality color graphics give you great flexibility in generating visual aids, but overuse of color in circle graphs can lessen their effectiveness.

- Use color to enhance the point you are making with your graph. As Figure 11–22 shows, use of the same shades to portray the same information on each circle graph allows easy

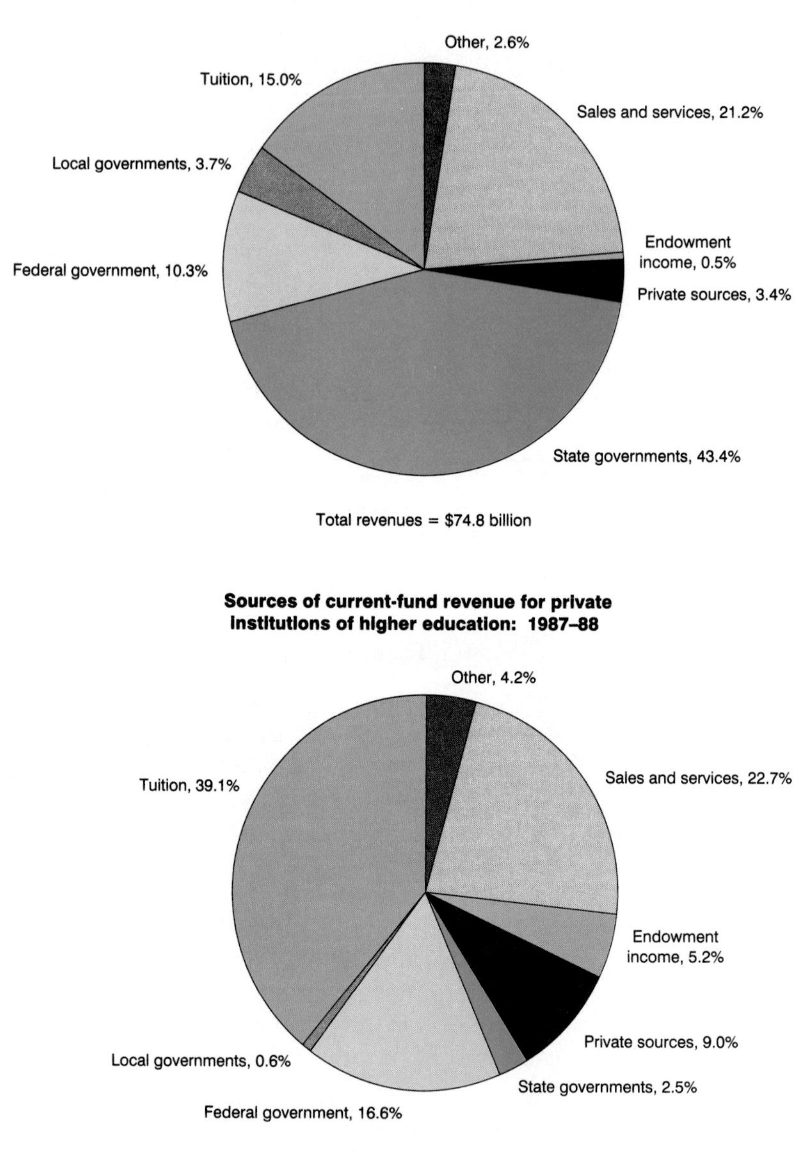

Figure 11–22 Comparative Circle Graphs: Note the use of shading to coordinate similar segments on both graphs.

Source: National Center for Education Statistics, *Digest of Education Statistics, 1991* (Washington, DC: U.S. Department of Education, Office of Educational Research and Improvement, 1991) 37.

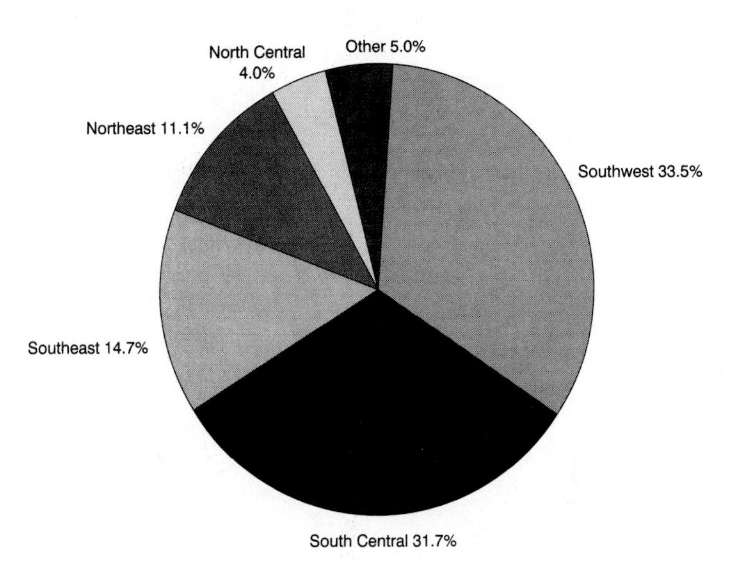

Figure 11–23 Black-and-White Segmented Circle Graph

visual comparison between corresponding segments of the two graphs.

- Watch for distortion when you use three-dimensional circle graphs.

Computer software allows you to draw three-dimensional bar graphs and circle graphs, but you must be sure that the perspective from which the graph is drawn does not distort the proportions. Figure 11–24 illustrates a three-dimensional circle graph that uses shading

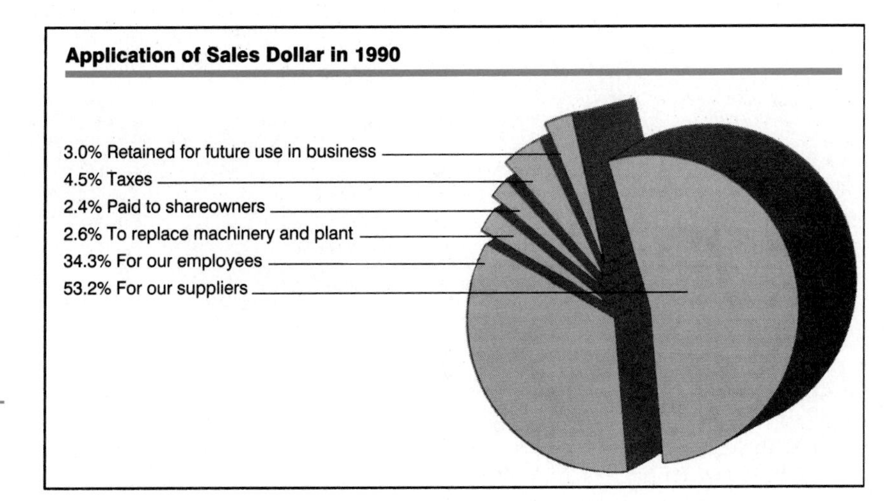

Figure 11–24 Three-Dimensional Circle Graph (No Distortion)

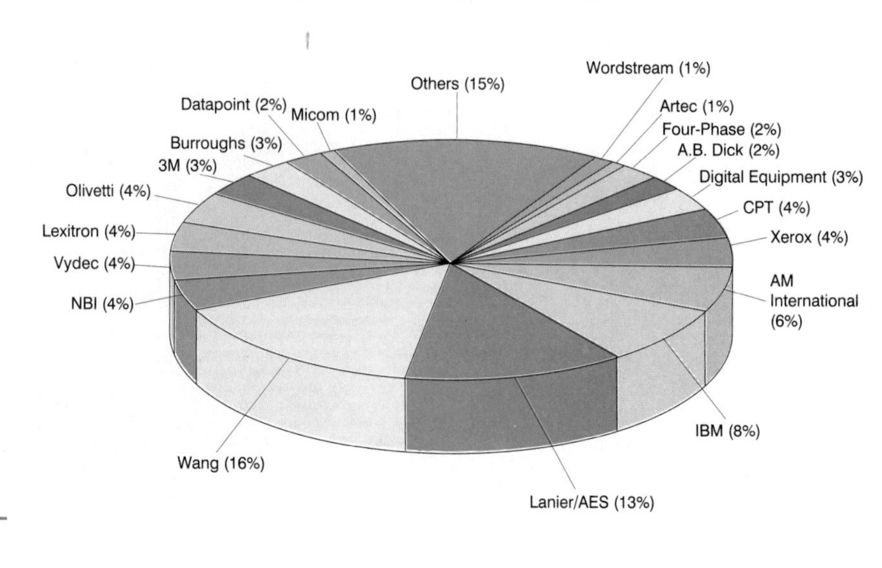

1979 U.S. DISPLAY WORD PROCESSOR SHIPMENTS
TOTAL KEYBOARDS—85,000

Figure 11–25 Three-Dimensional Circle Graph with Distortion

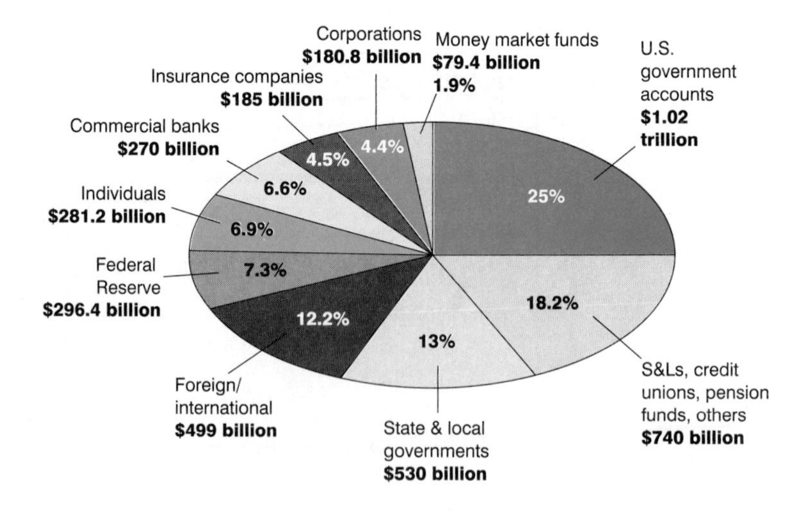

Source: U.S. Treasury Department

Figure 11–26 Circle Graph with Segments in Descending Order by Size

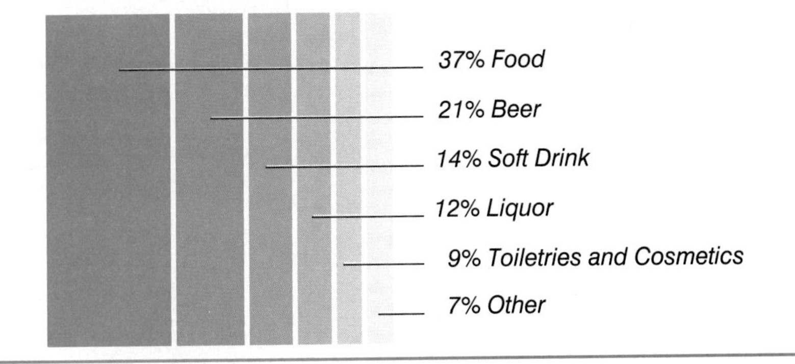

Figure 11–27 Divided Whole Bar Graph

and three-dimensional presentation effectively. In the circle graph in Figure 11–25, shading is used indiscriminately, there are too many segments, and the three-dimensional presentation distorts the size of the segments.

- Clearly label all segments.

Like all graphs, circle graphs must be clearly labeled. Whether identifications are placed inside or outside the circle, percentage figures and identifications should be horizontal, as in Figure 11–26, to preserve the comprehension of the graph. Figure 11–26 also illustrates a useful convention to follow: As you segment the graph, begin with the largest section in the upper right-hand quadrant. The remaining segments should be arranged clockwise in descending order.

Divided whole bars (as illustrated in Figure 11–27) are gaining in popularity because they are similar to divided bar graphs, as shown in Figures 11–19 through 11–21, which have been used extensively in technical reports for the past two decades.

Line Graphs

Line graphs are most often used to show relationships when more precision is required than bar graphs can show. Line graphs show trends well. Many times you will have choices between bar graphs and line graphs, with the information for both coming from a table, as we have illustrated in Figure 11–3.

- Label the x-axis and the y-axis clearly.

Like bar graphs, line graphs must have clearly labeled axes to show what variables you are comparing. See Figure 11–28. The independent

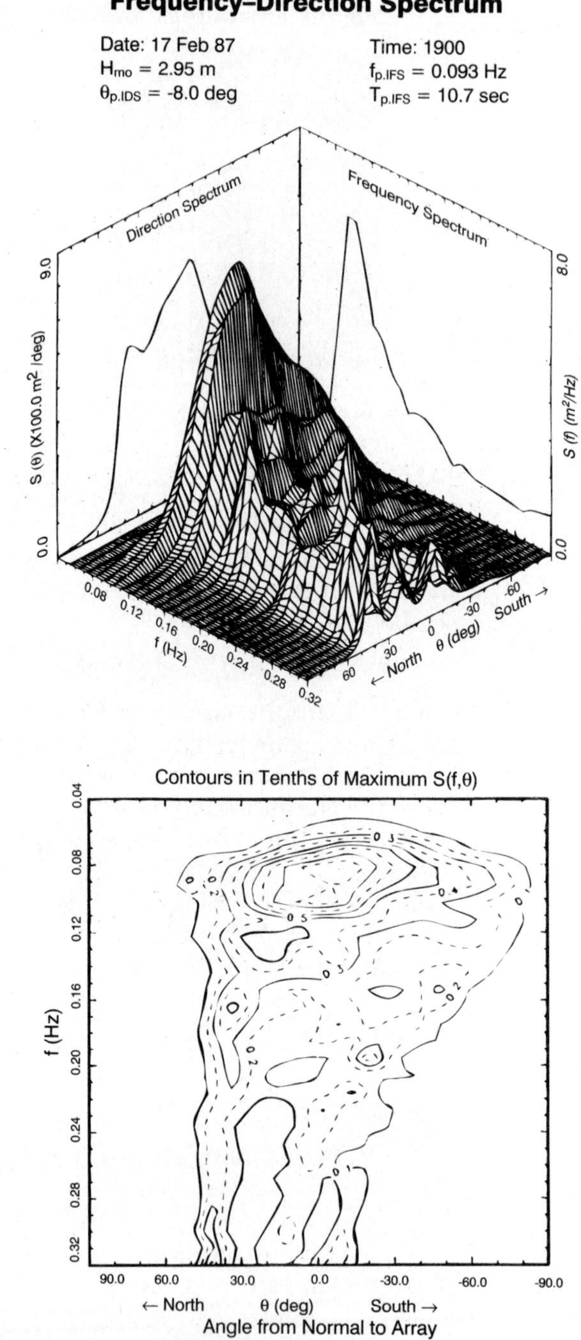

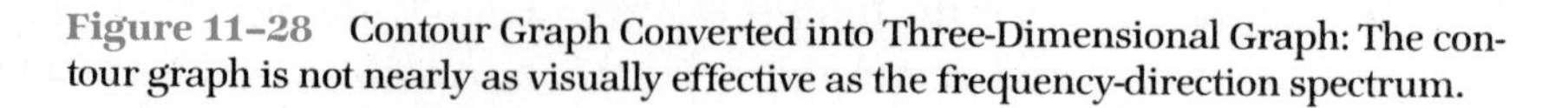

Figure 11–28 Contour Graph Converted into Three-Dimensional Graph: The contour graph is not nearly as visually effective as the frequency-direction spectrum.

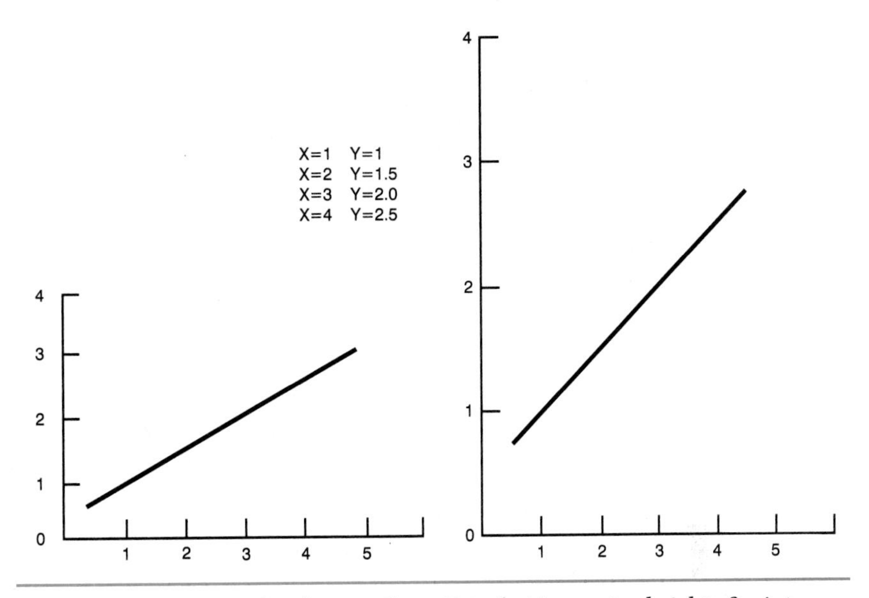

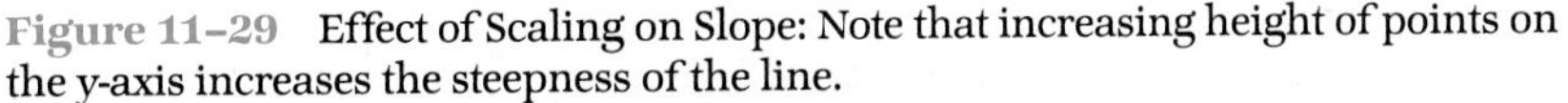

Figure 11–29 Effect of Scaling on Slope: Note that increasing height of points on the y-axis increases the steepness of the line.

variable is placed on the horizontal (x) axis, and the dependent variables are placed on the (y) axis and the (z) axis. Note that the two-dimensional contour graph is not nearly as visually effective as the computer-generated three-dimensional version.

- Choose the scale of each axis—x, y, and z—to show the appropriate steepness of the slope of the line.

The major difficulty in designing line graphs lies in choosing the spacing for both axes so that the steepness (slope) of the line accurately measures the actual trend suggested by the data. In Figure 11–29, both graphs depict the same data, but the slope changes drastically because of the different scales used in designing each set of axes. Computer graphics will allow you to adjust the intervals on the x- and the y-axis, but only you can tell whether the slope of your graph shows the trend depicted by your data.

- Avoid placing excessive lines on one graph.

Another issue in designing line graphs concerns the number of lines that may be imposed on one plane. When several lines occur on one graph and then overlap, the graph may become difficult to interpret, as illustrated in Figure 11–30. The number of lines on one axis should generally not exceed three, but the number you can impose will depend on clarity. If lines are spaced apart and do not overlap, you may be able to place more than three on the axis, as is shown in Figure 11–31.

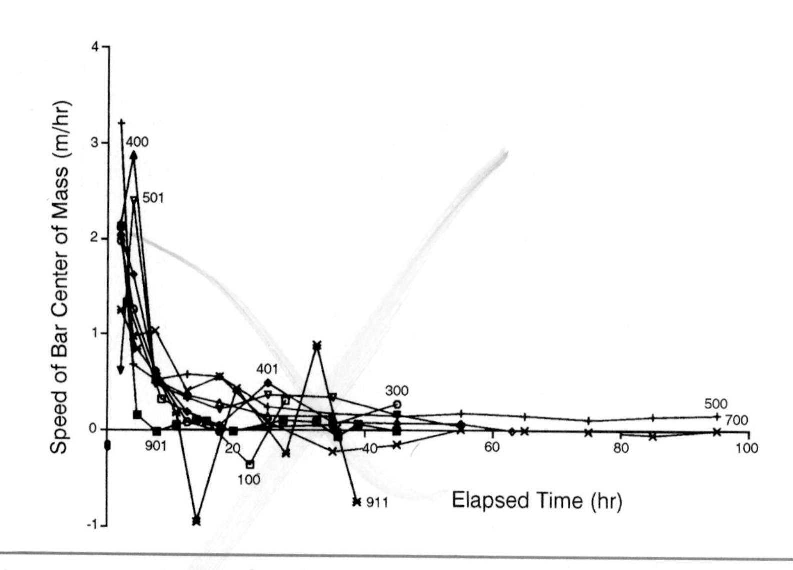

Figure 11–30 Line Graph with Excessive Overlapping Lines: Overlapping lines can reduce legibility.

Source: Nicholas C. Kraus, Coastal Engineering Research Center, Vicksburg, MS, *Beach Profile Change Measured in the Tank for Large Waves* (Washington, DC: Department of the Army, U.S. Army Corps of Engineers, 1990) 76.

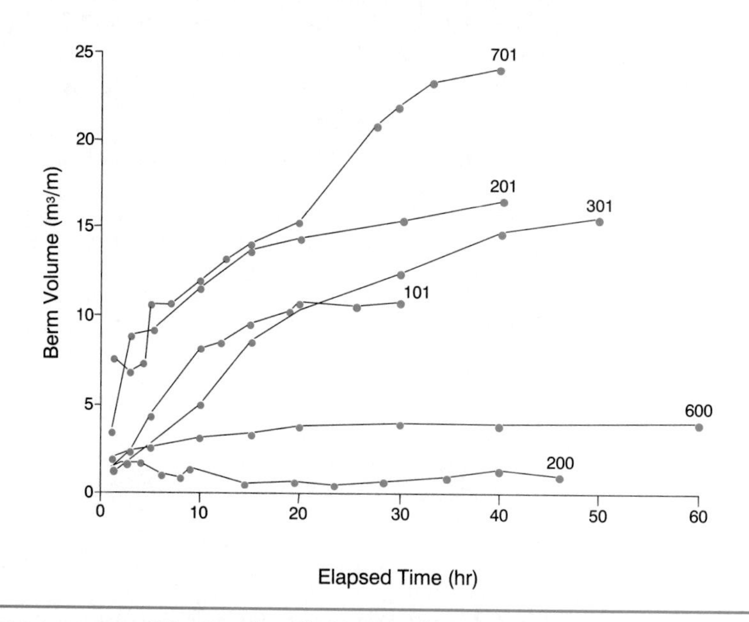

Figure 11–31 Line Graph with Six Lines that Do Not Reduce Comprehension—Minimal Overlapping

Source: Nicholas C. Kraus, Coastal Engineering Research Center, Vicksburg, MS, *Beach Profile Change Measured in the Tank for Large Waves* (Washington, DC: Department of the Army, U.S. Army Corps of Engineers, 1990) 74.

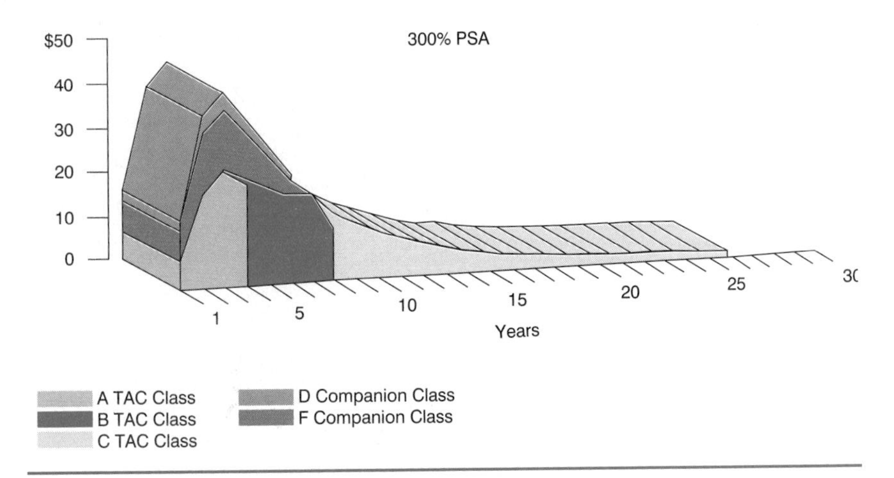

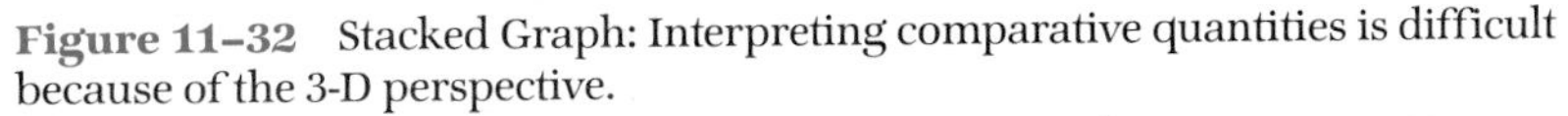

Figure 11–32 Stacked Graph: Interpreting comparative quantities is difficult because of the 3-D perspective.
Source: Fannie Mae, *Investing in REMIC Securities.* Washington, DC, 1992.

An interesting version of the line graphs is the surface, or stacked graph, which emphasizes the line by shading the area beneath it. Surface graphs can be very artistic, as shown in Figure 11–32, but the quantities they present are often hard to determine. The reader has to interpolate at any point to determine the difference and the amount pictured. In Figure 11–31, the need for the legend or key reduces the visual accessibility of the graph even more.

Combined Line and Bar Graphs

As Figures 11–33 and 11–34 illustrate, you can combine line graphs and bar graphs in the same visual aid, depending on the kind of information you want to display. The step graph, shown in Figure 11–34, combines features of the bar graph, stacked graph, and line graph to compare cash flow, earnings, and dividends per share. Figure 11–33 imposes a line graph on a bar graph to compare the company's cumulative total return with its annual total return.

Designing Other Forms of Graphics

Photographs, line drawings, representation drawings, and flowcharts are other useful types of graphics that allow you to help your reader visualize the idea you are presenting.

Photographs

As you have already observed in Figure 11–6, photographs may be used in technical reports to capture important detail, and perhaps to add

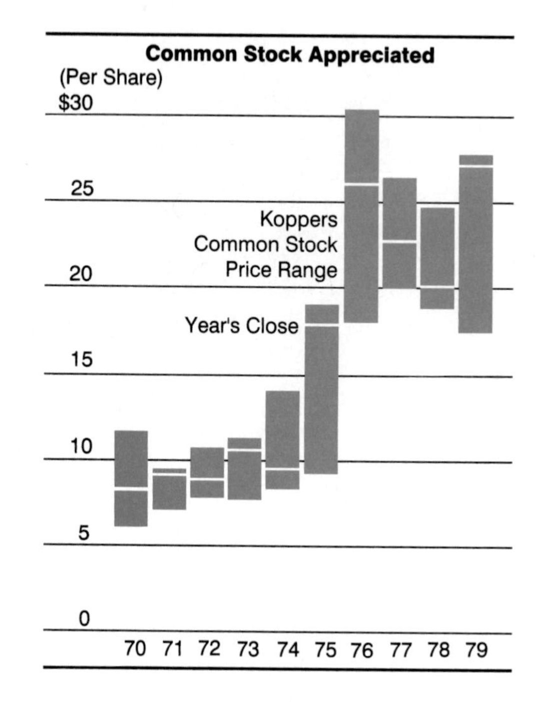

Figure 11–33 Graph that Combines Features of Line Graphs and Bar Graphs

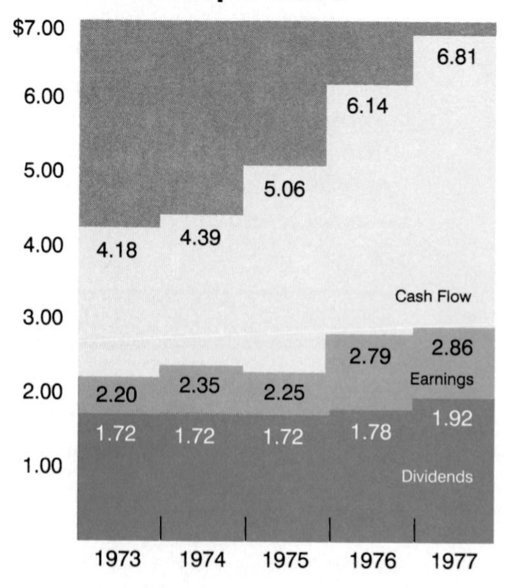

Figure 11–34 Graph that Combines Features of Line Graphs and Bar Graphs

realism. The close-up photograph of the termite-infested wood piling helps readers understand termite damage when they see it. In accident reports, photographs preserve details that may be essential in determining the cause of the damage or its extent. Because photographs add substantially to the cost of a report and must be of high quality to be effective, the photographs selected must clearly achieve their purpose.

Line Drawings

Line drawings are often preferable to photographs when you want to present only the details of an object or process. Unlike photographs, line drawings allow you to select what you want the reader to see. As you have seen in Figure 11-7, the line drawings of the ant and the termite help you see the differences in these two insects, which often look a great deal alike. Cutaway line drawings, such as the one of the Sealab in Figure 11-35, allow you to see the internal structure of the vehicle.

As Figure 11-36 shows, an exploded view of an object—in this case the rear drive mechanism of a proposed soap-box racer—allows a reader to see quickly the relationships among the components of the object.

However, drawings also enable you to help your reader understand processes. For example, Figure 11-37 uses a line drawing and a two-sentence description to show how solar-heated air enters a greenhouse.

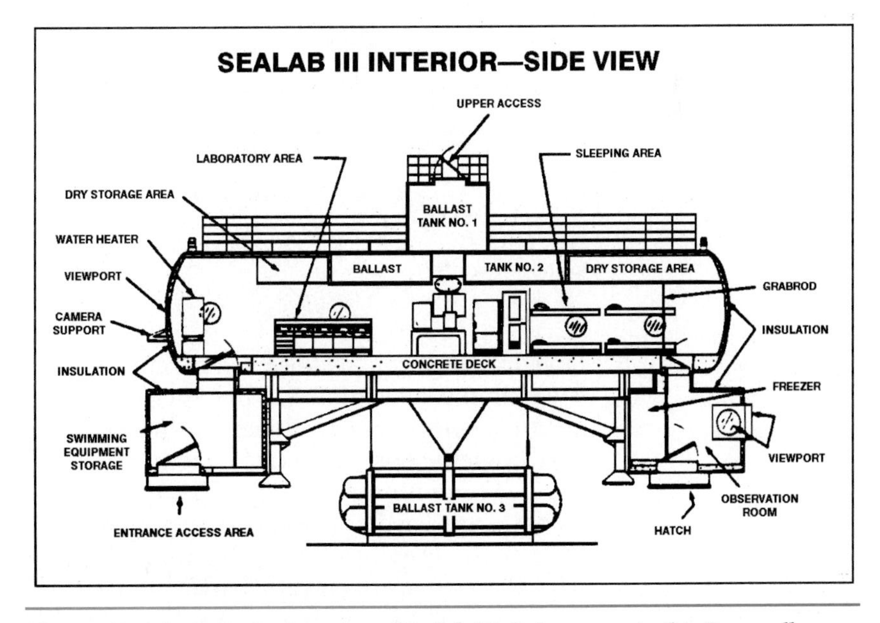

Figure 11–35 **Cutaway Drawing of Sealab III: Cutaways as in this figure allow readers to see inside objects.**

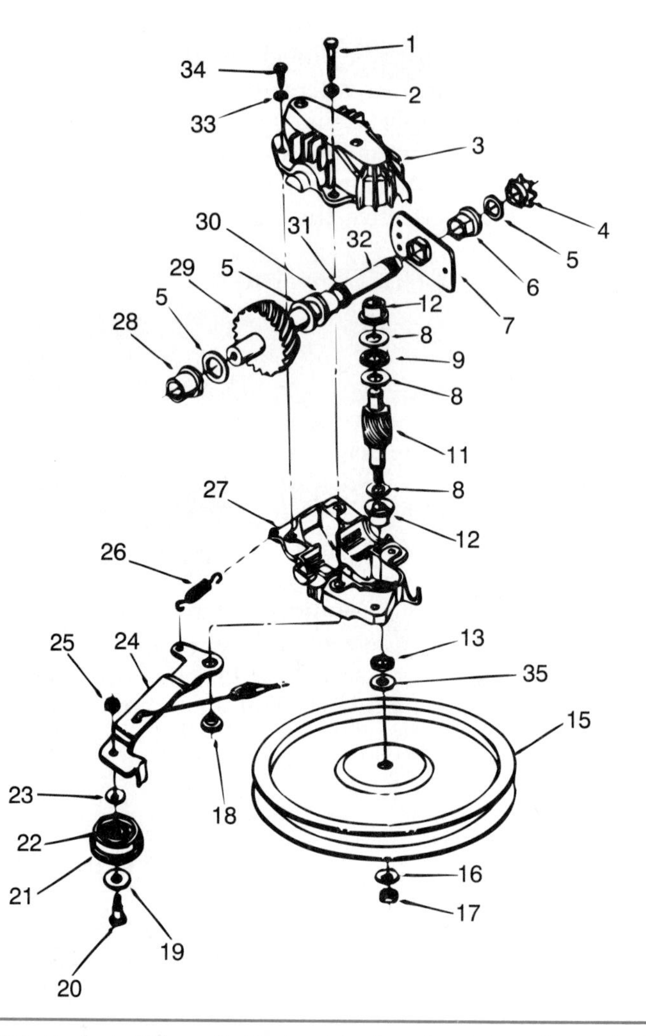

Figure 11–36 Exploded View of a Mechanism that Shows the Relationships among Parts

Line drawings can also be used to represent concepts discussed in text. For example, a technology forecaster in 1980 predicted the merging of personal computers, intelligent terminals, and word processing in the development of office systems (Figure 11–38). The speaker represented this merger with a conceptual drawing that shows how computer technology provides power tools for generating data and reports.

As these drawings illustrate, an important feature of line drawings (as well as all graphics) is effective use of labeling. From the drawing, the labels on various parts, and the title of the drawing, the reader should be able to determine what the drawing means.

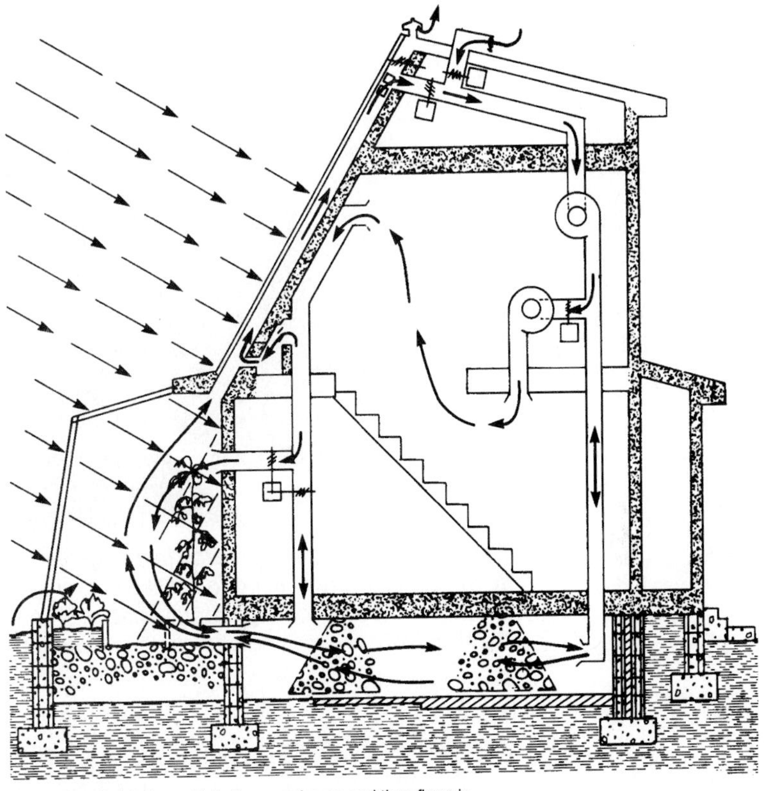

Solar-heated air is fanned into the greenhouse and then flows in the crushed rock for storage and reheating. From there, the air circulates through the residence and returns to the greenhouse for recycling (PN-6809).

Figure 11–37 Drawing Used to Show Process of Heating a Greenhouse by Solar Energy

Flowcharts

Flowcharts are drawings that show steps in a process to help a reader follow the process or understand the relationships among the steps. For example, the flowchart in Figure 11–8 was designed to help readers see the various levels of education in the United States with an emphasis on their progression. The flowchart in Figure 11–39 shows how a data base was developed by displaying the function of each step of the development process. This kind of flowchart is particularly useful when work in each category is developing at the same time and when the writer wants to show the concomitant development.

Algorithm flowcharts are often useful for showing decision-making processes. Figure 11–40 shows how decisions are made in the dredging operations conducted by the Department of the Army.

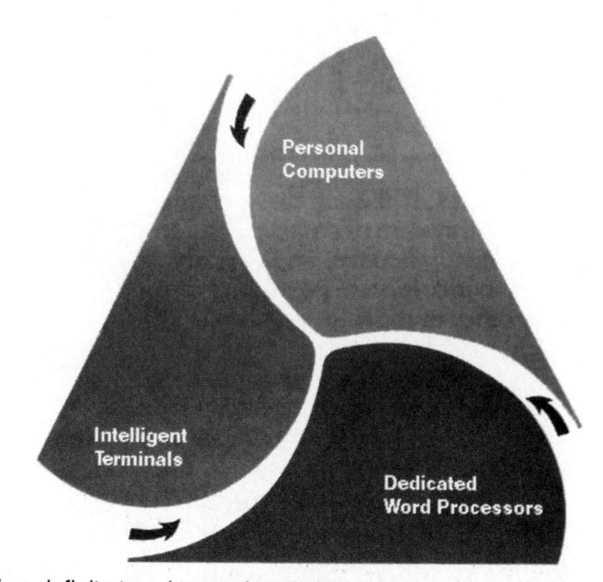

Figure 11–38 This conceptual drawing shows increasing integration of information technologies.

Source: Frances Hansen, "Trends in Office Automation: Implications for Communication Development," Lecture presented at Pi Omega Pi, University of Houston, March 1980.

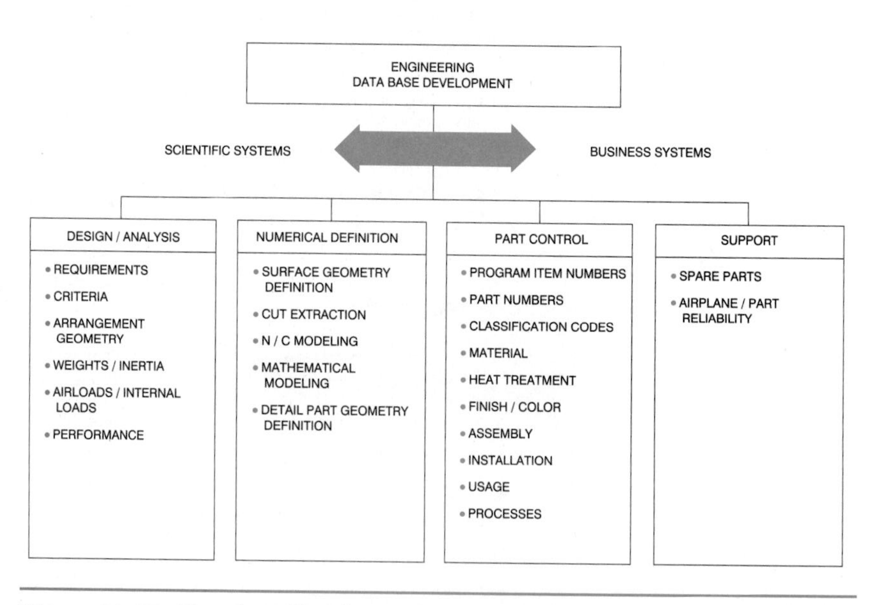

Figure 11–39 Flowchart Showing Entities Responsible for the Development of an Engineering Data Base

Source: E. Koch and R.L. Rees, *Analysis of Pressure Distortion Testing, Contractor Report* (Washington, DC: National Aeronautics and Space Administration, 1989) 57.

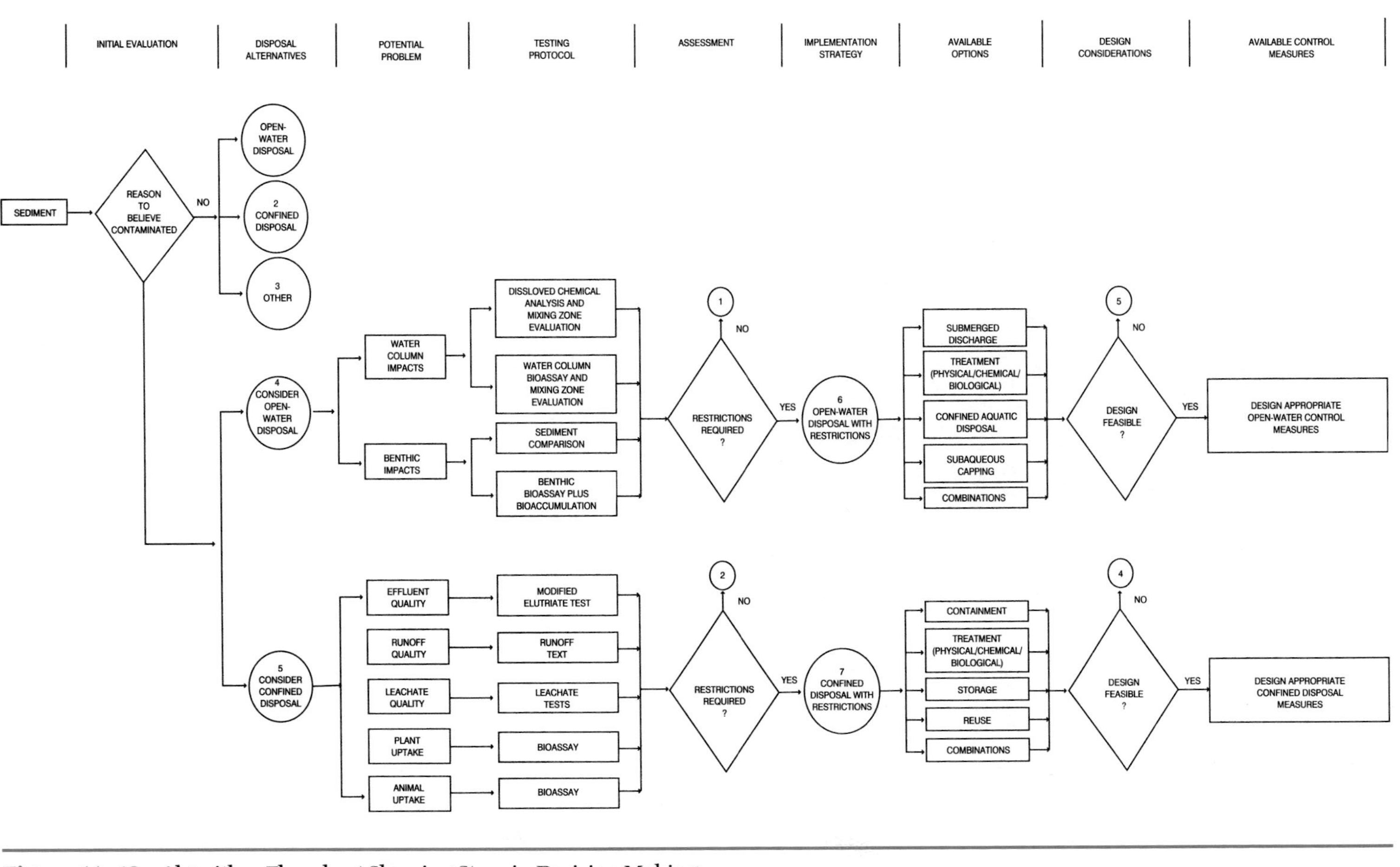

Figure 11–40 Algorithm Flowchart Showing Steps in Decision Making
Source: Charles R. Lee et al., *General Decisionmaking Framework for Management of Dredge Material, Example Application to Commencement Bay, Washington* (Seattle, WA: Department of the Army, Waterways Experiment Station, Corps of Engineers, June 1991) 41.

Integrating Text and Visuals

Many of the tables, bar graphs, line graphs, and line drawings in this chapter can stand alone. Effective titles and labeling often allow one visual to replace hundreds of words. In other situations, however, an accompanying verbal discussion helps a reader visualize a highly technical, abstract concept. Thus, an important technique in graphic design is coordination of the verbal and the visual presentation, as you have already seen in Figure 11–12. In this chapter, we want you to be able to see the differences in effective and ineffective graphics. However, we also integrate our examples with written guidelines for their selection and accurate use.

Designing Graphics for Accuracy and Effectiveness

In the previous sections on tables and graphs, we have briefly alluded to the importance of making graphics accurate as well as effective. The challenge of making your graphics present information accurately requires that you make careful choices as you design your graphics.

Using three-dimensional graphics is not the only way that information can be visually distorted. Frequently, as we discussed in Figure 11–29, the scale of the x- and y-axis can make a significant difference on how your information appears. In designing a graph, you should begin the x- and y-axis at 0, unless beginning at some other point will not distort information. For example, line graphs showing changes in the prime lending rate will usually begin at 7 and move upward to the highest point in history. The prime rate has not fallen below 5 for more than 50 years. Thus, a change in even 1% is significant, as shown in Figure 11–41.

In order to reduce the height of a bar or line graph without creating visual distortion, you can remove a portion of the bars, as shown in Figure 11–42, without changing the general trend that the bars indicate.

You avoid distorting the trend of the bar graph by choosing the intervals on the y-axis carefully. For example, examine Figure 11–43, the graph of Compaq Computer Corporation's stock prices for January and February 1993. If you were to redraw the graph to begin the origin at 0, the slope of the line would not change. The slope of the line, indicating the drop from $55 per share to 41 7/8 in one month is represented accurately on the graph. Similarly, Figure 11–44, showing changes in natural gas prices, uses interval differences of $.20, as this is a significant change in price.

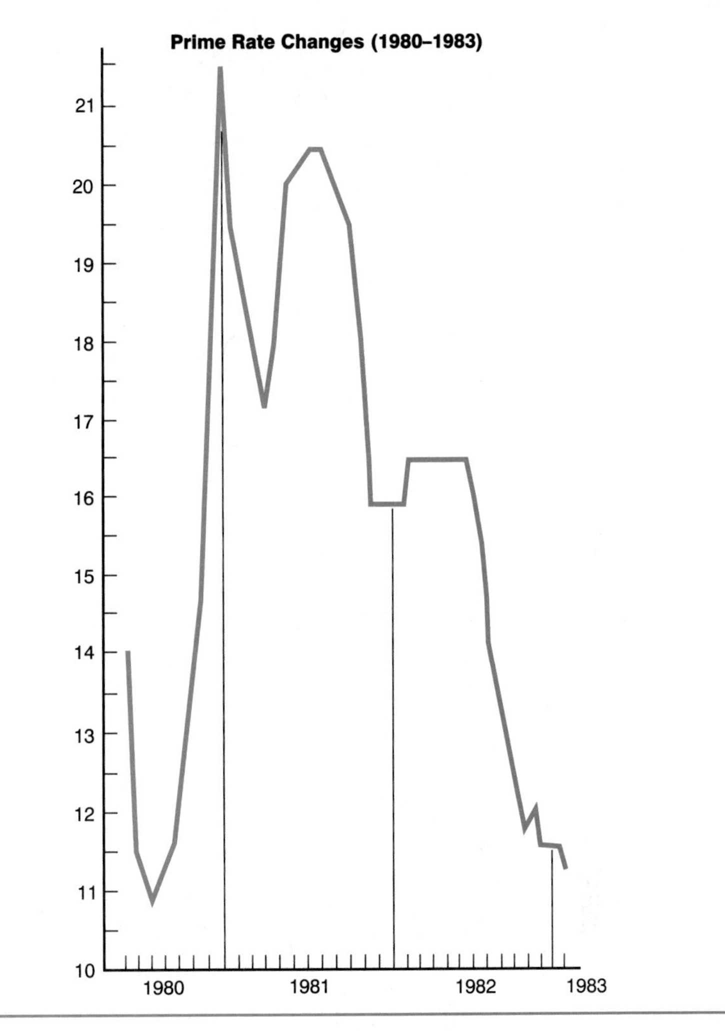

Figure 11–41 Suppressed Zero Line Graph with No Distortion

However, business publications often distort the intervals on the y-axis to emphasize extreme highs or lows in financial performance. Figure 11–45 graphs sharp declines in the Dow Jones industrial index over a short period—six hours—to convey the writer's view of the severity of this decline. Figure 11–46, with its y-axis scaling of $90–$270 million, is designed to show that inventories have not declined substantially, and new orders have not increased substantially.

In short, how you set up your y-axis will depend on the change you want your graph to show. As a writer, you have an ethical responsibility to present your graphic so that it represents the trend you want to show your reader without misrepresenting the facts.

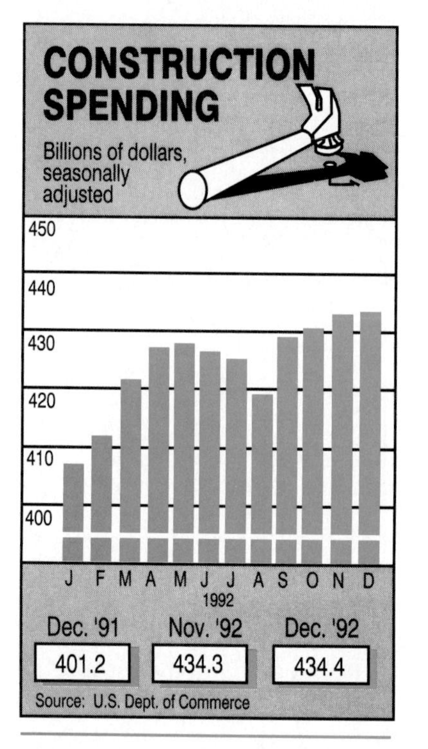

Figure 11–42 Bar Graph with Segment Removed to Show That the x-axis Has Been Distorted

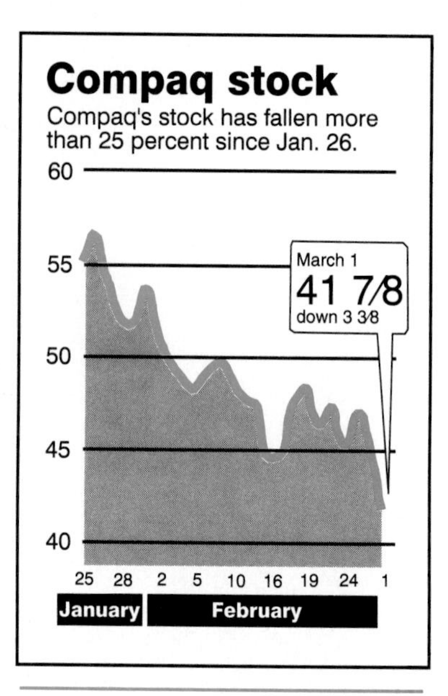

Figure 11–43 Line Graph with Suppressed Zero to Show Sharp Stock Price Drop

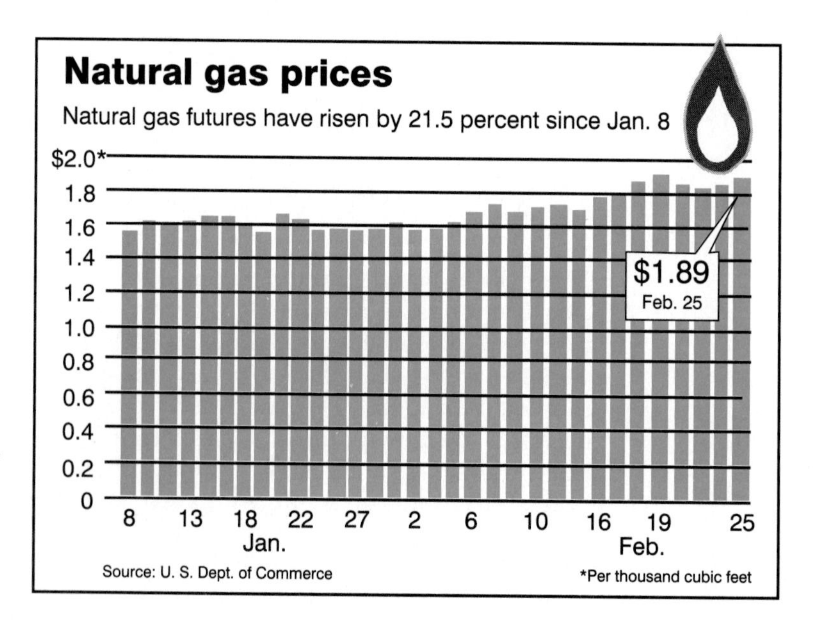

Figure 11–44 Bar Graph with x-axis Designed to Reflect Accurate Changes in Gas Prices

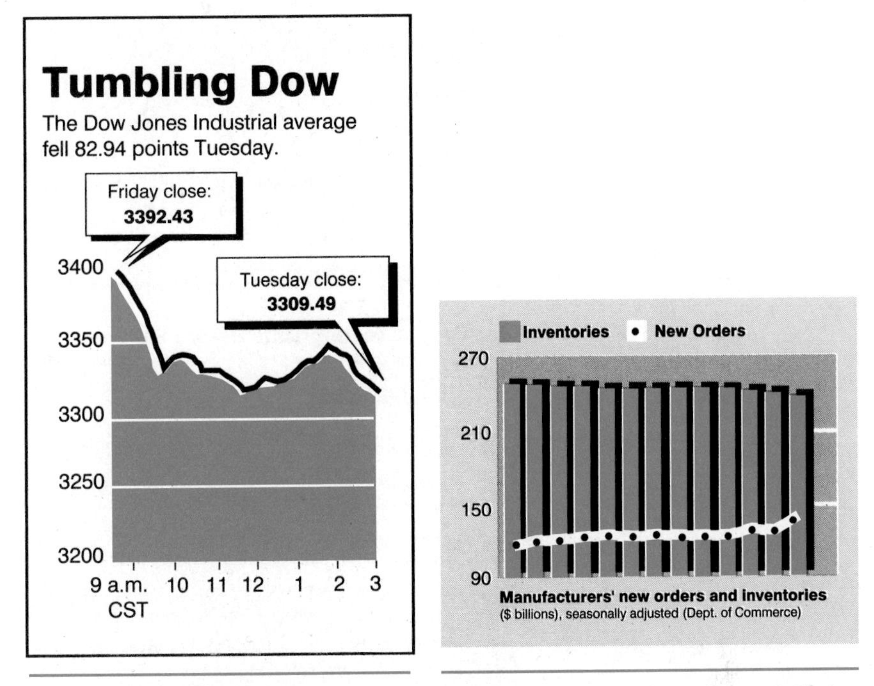

Figure 11–45 Suppressed Zero to Accentuate Dow Change

Figure 11–46 Suppressed Zero to Flatten Trends

Choosing the Best Graphic for the Information and the Audience

Much numerical information usually appears in tabular form, generated as computer printouts of raw data. After selecting the data from the printout, you then need to know how to display the information. You can begin with a table, then determine how best to display the information. Sometimes you have choices, as we show in Figure 11–46. To choose the best graphic use the help we provide:

- As you *plan* your report, consult the planning checklist on page 326.
- As you *generate* your graphics, keep the following summary (page 325) handy to guide you in your choice and development of visuals for accuracy and effectiveness.
- As you *evaluate* your graphics for possible revision, consult the revising checklist on pages 326–327.

Table IV: Testing Time Before Failure

Test no.	RPM	Hours–minutes
1	4,000	4:08
2	5,000	4:02
3	6,000	3:55
4	7,000	3:45
5	8,000	3:26
6	9,000	2:45
7	10,000	1:15

Bar Graph: Testing Time Before Failure

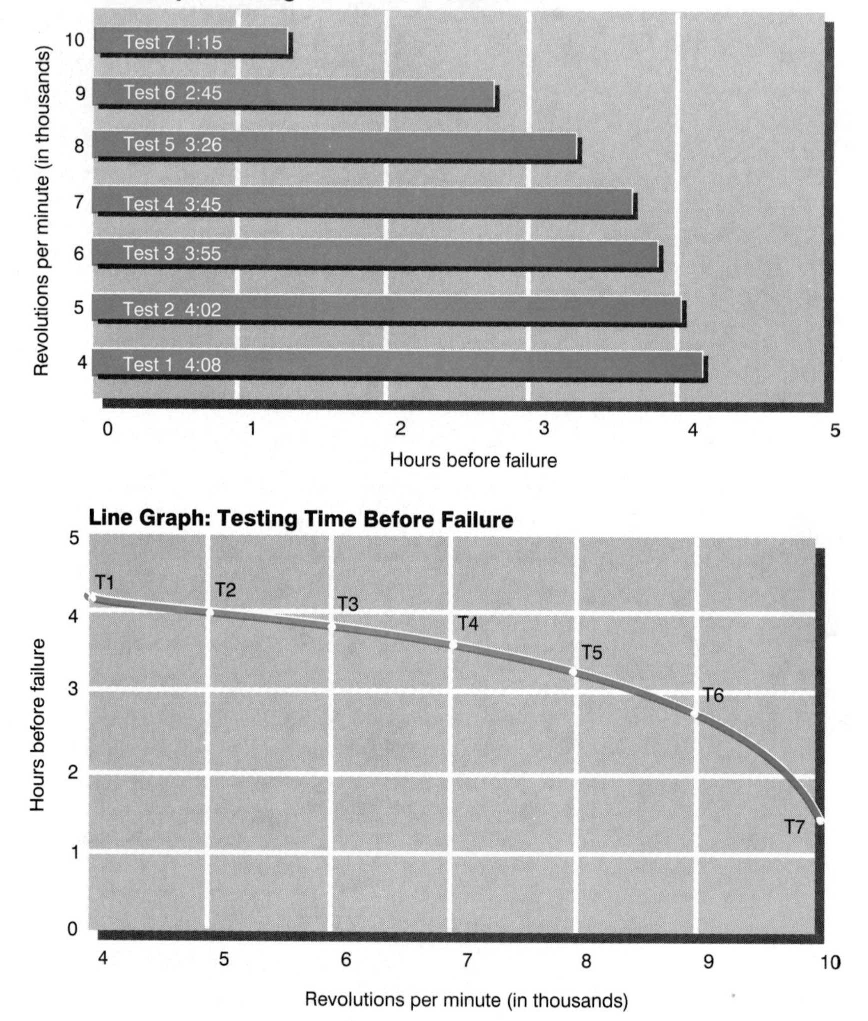

Figure 11–47 Table-Graph Relationship: The two graphs show clearly the relationship between rpm and time before failure. The line graph shows the relationship as a continuous variable and is well suited to technical and expert audiences.

Choosing Graphics Checklist

- Make sure the graphic communicates what you want it to convey quickly and accurately.
- Be sensitive for ways to use graphics to enhance the effectiveness of your reports. Choose the graphic that will do the following the best:
  - Clarify information
  - Summarize information
  - Emphasize a specific idea
- Use tables rather than bar graphs or line graphs if you have to present extensive data.
- If information requires your reader's immediate attention, look for ways to present this information graphically to help your reader see the urgency of your purpose.
- If your reader will not be motivated to read your report, graphics and visual design can encourage your reader. Seeing is easier than reading.
- Use tables if they do not contain too much information.
- Use flowcharts if the progression of ideas is complex.
- Use bar graphs and circle graphs if you want to show large comparisons.
- Use creative visuals to emphasize or portray important ideas, but use these sparingly.

DRAWING GRAPHICS

- Make the graphic aesthetically pleasing.
- Keep decoration from interfering with the message.
- Avoid overuse of color or shading.
- Select colors and shades carefully to show connections and similarities as well as differences.
- Do not put unnecessary information on any graphic. Eliminate any information not relevant to your point.
- Watch for graphics that are text-heavy.
- Avoid using too many typefaces and fonts. Avoid use of italics, open type, Olde English, and other typefaces that are difficult to read.

EVALUATING GRAPHICS FOR ACCURACY

- Check for distortion. You, not the computer, must make the final decision.
- If you choose three-dimensional graphics, check the accuracy of each segment to ensure that the size suggested is what the data suggest.
- Choose scales that are regular and logical.
- Keep the graphic as simple as possible. Avoid the tendency to overuse the technology available in computer programs. Excessive detail obscures or distorts the point of the graphic.

Planning and Revision Checklists

The following questions are a summary of the key points in this chapter, and they provide a checklist when you are planning and revising any document for your readers.

PLANNING A REPORT THAT WILL INCLUDE GRAPHICS

- How important are graphics to your presentation?
- How complex are your graphics likely to be?
- How expert is your audience in reading graphics?
- Based on your answers to the previous questions, how much prose explanation of your graphics are you likely to need?
- Do you have objects to portray?
- What do you have to illustrate about the objects?
- Do you want to draw attention to certain aspects of the objects and not to others?
- Will exploded or cutaway drawings of the objects serve your purpose?
- Would photographs of the objects add realism or drama to you report?
- Based on your answers to the previous questions about objects, what kinds of photos and drawings are you likely to need?
- Will any of your photos or drawings need a scale reference?
- Will any of your photos or drawings need annotation?
- Are you working with any concepts that can be best presented visually or in a combination of words and graphics?
- Do you have any definitions that should be presented visually in whole or in part?
- Do you have any processes or algorithms that should be depicted visually?
- Would a flowchart of any of your processes or algorithms aid your readers?
- Will you be presenting information on trends or relationships? Should some of the information be presented in tables and graphs?
- Do you have masses of statistics that should be summarized in tables?
- Would some informal tables help you present your data?
- Which are your readers most likely to comprehend: bar and circle graphs or line graphs?
- For each graph you plan, ask this question: Is this graph intended to give the reader a general shape of a trend, or should the reader be able to extract precise information from the graph?
- Would a pictogram add interest to your report?

REVISING REPORT GRAPHICS

- Are your graphics suited to your purpose and audience?
- Are your graphics well located and easy to find?
- Have you shown scale on your photos when necessary?
- Are you annotations horizontal? Are they easy to read and find?
- Will your readers grasp the concepts you have shown visually? Do your verbal and graphic elements complement each other?

- Will your readers easily follow any processes you have shown graphically?
- Do you have any blocks of data that should be converted to informal or formal tables?
- Are your tables and graphs properly titled and properly numbered?
- Are your tables and graphs simple, clear, and logical?
- Have you referred to your tables and graphs in your text?
- Have you, when necessary, included units of measure in your tables and graphs?
- Are the numbers in your tables aligned correctly? Whole numbers on right-hand digits? Fractional numbers on the decimal points?
- Have you acknowledged the sources for your tables and graphs?
- Are your graphs legible?
- Are your graphs attractive?
- When necessary, have you helped you readers to interpret your graphs with commentary or annotations?
- Do your graphs need a grid or hash marks for more accurate interpretation?
- Have you avoided the use of keys? If not, have you kept them simple?
- Have you plotted your graphs according to the conventions—independent variable horizontally, dependent variable vertically? If not, do you have a good reason for your arrangement?
- If you have used a suppressed zero, will it be obvious to your reader?
- Do your tables and graphs complement each other?

Exercises

1. Analyze the effectiveness of the following two graphics, both of which were designed to appear in reports to investors. In what ways could each be redrawn to eliminate distortion?

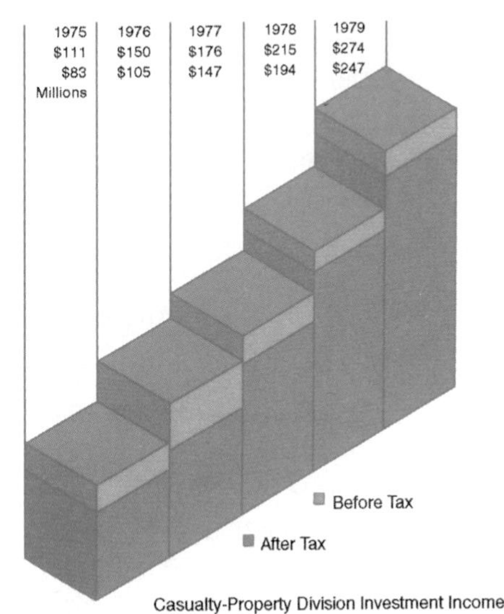

Casualty-Property Division Investment Income

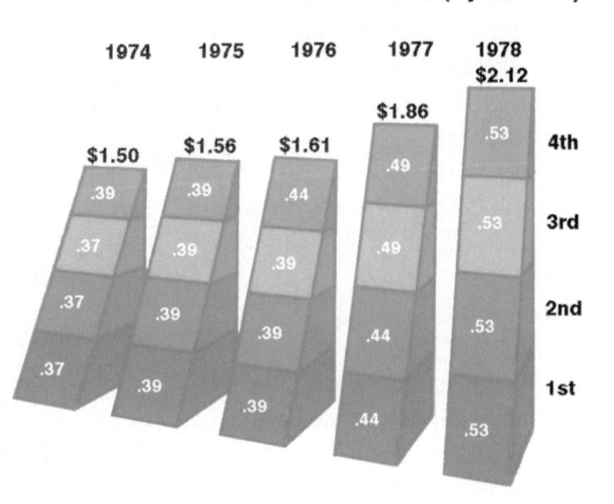

2. Mega Corp., which designs and test markets word processing programs, had the following sales figures for the years 1988, 1989, 1990, 1991, and 1992 in the six states in which it does business. The company has two divisions: Texas, Louisiana, and New Mexico comprise Division I; Oklahoma, Arkansas, and Kansas comprise Division II.

MEGA SALES 1988—INDIVIDUAL UNITS

State	CorrectWrite 1	CorrectWrite 2	CorrectWrite 3
Texas	904	741	318
Louisiana	713	751	722
New Mexico	823	755	720
Oklahoma	679	596	483
Arkansas	552	396	246
Kansas	327	219	435

MEGA SALES 1989—INDIVIDUAL UNITS

State	CorrectWrite 1	CorrectWrite 2	CorrectWrite 3
Texas	804	700	299
Louisiana	703	720	799
New Mexico	923	875	825
Oklahoma	655	543	450
Arkansas	502	380	236
Kansas	315	380	495

MEGA SALES 1990—INDIVIDUAL UNITS

State	CorrectWrite 1	CorrectWrite 2	CorrectWrite 3
Texas	740	738	466
Louisiana	803	800	789
New Mexico	843	769	790
Oklahoma	773	555	490
Arkansas	525	453	426
Kansas	337	287	488

MEGA SALES 1991—INDIVIDUAL UNITS

State	CorrectWrite 1	CorrectWrite 2	CorrectWrite 3
Texas	888	902	500
Louisiana	800	845	876
New Mexico	819	803	865
Oklahoma	723	505	390
Arkansas	500	353	326
Kansas	330	300	495

MEGA SALES 1992—INDIVIDUAL UNITS

State	CorrectWrite 1	CorrectWrite 2	CorrectWrite 3
Texas	940	838	766
Louisiana	930	900	889
New Mexico	634	567	590
Oklahoma	673	455	290
Arkansas	425	353	226
Kansas	237	187	288

Use line graphs, bar graphs, or circle graphs to show sales changes for the following:

A. Total sales of CW1, CW2, and CW3 for each year.

B. Comparison of total sales for Division I and Division II for each year.

C. Comparison of performance for Division I and Division II for each CorrectWrite version.

D. State which division has the largest percentage of gains in sales.

3. Redraw Figure 11-31 to remove the distortion among the bars.

4. Redraw Figures 11-14 and 11-15 to improve the comparisons suggested by the divided bars.

5. Reexamine Figure 11-18. Could you redesign this graphic so that it better presents the information?

6. Reexamine Figure 11-24. In what other way could you present the same information shown here?

7. Reexamine Figure 11-38. Could you design a more effective way to suggest the merging of word processors, desktop devices, and intelligent terminals?

8. Examine several technical publications in your field of study. Choose examples of good or bad graphics. Write a report that analyzes the effectiveness (or ineffectiveness) of these graphics. Be sure to integrate a copy of the graphic with your analysis of each.

Oral Exercises

See Chapter 18, "Oral Reports," before doing these exercises.

9. Assume that you are Director of Marketing for Mega Corporation. You receive a call from Megan Pierce, Vice President of Operations of Mega, who wants you to give an oral sales report to the board of directors. Your presentation should be approximately 10 minutes. Prepare your graphics as visual aids to explain the sales performance of Mega for 1988–1992.

10. Evaluate the effectiveness of the graphics in a journal in your major field of study. Prepare an oral presentation assessing the effectiveness of the graphics. Reproduce sample graphs, tables, drawings, and other graphic elements, and perhaps sample pages for use as overhead transparencies to explain your evaluation.

Report Exercise

See Chapter 14, "Development of Reports," and Chapter 15, "Development of Analytical Reports," before doing this exercise.

11. Assume that, after your sales presentation, Megan Pierce asks you to write a report to her. She wants to share your information with the Executive Committee. She also wants you to make projections for 1993 based on the 1988–1992 figures.

- Should the report be formal or informal?
- Based on your findings of the sales analysis, what kind of report will you write? How will you structure it? What elements will you include?

PART IV
APPLICATIONS

The first six chapters of Part IV put all the basic writing, design, and graphic techniques of the first three parts to work. The chapters of Part IV discuss correspondence, the job hunt, and reports such as feasibility reports, instructions, progress reports, and proposals. Chapters 14 and 15, in particular, show you how to deal with a variety of reports. Chapter 18 discusses how to use the principles and techniques of this book in oral reports. In short, Part IV covers most of the kinds of reports, written and spoken, that professionals create in every field.

CHAPTER 12
Correspondence

CHAPTER 13
The Strategies and Communications of the Job Hunt

CHAPTER 14
Development of Reports

CHAPTER 15
Development of Analytical Reports

CHAPTER 16
Instructions

CHAPTER 17
Proposals and Progress Reports

CHAPTER 18
Oral Reports

CHAPTER 12 CORRESPONDENCE

Even in this day of instant communication through telephones, fax machines, and electronic mail, typed—or, more likely, word processed—correspondence still plays a major role in getting an organization's work done. For example, executives may reach a decision in a telephone conversation or a teleconference, but that decision has to be documented. People may forget or incorrectly remember what the decision was. A letter or memo records the decision for everyone.

In fact, as we point out in Chapter 9, "Document Design," it's a myth that electronic technology is creating the paperless office. American business and government turn out more paper in the electronic age than ever before. Much of that paper consists of correspondence. On the job, you will often be responsible for a share of your organization's letters and memorandums.

We do not attempt in this chapter to give you a course in business correspondence. There are many fine books on the subject, some of them listed in our bibliography. However, we do cover the aspects of correspondence that will be most useful to you. We discuss the composition of business letters and memorandums, their style and tone, and their format. Then we instruct you how to write letters of inquiry, replies to letters of inquiry, letters of complaint and adjustment, and letter and memorandum reports.

Composing Letters and Memorandums

Composing letters and memorandums—memos, for short—is little different from composing any piece of writing as we describe the process in Chapter 2, "Composing." But a few points about topic, purpose, and audience might be worth reemphasizing because of their importance to good correspondence.

Topic and Purpose

People on the job use letters and memos to inform, to instruct, to analyze and evaluate, to argue—the whole range of communication activities that occupy people at work. Letters go to people outside the company or organization; memos go to people within. Any of the following would be a typical topic and purpose:

- Instructing secretaries in how to complete the new travel payment requests
- Convincing the boss that the office needs two new microcomputers
- Answering a complaint

Whatever your topic and purpose may be, be sure to announce them immediately. Nothing irritates a reader more than to read through several paragraphs of details with no idea of where the letter or memo is heading. Be directive. Tell your reader what your topic and purpose are with an opening like this one:

> Additional report writing responsibility in our office has increased our need for greater word processing capability. This memo describes the problems caused by the increased work load and shows the benefits of purchasing two microcomputers and a laser printer for the use of our staff.

The only exception to such a clear statement of topic and purpose may be when the recipient is likely to be hostile to your conclusions. We discuss this possibility and how to handle it when we discuss bad-news letters on pages 356–359.

Audience

In composing your letter or memo, consider your readers: Who are they and what do they know already? What is their purpose in reading your letter or memo?

Who Are the Readers? Letters and memos may be addressed to one person or to many. Even when you address a letter or memo to one person, you may send copies of it to many others. When such is the case, you have to think about the knowledge and experience of those receiving copies as well as those receiving the original. If those receiving copies lack background in the topic under discussion, you may have to take time to fill them in.

In another common situation, you may be explaining a technical problem and its solution to a colleague with technical knowledge equal to your own. However, you may be sending a copy to your boss, who lacks that technical knowledge. When such is the case, you would be wise to lead your discussion with what amounts to an executive summary. (See pages 250–252.) Fill the boss in quickly on the key points, and tell him or her the implications of what you are saying.

What Is the Reader's Purpose? You have a purpose in writing your letter or memo, but your reader has a purpose as well. Be sure to match your purpose to your reader's. For instance, your reader may be reading to evaluate your recommendation and accept or reject it. Your purpose must be to provide enough information to make that evaluation and decision possible.

In another situation, your purpose may be to explain why a project is running late. Your reader, somewhat skeptically no doubt, will be reading to determine whether your reasons are valid and acceptable. Provide enough information to justify your position. If you don't match your purpose to your reader's, you may not achieve the result you intended.

Style and Tone

In your correspondence, work for a clear style and a human tone.

Clear Style

Everything we say about clear and readable style in Chapter 5 goes doubly when you are writing letters and memos. Letters must be clear, so clear that the reader cannot possibly misunderstand them. Use short paragraphs, lists, clear sentence structure, and specific words. Above all, avoid pomposity and the cold formality of the passive voice.

Do not make your letters and memos a repository for all the clichés that writers before you have used. Avoid expressions like these:

- We beg to advise you that...
- We are in receipt of your letter that...
- It is requested that you send a copy of the requested document to our office.

There are hundreds of such expressions that weigh down business correspondence. To protect yourself from such prose pachyderms, remember our closing advice from Chapter 5: Ask yourself whether you would or could say in conversation what you have written. If you know you would strangle on the expression, don't write it. Restate it in simpler language, like this:

- We'd like you to know that...
- We received your letter that...
- Please send us a copy of your latest tax return.

Human Tone

To get a human tone into your correspondence, focus on the human being reading your letter or memo. Develop what has been labeled the *you-attitude.*

To some extent, the you-attitude means that your letters and memos contain a higher percentage of *you*'s than *I*'s. But it goes beyond this mechanical use of certain pronouns. With the you-attitude, you see things from your reader's point of view. You think about what the letter or memo will mean to the reader, not just what it means to you. We can illustrate simply. Suppose you have an interview scheduled with a prospective employer and, unavoidably, you must change dates. You could write as follows:

> Dear Ms. Moody:
>
> A change in my final examination schedule makes it impossible for me to keep our appointment on June 10.
>
> I am really disappointed. I was looking forward to coming to Los Angeles. It's a great inconvenience, but I hope we can work out an appointment. Please let me know when we can arrange a new meeting.

Now this letter is clear enough, and it may even get a new appointment for its writer. But it has the I-attitude, not the you-attitude. The persona projected is of a person who thinks only of himself or herself. The reader may be vaguely or even greatly annoyed. This next version will please a reader far more:

> Dear Ms. Moody:
>
> A change in my final examination schedule makes it impossible for me to keep our appointment on June 10. When I should have been talking with you, I'll be taking an exam in chemistry.
>
> I hope this change will not seriously inconvenience you. Please accept my regrets.
>
> Will you be able to work me in at a new time? Final exam week ends on June 12. Please choose any later date that is convenient for you.

Mechanically, the second letter contains more *you*'s, but more to the point, it considers the inconvenience caused the reader, not the sender. It also makes it easy for the reader to set up a new date.

Notice also that the second letter is a bit more detailed than the first. Many people have stressed the need for brevity in letters and memos, and certainly it is a good thing to be concise. But do not get carried away with the notion. Letters and memos are not telegrams. When they are too brief, they give an impression of brusqueness, even rudeness. Often, a longer letter or memo gives a better impression. Particularly avoid brevity when you must refuse people something or disappoint their expectations. People appreciate your taking time to explain in such a situation.

Our advice about style and tone can be summed up with what we call the four **C**'s of correspondence: Correspondence should be **clear, concise, complete,** and **courteous.** Sometimes, **concise** and **complete** may be in conflict. If in doubt, opt for completeness. Always remember the importance of being **courteous.** Taking time for the you-attitude is the good and human way to act. Luckily, it's also very good business.

Format

Almost any organization you join will have rules about its letter and memo formats. You will either have a secretary to do your correspondence, or you will have to learn the rules for yourself. In this section, we give you only enough rules and illustrations so that you can turn out a good-looking, correct, and acceptable business letter or memo. You will find this an especially valuable skill when you go job hunting.

Figures 12-1 and 12-2 illustrate the block and semiblock styles on nonletterhead stationery. Figures 12-3 and 12-4 illustrate the block style and the simplified style on letterhead stationery.

The chief difference between memos and letters is format. Figure 12-5 illustrates a typical memo format. Figure 12-6 illustrates the heading used for the continuation pages of a letter or a memo. We have indicated in these samples the spacing, margins, and punctuation you should use. In the text that follows, we discuss briefly the different styles and then give you some of the basic rules you should know about the parts of a letter or memo. Before continuing with the text, look at Figures 12-1 through 12-6. Particularly observe the spacing, placement, and punctuation of the various parts of a letter or memo.

For most business letters you may have to write, any of the styles shown would be acceptable. In letters of inquiry or complaint, where you probably do not have anyone specific in a company to address, we suggest the simplified style. For letters of application, we suggest the block or semiblock style without a subject line. Some people still find the simplified letter without the conventional salutation and complimentary close a bit too brusque. Unless you know for certain that the company you are applying to prefers the simplified form, do not take a chance with it.

If you are an amateur typist doing your own typing or word processing, we suggest the block style as the best style. All the conventional parts of a letter are included, but everything is lined up along the left margin. You do not have to bother with tab settings and other complications. Some people feel that a block letter looks a bit lopsided, but it is a common style that no one will object to. No matter which style you

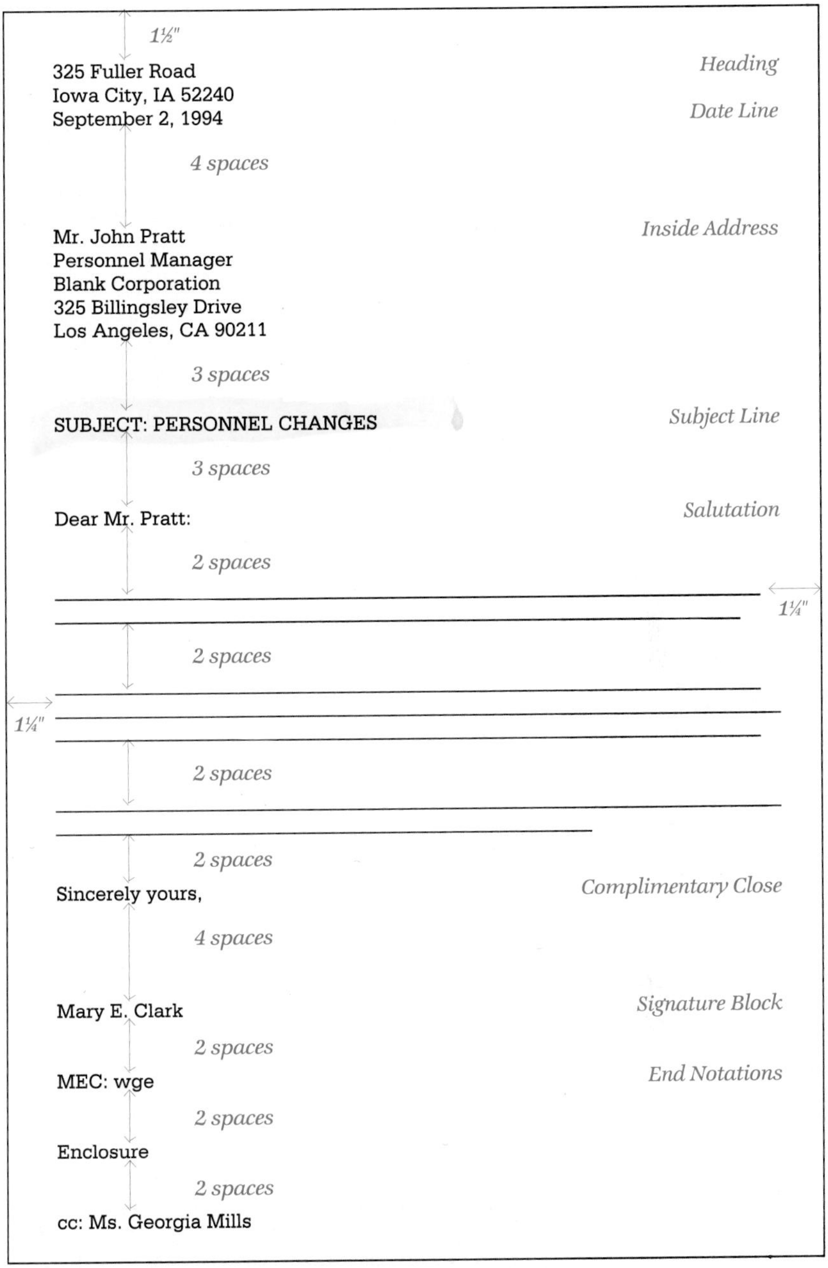

1½"

325 Fuller Road *Heading*
Iowa City, IA 52240
September 2, 1994 *Date Line*

4 spaces

Mr. John Pratt *Inside Address*
Personnel Manager
Blank Corporation
325 Billingsley Drive
Los Angeles, CA 90211

3 spaces

SUBJECT: PERSONNEL CHANGES *Subject Line*

3 spaces

Dear Mr. Pratt: *Salutation*

2 spaces

1¼"

2 spaces

1¼"

2 spaces

2 spaces

Sincerely yours, *Complimentary Close*

4 spaces

Mary E. Clark *Signature Block*

2 spaces

MEC: wge *End Notations*

2 spaces

Enclosure

2 spaces

cc: Ms. Georgia Mills

Figure 12–1 Block Letter on Nonletterhead Stationery

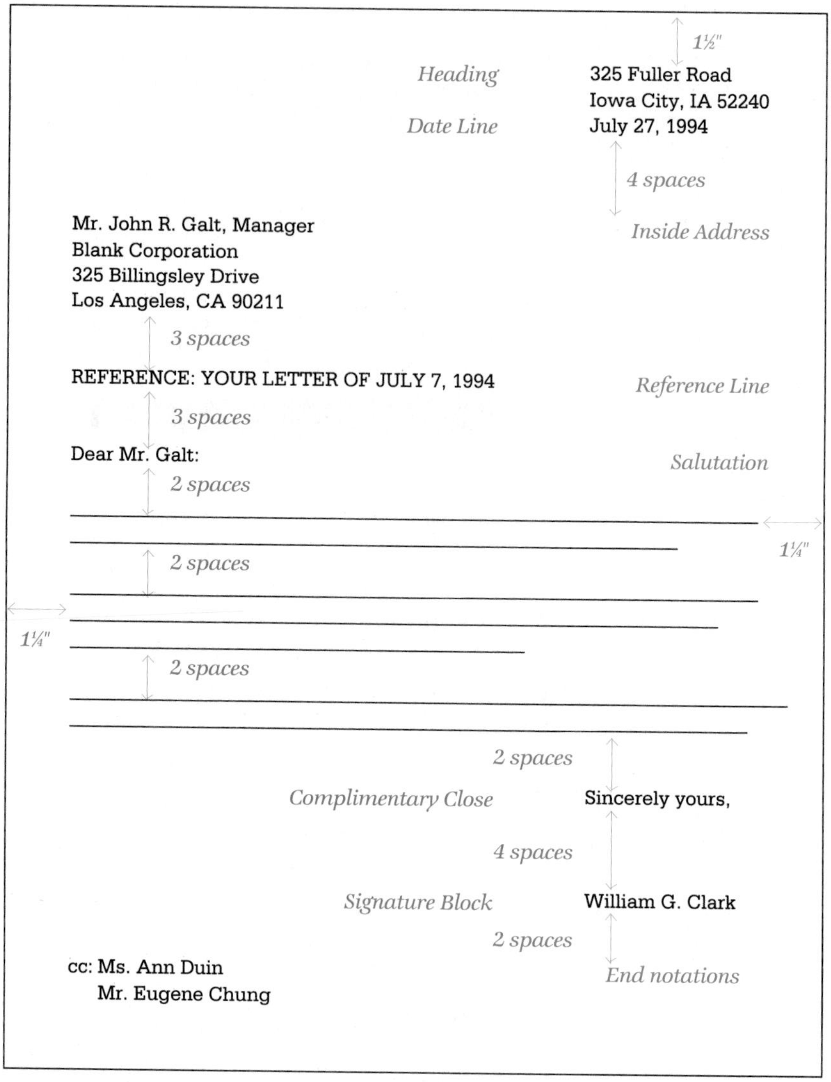

1½"

Heading 325 Fuller Road
Iowa City, IA 52240
Date Line July 27, 1994

4 spaces

Inside Address

Mr. John R. Galt, Manager
Blank Corporation
325 Billingsley Drive
Los Angeles, CA 90211

3 spaces

REFERENCE: YOUR LETTER OF JULY 7, 1994 Reference Line

3 spaces

Dear Mr. Galt: Salutation

2 spaces

1¼"

2 spaces

1¼"

2 spaces

2 spaces

Complimentary Close Sincerely yours,

4 spaces

Signature Block William G. Clark

2 spaces

cc: Ms. Ann Duin
Mr. Eugene Chung

End notations

Figure 12–2 Semiblock Letter on Nonletterhead Stationery

choose, leave generous margins, from an inch to two inches all around, and balance the first page of the letter vertically on the page. Because letters look more inviting with lots of white space, you should seldom allow paragraphs to run more than seven or eight lines.

Heading

When you do not have letterhead stationery, you will have to type your heading. In the semiblock style, the heading is approximately flush

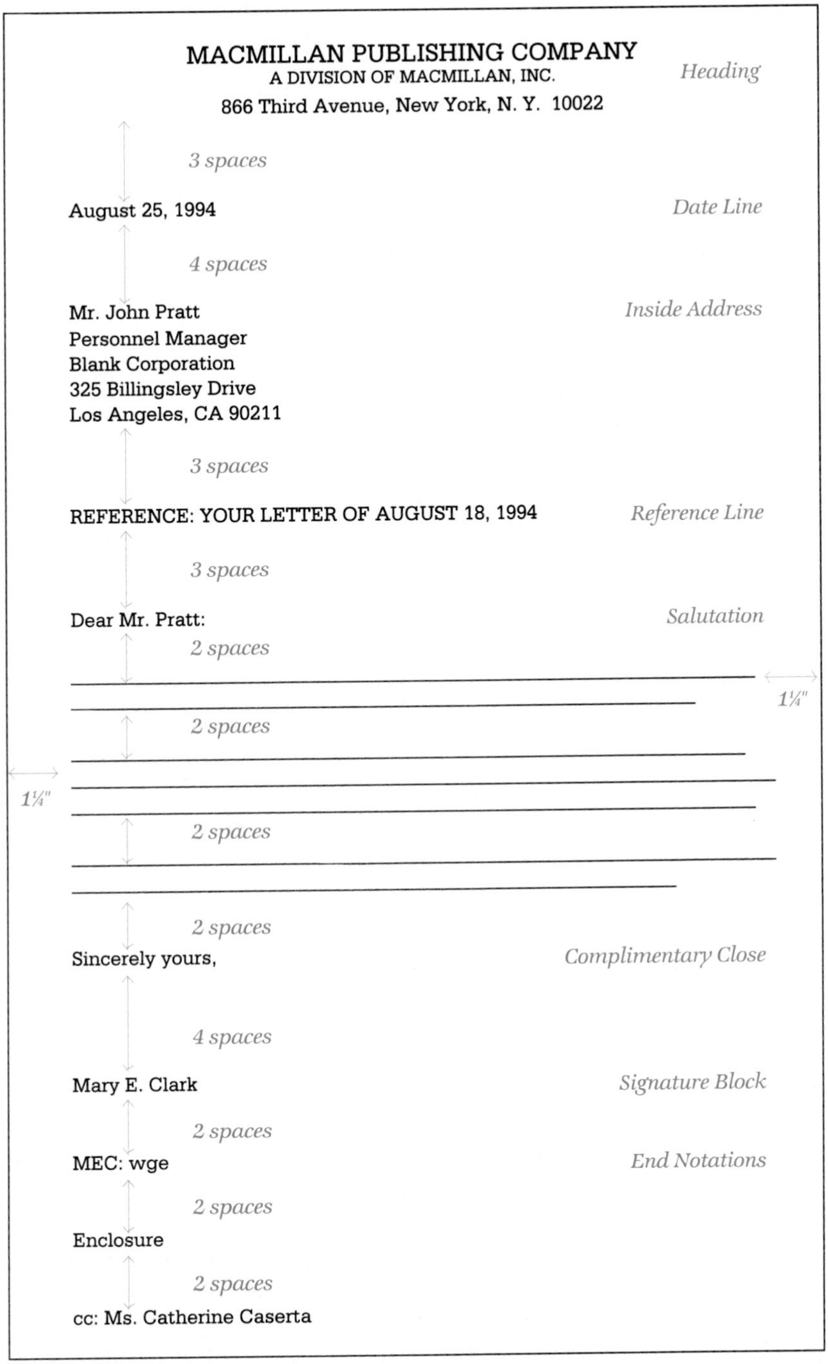

MACMILLAN PUBLISHING COMPANY
A DIVISION OF MACMILLAN, INC.
866 Third Avenue, New York, N. Y. 10022 — *Heading*

3 spaces

August 25, 1994 — *Date Line*

4 spaces

Mr. John Pratt — *Inside Address*
Personnel Manager
Blank Corporation
325 Billingsley Drive
Los Angeles, CA 90211

3 spaces

REFERENCE: YOUR LETTER OF AUGUST 18, 1994 — *Reference Line*

3 spaces

Dear Mr. Pratt: — *Salutation*

2 spaces

______________________________________ *1¼"*
____

2 spaces

____
1¼" ______________________________________
____

2 spaces

____
____

2 spaces

Sincerely yours, — *Complimentary Close*

4 spaces

Mary E. Clark — *Signature Block*

2 spaces

MEC: wge — *End Notations*

2 spaces

Enclosure

2 spaces

cc: Ms. Catherine Caserta

Figure 12–3 Block Style on Letterhead Stationery

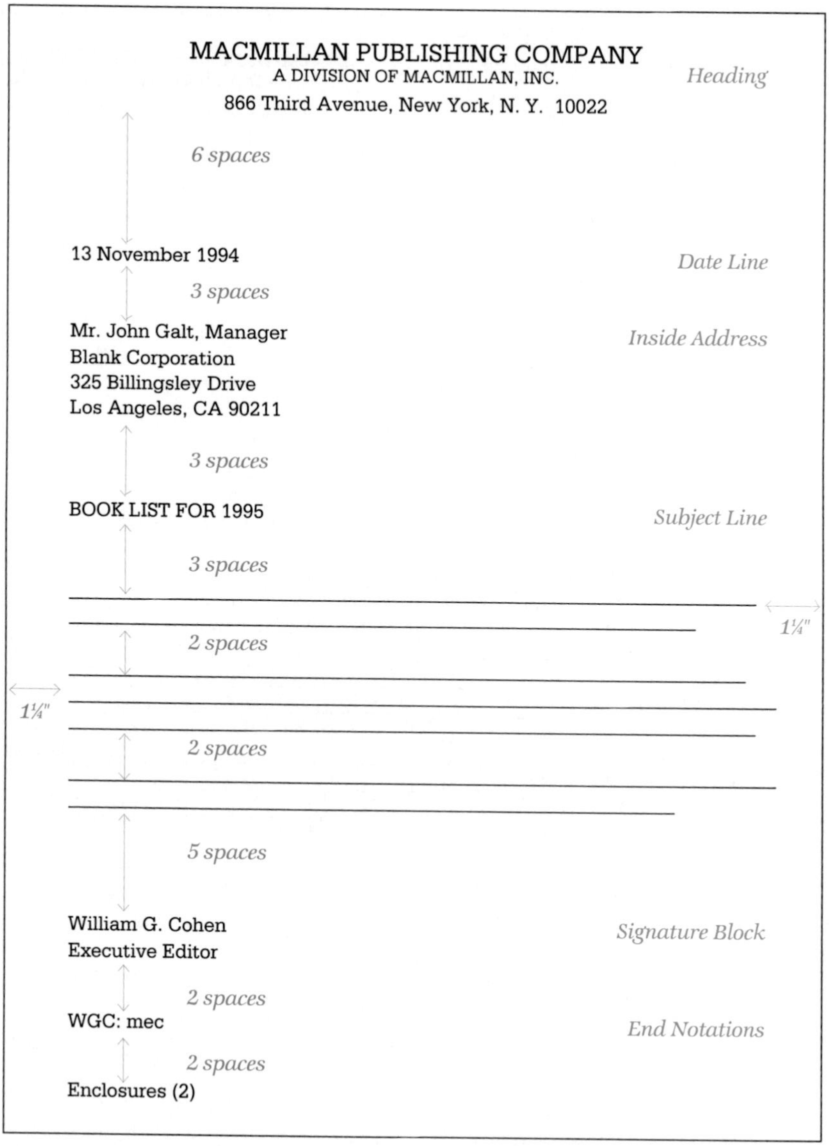

MACMILLAN PUBLISHING COMPANY
A DIVISION OF MACMILLAN, INC.
866 Third Avenue, New York, N. Y. 10022

Heading

6 spaces

13 November 1994

Date Line

3 spaces

Mr. John Galt, Manager
Blank Corporation
325 Billingsley Drive
Los Angeles, CA 90211

Inside Address

3 spaces

BOOK LIST FOR 1995

Subject Line

3 spaces

1¼"

2 spaces

1¼"

2 spaces

5 spaces

William G. Cohen
Executive Editor

Signature Block

2 spaces

WGC: mec

End Notations

2 spaces

Enclosures (2)

Figure 12–4 Simplified Style on Letterhead Stationery

right. In the other formats shown, the heading is flush left. Do not abbreviate words such as *street* or *road.* Write them out. You may abbreviate the names of states and provinces, however. Figure 12–7 lists the state abbreviations for the United States and province abbreviations for Canada. If you have business letterheads made up, have them printed on good quality white bond in a simple style. With word

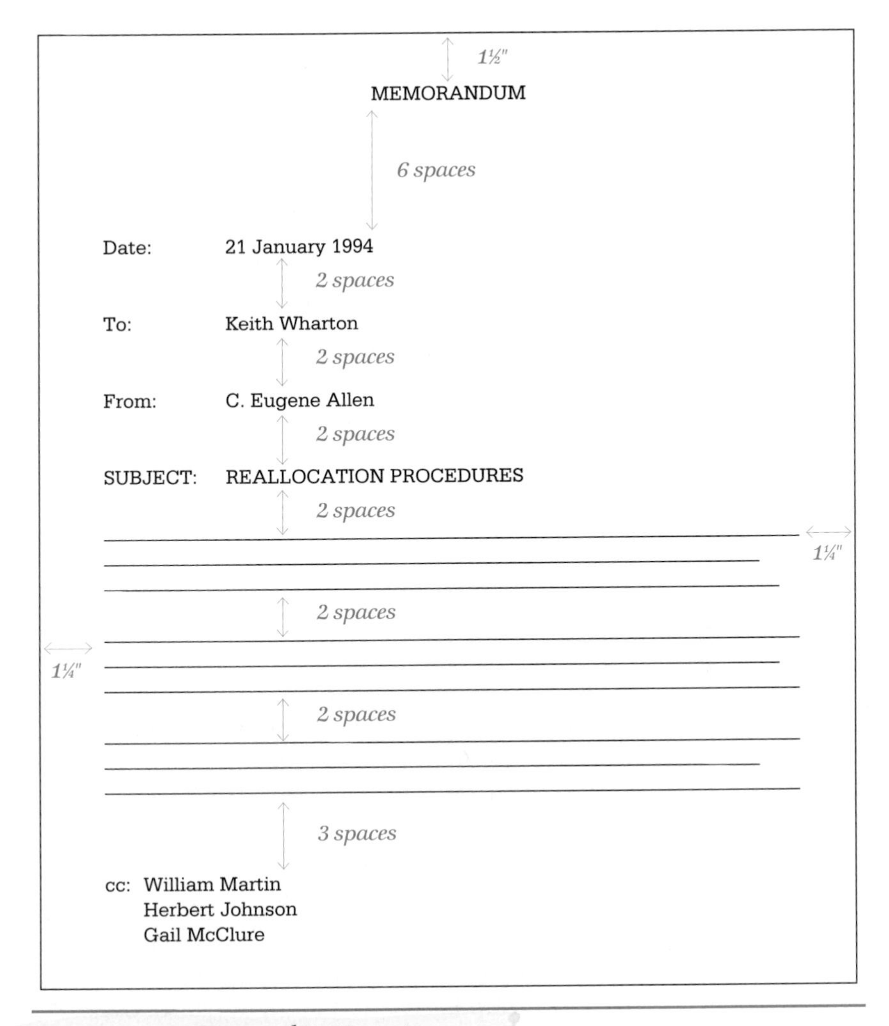

Figure 12–5 Memorandum Format

processing and a laser printer, you can print your own letterheads. We have two cautions if you choose to do so. Only the most expensive laser printers will turn out printing as crisp and sharp as commercial printing. Second, be careful not to design a too-elaborate letterhead. Use a simple, standard design.

Date Line

In a letter without a printed letterhead, the date line is part of your heading in the block, semiblock, and simplified styles. When you do have a printed letterhead, the date line is flush left in the block and simplified styles and approximately flush right in semiblock. Place it three

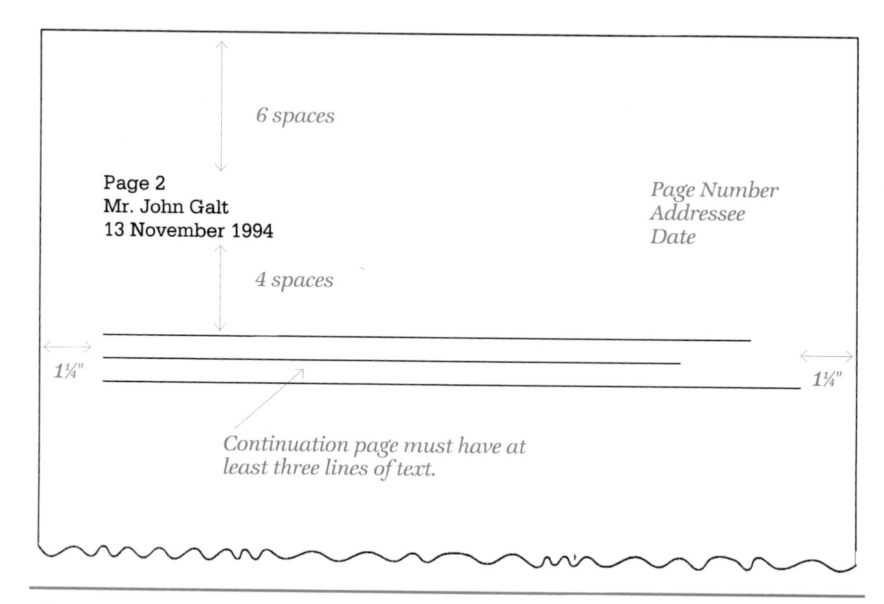

Figure 12–6 Continuation Page

to six spaces below a printed letterhead in a manner to help balance the letter vertically on the page. Write the date out fully either as in *June 3, 1994* or *3 June 1994.* Do not abbreviate the month or add *st* or *nd* (e.g., *1st, 2nd*) to the number of the day.

Inside Address

The inside address is placed flush left in all the formats shown. Make sure the inside address is complete. Follow exactly the form used by the person or company you are writing to. If your correspondent abbreviates *Company* as *Co.,* you should also. Use *S. Edward Smith* rather than *Samuel E. Smith* if that is the way Smith wants it. Do use courtesy titles such as *Mr., Ms., Dr.,* and *Colonel* before the name. The abbreviations usually used are *Mr., Ms.,* and *Dr.* (*Miss* and *Mrs.* are still in use, but *Ms.* is standard usage in the workplace.) Place one-word titles such as *Manager* or *Superintendent* immediately after the name. When a title is longer than one word, place it on the next line by itself. Do not put a title after the name that means the same thing as a courtesy title. For example, don't write *Dr. Samuel E. Smith, Ph.D.*

Attention Line

You may wish to write to an organizational address but also draw your letter to the attention of some individual. It's a way of saying, in effect, "Anyone there can answer this letter, but if Mr. Smith is there he is the

United States				Canada	
Alabama	AL*	Missouri	MO	Alberta	AB
Alaska	AK	Montana	MT	British Columbia	BC
American Samoa	AS	Nebraska	NE	Labrador	LB
Arizona	AZ	Nevada	NV	Manitoba	MB
Arkansas	AR	New Hampshire	NH	New Brunswick	NB
California	CA	New Jersey	NJ	Newfoundland	NF
Colorado	CO	New Mexico	NM	Nova Scotia	NS
Connecticut	CT	New York	NY	Northwest Territories	NT
Delaware	DE	North Carolina	NC	Ontario	ON
District of		North Dakota	ND	Prince Edward Island	PE
Columbia	DC	Ohio	OH	Quebec (Province de Quebec)	PQ
Florida	FL	Oklahoma	OK	Saskatchewan	SK
Georgia	GA	Oregon	OR	Yukon Territory	YT
Guam	GU	Pennsylvania	PA		
Hawaii	HI	Puerto Rico	PR		
Idaho	ID	Rhode Island	RI		
Illinois	IL	South Carolina	SC		
Indiana	IN	South Dakota	SD		
Iowa	IA	Tennessee	TN		
Kansas	KS	Texas	TX		
Kentucky	KY	Utah	UT		
Louisiana	LA	Vermont	VT		
Maine	ME	Virginia	VA		
Maryland	MD	Virgin Islands	VI		
Massachusetts	MA	Washington	WA		
Michigan	MI	West Virginia	WV		
Minnesota	MN	Wisconsin	WI		
Mississippi	MS	Wyoming	WY		

*Notice that both letters of the abbreviation are capitalized and that no period is used.

Figure 12–7 Geographic Abbreviations for the United States and Canada

best person to handle the matter." When you use an attention line, type it flush left two spaces below the inside address. Capitalize only the *A*. You can use a colon or not between *Attention* and the name:

Attention Mr. Frank Rookard

or

Attention: Mr. Frank Rookard

Reference Line

Use the reference line to refer to the letter or memo you are answering; for example,

REFERENCE: YOUR LETTER OF MAY 12, 1994

Place the reference line heading (sometimes abbreviated *RE.* or *REF.*) flush left in all styles. Type the heading and the reference line itself in capital letters. Generally, you will follow the heading with a colon, although sometimes the colon is omitted. A letter may have both a reference line and a subject line, or the two may be combined:

REFERENCE: YOUR LETTER OF 12 JULY 1994
APPLICATION FOR A MORTGAGE AT 452 LITTLE COMFORT ROAD

Subject Line

Place the subject line flush left in all the styles. In the block and semiblock styles, it is usually preceded by the heading *SUBJECT,* although sometimes you will see it with nothing before it. Generally, the heading is followed by a colon, but sometimes the colon will be omitted. Type the heading and the subject line in capital letters. In the simplified style, omit the heading, and type the line in all-capital letters.

If you are answering a letter that has a subject line, repeat the subject line from the original letter. If you are making up your own subject line, be sure it is complete enough to be useful. If, for example, you are reporting progress on an architectural design for a building at 452 Little Comfort Road, don't write merely "452 LITTLE COMFORT ROAD." Rather, write "PROGRESS REPORT ON THE ARCHITECTURAL DESIGN FOR THE BUILDING AT 452 LITTLE COMFORT ROAD."

Both subject lines and reference lines get your letter or memo off to a good start. They allow you to avoid clichéd openers like "With regard to your letter of April 5, 1994."

Salutation

Place the salutation flush left. Convention still calls for the use of *Dear.* Always use a name in the salutation when one is available to use. When you use a name, be sure it is in the inside address as well. Also, use the same courtesy title as in the inside address, such as *Dear Dr. Sibley* or *Dear Ms. McCarthy.* You may use a first-name salutation, such as *Dear Samantha,* when you are on friendly terms with the recipient. Follow the salutation with a colon.

What do you do when you are writing a company blindly and have no specific name to address? Some people use *Dear* followed by the name of the department being written to, such as *Dear Customer Relations Department.* Still others substitute a *Hello* or *Good Day* for the traditional salutation. One solution, perhaps the best, is to choose the simplified style, where no salutation is used.

In any case, do **not** begin letters with *Dear Person,* which is distasteful to many, *Dear Sir,* which is sexist, or *To Whom It May Concern,* which is old-fashioned.

Body

In typing the body of an average-length letter or memo, single-space between the lines and double-space between the paragraphs. In a particularly short letter, double-space throughout the body and use five-space indentations to mark the first lines of paragraphs. Avoid splitting words between lines. Never split a date or a person's name between two lines.

Complimentary Close

In the block style, place the complimentary close flush left. In the semiblock style, align the close with the heading (or with the date line in a letterhead letter). Settle for a simple close, such as *Sincerely yours* or *Very truly yours.* Capitalize only the first letter of the close and place a comma after the close.

Signature Block

Type your name four spaces below the complimentary close in the block and semiblock styles, five spaces below the last line in the simplified style. Use your first name, middle initial, and last name or, if you prefer, your first initial and middle name, as in *M. Lillian Smith.* We don't recommend the use of initials only, as in *M. L. Smith,* because this form puts your correspondents at a disadvantage. People who don't know you will not know whether to address you as *Dear Mr.* or *Dear Ms.*

If you have a title, type it below your name. Sign your name immediately above your typed name. Sign it legibly without fancy flourishes. Your signature and your typed name should agree. In memos it's customary to initial next to your typed name in the "From" line.

End Notations

Various end notations may be placed at the bottom of a letter or memo, always flush left. The most common ones indicate identification, enclosure, and carbon copy.

Identification The notation for identification is composed of the writer's initials in capital letters and the typist's initials in lowercase:

DHC:lnh

Enclosure The enclosure line indicates that additional material has been enclosed with the basic letter or memo. You may use several forms:

Enclosure
Enclosures (2)
Encl: Employment application blank

Copy The copy line informs the recipient of a letter or memo when you have sent a copy of the letter to someone else. The copy notation looks like this:

cc: Ms. Georgia Mills

See Figures 12-1 to 12-5 for proper spacing and sequence of these three notations.

Continuation Page

Use a continuation page or pages when you can't fit your letter or memo onto one page. Do not use letterhead stationery for a continuation page. Use plain bond of the same quality and color as the first page.

As shown in Figure 12-6, the continuation page is headed by three items: page number, name of addressee, and date.

When you have a continuation page, the last paragraph on the preceding page should contain at least two lines. The last continuation page must have at least three lines of text to accompany the complimentary close and the signature block.

Letters of Inquiry

As a student, businessperson, or simply a private person, you will often have occasion to write letters of inquiry. Students often overlook the rich sources of information for reports that they can tap with a few well-placed, courteous letters. Such letters can bring brochures, photographs, samples, and even very quotable answers from experts in the field the report deals with.

Sometimes companies solicit inquiries about their products through their advertisements and catalogs. In such cases your letter of inquiry can be short and to the point:

> Your advertisement on page 89 of the January 1994 Scientific American invited inquiries about your new Film-X developing process.
>
> I am a college student and president of a 20-member campus photography club. The members of the club and I would appreciate any information about this new process that you can send us.
>
> We are specifically interested in modernizing the film-developing facilities of the club.

As in the letter just quoted, you should include three important steps:

- Identify the advertisement that solicited your inquiry.
- Identify yourself and establish your need for the information.
- Request the information. Specify the precise area in which you are interested.

Obviously, in this step you also identify the area in which the company may expect to make a sale to you. You thus, in a subtle way, point out to the company why it is in its best interest to answer your inquiry promptly and fully: a good example of the you-attitude in action.

An unsolicited letter of inquiry cannot be quite so short. After all, in an unsolicited letter you are asking a favor, and you must avoid the risk of appearing brusque or discourteous. In an unsolicited letter you include five steps:

- Identify yourself.
- State clearly and specifically the information or materials you want.
- Establish your need for the requested information or materials.
- Tell the recipient why you have chosen him or her as a source for this information or material.
- Close courteously, but do **not** say "thank you in advance." Many feel that this phrase is presumptuous.

The first four steps may be presented in various combinations rather than as distinct steps, or in different order, but none of the steps should be overlooked.

Identification

In an unsolicited letter, identify yourself. We mean more here than merely using a title in your signature block. Rather, you should identify yourself in terms of the information sought. That is, you are not merely a student, you are a student in a dietetics class seeking information for a paper about iron enrichment of flour. Or you are a member of a committee investigating child abuse in your town. Or you are an engineer for a state department of transportation seeking information for a study of noise walls on city freeways. Certainly, the more prestigious your identification of yourself, the more likely you are to get the information you want, but do not misrepresent yourself in any way. Misrepresentation will usually lead to embarrassment.

Some years ago, one of the authors of this text wrote for information on bookshelves using his college letterhead instead of his private stationery. He was deluged with brochures that were followed up by a long-distance telephone call from a sales manager asking how many shelves the college was going to need: "Is the college expanding its library facilities?"

"No," the feeble answer came, "I just wanted to put up a few shelves in my living room."

Do not misrepresent yourself, but do honestly represent yourself in the best light you can. Most companies are quite good about answering student inquiries. They recognize in students the buyers and the employees of the future and are eager to court their goodwill.

Inquiry

State clearly what you want. Be very specific about what you want, and do not ask for too much. Avoid the shotgun approach of asking for "all available information" in some wide area.

Particularly, do not expect other people to do your work for you. Do not, for example, write to someone asking for references to articles on some subject. A little time spent by you in the library will produce the same information. On the other hand, it would be quite appropriate to write to a sociologist asking for a clarification or amplification of some point in a recent article that he or she had written. Science thrives on the latter kind of correspondence.

If you do need detailed information, put your questions in an easy-to-answer questionnaire form. If you have only a few questions, include them in your letter. Indent the questions and number them. Be sure to keep a copy of your letter. The reply you receive may answer your questions by number without restating them.

For a larger group of questions, or when you are sending the questionnaire to many people, make the questionnaire a separate attachment. Be sure in your letter to refer the reader to the attachment. Questionnaires are tricky. Unless they are presented properly and carefully made up, they will probably be ignored. Do not ask too many questions. If possible, phrase the questions for yes or no answers or provide multiple-choice options. Sometimes, meaningful questions just cannot be asked in this objective way. In such cases, do ask questions that require the respondent to write an answer, but try to phrase your questions so that answers to them can be short. Provide sufficient space for the answer expected. If you are asking for confidential information, stress that you will keep it confidential. The questionnaire should be typed, printed, or photocopied.

Need

Tell the recipient why you need the information. Perhaps you are writing a report, conducting a survey, buying a camera, or simply satisfying a healthy, scientific curiosity. Whatever your reason, do not be complicated or devious here. Simply state your need clearly and honestly. Often this step can best be combined with the identification step. If there is some deadline by which you must have an answer, say what it is.

Choice of Recipient

Tell the recipient of the letter why you have chosen him or her or the recipient's company as a source for the information. Perhaps you are writing a paper about stereo equipment and you consider this company to be one of the foremost manufacturers of FM tuners. Perhaps you have read the recipient's recent article on space medicine. You specifically want an amplification of some point. Obviously, this section is a good place to pay the recipient a sincere compliment, but avoid flattery or phoniness.

Point out any benefit the recipient may gain by answering your request. For example, if you are conducting a survey among many companies, promise the recipient a tabulation and analysis of results when the survey is completed.

Courteous Close

Close by expressing your appreciation for any help the receiver may give you. But do not use the tired, old formula, "Thanking you in advance, I remain ... " Later, do write a thank-you note even if all you get is a refusal. Who knows? The second letter may cause a change of heart.

Sample Letter

Dear Mr. Hanson:

Identification / Need

I am a second-year student at Florida Technological Institute. For a course in technical writing, I am writing a paper on the proper way to educate Americans in the use of the metric system. The paper is due on 3 March.

Choice of source

In the journals where I have been researching the subject, you are frequently mentioned as a major authority in the field. Would you be kind enough to give me your opinion about how metrics should be taught?

Specific information requested

The specific question that concerns me is whether metric measurements should be taught in relation to present standard measurements, such as the foot and pound, or whether they should be taught independently of other measurements. I see both methods in use.

Courteous close

Any help you can give me will be greatly appreciated, and I will, of course, cite you in my paper.

Replies to Letters of Inquiry

In replying to a letter of inquiry, be as complete as you can. Probably you will be trying to avoid a second exchange of letters of questions and answers. If you can, answer the questions in the order in which they appear in the inquiry. If the inquirer represents an organization and has written on letterhead stationery, you may safely assume that a file copy of the original letter has been retained. In that case, you don't have to repeat the original questions—just answer them. If the inquirer writes as a private person, you cannot assume that a copy of the letter has been kept. Therefore, you will need to repeat enough of the original question or questions to remind the inquirer of what has been asked.

You might answer the earlier letter about metric education in the following fashion:

Dear Ms. Montez:

Repetition of question

The question of how to educate people about the metric system concerns a great many educators. I suppose it's inevitable that those familiar with the present system will be tempted to convert metric measurements to ones they already know. For example, people will say, "A kilogram, that's about two pounds."

Answer to question

In my opinion, however, such conversion is not the best way to teach metrics. Rather, people should be taught to think in terms of what the metric measurement really measures, to associate it with familiar things. Here are some examples:

A paper clip is about a centimeter wide.

A dollar bill weighs about a gram.

A comfortable room is 20° C.

Water freezes at 0° C.

At a normal walking pace, we can go about five kilometers an hour.

These are the kinds of associations we have made all our lives with the present system. We need to do the same for metric measurements. As in learning a foreign language, we really learn it only when we stop translating in our heads and begin to think in the new language.

Source of additional information

I hope this answers your question. You can get valuable materials about metrics by writing to the U.S. Metric Association, 10245 Andasol Avenue, Northridge, CA 91325 (FAX: 818-368-7443).

Because the writer of the letter of inquiry wrote as a private person, she is, in this answer, reminded of the original question. The tone of the letter is friendly. The question is answered succinctly but completely. An additional source of information is mentioned, an excellent idea that should be followed when possible.

When you can, include as enclosures to your letter previously prepared materials that provide answers to the questions asked. If you do so, provide whatever explanation of these materials is needed.

Letters of Complaint and Adjustment

Mistakes and failures happen in business as they do everywhere else. Deliveries don't arrive on time or at all. Expensive equipment fails at a critical moment. If you're on the receiving end of such problems, you'll probably want to register a complaint with someone. Chances are good that you'll want an adjustment—some compensation—for your loss or inconvenience. To seek such an adjustment, you'll write a letter of complaint or, as it is also called, a claim letter.

26 Shady Woods Road
White Bear Lake MN 55101
28 October 1994

Customer Relations Department
Chapman Products, Inc.
1925 Jerome Street
Brooklyn, NY 11205

BROKEN SANDING BELTS

Product information

This past September, I purchased two Chapman sanding belts, medium grade, #85610, at the Fitler Lumber Company in St. Paul, Minnesota. I paid a premium price for the Chapman belts because of your reputation. However, the belts have proved to be unsatisfactory, and I am returning them to you in a separate package (at a cost to me of $2.36).

Problem

The belt I have labeled #1 was used only 10 minutes before it broke. The belt labeled #2 was used only 5 minutes when the glue failed and it broke.

Inconvenience caused

I attempted to return the belts to Fitler's for a refund. The manager refused me a refund and said I'd have to write to you.

I am disturbed on two counts. First, I paid a premium price for your belts. I did not expect an inferior product. Second, does the retailer have no responsibility for the Chapman products he sells? Do I have to write to you every time I have a problem with one of your products?

Motivation for fair adjustment

I'm sure that Chapman is proud of its reputation and will want to adjust this matter fairly.

John Griffin

John Griffin

Figure 12–8 Letter of Complaint

Letter of Complaint

Your attitude in a letter of complaint should be firm but fair. There is no reason to be discourteous. You'll do better in most instances if you write the letter with the attitude that the offending company will want to make a proper adjustment once they know the problem. Don't threaten to withdraw your business on a first offense. Of course, after repeated offenses you will seek another firm to deal with, and that should be made clear at the appropriate time.

Be very specific about what is wrong and about any inconvenience you suffered. Be sure to give any necessary product identification such as serial numbers. At the end of your letter, motivate the receiver to make a fair adjustment. If you know exactly what adjustment you want, spell it out. If not, allow the company you are dealing with to suggest an adjustment. Figure 12-8 shows you what a complete letter of complaint looks like. Note that it is in simplified format and addressed to the Customer Relations Department. You can safely assume that most companies have an office or an employee specifically responsible for complaints. In any case, if the letter is addressed in this manner, it will reach the appropriate person much more rapidly.

Letter of Adjustment

What happens at the other end of the line? You have received a letter of complaint and must write the adjustment letter. What should be your attitude? Oddly enough, most organizations welcome letters of complaint. They prefer customers who complain rather than customers who think "never again" when a mistake occurs and take their business elsewhere. Most organizations will go out of their way to satisfy complaints that seem at all fair. A skillful writer will use the adjustment letter as a means of promoting future business.

Letters of adjustment fall into two categories: granting of the adjustment requested or a refusal. The first is good news and easy to write. The second is more difficult, as indicated by its common name: the bad-news letter.

The Good-News Letter When you are granting an adjustment, be cheerful about it. Remember, your main goal is to build goodwill and future business. Follow these three steps:

- Begin by expressing regret about the problem or stating that you are pleased to hear from the customer—or both. Keep in mind our earlier comments about the you-attitude while writing an adjustment letter.
- Explain the circumstances that caused the problem. State specifically what the adjustment will be.
- Handle any special problems that may have accompanied the complaint and close the letter.

Figure 12-9 shows you such a letter.

The Bad-News Letter A letter refusing an adjustment is obviously more difficult to write. You want, if possible, to keep the customer's goodwill. You want at the very least to forestall future complaints. In

CHAPTER 12 • Correspondence

CHAPMAN PRODUCTS, INC.
1925 Jerome Street
Brooklyn, NY 11205

November 11, 1994

Mr. John Griffin
26 Shady Woods Road
White Bear Lake, MN 55101

Dear Mr. Griffin:

Expression of regret

Thank you for your letter of October 28. We're sorry that you had a problem with a Chapman product. But we are happy that you wrote to us about your dissatisfaction. We need to hear from our customers if we are to provide them satisfactory products.

Explanation of circumstances

The numbers on the belts you returned indicate that they were manufactured in 1984. Sanding belts, like many other products, have a "shelf life," and the belts you purchased had exceeded theirs. Age, heat, and humidity had weakened them.

Statement of adjustment

Mr. Griffin, we stand behind our products. Although we are sure that the belts you purchased were not defective when we shipped them, we wish to replace them for you. You are being shipped a box of 10 belts, medium grade. We're sure that these belts will live up to the Chapman name.

Handling of special problems

We also suggest that you look in the yellow pages of your telephone directory under "Hardware-Retail" for authorized Chapman dealers. We can only suggest to independent dealers how they should shelve and sell our products. We can exert more quality control with our own dealers. We know that you can find a Chapman dealer who will give you excellent service.

Sincerely,

Ahmad Attallah
Customer Service

AA/ay

cc: Fitler Lumber Company

Figure 12–9 Good-News Letter of Adjustment

stating your refusal, you must exercise great tact. Bad-news letters usually consist of five steps:

- Begin with a friendly opener. Try to find some common ground with the complainant. Express regret about the situation. Even though you may think the complaint is totally unfair, don't be discourteous. Incidentally, not everyone writes a letter as courteous as the one in Figure 12-8. Sometimes, people are downright abusive. If so, attempt to shrug it off. Just as you would not pour gasoline on a fire, don't answer abuse with abuse. Pour on some cooling words instead.
- Second, explain the reason for the refusal. Be very specific here and answer at some length. The very length of your reply will help to convince the reader that you have considered the problem seriously.
- Third, at the end of your explanation, state your refusal in as inoffensive a way as possible.
- Fourth, if you can, offer a partial or substitute adjustment.
- Finally, close your letter in a friendly way.

Companies selling products are not the only organizations that receive letters of complaint. Public service organizations do, also. The letter that follows illustrates a refusal from such an organization—a state department of transportation. In this case, a citizen had written stating that a curved section of highway near her home was dangerous. She requested that the curve be rebuilt and straightened. The reply uses the strategy that we have outlined.

In the case of this letter, the goal is not to keep a paying customer but to keep a taxpayer friendly. In either case, the strategy is to offer an honest, detailed, factual explanation in a cheerful way.

Dear Mrs. Ferguson:

Friendly opener

Thank you for your letter concerning the section of Trunk Highway (TH) 50 near your home. The Department of Transportation shares your concern about the safety of TH 50, particularly the section between Prestonburg and Pikeville, near which your home is located.

TH 50, as you mentioned, has many hills and curves and, because of the terrain, some steep embankments. However, its accident rate–3.58 accidents per million vehicle miles–is far from the worst in the state. In fact, there are 64 other state highways with worse safety records.

Reasons for refusal

We do have studies under way that will result ultimately in the relocation of TH 50 to terrain that will allow safer construction. These things take time, as I'm sure you know. We have to coordinate our plans with county and town authorities and the federal government. Money is assigned to these projects by priority, and based on its accident rate, TH 50 does not have top priority.

Accidents along TH 50 are not concentrated at any one curve. They are spread out over the entire highway. Reconstruction at any one location would cause little change in the overall accident record.

Statement of refusal

For all the above reasons we do not anticipate rebuilding any curves on TH 50. However, we are currently evaluating the need for guardrails along the entire length of the highway. Within a year we will probably construct guardrails at a number of locations. Most certainly, we will place a guardrail at the curve that concerns you. This should correct the situation to some extent.

Offer of substitution

Friendly close

We appreciate your concern. Please write to us again if we can be of further help.

The five-step strategy of the bad-news letter can be useful on many occasions. Any time you have to disappoint someone's expectations or you expect a hostile audience, consider using the bad-news approach.

Letter and Memorandum Reports

Many business reports run from two to five pages. They are too short to need the elements of more formal reports, such as title pages and tables of content. Usually, such a short report will be written as a letter or memo. All of the reports we discuss in Chapters 14 through 17 can be and often are written as memos or letters. All of the strategies discussed in Chapters 6, 7, and 8, used singly or in combination, are found in letter and memorandum reports. A common plan for either a letter report or a memo report calls for an introduction, a summary, a discussion, and an action step. Figures 12–10 and 12–11 illustrate how such a report can be presented as either a memo that stays within the organization or a letter that goes outside.

Introduction

Begin by telling the reader the subject and purpose of the report. Perhaps you are reporting on an inspection tour or summarizing the agreements reached in a consultation between you and the recipient. You may be reporting the results of a research project, or the beginning of one. Or you may be writing a progress report on a project that is under way but not completed. Whatever the subject and purpose are, state them clearly. If someone has requested the report, name the requester. Typically, the introduction in a letter or memo report will not have a heading.

Summary

In most cases, the summary will be an executive summary and will have a heading labeling it as a summary. (See pages 250–252.) It will

Southern Wire Company
Memorandum

Date: 15 January 1994

To: Louise Carson, President

From: Thomas Kehoe TK
Chief Engineer

Subject: Research in High-Temperature Superconductivity

Combination executive summary and introduction

Since technical breakthroughs in high-temperature superconductivity (HTS) in 1987, the federal government has been supporting research in HTS at an increasing rate. In the next 6 years, the government will fund a number of joint ventures in HTS between government laboratories and private industry to prepare for a market in HTS technology it estimates to be worth $12 billion by 2000.

Wire fabrication will be a major research area and a major piece of the HTS market. If we are to afford significant research and to maintain a market share in HTS, we must act quickly to seek a joint venture with a government laboratory. To that end, this report, after a brief discussion of HTS technology and government policy, recommends that management set up a small task force to explore and report on the funding possibilities and to produce a model proposal for management's consideration.

Discussion

In ordinary conductivity, electrons meet resistance in passing through a conductor, such as copper wire, and, therefore, lose efficiency. In superconductivity, electrons meet little resistance in passing through the conductor, and efficiency increases. Low-temperature superconductivity has existed for about 30 years, but at extremely low temperatures of 0 Kelvin (K), about -464° Fahrenheit (F). Expensive liquid helium is required for such temperatures. High-temperature superconductivity operates at about 125K (-234° F). Widely available liquid nitrogen will produce such a temperature and provide an inexpensive and reliable refrigeration system. The cost savings are shown in the following table that compares the life-cycle savings of HTS with those of low-temperature and conventional systems in several applications.

Figure 12–10 Memorandum Report
Source: Adapted from material in Solar Energy Research Institute, *Superconductivity for Electric Power Systems* (Washington, DC: U.S. Department of Energy, 1989).

Advantages of tables:
- Provide white space
- Make comparisons easy to see

Page 2
Louise Carson, President
15 January 1994

Life-Cycle Cost Savings (%) with High-Temperature Superconductors

	Compared with Low-Temperature System	Compared with Conventional System
Generator (300 megawatt)	27	63
Transformer (1,000 megavolt-ampere)	36	60
Transmission Line (10,000 megavolt-ampere, 230-kilovolt)	23	43
Motors	11	21
Magnetic Separators	15	20

The lower cost and higher efficiency of HTS will have a tremendous effect on virtually all applications of electrical power, such as motors, generators, heat pumps, transmission lines, magnetic storage, and computing. The Department of Energy (DOE) estimates the market to be $12 billion by 2000.

The federal government is funding research in all aspects of HTS at the rate of about $200 million a year. DOE has established three pilot programs to make the resources of the national laboratories at Argonne, Los Alamos, and Oak Ridge available to private industry and is seeking joint ventures in research and development. Furthermore, companies acting in such joint ventures can obtain the patents and licenses needed to go commercial with HTS.

The major "so-what"

As a wire fabricator, we clearly have a major stake in HTS. If we do not move swiftly and surely to develop wire suitable for HTS, we run the risk of losing a share of what will be a major market.

Figure 12-10 (continued).

Page 3
Louise Carson, President
15 January 1994

Action

Recommendations

I recommend that management establish a small task force composed of myself and one representative each from marketing, finance, and the legal department and charge it with three tasks:

- Contact DOE to obtain information on federal funding for joint ventures.
- Report on the funding and research possibilities available to us.
- Develop a model proposal for a joint venture between us and DOE.

Copies for those directly concerned

cc: Joseph Roberts
Vice President, Marketing

Emmie Caraballo
Chief Financial Officer

Thomas Scanlan
Director, Legal Department

Figure 12-10 (continued).

ENERGY CONSULTING

2221 K Street NW
Washington, DC 20007

15 January 1994

Ms. Louise Carson, President
Southern Wire Company
651 River Street
Savannah, GA 31455

SUBJECT: REPORT ON HIGH-TEMPERATURE SUPERCONDUCTIVITY RESEARCH

Dear Ms. Carson:

Reason for report

Combination executive summary and introduction

At your request Energy Consulting has looked into high-temperature superconductivity (HTS) and the funding possibilities for Southern Wire Company in HTS. We can report favorable news. Since technical breakthroughs in HTS in 1987, the federal government has been supporting research in HTS at an increasing rate. In the next 6 years, the government will fund a number of joint ventures in HTS between government laboratories and private industry to prepare for a market in HTS technology it estimates to be worth $12 billion by 2000.

Wire fabrication will be a major research area and a major piece of the HTS market. If Southern Wire is to conduct significant research and maintain a market share in HTS, you should act quickly to seek a joint venture with a government laboratory. To that end, this report, after a brief discussion of HTS technology and government policy, recommends that Southern Wire management set up a small task force to explore and report on the funding of possibilities and to produce a model proposal for management's consideration.

Figure 12–11 Letter Report
Source: Adapted from material in Solar Energy Research Institute, *Superconductivity for Electric Power Systems* (Washington, DC: U.S. Department of Energy, 1989).

Page 2
Louise Carson, President
15 January 1994

Discussion

In ordinary conductivity, electrons meet resistance in passing through a conductor, such as copper wire, and, therefore, lose efficiency. In superconductivity, electrons meet little resistance in passing through the conductor and efficiency increases. Low-temperature superconductivity has existed for about 30 years, but at extremely low temperatures of 0 Kelvin (K), about -464° Fahrenheit (F). Expensive liquid helium is required for such temperatures. High-temperature superconductivity operates at about 125K (-234° F). Widely available liquid nitrogen will produce such a temperature and provide an inexpensive and reliable refrigeration system. The cost savings are shown in the following table that compares the life-cycle savings of HTS with those of low-temperature and conventional systems in several applications.

Advantages of tables:
- Reduce wordage
- Provide relief for reader

Life-Cycle Cost Savings (%) with High-Temperature Superconductors

	Compared with Low-Temperature System	Compared with Conventional System
Generator (300 megawatt)	27	63
Transformer (1,000 megavolt-ampere)	36	60
Transmission Line (10,000 megavolt-ampere, 230-kilovolt)	23	43
Motors	11	21
Magnetic Separators	15	20

The lower cost and higher efficiency of HTS will have a tremendous effect on virtually all applications of electrical power, such as motors, generators, heat pumps, transmission lines, magnetic storage, and computing. The Department of Energy (DOE) estimates the market to be $12 billion by 2000.

Figure 12-11 (continued).

Page 3
Louise Carson, President
15 January 1994

The federal government is funding research in all aspects of HTS at the rate of about $200 million a year. DOE has established three pilot programs to make the resources of the national laboratories at Argonne, Los Alamos, and Oak Ridge available to private industry and is seeking joint ventures in research and development. Furthermore, companies acting in such joint ventures can obtain the patents and licenses needed to go commercial with HTS.

As a wire fabricator, Southern Wire has a large stake in HTS. If you do not move swiftly and surely to develop wire suitable for HTS, you run the risk of losing a share of what will be a major market.

Action

I recommend that management establish a small internal task force composed of one representative each from engineering, marketing, finance, and the legal department and charge it with three tasks:

List format for recommendations

- Contact DOE to obtain information on federal funding for joint ventures.
- Report on the funding and research possibilities available to Southern Wire.
- Develop a model proposal for a joint venture between Southern Wire and DOE.

If you would like Energy Consulting to work with the task force or to help Southern Wire in working with DOE, we would be happy to do so.

Sincerely yours,

Walter Mazura
Vice President and Director

WM: ekg

Figure 12-11 (continued).

emphasize the things in the report important to an executive's decision-making process and state clearly any conclusions, recommendations, and decisions that have been reached. Often, the functions of the introduction and the executive summary are combined. When this is the case, the combination usually will not have a heading.

Discussion

Give your discussion a heading. You might label it simply as "Discussion," or perhaps call it "Findings," "Results and Discussion," or the like. Develop your discussion using the same techniques and rhetorical principles that you would use in a longer report. Remember to consider your audience as you do in longer reports. For example, if you are writing to an executive, do not fill your letter or memo with jargon and technical terms.

If you must report a mass of statistics, try to round them off. If absolute accuracy is necessary, perhaps you can give the figures in an informal table. Some word processing programs now make it possible to incorporate graphs into memos and letters. If you have such a capability, take advantage of it by following the principles developed in Chapter 11, "Graphical Elements." You may also find listing a useful technique in letters and memos. (See pages 88–89 and 198–202.) Use subheadings just as you would in a longer report. (See pages 209–215.)

Action

If your letter or memo report recommends action on someone's part, include an action section with an "Action" or "Recommendation" heading. In this section, state—or sometimes, more diplomatically, suggest—what that action should be.

Planning and Revision Checklists

You will find the planning and revision checklists that follow Chapter 2, "Composing" (pages 38-39), Chapter 4, "Writing for Your Readers" (pages 81-82), and inside the front cover .valuable in planning and revising any presentation of technical information. The following questions specifically apply to correspondence. They summarize the key points in this chapter and provide a checklist for planning and revising.

PLANNING

General

- What is your topic and purpose?
- Who are your primary readers? Secondary readers? Do they have different needs? How can you satisfy all your readers?
- Why will your readers read your letter or memo?
- What format will best suit the situation?
- Do you have all the information you need to address your letter or memo? Names, title, addresses, and so forth?

Letters of Inquiry

- In what capacity are you making your request?
- What specifically do you want?
- Why do you need what you are requesting?
- Why are you making your request to this particular source?

Replies to Letters of Inquiry

- Has the inquirer written to you as a private person or as a member of an organization?
- Can you answer the inquirer's questions point by point?
- Do you have additional information or material that the inquirer would find useful?

Letters of Complaint

- What specifically is the problem?
- What inconvenience or loss has the problem caused you?
- Do you have all the product information you need?
- What is the adjustment you want?
- How can you motivate the recipient to make a fair adjustment?

Letters of Adjustment

- Is the letter of adjustment good news or bad news?
- If good news, what circumstances caused the problem? Are there special problems that must be handled? What adjustment has been asked for? What adjustment can you offer?
- If bad news, what are the circumstances for refusing the request for adjustment?
- Is there a substitute that can be offered in place of the requested adjustment?

Letter and Memorandum Reports

- What is the occasion, subject, purpose, and audience for the report?
- Would lists, tables, and graphs help your discussion?
- What conclusions, recommendations, and decisions do you have to report?
- What action should be the outcome of the report?

REVISION

General

- Have you stated your topic and purpose clearly in the first sentence or two of your letter or memo?
- Have you met your reader's purpose?
- Have you avoided jargon and clichés?
- Does your correspondence demonstrate a you-attitude?
- Is your letter or memo clear, concise, complete, and courteous?
- Have you used your chosen format correctly? Checked for correct punctuation and spacing?

Letters of Inquiry

- Does your letter of inquiry identify you, state specifically what you want, establish your need for the information or material you request, tell the recipient why you have chosen him or her as a source, and close graciously?

Replies to Letters of Inquiry

- If you are replying to a letter of inquiry, have you answered it as completely as need be? Is there additional information you could refer to or additional material you could send?

Letters of Complaint

- Do you have a firm but courteous tone in your letter of complaint? Are you specific about the problem and about the adjustment you seek? Have you motivated the recipient to grant your request?

Letters of Adjustment

- Is your letter of adjustment courteous? Is the you-attitude evident?
- If your letter of adjustment is a bad-news letter, have you explained your refusal in an honest, detailed way? Have you expressed the refusal clearly but courteously? Have you offered a substitute adjustment? Have you opened and closed the letter in a friendly way?

Letter and Memorandum Reports

- Is your letter or memorandum report well introduced and summarized? Do your headings reflect the content of the report? Have you used lists, tables, and graphs if they would help? Are your conclusions and recommendations clearly stated? Is any action needed or desired clearly stated?

Exercises

1. Write an unsolicited letter of inquiry to some company asking for sample materials or information. If you really need the information or material, mail the letter, but do not mail it as an exercise.
2. Imagine that you are working for a firm that provides a service or manufactures a product you know something about. Someone has written the firm a letter of inquiry asking about the service or product. Your task is to answer the letter.
3. Think about some service or product that has recently caused you dissatisfaction. Find out the appropriate person or organization to write to and write that person or organization a letter of complaint.
4. Swap your letter of complaint written for Exercise 3 for the letter of complaint written by another member of the class. Your assignment is to answer the other class member's letter. You may have to do a little research to get the data you need for your answer.
5. Write a memo to some college official or to an executive at your place of work. Many of the papers you have written earlier in your writing course are probably suitable for a memorandum format. Or you could choose some procedure, such as college registration, and suggest a new and better procedure. Perhaps your memo could be to an instructor suggesting course changes. A look at Figure 4-1 on page 60 could also suggest a wide range of topics and approaches that you could use in a memo.

CHAPTER 13 THE STRATEGIES AND COMMUNICATIONS OF THE JOB HUNT

If you are a college student or a recent college graduate, unfortunately, you can't take it for granted that you will get a good job right out of college. In some cases, you may be competing with experienced workers for the same job. To help you in this difficult environment, in this chapter we cover the major steps of the job hunt: preparation, letters of application, resumes, and interviews.[1]

Preparation

What you can take for granted is that job hunting is nearly a full-time occupation. If you are still in college, it has to be at least a part-time occupation. Most professional jobs require that you follow a regular schedule and work 40 to 50 hours a week. Job hunting also requires that you schedule your time around various activities and that you, if you are still in college, spend at least 15 to 20 hours a week in the hunt. Your first task is to prepare for the hunt. That involves **self-assessment, gathering information about possible jobs,** and **networking.**

Self-Assessment

The goal of **self-assessment** is twofold. First and most important, you want to avoid pounding a square peg (you) into a round hole (the wrong job). You want to determine what jobs among the many possible would suit you the best: What kind of work can you do well and what kind of work pleases you? Secondly, in the job hunt, you'll be creating resumes, completing applications, and answering interview questions. Self-assessment will help you to list necessary details, such as dates, names, and job responsibilities, about past work and educational experiences.

In your self-assessment, you may ask yourself the following questions: What are my strengths? What are my weaknesses? How well have I performed in past jobs? Have I shown initiative? Have I improved procedures? Have I accepted responsibility? Have I been promoted or given a merit raise? How can I present myself most attractively? What skills do I possess that relate directly to what the employer seems to need? How and where have I obtained those skills? To perform your self-assessment in a serious and systematic way, use the questionnaire provided in Figure 13–1.

When you have completed the questionnaire, you will have a good record of your work and educational experiences. You should have a good idea of what skills you have and what you like to do.

Using that information, match jobs and careers that you might have to your skills and interests. Three U.S. Department of Labor publications can help you make this match:

- *Guide for Occupational Exploration.* Lists more than 12,000 occupations and guides you in relating your skills and interests to possible careers.
- *Occupational Outlook Handbook.* Describes the educational requirements, duties, and job prospects for most occupations.
- *Dictionary of Job Titles.* Describes in detail more than 12,000 occupations.

College libraries and placement offices should have these books. You can find them in the state employment service offices listed in Figure 13–2. You can order them and any other U.S. government publication from

Work Experience (Use a sheet like this for each position you have held, including military service.)

Company: ______________________________
Address: ______________________________
Supervisor's Name and Title: ______________________________
Dates of Employment: ______________________________
Position(s)/Title(s)/Military Rank: ______________________________

Duties and responsibilities: ______________________________

Accomplishments (including awards or commendations): ______________________________

Skills, Knowledge, and Abilities Used: ______________________________

Duties Liked and Disliked: ______________________________

Reason for Leaving: ______________________________

Education and Training

School, College, University	Dates of Enrollment	Major	Degree or Certificate	Date	G.P.A.

Career-Related Courses: ______________________________

Scholastic Honors, Awards, and Scholarships: ______________________________

College Extracurricular Activities: ______________________________

Other Training (include courses sponsored by the military, employers, or professional associations, etc.): ______________________________

Courses, Activities Liked and Disliked: ______________________________

Skills, Knowledge, and Abilities Learned: ______________________________

Figure 13–1 Self-Assessment Questionnaire
Source: U.S. Department of Labor, *Job Search Guide: Strategies for Professionals* (Washington, DC: GPO, 1993).

Professional Licenses: ______________________

Personal Characteristics (e.g., organizational ability, study habits, social skills, like to work alone or on a team, like or dislike public speaking, detail work): ______________________

Personal Activities

Professional (association memberships, positions held, committees served on, activities, honors, publications, patents, etc.): ______________________

Community (civic, cultural, religious, political organization memberships, offices or positions held, activities, etc.): ______________________

Other (hobbies, recreational activities and other personal abilities and accomplishments): ______________________

Overall Assessment

Take a look at all the work sheets you have completed: Work Experience, Education and Personal Activities. Considering all you have done, list your strengths and positive attributes in each of the areas below.

Skills, Knowledge and Abilities: ______________________

Accomplishments: ______________________

Personal Characteristics: ______________________

Activities Performed Well: ______________________

Activities Liked: ______________________

Figure 13-1 (continued)..

Superintendent of Documents
P.O. Box 371954
Pittsburgh, PA 15250-7954
Phone: (202) 783-3238
Fax: (202) 512-2250

What if at the end of your self-assessment you either can't decide what you want to do or, worse, have decided that your college major is in a field that no longer interests you? In either case you might seek professional job counseling. Many college placement offices offer such help. You can also find counseling services at the state employment service offices listed in Figure 13–2.

If you are reasonably sure of your career direction, your next step is to find where the jobs are that will lead in that direction.

Information Gathering

Two good ways exist to discover and gather information about companies and organizations that offer the jobs you are seeking. One way, often overlooked, is networking. We discuss that method in the next section. The other way is researching the many publications that contain information about where the jobs are, how much they pay, and how big a future such jobs have. Figure 13–3 lists many such sources with brief descriptions of what you'll find in them. Most of these sources will be available in college libraries and placement offices.

In addition to the sources listed in Figure 13–3, consult magazines and newspapers that regularly carry business news such as *Forbes, The Wall Street Journal,* and *Business Week.* To see what has recently appeared in the business press about a company, see the *F & S Index of Corporations and Industries* and the *Business Periodicals Index.* For general coverage, see *The New York Times Index* and the *Reader's Guide to Periodical Literature.* Analyze the company to see what achievements it is most proud of and to determine its goals.

A major source of information is the *CPC Annual,* published each year by the College Placement Council and found in all college placement offices. The *CPC Annual* covers most major companies and many governmental organizations, providing information about their products, services, locations, job opportunities, and even their management philosophy. For example, a recent entry for the Jet Propulsion Laboratory in Pasadena, California, states the following:

> The emphasis here is on a creative engineering and science environment with a free exchange of innovative ideas and opinions. Management philosophy is to develop staff members to their highest level of maturity in an atmosphere of involvement. The work is exciting, and creative expression is encouraged.

State Employment Service Offices

ALABAMA
Employment Service,
Dept. of Ind. Rel.
649 Monroe Street
Montgomery, AL 36130
(205) 261-5364

ALASKA
Employment Service
Empl. Sec. Div.
P.O. Box 3-7000
Juneau, AK 99602
(907) 465-2712

ARIZONA
Department of Economic Security
P.O. Box 6123
Site Code 730A
Phoenix, AZ 85005
(602) 542-4016

ARKANSAS
Employment Security Division
P.O. Box 2961
Little Rock, AR 72203
(501) 371-1683

CALIFORNIA
Job Service Division
Empl. Dev. Dept.
800 Capitol Mall
Sacramento, CA 95814
(916) 322-7318

COLORADO
Employment Programs
Div. of Empl. & Trng.
251 East 12th Avenue
Denver, CO 80203
(303) 866-6180

CONNECTICUT
Job Service
CT Labor Department
200 Folly Brook Blvd.
Wethersfield, CT 06109
(203) 566-8818

DELAWARE
Employment & Trng. Div.
DE Department of Labor
P.O. Box 9029
Newark, DE 19711
(302) 368-6911

DISTRICT OF COLUMBIA
Office of Job Service
Dept. of Empl. Services
500 C Street, NW, Rm. 317
Washington, D.C. 20001
(202) 639-1115

FLORIDA
Dept. of Labor & Empl. Sec.
1320 Executive Center Cir.
300 Atkins Building
Tallahassee, FL 32101
(904) 488-7228

GEORGIA
Employment Service
148 International Blvd, N
Room 400
Atlanta, GA 30303
(404) 656-0380

HAWAII
Employment Service Division
Dept. of Labor & Ind. Rel.
1347 Kapiolani Blvd.
Honolulu, HI 96814
(808) 548-6468

IDAHO
Operations Div. Emp. Svc.
Dept. of Empl.
317 Main Street
Boise, ID 83735
(208) 334-3977

ILLINOIS
Employment Services
Employment Security Division
910 S. Michigan Avenue
Chicago, IL 60605
(312) 793-6074

INDIANA
E.S., Employment
Security Div.
10 North Senate Avenue
Indianapolis, IN 46204
(317) 232-7680

IOWA
Job Service Program Bureau
Department of Job Service
1000 East Grand Avenue
Des Moines, IA 50319
(515) 281-5134

KANSAS
Div. of Employment & Training
Dept. of Human Resources
401 Topeka Avenue
Topeka, KS 66603
(913) 296-5317

KENTUCKY
Dept. for Employment Services
275 E Main Street, 2nd Floor
Frankfort, KY 40621
(502) 564-5331

LOUISIANA
Employment Service
Office of Employment Security
P.O. Box 94094
Baton Rouge, LA 70804-9094
(504) 342-3016

MAINE
Job Service Division
Bureau of Employment Security
P.O. Box 309
Augusta, ME 04330
(207) 289-3431

MARYLAND
MD Department of Employment &
Economics Development
1100 North Eutaw St., Rm. 701
Baltimore, MD 21201
(301) 383-5353

MASSACHUSETTS
Div. of Employment Security
Charles F. Hurley Building
Government Center
Boston, MA 02114
(617) 727-6810

MICHIGAN
Bureau of Employment Service,
Employment Security Commission
7310 Woodward Avenue
Detroit, MI 48202
(313) 876-5309

MINNESOTA
Job Service & UI Operations
690 American Center Bldg.
150 East Kellogg
St. Paul, MN 55101
(612) 296-3627

MISSISSIPPI
Employment Service Division
Employment Service Commission
P.O. Box 1699
Jackson, MS 39215-1699
(601) 354-8711

MISSOURI
Employment Service
Division of Employment Security
P.O. Box 59
Jefferson City, MO 65104
(314) 751-3790

MONTANA
Job Service/Employment
and Training Division
P.O. Box 1728
Helena, MT 59624
(406) 444-4524

Figure 13–2 State Employment Service Offices
Source: U.S. Department of Labor, *Tips for Finding the Right Job* (Washington, DC: GPO, 1992) 26–27.

NEBRASKA
Job Service
NE Dept. of Labor
P.O. Box 94600
Lincoln, NE 68509
(402) 475-8451

NEVADA
Employment Service
Employment Security Department
500 East Third Street
Carson City, NV 89713
(702) 885-4510

NEW HAMPSHIRE
Employment Service Bureau,
Department of Employment
Security
32 South Main Street
Concord, NH 06391
(603) 224-3311

NEW JERSEY
NJ Department of Labor
Labor & Industry Bldg.,
CN 058
Trenton, NJ 08625
(609) 292-2400

NEW MEXICO
Employment Service Employment
Security Department
P.O. Box 1928
Albuquerque, NM 87103
(305) 841-8437

NEW YORK
Job Service Division
NY State Department of Labor
State Campus
Building 12 g.
Albany, NY 12240
(518) 457-2612

NORTH CAROLINA
Employment Security Commission
of North Carolina
P.O. Box 27625
Raleigh, NC 27611
(919) 733-7522

NORTH DAKOTA
Employment & Training Division
Job Service North Dakota
P.O. Box 1537
Bismarck, ND 58502
(701) 224-2842

OHIO
Employment Service Division
Bureau of Employment Services
145 S. Front Street, Rm. 640
Columbus, OH 43215
(614) 466-2421

OKLAHOMA
Employment Service
Employment Security
Commission
Will Rogers Memorial Ofc. Bldg.
Oklahoma City, OK 73105
(405) 521-3652

OREGON
Employment Service
OR Employment Division
875 Union Street, N.E.
Salem, OR 97311
(503) 378-3212

PENNSYLVANIA
Bureau of Job Service
Labor 7 Industry Building
Seventh & Forster Streets
Harrisburg, PA 17121
(717) 787-3354

PUERTO RICO
Employment Service Division
Bureau of Employment Security
505 Munoz Rivera Avenue
Hato Rey, PR 00918
(809) 754-5326

RHODE ISLAND
Job Service Division
Dept. of Employment Security
24 Mason Street
Providence, RI 02903
(401) 277-3722

SOUTH CAROLINA
Employment Service
P.O. Box 995
Columbia, SC 29202
(803) 737-2400

SOUTH DAKOTA
SD Department of Labor
700 Governors Drive
Pierre, SD 57501
(605) 773-3101

TENNESSEE
Employment Service
Dept. of Employment Security
503 Cordell Hull Building
Nashville, TN 37219
(615) 741-0922

TEXAS
Employment Service
Texas Employment Commission
12th & Trinity, 504BT
Austin, TX 78778
(512) 463-2820

UTAH
Employment Services/Field Oper.
Department Employment
Security
174 Social Hall Avenue
Salt Lake City, UT 84147
(801) 533-2201

VERMONT
Employment Service
Dept. of Employment and Training
P.O. Box 488
Montpelier, VT 05602
(802) 229-0311

VIRGINIA
Employment Service
VA Employment Commission
P.O. Box 1258
Richmond, VA 23211
(804) 786-7097

VIRGIN ISLANDS
Employment Service
Employment Security Agency
P.O. Box 1090
Charlotte Amalie, VI 00801
(809) 776-3700

WASHINGTON
Employment Security Department
212 Maple Park
Olympia, WA 98504
(206) 753-0747

WEST VIRGINIA
Employment Service Division
Dept. of Employment Security
112 California Avenue
Charleston, WV 25305
(304) 348-9180

WISCONSIN
Job Service
P.O. Box 7905
Madison, WI 53707
(608) 266-8561

WYOMING
Employment Service
Employment Security Commission
P.O. Box 2760
Casper, WY 82602
(307) 235-3611

National Office
United States Employment
Service
200 Constitution Ave. NW
Washington, DC 20210
(202) 539-0188

Figure 13-2 (continued).

- *The Job Hunter's Guide to 100 Great American Cities.* (Brattle Communications, Latham NY). Rather than concentrating on a particular locale, this guide gives the principal-area employers for 100 of America's largest cities.

- *Macrae's State Industrial Directories.* (New York, NY). Published for 15 Northeastern states. Similar volumes are produced for other parts of the country by other publishers. Each book lists thousands of companies, concentrating almost exclusively on those that produce products, rather than services. They include a large number of small firms, in addition to the larger ones listed in many other guides.

- *National Business Telephone Directory.* (Gale Research, Detroit, MI). An alphabetical listing of companies across the United States, with their addresses and phone numbers. It includes many smaller firms (20 employees minimum).

- *Thomas Register.* (New York, NY). Lists more than 100,000 companies across the country. Contains listings by company name, type of product made, and brand name of product produced. Catalogs provided by many of the companies also are included.

- *America's Fastest Growing Employers.* (Bob Adams Inc., Holbrook, MA). Lists more than 700 of the fastest-growing companies in the country. Also gives many tips on job hunting.

- *The Hidden Job Market: A Guide to America's 2000 Little-Known Fastest Growing High-Tech Companies.* (Peterson's Guides, Princeton, NJ). Concentrates on high-tech companies with good growth potential.

- *Dun & Bradstreet Million Dollar Directory.* (Parsippany, NJ). Provides information on 180,000 of the largest companies in the country. Gives the type of business, number of employees and sales volume for each. It also lists the company's top executives.

  An abbreviated version of this publication also exists, which gives this information for the top 50,000 companies.

- *Standard & Poor's Register of Corporations, Directors and Executives.* (New York, NY). Information similar to that in Dun and Bradstreet's directory. Also contains a listing of the parent companies of subsidiaries and the interlocking affiliations of directors.

- *The Career Guide-Dun's Employment Opportunities Directory.* (Parsippany, NJ). Aimed specifically at the professional job seeker. Lists more than 5,000 major U.S. companies that plan to recruit in the coming year. Unlike the other directories from Standard and Poor and Dun and Bradstreet, this guide lists personnel directors and gives information about firms' career opportunities and benefits packages. Also gives a state-by-state list of headhunters and tips on interviewing and résumé writing.

Figure 13–3 Job Information Sources

Source: U.S. Department of Labor, *Job Search Guide: Strategies for Professionals* (Washington, DC: GPO, 1993) 24–25.

The *CPC Annual* provides the addresses you'll need to contact the organizations listed. Also included in each *Annual* is advice on interviewing and the correspondence of the job hunt.

If you are interested in federal employment, seek out *Federal Career Opportunities* and the *Federal News Digest.* The state employment service offices listed in Figure 13-2 can conduct a computerized search of federal job opportunities for you. Also, the state employment offices offer help in locating state jobs.

To find employment and wage data around the country, call one of the eight regional offices of the Bureau of Labor Statistics listed in Figure 13-4. The New York office offers recorded information 24 hours a day. You can get similar information from the State Occupational Information Coordinating Committees located in all 50 states. You can find the addresses and phone numbers for each state committee in the *Occupational Outlook Handbook.*

We cannot emphasize enough how important this preparation is. Most job seekers do not even know such rich information sources exist. Many who do know do not take advantage of them. You now know what the sources are. The rest is up to you.

Networking

Networking is a way of finding both job information and job opportunities. Networking starts with broadcasting the news that you are looking for a job. Whom do you tell? Start with family and friends, including grandparents, aunts, uncles, cousins, and in-laws. Talk to professors, clergy, and favorite teachers. Don't overlook family doctors, dentists, lawyers, bankers, barbers, and hairdressers. If there are professional associations that cover fields you are interested in, join them. For

Boston	(617) 565-2327
New York	(212) 337-2400
Philadelphia	(215) 596-1154
Atlanta	(404) 347-4416
Chicago	(312) 353-1880
Dallas	(214) 767-6970
Kansas City	(816) 426-2481
San Francisco	(415) 744-6600

Figure 13-4 Regional Offices of the Bureau of Labor Statistics

Source: U.S. Department of Labor, *Job Search Guide: Strategies for Professionals* (Washington, DC: GPO, 1993) 24.

information about such organizations, see *Career Guide to Professional Organizations.* If there are local chapters of such organizations within a reasonable distance, attend their meetings.

Visit or phone local businesses or organizations, particularly those that have jobs you might be interested in, and ask for the names of people who have or manage jobs like the ones you want. Call such people and ask if you can come by for a talk. Make it clear that you are not looking for a job interview but rather an informational interview to seek advice about looking for work or how to prepare for a certain kind of job.

If you have a name you can use, such as a relative or teacher, it will help in making the initial contact. But don't let not having a name deter you. Try to make the contact anyway. Expect a lot of people to put you off, but you will likely be pleasantly surprised at how many people will talk to you. People like to give advice, and, in particular, they like to talk about their occupation. Furthermore, organizations are on the lookout for enterprising self-starters. In calling on the organization, you are showing yourself to be such a person. In these interviews, you are the interviewer. To help people remember you, have business cards made up that look like the one in Figure 13-5. If you have a profession, such as computer programmer, add it to the card beneath your name. To help yourself remember who you have seen and what you have learned, keep good records as you proceed. Write down names (accurately spelled), addresses, and phone numbers. Record what seems to be important to the people you talk to about work in general and the specific work that interests you.

All this may strike you as an informal way of looking for work, and it is. But research, as the chart in Figure 13-6 shows, says that most people find work in precisely this way. Partly, this is true because many jobs never reach a formal search. They are filled through contacts of the kind we are telling you to develop. Also, two-thirds of all jobs are in

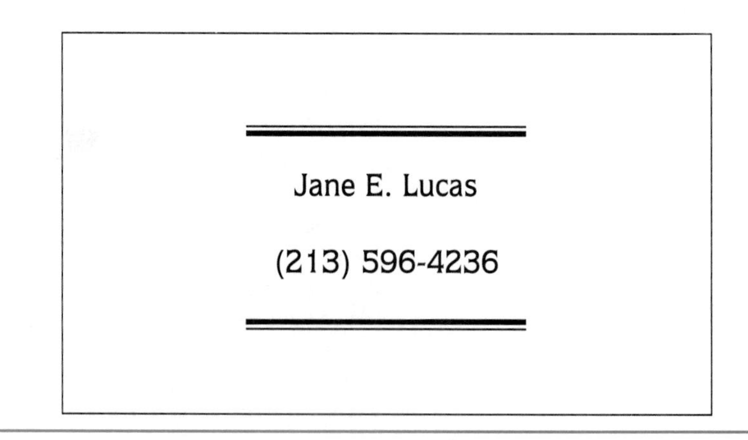

Figure 13-5 Business Card

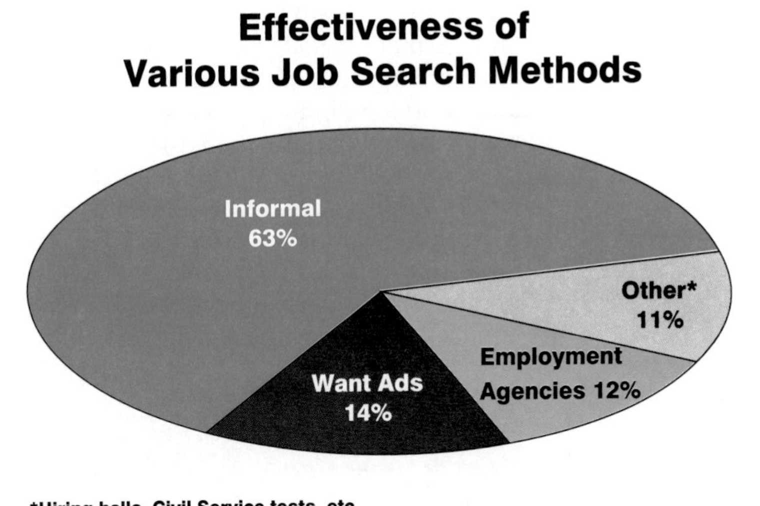

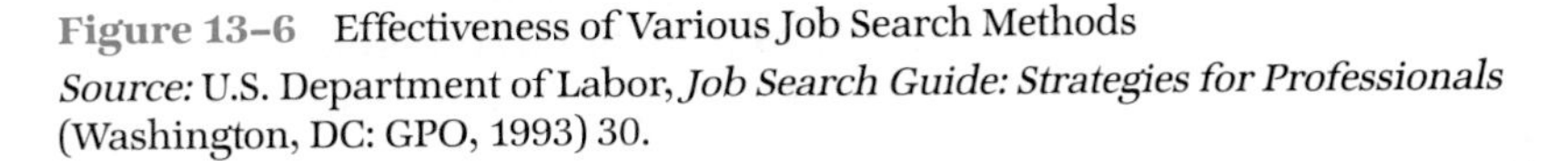
Figure 13–6 Effectiveness of Various Job Search Methods
Source: U.S. Department of Labor, *Job Search Guide: Strategies for Professionals* (Washington, DC: GPO, 1993) 30.

companies with fewer than 25 employees.[2] Small companies often don't participate in formal searches such as sending recruiters to college campuses. You have to search them out, and networking is probably the best way to do it.

More formally, you can network with organizations specifically set up to help people find work. Foremost among these, if you are a college student, is your placement office. Many large firms regularly call on college placement offices to seek new employees. The placement office schedules campus interviews for graduating students, and many offices have job fairs once or twice a year. Job fairs, to which many companies send representatives, are good places to gather information and to network. In addition, the placement office maintains a library of books about how to seek employment and usually has a file of brochures and articles about companies and organizations that might interest you.

If you have already graduated from college, ask whether your college alumni association offers help in seeking employment. Many do. Other formal possibilities are private employment agencies and public employment agencies. Every state has an employment office. We list the state offices with their phone numbers and addresses in Figure 13–2.

You can find published information about job openings in several sources. Your local and regional newspapers will carry want ads. You can widen the search by regularly reading *The National Business*

Employment Weekly and *National Ad Search.* As their titles indicate, both of these weekly papers offer nationwide coverage. Both may be available in your college library or placement office. If you have a computer and a modem you can use *Adnet Online,* which carries national want ads and updates them twice a week. You can access *Adnet Online* from the following information services: America Online, Bix, Compuserve, Genie, PC-Link, Promenade, and Prodigy.

Hunting a job is a hard job in itself. There is a great deal of help out there, but you can't be passive. You have to actively search out job information and opportunities. Another way to do that is with letters of application and resumes, which we discuss next.

The Correspondence of the Job Hunt

In some cases, the first knowledge prospective employers will have of you is the **letter of application** and **resume** that you send to them. A good letter of application, sometimes called a *cover letter,* and resume will not guarantee that you get a job, but bad ones will probably guarantee that you do not. In this section we'll describe how to prepare letters of application and resumes, and then discuss several follow-up letters needed during the job hunt.

Letter of Application

Plan the mechanics of your **letter of application** carefully. Buy the best quality white bond paper. This is not the time to skimp. Plan to type or word process your letter, of course, or have it done. Use a standard typeface; do not use italics. Make sure your letter is mechanically perfect, free from erasures and grammatical errors. Be brief but not telegraphic. Keep the letter to one page. Don't send a letter duplicated in any way. Accompany each letter with a resume. We have more to say about resumes later.

Pay attention to the style of the letter and the resume that accompanies it. The tone you want in your letter is one of self-confidence. You must avoid both arrogance and humility. You must sound interested and somewhat eager, but not fawning. Do not give the impression that you must have the job, but do not seem uncaring about whether you get the job.

When describing your accomplishments in the letter and resume, use action verbs. They help to give your writing brevity, specificity, and force. For example, don't just say that you worked as a sales clerk. Rather, tell how you maintained inventories, sold merchandise, prepared displays, implemented new procedures, and supervised and trained new clerks. Here's a sampling of such words:

administer	edit	oversee
analyze	evaluate	plan
conduct	exhibit	produce
create	expand	reduce costs
cut	improve	reorganize
design	manage	support
develop	operate	was promoted
direct	organize	write

You cannot avoid the use of *I* in a letter of application, but take the you-attitude as much as you can. Think about what you can do for the prospective employer. The letter of application is not the place to be worried about salary and pension plans. Above all, be mature and dignified. Forget about tricks and flashy approaches. Write a well-organized, informative letter that highlights the skills your analysis of the company shows it desires most. We will discuss the application letter in terms of a beginning, a body, and an ending.

The Beginning Beginnings are tough. Do not begin aggressively or cutely. Beginnings such as "WANTED: An alert, aggressive employer who will recognize an alert, aggressive young forester" will usually send your letter into the wastebasket. If it is available to you, a bit of legitimate name dropping is a good beginning. Use this beginning only if you have permission and if the name dropped will mean something to the prospective employer. If you qualify on both counts, begin with an opener like this:

> Dear Ms. Marchand:
>
> Professor John J. Jones of State University's Food Science faculty has suggested that I apply for the post of food supervisor. In June I will receive my Bachelor of Science degree in Food Science from State University. Also, I have spent the last two summers working in food preparation for Memorial Hospital in Melbourne.

Remember that you are trying to arouse immediate interest about yourself in the potential employer. Another way to do this is to refer to something about the company that interests you. Such a reference demonstrates that you have done your homework. Then try to show how some preparation on your part ties you into this special interest. See Figure 13–7 for an example of such an opener.

Sometimes the best approach is a simple statement about the job you seek accompanied by something in your experience that fits you for the job, as in this example:

> Your opening for a food supervisor has come to my attention. In June of this year, I will graduate from State University with a Bachelor of Science in Food Science. I

635 Shuflin Road
Watertown, CA 90233
March 23, 1994

Mr. Morrell R. Solem
Director of Research
Price Industries, Inc.
2163 Airport Drive
St. Louis, MO 63136

Dear Mr. Solem:

Opener showing knowledge of company

I read in the January issue of Metal Age that Dr. Charles E. Gore of your company is conducting extensive research into the application of X-ray diffraction to problems in physical metallurgy. I have conducted experiments at Watertown Polytechnic Institute in the same area under the guidance of Professor John J. O'Brien. I would like to become a research assistant with your firm and, if possible, work for Dr. Gore.

Specific job mentioned

Highlights of education

In June, I will graduate from WPI with a Bachelor of Science degree in Metallurgical Engineering. At present, I am in the upper 25 percent of my class. In addition to my work with Professor O'Brien, I have taken as many courses relating to metal inspection problems as I could.

Highlights of work experience

For the past two summers, I have worked for Watertown Concrete Test Services where I have qualified as a laboratory technician for hardened concrete testing. I know how to find and apply the specifications of the American Society for Testing and Materials. This experience has taught me a good deal about modern inspection techniques. Because this practical experience supplements the theory learned at school, I could fit into a research laboratory with a minimum of training.

Reference to resume

You will find more detailed information about my education and work experience in the résumé enclosed with this letter. I can supply job descriptions concerning past employment and the report of my X-ray diffraction research.

Request for interview

In April, I will attend the annual meeting of the American Institute of Metallurgical Engineers in Detroit. Would it be possible for me to talk with some member of Price Industries at that time?

Sincerely yours,

Jane E. Lucas

Jane E. Lucas

Enclosure

Figure 13–7 Letter of Application

> have spent the last two summers working in food preparation for Memorial Hospital in Melbourne. I believe that my education and work experience qualify me to be a food supervisor on your staff.

Be specific about the job you want. As the vice-president of one firm told us, "We have all the people we need who can do anything. We want people who can do something." Quite often, if the job you want is not open, the employer may offer you an alternative one. However, employers are not impressed with vague statements such as, "I'm willing and able to work at any job you may have open in research, production, or sales."

The Body In the body of your letter, select items from your education and experience that show your qualifications for the job you seek. Remember always, you are trying to show the employer how well you will fit into the job and the organization.

In selecting your items, it pays to know what things employers value the most. One thorough piece of research[3] shows that for recent college graduates, employers give priority as follows:

First priority

Major field of study
Academic performance
Work experience
Plant or home-office interview
Campus interview

Second priority

Extracurricular activities
Recommendations of former employer
Academic activities and awards

Third priority

Type of college or university attended
Recommendations from faculty or school official

Fourth priority

Standardized test scores
In-house test scores
Military rank

Try to include information from the areas that employers seem to value the most, but emphasize the areas in which you come off best. If your grades are good, mention them prominently. If you stand low in your class—in the lowest quarter, perhaps—maintain a discreet silence. Speak to the employer's interests, and at the same time highlight your own accomplishments. Show how it would be to the employer's advantage to hire you. The following paragraph, an excellent example of the you-attitude in action, does all these things:

> I understand that the research team of which I might be a part works as a single unit in the measurement and collection of data. Therefore, team members need a general knowledge and ability in fishery techniques as well as a specific job skill. I would like to point out that last summer I worked for the Department of Natural Resources on a fish population study. On that job, I gained electro-fishing and seining experience and learned how to collect and identify aquatic invertebrates.

By being specific about your accomplishments, you avoid the appearance of bragging. It is much better to say, "I was president of my senior class," than to say, "I am a natural leader."

One tip about job experience: The best experience is that which relates to the job you seek, but mention job experience even if it does not relate to the job you seek. Employers feel that a student who has worked is more apt to be mature than one who has not.

Do not forget hobbies that relate to the job. You are trying to establish that you are interested in, as well as qualified for, the job.

Do not mention salary unless you are answering an advertisement that specifically requests you to. Keep the you-attitude. Do not worry about pension plans, vacations, and coffee breaks at this stage of the game. Keep the prospective employer's interests in the foreground. Your self-interest is taken for granted.

If you are already working and not a student, construct the body of your letter in much the same fashion. The significant difference is that you will emphasize work experience more than college experience. Do not complain about your present employer. Such complaints will lead the prospective employer to mistrust you.

In the last paragraph of the body, refer the employer to your enclosed resume. Mention your willingness to supply additional information such as references, letters concerning your work, research reports, and college transcripts.

The Ending The goal of the letter of application is an interview with the prospective employer. In your ending, request this interview. Your request should be neither humble nor overaggressive. Simply indicate that you are available for an interview at the employer's convenience and give any special instructions needed for reaching you. If the prospective employer is in a distant city, indicate (if you can) some convenient time and place where you might meet a representative of the company, such as the convention of a professional society. If the employer is really interested, you may be invited to visit the company at its expense.

The Complete Letter Figure 13-7 shows the complete letter of application. Take a minute to read it now. The beginning of the letter shows that the writer has been interested enough in the company to investigate it. The desired job is specifically mentioned. The middle portion highlights the course work and work experience that relate directly to the job sought. The close makes an interview convenient for the employer to arrange.

The word processor is a great convenience when you are doing application letters. It allows you to store basic paragraphs of your letter that you can easily modify to meet the needs and interests of the organization you are writing to. Such modification for each organization is truly necessary. Personnel officers read your letter and its accompanying resume in about 30 seconds or so. If you have not grabbed their interest in that time, you are probably finished with that organization.

The Resume

With your letter of application, enclose a **resume** that provides your prospective employer with a convenient summary of your education and experience. To whom should you send letters and resumes? When answering an advertisement, follow whatever instructions are given there. When operating on your own, send them, if at all possible, to the person within the organization for whom you would be working; that is, the person who directly supervises the position. This person normally has the power to hire for the position. Your research into the company may turn up the name you need. If not, don't hesitate to call the company switchboard and ask directly for the name and title you need. Write to human resource directors only as a last resort. Whatever you do, write to *someone* by name. Don't send "To Whom It May Concern" letters on your job hunt. It's wasted effort. Sometimes, of course, you may gain an interview without having sent a letter of application—for example, when recruiters come to your campus. When you do, bring a resume with you and give it to the interviewer at the start of the interview. He or she will appreciate this help tremendously. Furthermore, the resume gives the interviewer a point of departure for questions, which often helps to structure the interview to your best advantage.

Research indicates that, as in the letter of application, good grammar, correct spelling, neatness and brevity—ideally only one page—are of major importance in your resume.[4] If you have wide experience to report, you may need to use smaller type to fit it on one page. But don't go any smaller than 10-point type or your resume will be too hard to read and too cramped looking.

You have several alternatives for producing your resume. You can type it and have it photoreproduced. You can have it printed by a commercial printer. You can word process it if you have access to a letter-quality printer. Word processing is probably the best alternative because it allows you to modify your resume for different employers.

Use good paper in a standard color such as white or off-white. It's best if your letters and resumes are on matching paper. Make the resume good-looking. Leave generous margins and white space. Use distinctive headings and subheadings. The samples in Figures 13-8, 13-9, and 13-10 provide you with good models of what a resume should look like. Take a look at them now before you read the following comments.

The resumes shown are in the three most used formats: **chronological, functional,** and **targeted.** All have advantages and disadvantages.

Chronological Format The advantages of a chronological resume, Figure 13-8, are that it's traditional and acceptable. If your education and experience show a steady progression toward the career you seek, the chronological resume portrays that progression well. Its major disadvantage is that your special capabilities or accomplishments may sometimes get lost in the chronological detail. Also, if you have holes in your employment or educational history, they show up clearly.

Address In a chronological resume, put your address at the top. Give your phone number and don't forget the area code. If you have a fax number or E-mail address, include that as well.

Career Objective A career objective entry specifies the kind of work you want to do and sometimes the industry or service area where you want to do it, something like this:

> Work in food service management.

or like this:

> Work in food service management in a metropolitan hospital.

If you are going to have only one resume printed or photocopied, you would be wise to omit the career objective. It may not be appropriate for every potential employer you send your resume to. You can always state your career objective in your letter of application. On the other hand, if you are using word processing for your resume and can modify it for different employers, including a career objective is a good idea. Place the career objective entry immediately after the address and align it with the rest of the entries, as shown in Figure 13-8.

Education For most students, educational information should be placed before business experience. People with extensive work experience, however, may choose to put that first. List the colleges or universities you have attended in reverse chronological order; in other words,

Framentary sentences used

Highlights of education

Special activities

Honors and activities

Special work skills

Summary of early employment

RÉSUMÉ OF JANE E. LUCAS

635 Shuflin Road
Watertown, California 90233
Phone: (213) 596-4236

Education 1993–1995	WATERTOWN POLYTECHNIC INSTITUTE WATERTOWN, CALIFORNIA
	Candidate for Bachelor of Science degree in Metallurgical Engineering in June 1995. In upper 25% of class with GPA of 3.2 on 4.0 scale. Have been yearbook photographer for two years. Member of Outing Club, elected president in senior year. Elected to Student Intermediary Board in senior year. Oversaw promoting and allocating funds for student activities. Wrote a report on peer advising that resulted in a change in college policy. Earned 75% of college expenses.
1991–1993	SAN DIEGO COMMUNITY COLLEGE SAN DIEGO, CALIFORNIA
	Received Associate of Arts degree in General Studies in June 1993. Made Dean's List three of four semesters. Member of debate team. Participated in dramatics and intramural athletics.
Business Experience 1993–1994 Summers	WATERTOWN CONCRETE TEST SERVICES WATERTOWN, CALIFORNIA
	Qualified as laboratory technician for hardened concrete testing under specification E329 of American Society for Testing and Materials (ASTM). Conducted following ASTM tests: Load Test in Core Samples (ASTM C39), Penetration Probe (ASTM C803-75T), and the Transverse Resonant Frequency Determination (ASTM C666-73). Implemented new reporting system for laboratory results.
1993–1995	WATERTOWN ICE SKATING ARENA WATERTOWN, CALIFORNIA
	During academic year, work 15 hours a week as ice monitor. Supervise skating and administer first aid.
1989–1993	Summer and part-time jobs included newspaper carrier, supermarket stock clerk, and salesperson for large department store.
Personal Background	Grew up in San Diego, California. Travels include Mexico and the eastern United States. Can converse in Spanish. Interests include reading, backpacking, photography, and sports (tennis, skiing, and running). Willing to relocate.
References	Personal references available upon request

February 1995

Figure 13–8 Chronological Resume

RÉSUMÉ OF JANE E. LUCAS

635 Shuflin Road
Watertown, California 90233
Phone: (213) 596-4236

Education
1995

Candidate for degree in Metallurgical Engineering from Watertown Polytechnic Institute in June 1995.

Academic, work, and extracurricular activities categorized by capabilities

Technical

- Qualified as laboratory technician for hardened concrete testing under specification E329 of American Society for Testing and Materials (ASTM).
- Conducted following ASTM tests: Load Test in Core Samples (ASTM C39), Penetration Probe (ASTM C803-75T), and the Transverse Resonant Frequency Determination (ASTM C666-73).
- Will graduate in upper 25% of class with a GPA of 3.2 on a 4.0 scale.

People

- Elected President of Outing Club.
- Elected to Student Intermediary Board.
- Oversaw promoting and allocating funds for student activities.
- Participated in dramatics and intramural athletics.

Communication

- Wrote a report on peer advising that resulted in a change in Institute policy.
- Participated in intercollegiate debate.
- Worked two years as yearbook photographer.
- Completed courses in advanced speaking, small group discussion, and technical writing.

Summary of work experience in reverse chronological order

Work Experience

1993, 1994
Summers

- Watertown Concrete Test Services, Watertown, California: laboratory technician.

1993–1995

- Watertown Ice Skating Arena, Watertown, California: ice monitor.

1989–1993

- Summer and part-time jobs included newspaper carrier, supermarket stock clerk, and salesperson for large department store.
- Earned 75% of college expenses.

References

References available upon request.

February 1995

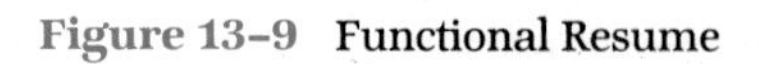

Figure 13–9 Functional Resume

RÉSUMÉ OF JANE E. LUCAS

635 Shuflin Road
Watertown, California 90233
Phone: (213) 596-4236

Job Objective Research assistant in a testing or research laboratory.

Education
1995 Candidate for degree in Metallurgical Engineering from Watertown Polytechnic Institute in June 1995.

Capabilities listed separately

Capabilities
- Find and apply specifications of the American Society for Testing and Materials (ASTM).
- Conduct X-ray diffraction tests.
- Work individually or as a team member in a laboratory setting.
- Take responsibility and think about a task in terms of objectives and time to complete.
- Report research results in both written and oral form.
- Communicate persuasively with nontechnical audiences.

Achievements that support capabilities listed

Achievements
- Will graduate in upper 25% of class with a GPA of 3.2 on a 4.0 scale.
- Earned 75% of college expenses.
- Qualified as laboratory technician for hardened concrete testing under specification E329 of ASTM.
- Conducted major ASTM tests and implemented new reporting system for laboratory results.
- Completed courses in advanced speaking, small group discussion, and technical writing.
- Wrote a report on peer advising that resulted in a change in Institute policy.

Summary of work experience in reverse chronological order

Work Experience

1993, 1994 Summers
- Watertown Concrete Test Services, Watertown, California: laboratory technician.

1993–1995
- Watertown Ice Skating Arena, Watertown, California: ice monitor.

1989–1993
- Summer and part-time jobs included newspaper carrier, supermarket stock clerk, and salesperson for large department store.

References References available upon request.

February 1995

Figure 13–10 Targeted Resume

list the school you attended most recently first, the one before second, and so on. Do not list your high school.

Give your major and date or expected date of graduation. Do not list courses, but list anything that is out of the ordinary, such as honors, special projects, and emphases in addition to the major. Extracurricular activities also go here.

Business Experience As you did with your educational experience, put your business experience in reverse chronological order. To save space and to avoid the repetition of *I* throughout the resume, use phrases rather than complete sentences. The style of the sample resumes makes this technique clear. As we advise you to do in the letter of application, emphasize the experiences that show you in the best light for the kinds of jobs you seek. Use active verbs in your descriptions. Do not neglect less important jobs of the sort you may have had in high school, but use a summary approach for them. You would probably put college internships and work-study programs here, although you might choose to put them under education. If you have military experience, put it here. Give highest rank held, list service schools attended, and describe duties. Make a special effort to show how military experience relates to the civilian work you seek.

Personal Background You may wish to provide personal information about yourself. Personal information can in a subtle way point out desirable qualities you possess. Recent travels indicate a broadening of knowledge and probably a willingness to travel. Hobbies listed may relate to the work sought. Participation in sports, drama, or community activities indicates a liking for working with people. Cultural activities indicate you are not a person of narrow interests.

If you indicate you are married, you might want to say that you are willing to relocate. Don't say anything about health unless you can describe it as excellent. Because of state and federal laws concerning fair employment practices, certain information should not appear in your resume. Do not mention handicaps and do not include a photograph.

References You have a choice with references. You can list several references with addresses and phone numbers or simply put in a line that says "References available on request." Both methods have an advantage and a disadvantage. If you provide information on references, a potential employer can contact them immediately, but you use up precious space that might be better used for more information about yourself. Conversely, if you don't provide the reference information, you save the space but put an additional step between potential employers and information they may want. It's a judgment call, but, on balance, we favor saving space by omitting the reference information. Your first goal is to interest the potential employer about yourself. If that happens then it will not be difficult to provide the reference information at a later time.

In any case, have at least three references available. Choose from among your college teachers and past employers people who know you well and are likely to say positive things about you. Get their permission, of course. Also, it's a smart idea to send them a copy of your resume. If you can't call on them personally, send them a letter that requests permission to use them as a reference, reminds them of the association with you, and sets a time for their reply, like this:

> Dear Ms. Pickford:
>
> In June of this year, I'll graduate from Watertown Polytechnic Institute with a B.S. in metallurgical engineering. I'm getting ready to look for work. May I have permission to use you as a reference?
>
> During the summers of 1993 and 1994, I worked as a laboratory technician in your testing lab at Watertown. They were good summers for me, and I qualified, with your help, to carry out several ASTM tests.
>
> I will send my resume out to some companies by 15 March and will need your reply by that time. I enclose a copy of my resume for you, so that you can see what I've been doing.
>
> Thanks for all your help in the past.

Date Line At the bottom of the resume, place the date—month and year—in which you completed the resume.

Functional Format A main advantage of the functional resume (Figure 13-9) is that it allows you to highlight the experiences that show you to your best advantage. Extracurricular experiences show up particularly well in a functional resume. Its major disadvantages are that you don't show as clearly a steady progression of work and education. Also, the functional resume is a newer format than the chronological and some employers may not find it as acceptable as the older format.

The address portion of the functional resume is the same as in the chronological. After the address, you may provide a job objective line if you like.

For education, simply give the school from which you received your degree, your major, and your date of graduation.

The body of the resume is essentially a classification. You sort your experiences—educational, business, extracurricular—into categories that reveal capabilities related to the jobs you seek. Remember that in addition to professional skills, employers want good communication skills and good people skills. Possible categories are *technical, professional, team building, communication, research, sales, production, administration,* and *consulting.*

The best way to prepare a functional resume is to brainstorm. Begin by listing some categories that you think might display your experiences well. Brainstorm further by listing your experiences in those categories. When you have good listings, select the categories and experiences that show you in the best light. Remember, you don't have to display everything you've ever done, just the things that might strike a potential employer as valuable.

Finish off the functional resume with a brief reverse chronological work history and a date line, as in the chronological resume.

Targeted Format The main advantage of the targeted resume (Figure 13-10) is also its main disadvantage. You zero in on one goal. If you can achieve that goal, fine, but the narrowness of the approach may block you out from other possibilities. The targeted resume displays your capabilities and achievements well, but like the functional resume, it's a newer format that may not have complete acceptance with all employers.

The address and education portions of the targeted resume are the same as in the functional resume. The whole point of a targeted resume is to aim at a specific job objective. Therefore, you express your job objective as precisely as you can.

Next, list your capabilities that match the job objective. Obviously, you have to understand the job sought to make the proper match. Capabilities are things you could do if called upon to do so. To be credible, they must be supported by the achievements or accomplishments that are listed next in your resume. Finish off the targeted resume with a reverse chronological work history and a date line, as in the functional resume.

Brainstorming is a good way to discover the material you need for your targeted resume. Under the headings *capabilities* and *achievements*, make as many statements about yourself as you can. When done, select the statements that best relate to the job sought.

General Advice Look again at the three resumes in Figures 13-8, 13-9, and 13-10. Notice that all three use informative headings and allow enough white space to be inviting. Figure 13-8 is done with standard typewriter print. Figures 13-9 and 13-10 show the possibilities if you have a word processor with different typefaces or if you have your resume set by a printer. The addition of the boldface type in the headings provides a more professional polish. We strongly recommend such printing if you can manage it.

Follow-Up Letters

Write follow-up letters (1) when you have had no answer to your letter of application in two weeks; (2) after an interview; (3) when a company refuses you a job; and (4) to accept or refuse a job.

No Answer When a company has not answered your original letter of application, write again. Be gracious, not complaining. Say something like this:

> Dear Mr. Souther:
>
> On 12 April I applied for a position with your company. I have not heard from you, so perhaps my original letter and resume have been misplaced. I enclose copies of them.
>
> If you have already reached some decision concerning my application, I would appreciate your letting me know.
>
> I look forward to hearing from you.
>
> Sincerely yours,

After an Interview Within a day, follow up your interview with a letter. Such a letter draws favorable attention to yourself as someone who understands business courtesy and good communication. Express appreciation for the interview. Draw attention to any of your qualifications that seemed to be important to the interviewer. Express your willingness to live with any special conditions of employment such as relocation. Make clear you want the job and feel qualified to do it. If you include a specific question in your letter, it may hasten a reply. Your letter might look like this one:

> Dear Ms. Marchand:
>
> Thank you for speaking with me last Tuesday about the food supervisor position you have open.
>
> Working in a hospital food service relates well to my experience and interests. The job you have available is one I am qualified to do. A feasibility study I am currently writing as a senior project deals with a hospital food service's ability to provide more varied diets to people with restricted dietary requirements. May I send you a copy next week when it is completed?
>
> I understand that my work with you would include alternating weekly night shifts with weekly day shifts. This requirement presents no difficulty for me.
>
> Tuesdays and Thursdays are best for me for any future interviews you may wish, but I can arrange a time at your convenience.
>
> Sincerely yours,

After a Job Refusal When a company refuses you a job, good tactics dictate that you acknowledge the refusal. Express thanks for the time spent with you, and state your regret that no opening exists at the present time. If you like, express the hope that they may consider you in the future. You never know; they might.

Accepting or Refusing a Job Writing an acceptance letter presents few problems. Be brief. Thank the employer for the job offer and accept the job. Settle when you will report for work and express pleasure at the prospect of working for the organization. A good letter of acceptance might read as follows:

> Dear Mr. Solem:
>
> Thank you for offering me a job as research assistant with your firm. I happily accept. I can begin work by 1 July as you have requested.
>
> I look forward to working with Price Industries and particularly to the opportunity of doing research with Dr. Gore.
>
> Sincerely yours,

Writing a letter of refusal can be difficult. Be as gracious as possible. Be brief but not so brief as to suggest rudeness or indifference. Make it clear that you appreciate the offer. If you can, give a reason for your refusal. The employer who has spent time and money in interviewing you and corresponding with you deserves these courtesies. And, of course, your own self-interest is involved. Some day you may wish to reapply to an organization that for the moment you must turn down. A good letter of refusal might look like this one:

> Dear Ms. White:
>
> I enjoyed my visit to the research department of your company. I would very much have liked to work with the people I met there. I thank you for offering me the opportunity to do so.
>
> However, after much serious thought, I have decided that the research opportunities offered me in another job are closer to the interests I developed at the University. Therefore, I have accepted the other job and regret that I cannot accept yours.
>
> I appreciate the courtesy and thoughtfulness that you and your associates have extended me.
>
> Sincerely yours,

Interviewing

The immediate goal of all your preparation and letter and resume writing is an interview with a potential employer. Interviews come about in various ways. If you network successfully, you will obtain interviews that allow you to ask questions of people already on the job. Although these are information-seeking interviews, if you impress the person you are talking to, they may turn into interviews that assess your potential as an employee. Your letters and resumes may get you interviews. Recruiters may come to your campus and schedule screening interviews with graduating students. As their name suggests, screening interviews are preliminary interviews from which the recruiters choose people to continue the process.

Continuing the interview process often means multiple interviews at the organization's headquarters. In one day you might interview with human resources people, the person who would be your boss if you were employed, and perhaps his or her boss. If you make it to the point where you are offered a job, you will probably have an interview in which you negotiate the details of your job, salary, and benefits. All this can be quite stressful. The better prepared you are, the easier it will go. The screening and follow-up interviews follow a somewhat similar pattern. We discuss them first and then give you some advice about negotiation.

The Interview

If you have prepared properly, you should show up at the interview knowing a good deal about the organization. A comment by one interviewer emphasizes the importance of this:

> It's really impressive to a recruiter when a job candidate knows about the company. If you're a national recruiter, and you've been on the road for days and days, you have no idea how pleasant it is to have a student say, "I know your company is doing such and such, and has plants here and here, and I'd like to work on this particular project." Otherwise I have to go into my standard spiel, and God knows I've certainly heard myself give that often enough.[5]

For interviews, you should be well groomed and dressed in a conservative suit or dress. Arrive at the place of the interview early enough to be relaxed. Shake hands firmly but not aggressively, and make eye contact. Give the interviewer a copy of your resume, and, when the interviewer sits down, sit down comfortably. Body language is important. Be neither rigid nor slouching.

Most interviews follow a three-part pattern. To begin with, the interviewer may generate small talk designed to set you at ease. Particularly at a screening interview, the interviewer may give some information about the company. This is a good chance for you to ask some questions that demonstrate that you have done your homework about the company. But don't force your questions on the interviewer.

Most of the interview will be taken up with questions aimed at assessing your skills and interests and how you might be of value to the organization. A well-done self-assessment obviously is a necessity in answering such questions. Figure 13–11 lists some of the commonly asked questions. If you prepare answers for them, you should be able to handle most of the questions you are likely to receive. In your answers, relate always to the organization. To the question, "What do you want to be doing five years from now?" the answer "Running my own consulting firm" might bring the interview to an early close.

- What can you tell me about yourself?
- What are your strengths and weaknesses?
- What do you want to be doing five years from now?
- Do you know much about us? Why do you want to work for us?
- We're interviewing ten people for this job. Why should you be the one we hire?
- What in your life are you most proud of?
- Here is a problem we had (interviewer describes problem). How would you have solved it?
- If you won ten million dollars in a lottery, how would you spend the rest of your life?
- Why do you want a career in your chosen field?
- Which school subjects interested you the most (least)? Why?

Figure 13–11 Frequently Asked Interview Questions

To the question, "What can you tell me about yourself?" the interviewer really doesn't expect an extended life history. This question provides you the opportunity to talk about your work and educational experiences and your skills. Try to relate your skills and experience to the needs of the organization. Don't overlook those people and communication skills essential to nearly every professional job. In your answer to this and other questions, be specific in your examples. If you say something like "I have good managerial skills," immediately back it up with an occasion or experience that supports your statement.

The question "Why do you want to work for us?" allows you to display what you have learned about the organization. In answering this question, you should again show that what you have to offer meshes with what the company needs.

The question about how you would spend your life if you won ten million dollars is an interesting one. "Lolling around the beach" is obviously the wrong answer, but so might be "I'd continue to work for your corporation." What the question is intended to get at are those worthwhile things in your life that you really enjoy doing and would do even without pay. Building houses for "Habitat for Humanity" or setting up sports programs for inner-city youths might qualify as good answers.

In answering questions about your strengths and weaknesses, be honest, but don't betray weaknesses that could eliminate you from consideration. "I can't stand criticism," would likely finish you off. "Sometimes, I don't know when to quit when I'm trying to solve a problem," given as a weakness could be perceived as a strength.

In the last part of the interview you will likely be given a chance to ask some questions of your own. It's a good time to get more details about the job or jobs that may be open. Ask about the organization's goals. "What is the company most proud of?" is a good question. Don't ask these questions just to ask questions. The interview is a good time for you to find out if you really want to work for an organization. Not every organization is going to be a good fit for what you have to offer and what you want to do. Unless the interviewer has raised the question of salary and benefits, don't ask questions about these matters.

If you really want to go to work for the organization, make that clear before the interview ends. But don't allow your willingness to appear as desperation. At some point in the interview, be sure to get the interviewer's name (spelled correctly!), title, address, phone number, and fax number (usually, you can simply ask the interviewer for his or her card). You'll need them for later correspondence. When the interviewer thanks you for coming, thank him or her for seeing you and leave. Don't drag the interview out when it's clearly over.

Negotiation

Interviewers seldom bring up salary and benefits until either they see you as a good prospect, or until they are sure they want to hire you. If they offer you the job, the negotiation is sometimes done in a separate interview. For example, your future boss may offer you the job and then send you to negotiate with the human resources staff.

Sometimes, the negotiator may offer you a salary. At other times, you may be asked to name a salary. How do you know what to accept or what to ask for? You may have received useful salary information through your networking activities. Also, you can find such information through research. Figure 13-12 lists some books that will provide the information you need. Your research in these books will give you not a specific salary but a salary range. If asked to name a salary, do not ask for the bottom of the range. Ask for as near the top as you reasonably can. The negotiator will respect you for your knowing what you are worth. However, balance the compensation package—vacations, pension plans, health care, educational opportunities, and so forth—against the salary. Some compensation packages are worth a good deal of money and may allow you to take a lower salary.

Before and After the Interview

If you have not participated in job interviews before, you should practice them. Get together with several friends. Using the information you have gathered, roleplay several interviews. As two of you play interviewer and interviewee, the others act as observers. They should look for strengths and weaknesses in your answers, diction, grammar, and body language. The members of the group need to appraise each other honestly. Practice until you feel comfortable with the process.

When interviews are over, write down your impressions as soon as you can. How did your clothes compare to the interviewer's? Were there unexpected questions? How good were your answers? What did you

- *State and Metropolitan Area Data Book.* Published by the U.S. Department of Commerce. Compiles statistical data from many public and private agencies. Includes unemployment rates, rate of employment growth, and population growth for every state. Also presents a vast amount of data on employment and income for metropolitan areas across the country.
- *White Collar Pay: Private Goods-Producing Industries.* Produced by the U.S. Department of Labor's Bureau of Labor Statistics. Good source of salary information for white collar jobs.
- *1991 AMS Office, Professional and Data Processing Salaries Report.* (Administrative Management Society, Wash. DC). Salary distributions for 40 different occupations, many of which are professional. Subdivided by company size, type of business, region of the country, and by 41 different metropolitan areas.
- *American Salaries and Wages Survey.* (Gale Research, Detroit, MI). Detailed information on salaries and wages for thousands of jobs. Data is subdivided geographically. Also gives cost-of-living data for selected areas, which is very helpful in determining what the salary differences really mean. Provides information on numbers employed in each occupation, along with projected changes.
- *American Almanac of Jobs and Salaries.* (Avon Books, NY). Information on wages for specific occupations and job groups, many of which are professional and white collar. Also presents trends in employment and wages.

Figure 13–12 Sources for Salary Ranges

Source: U.S. Department of Labor, *Job Search Guide: Strategies for Professionals* (Washington, DC: GPO, 1993) 62–63.

learn about the organization? What did you learn about a specific job or jobs? Did anything make you uncomfortable about the organization? Do you think you would fit in there? By the next day, get a thank-you note (letter or fax) off to the interviewer. (See page 395.)

Planning and Revision Checklists

You will find the planning and revision checklists that follow Chapter 2, "Composing" (pages 38-39), and Chapter 4, "Writing for Your Readers" (pages 81-82), and inside the front cover valuable in planning and revising any presentation of technical information. The following questions specifically apply to the job hunt. They summarize the key points in this chapter and provide a checklist for planning and revising.

PLANNING AND REVISION: PREPARATION

- Have you completed the self-assessment questionnaire in Figure 13-1?
- Do you have a complete record of your past job and educational experiences?
- Do you know your strengths and weaknesses, your skills, and your qualifications for the jobs you seek? Do you have clear career objectives?
- Do you need professional career counseling?
- Have you researched the sources in Figures 13-2 and 13-3 to find job information?
- Have you found organizations that fit your needs?
- Have you started your networking?
- Have you had business cards made?
- Have you called on businesses and set up informational interviews?
- Have you joined a professional organization relevant to your career field?
- Have you located agencies and individuals who can help you in your job hunt?
- Have you kept good records of your networking?

PLANNING: CORRESPONDENCE OF THE JOB HUNT

- For your letter of application:
  - Do you have the needed names and addresses?
  - What position do you seek?
  - How did you learn of this position?
  - Why are you qualified for this position?
  - What interests you about the company?
  - What can you do for the organization?
  - Can you do anything to make an interview more convenient for the employer?
  - How can the employer reach you?
- For your resume:
  - Do you have all the details needed of your past educational and work experiences? Dates, job descriptions, schools, majors, degrees, extracurricular activities, and so forth?
  - How will you prepare your resume? Type and photocopy? Commercial printing? Word processing?
  - Which resume form will suit your experience and capabilities best? Chronological? Functional? Targeted? Why?
  - In a functional resume, which categories would best suit your experience and capabilities?
  - Do you know enough about the job sought to use a targeted resume?
  - Do you have three references and permission to use their names?

- What follow-up letters do you need? No answer to your application? Follow up an interview? Respond to a job refusal? Accept or refuse a job?

REVISION: CORRESPONDENCE OF THE JOB HUNT

- Do your letter of application and resume reflect adequate preparation and self-assessment?
- Are your letter of application and resume completely free of grammatical and spelling errors? Are they well designed and good-looking?
- For the letter of application:
  - Have you the right tone, self-confident without arrogance?
  - Does your letter show how you could be valuable to the employer? Will it raise the employer's interest?
  - Does your letter reflect a sure purpose about the job you seek?
  - Have you highlighted the courses and work experience that best suit you for the job you seek?
  - Have you made it clear you are seek ing an interview and made it convenient for the employer to arrange one?
- For the resume:
  - Have you chosen the resume type that best suits your experiences and qualifications?
  - Have you limited your resume to one page?
  - Have you put your educational and work experience in reverse chronological order?
  - Have you given your information in phrases rather than complete sentences?
  - Have you used active verbs to describe your experience?
  - Does the personal information presented enhance your job potential?
  - Do you have permission to use their names from your references?
  - If you are using a functional resume, do the categories reflect appropriately your capabilities and experience?
  - Has your targeted resume zeroed in on an easily recognizable career objec tive? Do the listed achievements support the capabilities listed?
- Are the follow-up letters you have written gracious in tone?
- Have you followed up every interview with a letter? Does that letter invite further communication in some way?
- Does your letter of acceptance of a job show an understanding of the necessary details, such as when you report to work?
- Does your letter of refusal make clear that you appreciate the offer and thank the employer for time spent with you?

PLANNING: INTERVIEW

- Have you found out as much about the organization as you can? Its products, goals, locations, and so forth?
- Have you practiced interviewing, using the questions in Figure 13–11?
- Have you the proper clothes for interviewing?

- Do you have good questions to ask the interviewer?
- Do you know the salary range for the jobs you seek?

REVISION: INTERVIEW

- How did your clothes compare to the interviewer's?
- Did you answer questions well? Which answers need improving?
- Did the interviewer ask any unexpected questions?
- How did the interviewer respond to your questions? Did he or she seem to think they were relevant?
- How well do you think you did? Why?
- Do you think your career goals and this organization are a good fit?

Exercises

1. Work out a schedule for your job hunt. Allocate time and set dates for completion of the following stages:
   - Self-assessment
   - Job information
   - Networking under way
   - Letter of application
   - Resume
   - Interview practice
2. Complete for yourself a summary of your self-assessment. Complete and turn in to your instructor a summary of your job information search, a networking plan, a sample letter of application and resume, and a summary of the salary ranges for entry-level jobs in your field.
3. In groups of four, plan and carry out practice interviews as described on page 399. Everyone in the group should get the chance to play interviewer and interviewee once.
4. Write a letter to some organization applying for full- or part-time work. Brainstorm and work out in rough form the three kinds of resumes: chronological, functional, and targeted. Choose the one that suits your purposes best, and work it into final form to accompany your letter. It may well be that you are seeking work and can write your letter with a specific organization in mind.

CHAPTER 14 DEVELOPMENT OF REPORTS

The Variable Nature of Reports

Although reports may generally be classified as informative or analytical, many reports that you write as part of your job will not fall neatly into one of these two categories. In fact, your greatest challenge as a writer will be to determine how to design a report that is best suited for whatever assignment you receive. A report assignment may not call precisely for an information report, a recommendation report, a feasibility study, or an empirical research report. For example, you may be told to "write a report summarizing what we have done in our meetings this week"; "write a report discussing our major problems in choosing the best toxic clean-up plan for this neighborhood of 200 single-family houses"; "write a report explaining to our client the cheapest way we have found for providing utilities"; "explain the best design for a pump that will remove 300–500 gallons of flood water per hour from golf course fairways located along creeks." Each of these situations incorporates characteristics of the different report types listed above. The question, then, is how do you go about designing any report, for any purpose?

To help you answer that question, we provide three sample reports in this chapter, each developed to meet a different situation. Study and analysis of each report will show you how the writers developed their reports to meet their situations. Applying what you learn will help you write your reports to meet your situations.

Any type of report can be presented as a formal report or a letter or memo report.

Formal reports include all or most of the elements discussed in Chapter 10—transmittal letter, title page, list of illustrations, abstracts, foreword, summary, glossary, and appendices. Figures 15-2 and 15-4 in Chapter 15 illustrate two variations of the formal report format.

Letter and memo reports are short reports seldom more than five pages long. Because they are shorter, they do not need the complex format of formal reports. Memo reports remain within the organization; letter reports go outside the organization.You can find examples of letter and memo reports in Chapter 12, "Correspondence, " and Chapter 15, "Development of Analytical Reports." Figures 14-1 and 14-2 are memo reports.

Liability and Report Writing

Before discussing report design, we want to stress that all reports carry legal responsibility. Reports, like letters and instructions, can be used as legal instruments. They document your activities as an employee and suggest your competence in pursuing your work. For those reasons, your reports should always be carefully written. They should be accurate, mechanically correct, effectively designed, and visually pleasing. In short, they should testify to your competence, your ability to think logically and precisely. When you write a report to a client, he or she will expect your information to be accurate. The client will expect to be able to make financial and technical decisions based on the accuracy and clarity of your reports. If your work is not clear, if it is inaccurate, the client can hold you legally responsible for any costs incurred as a result of your reports.

Because reports become permanent records of your work, you want to be sure that if they are read by litigants, by readers you do not know, or by anyone at any time, your reports present a positive picture of you and the professional manner in which you approach your work.

Producing the kind of document that achieves its purpose, a communication that you can send with confidence, begins and ends with the composing process, which we emphasize throughout this book. As you analyze audiences, your relationship with these audiences, the context that prompted your report, and the contexts in which it is being written and received, always look for ways in which your report could be misconstrued. As you plan, organize, write, revise, and edit, be sure to look for other ways your message could be interpreted. When necessary, revise to avoid any misinterpretation. Care in developing every document you write is the best defense against your writing being used to discredit you or your organization.

General Structure of Reports

Writing effective reports involves four considerations, which we highlight from the general composing process:

- Understanding what you want your readers to know
- Understanding their perspective on the information you will present
- Applying your knowledge of how people read and process information to the development and presentation of the message
- Choosing content, style, and tone that are suitable for your audience, the message, and your relationship to both in the organizational context

In short, good report writing begins with good planning. Because employees are often overwhelmed with paperwork, your reports—everything you write—needs to be designed to encourage audiences to read them. Therefore, you will want to carefully apply the principles covered in Chapter 9 on document design and Chapter 10 on the main design elements of reports. The following situation explains the background, rationale, and purpose of Figure 14–1, a good example of a document intended to report information.

Situation 1

Graham and Simpson is an architectural consulting firm that specializes in analyzing construction problems. The membership of a historic New England church has hired Graham and Simpson to determine why the chimneys of the church are leaking and to recommend what should be done about the problem. First Church is, of course, interested in the cost of correcting the problems. Tim Fong, the managing engineer on the First Church project, writes an internal report to Eben Graham, one of the principals of Graham and Simpson, who is dealing directly with the church committee on repairs. Tim's main purpose is to let Eben Graham know what the team has done on the First Church project so that Graham can phone the chairperson of the church committee about the progress of the study. Eben Graham requires project reports—periodic and final—from all managing engineers on all projects so that he knows what's going on and can inform clients regularly. In a sense, this is an internal status report that shows you the characteristics of a good informal memo report. Refer to Figure 14–1 as you review these characteristics.

TO: Eben Graham DATE: August 20, 1994

FROM: Tim Fong

SUBJECT: Status of Investigation of Masonry Deterioration and Leakage, The First Church, Nashua, NH

Jane Hazel, James Portales, and I have examined two of the chimneys of The First Church. The following points should aid you in preliminary conversations with First Church to inform them of the extent of damage. We will have a final letter report for the Church by 10/1/94 that will include repair estimates for all three chimneys and the roof.

Conclusions

The two chimneys we have examined are badly deteriorated and require extensive repairs to stop the leakage and to eliminate an increasing danger of falling debris.

(1) Leakage through faulty flashing can be stopped easily and effectively by repairing the flashings.

(2) Leakage through the stone masonry cannot be stopped effectively because the chimney masonry is badly deteriorated. The stones in the chimneys have shifted, thereby enlarging the stone joints and creating horizontal surfaces that catch water. Water entering the masonry can bypass the lead counterflashing internally and enter the building. Freeze-thaw action on the wet stone masonry has slowly pushed the stones apart. The stones in the chimneys are not mechanically tied together to prevent movement.

(3) Unless major repairs are made to the two chimneys and the bell tower, significant deterioration will occur in wood decking and masonry. The damage will cause dangerous conditions from falling debris. Eventually, the structural safety of the church will be threatened.

Recommendations

(1) We recommend removal of any chimney not needed. The work should include removal of the chimneys to below the roof level, repair of all deteriorated wood about the chimney, installation of a new roof deck, and installation of new roofing slate to match the existing roof.

(2) Any chimney scheduled to remain should be demolished and

Figure 14–1 Technical Memorandum Report

Eben Graham -2- August 20, 1994

reconstructed with the exterior stones. Reconstruction should include new clay flue liners, steel reinforcing in the horizontal joints of the stone masonry, and new stainless steel or copper through-wall flashings. The roofing around the chimney should be repaired with new metal base flashing of the same metal used for the through-wall flashing.

(3) The wood components of the deck should be thoroughly examined and replaced or repaired, as required by the rotted conditions. Repairs should include exploratory operations into hidden conditions to ensure all deteriorated wood is repaired.

(4) Reconstruction of the interior finishes on the church should be delayed until all repair work is complete and tested.

Field Observations

We have examined both the interior and the exterior of both chimneys.

Chimney No. 1

Chimney 1 has extensive problems with wood and masonry.

<u>Interior Observations</u>

* The chimney is constructed of brick below the roof deck.

* The underside of the roof deck and joists are water stained on four sides of the chimney. The heaviest staining occurs on the north side of the chimney, which is the low point of the surrounding roof.

* A gutter has been installed at the low side of the chimney to catch leakage water and direct it to trash barrels.

* The low header in the roof deck framing on the north side of the chimney opening is wet and extensively rotted, allowing easy insertion of a two-inch pocket knife. Large pieces of the rotted wood are easily removed by hand.

* The remaining structural components surrounding the chimney are sound, based on our visual examination from the attic space.

Figure 14-1 (continued)

Eben Graham -3- August 20, 1994

<u>Exterior Observations</u>

* The chimney is capped with roofing cement and fabric membrane (See Photo 1).

* The mortar joints have been coated with a caulking material (See Photo 1 and Photo 2).

* A cricket made of copper sheet metal is located on the roof above the chimney (See Photo 3). There is one small crack in a soldered seam on the ridge of the cricket.

* After lifting the counterflashing on the cricket, we found a large hole in the membrane (See Photo 3). Water draining over the roof and onto the cricket enters through this hole into the building.

* Some of the granite blocks in the chimney have shifted, and the mortar joints are open slightly.

Chimney No. 2

Our examination shows that Chimney 2 is also in a highly deteriorated state.

<u>Interior Observations</u>

* The roof deck and structural components surrounding the chimney are stained from water leakage, but all components are sound, as far as we can tell without removing any decking.

<u>Exterior Observations</u>

* The chimney is capped with roofing cement and fabric membrane.

* The mortar joints have been coated with a caulking material.

* The flue lining is brick, which has deteriorated. The mortar joints of the lining are heavily eroded (See Photo 4).

* The granite blocks that form the exterior width of the chimney have shifted outward, leaving many of the horizontal and vertical stone joints open (See Photo 5).

Figure 14-1 (continued).

Eben Graham -4- August 20, 1994

* Some horizontal stone joints are completely open. We examined these joints at the corners of the chimney for the presence of metal ties between stones. We found no ties.

* We saw no defects in the metal flashings at the base of the chimney.

General Observations of the Roof Structure

We are currently examining the roof surfaces of The First Church to determine the presence of deterioration resulting from the leakage about the chimneys. We will send results of these findings to you by 9/25/94 along with photos that support our observations.

We still need to examine Chimney No. 3 carefully, but our visual observations suggest that the problems with No. 3 are similar to those of No. 1 and No. 2.

We plan to provide photos for Chimney No. 3 as well as sections of the roof, if visual inspection of the roofing near No. 3 suggests that water damage has occurred.

Please contact

Jane Hazel about interior observations (ext. 2165)
James Portales about exterior observations (ext. 2171).

I will be in Orono for the structural design conference until August 25.

Figure 14-1 (continued).

1. *State your subject of your informal report clearly.* All reports need clear titles. Informal reports are no exception. A busy reader with a dozen items to read will often make a decision to read an internal company report based on the subject line. For example,

SUBJECT: Description of First Church Masonry, Nashua, NH

is not as effective as

SUBJECT: Status of Investigation of Masonry Deterioration and Leakage, First Church, Nashua, NH

The recruiting meeting subject line below tells the reader the purpose of the report and the meeting where the report will be discussed. Adding the date of the meeting is also useful in aiding the reader's memory.

If a report requires action by readers, you may want to add that immediately below the subject line:

SUBJECT: Recruiting Meeting—July 1, 1994—Change in Hiring Policies for Part-Time Research Technicians.

ACTION REQUIRED: Attached forms must be completed by all departments by June 15, 1994.

2. *In a memo report, attempt to begin with the main information.* That way, a reader who sees nothing but the subject line and the first page will get the point of your message.

In Figure 14-1, the subject line is followed immediately by a short introduction. The conclusions and the recommendations based on those conclusions explain precisely what needs to be done to repair the chimneys at the church.

3. *Keep additional paragraphs short to make the content easier to read, but be sure to provide all explanation needed to support the information in the opening paragraph.*

4. *Use design principles to highlight information.* In Figure 14-1, the supporting part of the report is divided into two main headings—Chimney 1 and Chimney 2. Main problems with each chimney can be seen at a glance because of the listing arrangement.

5. *Be sure to tell readers where to get additional information.* Figure 14-1 includes the names and phone numbers of the two engineers working on the First Church project so that Graham can contact the appropriate individual if he has questions about the findings in the report.

Figure 14-2 illustrates an information report. It is a tax research report written for tax accountants in an accounting firm who need regular notification about changes in the Internal Revenue Code. The following description of the context in which the report is written will help you understand why this report is designed the way it is.

Situation 2

Walker & Walker is a tax accounting firm that deals only with corporate and individual tax problems. As a result, the firm has its own research department that prepares reports for its tax specialists. The research department provides these reports to inform tax preparers about every change in the Internal Revenue Code and its application by court cases or letter rulings from the IRS. Because the research department may disseminate as many as 200 of these research reports a year, they must be designed for rapid reading: Employees must know which reports are applicable to their current clients so they can know which to read carefully and which to scan for basic information.

The example tax research report in Figure 14-2 can be read selectively, depending on how much the reader wants or needs to know about this particular tax research topic:

- The heading indicates that the report does not require the readers or their clients to take some action.
- The specific, descriptive *subject line,* often indicated by SUBJECT, SUBJ, or RE (Regarding), helps readers know whether they need to read further and whether the topic pertains to work they are currently doing.
- The *summary,* which may also be called a DIGEST or BRIEF, gives the gist of the report but also includes topics to be discussed.
- The *discussion* section is partitioned with boldface headings. Readers can read part or all of the report because it is visually partitioned to reveal discrete sections of the report.

Determining Report Structure

Many times, as you can see from Figures 14-1 and 14-2, reports may not need all the design elements discussed in Chapter 10. If they are designed for internal communication, they can be set up in informal report format (Figures 14-1 and 14-2, for example), even though they may be several pages long. Whatever the length of the report, you will want to organize it so that it will achieve its intended goals and use a logical system of headings to reveal the content to the reader, as discussed in Chapter 9.

Clearly stated subject line

Summary of content

Heading, subheading, listing, and highlighting help readers see the content

DECEMBER 7, 1992
TAX 92-276

FOR INFORMATION ONLY — NO ACTION REQUIRED

TO: ALL TAX STAFF

FROM: SHERYL COSTA

RE: FINAL REGULATIONS UNDER IRC SEC. 6694 — UNDERSTATEMENT OF TAXPAYER'S LIABILITY BY AN INCOME TAX RETURN PREPARER

DIGEST

The final regulations under Section 6694 were issued on December 31, 1991. These regulations are the final step in the Service's effort to substantially modify the analysis made under former IRC Section 6694 as to what constitutes valid legal authority, how that authority is weighed, and what standards return preparers must meet to avoid the various preparer penalties. The purpose of this internal report is to remind all tax staff of the provisions of this Section and to highlight those portions that are particularly important in avoiding unnecessary assessment of any penalties on Walker & Walker or our employees.

DISCUSSION

PREPARER PENALTY FOR UNDERSTATEMENTS DUE TO UNREALISTIC POSITIONS (SEC. 6694(a))

Effective for returns prepared after December 31, 1991, if a preparer files a return knowing a position does not have a **realistic possibility** of success if challenged administratively or litigated, and the position is not disclosed, the preparer is subject to a **$250 penalty**.

Realistic Possibility Standard

In Notice 90-20, the IRS stated its view that the new "realistic possibility" standard of Sec. 6694(a) is a stricter standard than under former law. A position will be considered to have a realistic possibility of being sustained on its merits if a reasonable and well-informed analysis by a person knowledgeable in the tax law would lead such a person to conclude that the position has approximately a **one in three, or greater**, likelihood of being sustained on its merits. The regulations further provide that the analysis prescribed by **Reg. Sec. 1662-4(d)(3)(ii)** for purposes of

Figure 14–2 Research Report

All Tax Staff -2- December 7, 1992

determining whether substantial authority is present applies for purposes of determining whether the realistic possibility standard is satisfied, and only the authorities specified in **Reg. Sec. 1662-4(d)(3)(iii)** are to be considered in the analysis.

Adequate Disclosure

An adequate disclosure will prevent imposition of a Sec. 6694(a) penalty **unless** the position is frivolous. The final regulations set forth different disclosure rules for signing and nonsigning preparers. **Signing** preparers must disclose on the return or claim for refund and will be protected only if the disclosure is made on **Form 8275** or **Form 8275-R**, if appropriate, or in accordance with an annual revenue procedure. **Nonsigning** preparers generally will meet the disclosure requirements if they inform the taxpayer or another preparer that disclosure is necessary, or if the position is adequately disclosed on the return or claim for refund.

Reasonable Cause

For a position that does not meet the realistic possibility standard, the preparer must show that the understatement was due to **reasonable cause** and the preparer acted in **good faith**. The regulations provide that the reasonable cause and good faith determination will be made by considering all the relevant facts and circumstances, including:

(1) the **nature** of the error causing the understatement;
(2) the **frequency** of errors;
(3) the **materiality** of errors;
(4) the preparer's **normal office practice**; and
(5) the extent to which the preparer **reasonably relies** on the advice of, or schedules prepared by, another preparer.

PREPARER PENALTY FOR WILLFUL OR RECKLESS CONDUCT (SEC. 6694(b))

Effective for returns prepared after December 31, 1991, if an understatement is attributable to a **willful** attempt by a preparer to understate a client's liability, or to the preparer's **reckless** or intentional disregard of the tax laws or regulations, then the preparer is subject to a **$1,000 penalty** under Sec. 6694(b). Again, disclosure on Form 8275 may prevent this penalty, but not if the position being disclosed is frivolous.

Figure 14-2 (continued).

All Tax Staff -3- December 7, 1992

A preparer is **reckless** in not knowing of a rule or regulation if the preparer makes little or no effort to determine whether a rule or regulation exists, under circumstances that demonstrate a substantial deviation from the standard of conduct a reasonable preparer would observe in the situation. The amount of penalty is reduced by the amount of penalty paid by the preparer for a Sec. 6694(a) penalty. Therefore, the total penalty under Sec. 6694(a) and Sec. 6694(b) for a single infraction is **$1,000**.

PROCEDURAL RULES (SEC. 6694(c))

If within **30 days** after the notice and demand for payment of the penalty, the preparer pays at least **15 percent** of the penalty amount and files a claim for refund, he will not be required to pay the remaining penalty amount until resolution of the refund claim or any suit is brought by the tax preparer in district court. If the claim is denied, suit must be brought within **30 days** or collection of the balance will ensue.

ABATEMENT OF PENALTY (SEC. 6694(d))

When a final administrative determination or a final judicial decision is reached concluding that no understatement or liability exists, the penalty assessment will be abated and any amount paid will be refunded.

Please do not hesitate to consult a superior regarding this or any other penalty if you are unsure of a position you are taking on a return.

Figure 14-2 (continued)

Determining Internal Development of Reports

Reports will usually begin with an introduction that briefly states the following:

- The purpose of the report
- The reason the report was written
- Scope of the report—what will be covered, as in Figure 14–2
- Perhaps a short summary of the main ideas covered in the discussion, also shown in the digest in Figure 14–2

As discussed in Chapter 10, the extensiveness of the introduction depends on your readers' needs: Will they expect the report? Do they understand the circumstances surrounding the report (why it was written)? Remember: Readers have to be prepared for the information they will read before they are ready to process it.

The introduction to a report prepares the reader for the content to be presented. A reader approaching a report for the first time will ask several questions: What is this report about? What does it cover? Do I need to read it? How does it affect me? The introduction should answer those questions. Thus, any report introduction should stand alone. It should make sense to a reader who is seeing the report for the first time. The length of the introduction is determined by what your reader needs to know before moving to the conclusions or the discussion.

A report may also require a summary that gives the main highlights of the entire report. Or, the report may use a combined introduction/summary, a short paragraph that states the purpose of the report and the main ideas presented. Figure 14–2 uses this configuration. Figure 14–1 uses a brief introduction and separate sections containing detailed conclusions and recommendations to let the senior engineer know the results of the analysis of the masonry.

The discussion section presents each main category of information specified in the introduction, as shown in Figures 14–1 and 14–2.

In short use an introduction, introduction/summary, summary, conclusions, recommendations, and discussions as purpose, audience, and situation call for them.

Situation 3

Wendy Smith is a graduate assistant for the Instructional Materials Services (IMS) of the Agricultural Education Department at Texas A&M. The IMS is responsible for developing the curriculum used in public school agricultural and technical classes in the state of Texas. The curriculum developed by the IMS consists of little pamphlets called topics. These topics are disseminated separately, but sometimes all the topics from a course are bound together in a notebook to form a text. Teachers choose which units they will cover.

APPLIED ENTOMOLOGY: INTEGRATED PEST MANAGEMENT

8421-A

DEFINING INTEGRATED PEST MANAGEMENT AND RECOGNIZING ITS DEVELOPMENT AND IMPORTANCE

INTRODUCTION

Insects are amazing creatures. They can thrive almost anywhere, and they exist in incredibly large numbers. Many millions of insects can live on one acre of land. Scientists have identified more than one million insect species. Some scientists estimate that there are more than 30 million different insect species. That is more than any other type of animal. More than one thousand insect species can live in a back yard.

Insects help humans and animals in many different ways. Honey bees provide honey and wax. Even more importantly, honey bees pollinate* many fruit and vegetable crops, including apples, pears, watermelons, and others. Certain insects help control weeds and other insect pests by eating them. Some insects eat dead plants and animals and help recycle nutrients in the ecosystem.

Insects can also harm humans and animals. Some insects can ruin agricultural crops, food, wood, or clothing. Some also transmit diseases and parasites, irritate humans and animals, and sting or bite. Insect pests cause millions of dollars worth of damage and much human misery. People must manage harmful insects to minimize their damage.

People can manage insect pests several ways. Researchers have developed many insecticides. These chemical products kill insects and are widely used to control insect pests. Insecticides have often been very effective. For instance, soldiers in World War II used the insecticide DDT to kill disease-carrying lice. This saved millions of soldiers' lives. If not used wisely, however, insecticides can pollute water supplies and poison humans and other organisms. Insecticides should not be the only insect control method producers use. People must also use other methods to manage insect pests, such as crop rotation or destroying crops soon after harvest. The Integrated Pest Management program for the Sweet Potato Whitefly, illustrated below, uses several control methods.

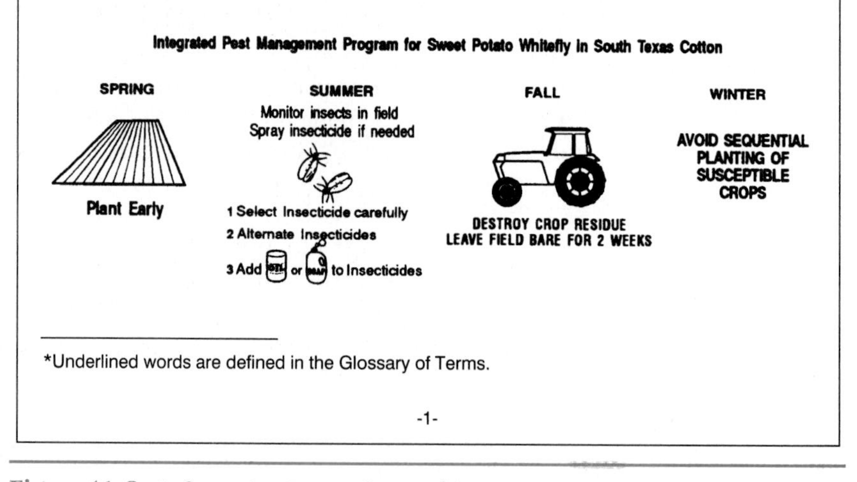

*Underlined words are defined in the Glossary of Terms.

-1-

Figure 14–3 Information Report for Teaching

Source: Instructional Materials Service, Texas A&M University, College Station, TX 77843.

Effective insect management requires using various procedures and steps in an Integrated Pest Management program. This topic answers three questions: What is Integrated Pest Management? How was it developed? Why is it important?

WHAT IS INTEGRATED PEST MANAGEMENT?

In applied entomology, Integrated Pest Management is a pest population management system that anticipates and prevents insect pests from reaching damaging levels by using all suitable tactics. An Integrated Pest Management system combines insect control methods to prevent economic loss while protecting human health and the environment. This combination of methods keeps an insect pest from causing too much damage to its host or habitat. [For the remainder of this topic's discussion, **IPM** will be used for **I**ntegrated **P**est **M**anagement.]

The success of an IPM program hinges on its five basic principles:

1. IPM considers the entire ecosystem. The program's manager considers everything known about an insect pest when planning an IPM program and making IPM decisions. The manager must consider the insect pest's host, environment, and life cycle, as well as weather conditions and the effect of control measures on other organisms. IPM tries not to disrupt the ecosystem.

2. IPM relies on naturally occurring controls as much as possible. These natural controls include enemies of the insect pest that eat, parasitize, or infect it, and weather conditions that kill the insect pest or slow its rate of reproduction.

3. IPM monitors insect pest populations or damage to determine when to use control methods. IPM does not completely eliminate the insect pest population. The key point is the number of individual insects found. IPM keeps the insect population at the level necessary to prevent economic loss. This population level is called the economic threshold (see Topic #8421-C).

4. IPM uses various insect control tactics (see Topic #8421-B). Using a variety of insect-control methods is more successful than using only one. This is because different methods may kill the insect pest at different times in its life cycle. Also, insects can develop resistance to insecticides when they are the only control method used. Insect resistance can result in a control failure when insecticides are applied. Overuse of insecticides can also allow other insects to become insect pests.

5. IPM always considers the cost and effectiveness of a control tactic. A tactic that works very well may be too expensive for a producer or home owner to afford. The IPM operator then must find another solution to the pest problem. The IPM operator weighs a control tactic's cost against the value of the resource to be protected and considers the effect that the tactic will have on the ecosystem.

HOW WAS IPM DEVELOPED?

IPM is a new term for an old approach. The concepts and ideas of IPM developed over many years. In 1880, S.A. Forbes recommended that entomologists apply the principles of ecology to insect control. In the early 20th century, producers used a combination of farming practices and field cleanliness to combat the boll weevil. By 1925, cotton growers regularly counted insects in cottonfields to decide when to apply insecticides. Organized programs to manage pea aphids in the Northwestern United States were in place by 1935. Several scientists defined the concepts of IPM throughout the late 1950s and early 1960s.

-2-

Figure 14-3 (continued)

Researchers continue to refine IPM. Each IPM program is unique to the individual insect pest and its situation. IPM concepts apply not only to insect pests of livestock, urban areas, and human health, but also to various organisms that cause plant diseases. IPM will continue to grow in popularity and use as scientists develop programs for more insect pests.

WHY IS IPM IMPORTANT IN APPLIED ENTOMOLOGY?

IPM is important because it is effective, and it protects human health and the environment. IPM programs prevent insect populations from causing economic loss. Many IPM programs reduce the need for insecticides. This can help prevent control failures and keep insecticides effective against insect pests. Less insecticide use also helps prevent other insects from becoming insect pests.

SUMMARY

Insects are amazing creatures that exist almost everywhere. Many are helpful, but many are pests. People control insect pests to minimize the damage they cause. In recent times, insects have usually been controlled with insecticides. However, the misuse of insecticides can pollute water supplies and poison humans and other organisms. Overuse of insecticides can cause insects to become resistant to them and cause new insect pest problems to develop. IPM uses a variety of insect control methods in an organized program that prevents economic loss without harming humans or the environment.

References:
Borror, Donald J., Charles A. Triplehorn, and Norman F. Johnson. *An Introduction to the Study of Insects*, Harcourt Brace Jovanovich Publishers: Orlando, Florida, 1989.
Pfadt, Robert E. (editor). *Fundamentals of Applied Entomology*, Macmillan Publishing Company: New York, New York, 1985.

Glossary of Terms

Applied entomology — takes knowledge about insects and uses it to control insect pests
Control — in IPM, to reduce an insect population below economically damaging levels
Control failure — an insect control method that does not reduce an insect population
Ecology — the study of the relationships between living things and their environment
Economic loss — a loss of profit greater than the cost of insect control measures
Ecosystem — all the living and nonliving parts of a given area
Entomologist — a person who studies insects and their relatives
Habitat — the immediate area in which an organism lives (a particular plant, a pond, etc.)
Host — the organism in or on which a parasite lives (e.g., the plant on which an insect feeds)
Insecticide — a chemical substance used to kill and control insects
Integrated Pest Management — a program of insect management that is economically and environmentally acceptable
Manage — in IPM, to maintain an insect population below a damaging level
Monitor — to check on; to observe; to watch over
Parasite — any animal or plant that lives in or on and at the expense of another organism
Parasitize — to live in or on and at the expense of another organism
Pollinate — in flowering plants, to transfer pollen from an anther to a stigma; can be done by insects
Resistance — regarding insects, the ability to overcome a control tactic; usually the ability to withstand an insecticide
Tactic — in IPM, a specific method of insect control

-3-

Figure 14-3 (continued)

The purpose of these pamphlets is to help students on the high school level understand agricultural topics—in this case, the integrated pest management concept. Students must learn to manage pests effectively within today's pest control legislation. Figure 14–3 is one of three topics written by Wendy that describes integrated pest management for Instructional Material Service's course on Applied Entomology.

Figure 14–3 illustrates a third kind of report, the informative teaching report. Even though this report does not appear in memorandum format, its structure is similar to that in Figures 14–1 and 14–2.

1. It has a clear title that indicates the information covered in the report.
2. It has an introduction that states the purpose of the report, the significance of the topic, and the issues that will be addressed.
3. It uses document design techniques and visual aids to highlight information.
4. Terms are properly defined; paragraphs and sentences are concise to make the report easy to read.

Note, however, that the writer places the summary at the end to draw together the important points discussed. You will want to examine the context for which this report was developed to help you understand the rationale for its particular design.

The three reports in this chapter, though they lack the complexity of more analytical reports, provide a good base for the study of report development. They illustrate well that the style, content, organization, and design of your report depend on your situation.

Planning and Revision Checklist

Because of the variable nature of reports, the general planning and revision checklist inside the front cover will serve best for this chapter.

Exercises

1. The following report is a poorly organized and formatted information report. Redesign the report to reflect good document design, effective style, and organization.

PLEASE READ AND INITIAL
TO: Structural Engineering Group
FROM: Program Planning Committee
DATE: September 7, 1993
RE: Construction Engineering Conference

The Construction Engineering Conference, which will be held October 28–30 at the Lancaster Hotel, will be attended by approximately 260 engineers. To give all of you some idea about how the conference will proceed, we want you to be aware of the following items.

The SE group should be on hand throughout the conference to represent our company, since we are the host company this year, and to answer any questions from participants. For our company, our group contacts will be Jim Mahann and Joanna Sturges. Their conference office will be in Suite 104. They will begin working from this office October 26.

Breakout rooms will be available for the second part of all sessions. Phones and fax machines will be available in all breakout rooms from 9:00–5:00 each day. House phones will be located only at the end of the first and second floor ballroom areas.

Structural software will be displayed by nine vendors all day October 29 and until noon October 30. Four book companies will also have displays of recent publications on structural topics. Computer consultants from IBM, Compaq, and Apple will also be on hand to discuss compatibility issues and to discuss their own products. Material for the conference packets has already begun arriving. Each folder will have brochures from at least a dozen new products.

Participants will arrive by mid-afternoon October 28. Everyone from our group is to be available to greet incoming guests. Room assignments have already been made for all guests, but if anyone has questions, be sure they are sent to Jim and Joanna. Our company will host happy hour beginning at 4:30 in the Cavanaugh Room. A buffet supper, hosted by MERK Structural Group in Kansas City, will be held in the same room. Food will be available at 6:00. The film of the Paratex Building structural failure will be shown at 8:00 in the Segram Room for those who have not had the opportunity to see this film.

Other information: Check cashing services are available with proper identification at the Manager's Office, Room 6. A list of suggested entertainment places is being prepared. You will receive copies for distribution.

Both days of the conference will feature both plenary and concurrent sessions. Two plenary sessions will be on the 28th and one on the 29th. Dr. Milo Nyuen and Dr. Hee Wong from MIT will conduct the 28th plenary sessions. Dr. Phil Rollins of Ohio State will be the principal speaker for the 29th. The exact schedule will be published soon. Everyone will receive a complete packet of information. Coffee and refreshments will be available during the morning and afternoon session breaks.

That's about it for now. Any questions or suggestions are welcome. Call Megan and Tracy at 2133 and 5623.

2. Write a memo report to your instructor presenting and discussing three topics you are considering for your major report.

CHAPTER 15 DEVELOPMENT OF ANALYTICAL REPORTS

Although the design of the elements of both information and analytical reports is the same, analytical reports go beyond reporting information: They analyze information or data. From the analysis, the writer may evaluate information, draw conclusions, and perhaps recommend action based on those conclusions. The type of report you write and the extensiveness of the analysis required depend on the organizational issue that is the topic and purpose of the report.

Analytical reports often defy rigid classification, but for the purpose of learning to write analytical reports, we can classify them into the following types.

If the analysis leads to a recommendation, the report may be called a **recommendation report.** If the analysis emphasizes evaluation of personnel, data, or perhaps options, the report may be called an **evaluation report. Feasibility reports** analyze a situation to determine the best solution to questions surrounding the situation or problem. Empirical research reports explain research that has been conducted, report the results, analyze the results, and sometimes recommend further work or research that is needed.

Many reports both inform and analyze: **progress reports,** which will be discussed in Chapter 17, describe and evaluate the work that has been done on a project, the cost, and problems encountered; **trip reports** document the information gathered on a trip, evaluate this information, and may suggest action based on the report findings. **Personnel reports** describe an employee's performance, analyze its effectiveness, and estimate the employee's potential for promotion. **Economic justification reports** explain the cost of a project or action and then argue for the cost-effectiveness of the project.

In this chapter, we will discuss the design of four of these types: the analytical report in general, the recommendation report, the feasibility study, and the empirical research report. Each report type will be accompanied by illustrations so that you can see how design works in specific situations. However, you should always remember that report guidelines are just that—guidelines. Your report should respond to the needs of your content and your audience.

Analytical Reports

Isolating analytical reports for study is useful for several reasons. Analytical reports usually have a more complex design than information reports. Analytical reports usually require a more extensive introduction than is needed for the information report. Analytical reports begin with an introduction that may include:

- The purpose of the report
- The reason the report was written
- The scope of the report—what issues will be and will not be covered and not covered
- The procedure for investigating (analyzing) the topic of the report

After the introduction, the report can be developed in several ways:

Introduction
Conclusion—results of the analysis
Recommendation—if required by the investigation
Presentation of information
Criteria for evaluation
Discussion/evaluation of information

Or,

Introduction
Presentation of information
Criteria for evaluation
Discussion/evaluation of information
Conclusion—results of the analysis
Recommendation—if required by the investigation

The organization you choose should depend on your reader or the format used by your company.

Use the first plan—placing the conclusions of the analysis first—if your reader is most interested in the conclusions.

Use the second plan—presenting the data or information, evaluating it, then presenting the conclusions and any recommendations you have—if the analysis, rather than the conclusions, is the focus of your report.

Figure 15-1, based on Situation 1, illustrates a report that emphasizes analysis rather than any firm conclusion. For that reason, the writer does not include a summary or a conclusion at the beginning, but places the less-than-definitive conclusion at the end. Note that in this report, the writer gives several sentences of background, as several readers besides the main reader will receive the report. These readers may not be as familiar with the context of the report as the main reader.

Do not place the conclusion first if you think your reader may misunderstand or object to the conclusion; present the criteria for analysis, the analysis, and then the evaluation from which the conclusion and any recommendations evolve. Note that in Figure 15-2, based on Situation 2, the consultant writing the report makes clear how the conclusions have been determined. He does so in the letter of transmittal and in the summary, which explains the background and the nature of the problem before proceeding to a brief statement of the recommendations, which are fully explained at the end of the discussion. See Situation 2, page 447.

Situation 1 (Figure 15-1)

Melanie Pierce is an international market analyst for REVAC, a pipeline company that is considering expanding its sales into European markets. This expansion will require major revisions in REVAC's financial operations to handle marketing finances. Her report to the ENR (Engineering Network Reorganization) Group has no conclusion at this point, but Pierce does want to inform the group about the two options being considered for dealing with the proposed international expansion.

TO: ENR Group DATE: March 11, 1994

FROM: Melanie Pierce

SUBJECT: Alternatives for Restructuring the FACG Team

Introduction

Background of the report; the problem that needs analysis

Recent decisions to reposition REVAC's main product line have resulted in the decision by the VP-Operations to refocus the direction of the Financial Analysis Consulting Group (FACG). Diversification of REVAC products into European markets requires new types of financial analysis be available for sales, marketing, as well as product development.

My office was asked to analyze two options considered by the GOC during its fall decision to establish European markets with the goal of achieving a wider range of financial information.

Answers being considered

1. Create a separate International Financial Analysis Office, which would report directly to the VP-Operations.

2. Create an international financial group within the existing FACG.

Current FACG Situation

Present status of the organization as it relates to the problem

The FACG has not hired new financial personnel for three years. During this time, two have left the group. Their responsibilities were absorbed by existing analysts, whose primary responsibility was to monitor U.S. markets. The recent shift of Marvin Perry to operations manager for general organizational planning makes a replacement for him mandatory. The upshot of the situation is that FACG cannot be responsible for more analytical data, particularly European data, without adding additional staff.

Option 1: Establishment of new FACG International Group

Analysis of Option 1

Creating a separate office for international financial analysis would avoid overloading the current FACG. European financial data must be considered no later than 10/1/95 to be of value to design and pre-market planning. Personnel reports that three Euro-specialists, whose contract expires at M-L 8/15/95 will hear resituation offers to REVAC. Should

Figure 15–1 Analytical Report, Situation 1

ENR Group - 2 - March 11, 1994

they begin work 9/1, they could make the first cost estimates available by the 10/1 deadline. All three are experienced overseas financial specialists who would need minimal instruction in what kind of financial information the VP-O needs to launch the Euromarket. The VP-O's administrative staff would be responsible for the new group's set up.

Option 2: Expansion of FACG

Analysis of Option 2

Space is currently available in the FACG office suite on the 41st floor, if Eurospecialists with U.S. background could be found who would work with the current FACG team. Personnel has been interviewing new M.S.-International Finance graduates at Stanford and is ready to recommend those who meet qualifications. New graduates must have educational backgrounds that equip them to work with either U.S. or Euro-markets. Two recent M.B.A.s are under consideration to fill FACG vacancies to enable the office to assume a full range of responsibility.

Comparison of Current Alternatives

Comparison of Option 1 with Option 2

Replacing FACG management personnel because of the stringent qualifications required of these select analysts usually takes at least six months after a position vacancy has been announced. Assuming that estimate is on track, a replacement for Marvin Perry and one other experienced analyst will not likely happen until 1/1/96 at the earliest. None of the college students interviewed so far will be available before 1/1. After start date, new analysts will still have to have time to take the required SEC exams, which means that their effectiveness date to REVAC begins about 7/1/96.

While expanding the current FACG with new hires is the most cost effective move, the effective time makes the value of recruiting new college graduates less than desirable. In contrast, the cost of resituating the M-L team to REVAC would increase start-up costs for the unit by 60% and require space availability near the VP-O office. However, the fact that these analysts are fully SEC certified and ready to begin work with only minimal instruction suggests that this expenditure is worth considering.

Status as of 4/5/94

What actions need to be taken now

Cost requirements for start-up of the new unit need to be specified. The VP-O's Executive Committee authorized the creation of the international unit apparently on the assumption that salaries would be commen-

Figure 15-1 (continued).

ENR Group - 3 - March 11, 1994

surate with those of existing FACS members. Since that is not the case, cost guidelines vs. the importance of the 10/1 start time need to be clarified.

If expansion of the FACG unit is selected because of cost, Planning needs to be made aware that Euro-market data analysis will be delayed at least eight months. If this option is chosen, Personnel needs to be informed immediately to step up recruiting efforts with an emphasis on candidates with international background. In any case, salary structure needs to be clearly specified in the event that a range of individuals becomes available.

pc: File 9104.22A
S. Smith
P. Sanchez
N. Silanski

Figure 15-1 (continued).

FIRST LINE OIL SPILL RESPONSE RECOMMENDATIONS
FOR BRITISH PETROLEUM'S
MARCUS HOOK FACILITY

J. Larry Payne
R. J. Meyers and Associates
Environmental Engineering and Consulting

November 25, 1990

Figure 15–2 Analytical Report, Situation 2

November 24, 1990

British Petroleum Oil Company Ltd.
520 Smith Street
Marcus Hook, Pa. 20190

Attention: R. G. Rolan
Environmental Coordinator

Subject: First Line Oil Spill Response Recommendations For British Petroleum's Marcus Hook Facility

Dear Bob:

Report purpose & central recommendation

Attached is our final report of first line oil spill response recommendations for your personnel at the Marcus Hook facility. Our analysis indicates that containment booming can be effective if deployment techniques are adapted to swift water conditions.

Summary arguments for the recommendation

We have verified with Mr. Michael Flaherty of the EPA's Chemical Countermeasures Division that dispersant application within the Delaware River is not a viable response option. Therefore, containment booming as a response option is mandated by Annex X of the National Contingency Plan.

During our site assessment the first week of November, Clyde Strong and I worked with B. P. personnel to develop the angled method of boom deployment outlined in the report. We are confident that this strategy will work well for your personnel.

You will also note our recommendations to purchase an additional 3,500 feet of boom and 1 extra tow vessel. Since you have 1,500 feet of SLICKBAR boom, we've suggested that you stick with this brand. The SLICKBAR boom will cost less than half what a compactible boom will run and it will easily meet your needs based on our field trials. Another tow vessel similar to your MONARK will also be required.

Based on our analysis, B. P.'s capital equipment investment will be less than your $200,000 ceiling (see the report for our estimates). The only other expenses you will need to consider are training costs to bring your personnel up to speed on the new deployment method. We'll be happy to discuss this with you in detail and submit a detailed proposal after the first of the year.

Further information supporting the implementation of the solution

Finally, if you need information on equipment purchases give me a call. Should you decide to stay with SLICKBAR we'll need to get you in touch with John Sullivan and Russ Blair in Connecticut. If you want to consider a compactible boom, you'll need to contact Frank Meyer with Kepner Plastics in Torrance, California.

Figure 15-2 (continued).

Note that the style of the transmittal letter is informal and personal. Sentences are concise and direct. The transmittal offers a brief summary of the report.

R. G. Rolan - 2 - November 24, 1990

Thanks again for your hospitality and assistance in setting up the equipment trials. If you have any questions, please give us a call. If holidays intervene, call me at home (409) 779-8899.

Best regards,

J. Larry Payne

J. Larry Payne

JLP/mlp

enc:

Figure 15-2 (continued).

TABLE OF CONTENTS

Figure 15-2 (continued).

Report subject

Report purpose

INTRODUCTION

Product volume and proximity to environmentally sensitive wetlands render B. P.'s Marcus Hook facility susceptible to major oil spill difficulties. Viability of the only possible first line response measure, containment booming, is severely compromised by swift currents in the Delaware River. This study provides the following:

1. A discussion of the background and nature of the problem.
2. An analysis of conventional booming methods and their limitations.
3. Recommended alterations in booming techniques to enhance B. P.'s first spill response effectiveness.

Recommendations made by R. J. Meyers and Associates, Inc., comply with all pertinent local, state, and federal spill response regulations. Implementing these recommendations will significantly reduce B. P.'s spill liability and cleanup costs.

SUMMARY: OIL SPILL RESPONSE AT B. P.'s MARCUS HOOK FACILITY

Background Information

The British Petroleum (B. P.) facility at Marcus Hook lies in one of the busiest product corridors in the U.S. Maximum daily throughput capacity is 75,000 barrels of sweet crude oil. The average monthly processing rate exceeds 1,000,000 barrels. Throughput volume alone signifies considerable potential for a major oil spill.

In addition, B. P.'s facility sits in the midst of numerous environmentally sensitive areas that line the Delaware River. River currents running between 1.5–2.5 knots are capable of driving an uncontained oil spill into those wetlands adversely affecting wildlife populations in Pennsylvania, New Jersey, and Delaware.

During the past decade B. P. has experienced 3 major spills at the Marcus Hook facility as well as a number of lesser incidents. Cleanup costs for these spills (excluding federal penalties, habitat restoration, and third party damage claims) have ranged from an estimated low of $75 to an estimated high of $400 per barrel of oil spilled.

-1-

Figure 15-2 (continued).

Nature of the Problem

The throughput volume, the environmentally sensitive location of the facility, and the swift river currents in the area require an immediate and effective first line response strategy.

First line spill response methods available to B.P. emergency crews are limited to containment booming operations. While large spills at sea can be dispersed with chemical agents or burned with incendiary devices, environmental factors and regulatory restrictions preclude these actions in the Marcus Hook area.

Major Implications

An evaluation of pertinent factors indicates that containment booming of an oil spill at B.P.'s Marcus Hook facility constitutes the only first line response action open to B.P. spill response personnel.

However, current speeds in the Delaware River (exceeding 1.5 knots) are not compatible with conventional containment booming strategies. Quite simply, traditional methods of containment booming fail catastrophically in such currents. If containment booming is to be effective, some means of adapting traditional methods to swift water situations must be devised.

Field studies conducted at the Marcus Hook facility in early November, 1990, demonstrated that conventional booming strategies could be modified to accommodate the local swift currents. Additional work in the area of angling boom directly into currents exceeding .75 knot is expected to further improve swift water containment capabilities.

Summary of Recommendations

Brief statement of recommendation discussed in detail on pp. 11–12.

1. B.P. emergency response teams should receive prompt training in swift current boom deployment tactics.
2. 5,000 feet of round boom should be stockpiled at 2 locations for rapid deployment. (This requires purchasing 3,500 feet of boom at an estimated cost of $105,000.00.)
3. Two tow vessels should be equipped and available on a 24 hour basis for boom deployment. (This requires purchasing 1 vessel at an estimated cost of $56,500.00.)
4. Follow-up drills and hands-on training should be conducted regularly to improve B.P.'s response capabilities.

-2-

Figure 15-2 (continued).

Description of how boom containment works

DETAILED DISCUSSION

Basic Boom Design

Spill containment booms are designed to contain and control the movement of oil floating on the water's surface. Booms are not effective with emulsions, sinking products, or water-soluble materials. A floatation unit or freeboard corrals the floating slick on the surface while a flexible skirt or draft prevents oil from passing beneath the boom (see Figure 1).

Verbal discussion correlates with the visual aid

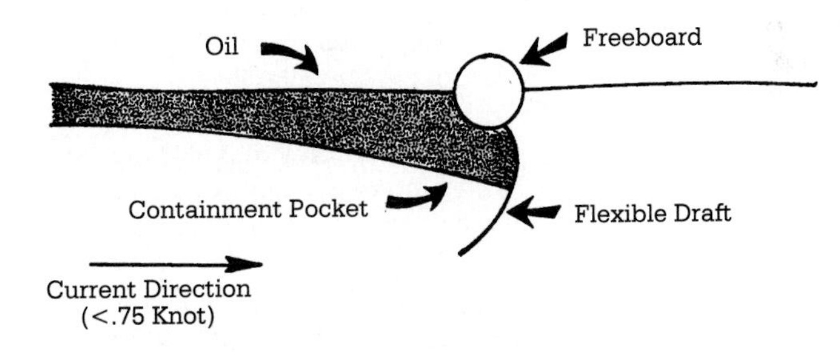

Figure 1. Cross Sectional View of Round Boom in Low Current

Most containment booms rely upon round freeboard designs with floatation units ranging in diameter from 6 to 10 inches. Experience has shown that booms having round freeboards within this size range sufficiently limit splashover (oil sloshing over the top of the boom) in windy conditions. Field observations and informal boom tests conducted at the Environmental Protection Agency's Oil and Hazardous Material Simulated Environmental Test Tank (OHMSETT) in Leonardo, New Jersey, seem to support the contention

-3-

Figure 15-2 (continued).

that a round freeboard design helps limit splashover largely because of the reserve buoyancy inherent in the configuration.

Similar tests at OHMSETT indicate that boom drafts of between 12 and 16 inches in length work best to restrict carryunder (oil being entrained and carried beneath the boom by swift currents). Most boom drafts consist of flexible laminated plastic skirts that are attached to the freeboard. In low current situations (<.75 knot) the skirt flexes with the current to form a cup-like containment pocket directly beneath the freeboard that aids containment and restricts carryunder (see Figure 1).

Many booms are manufactured from closed cell plastic compounds that are resistant to hydrocarbons, industrial solvents, ultraviolet rays (sunlight), and salt water. In addition, booms should retain their flexibility during cold weather. Other major design considerations include:

1. Ease of deployment
2. Cost and availability
3. Length and weight per section or linear foot
4. Compactibility and storage requirements
5. Maintenance and repair requirements
6. General shape and configuration

Compactible or "compressible" booms are designed to fold up in an accordion fashion to facilitate storage. These types require little storage space compared to rigid booms and are characteristically easy to deploy. As the compressed sections are pulled into the water by a tow vessel, they expand and inflate themselves with ambient air. Compactible booms are becoming increasingly popular and demonstrate excellent wave conformity capabilities.

-4-

Figure 15-2 (continued).

Visual aid is introduced then discussed.

Modes of Boom Failure

Containment booms generally fail in one of two ways. In unprotected open water areas, oil may escape by splashing over the freeboard. Surface chop conditions usually result in minimal splashover. However, sustained winds in excess of 20 knots may result in catastrophic boom failure. Fortunately, B. P.'s Marcus Hook facility is situated in a well protected area with little threat of splashover.

However, booms are also subject to failure by carryunder. Carryunder, as noted earlier, is a failure mode whereby oil becomes entrained in swift water and is forcefully carried beneath the boom's draft later surfacing downstream (see Figure 2).

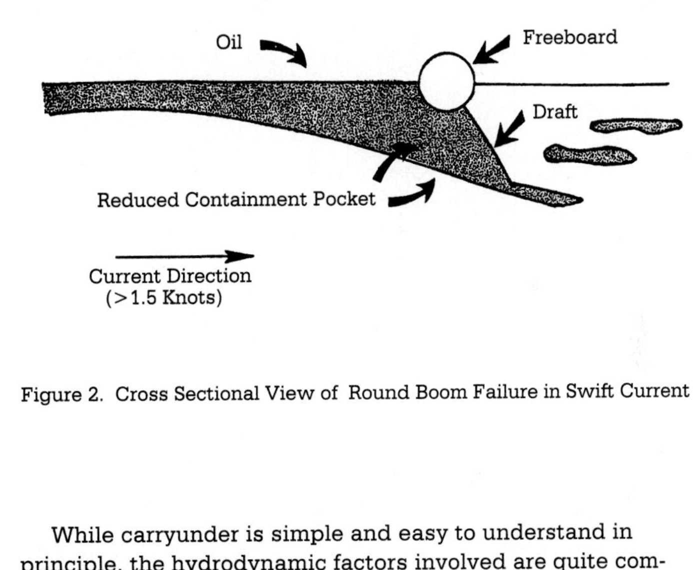

Figure 2. Cross Sectional View of Round Boom Failure in Swift Current

While carryunder is simple and easy to understand in principle, the hydrodynamic factors involved are quite complex. As Figure 2 illustrates, swift water causes the boom's flexible draft to plane or roll with the current such that most of the containment pocket is lost. Consequently, some oil is pulled down the face of the boom directly into the swift

-5-

Figure 15-2 (continued).

Methods of countering potential problems with boom containment

current. As the oil reaches the bottom of the boom's draft, water rushing past the skirt exerts a sort of venturi effect that actually pulls the oil beneath and past the boom.

A seemingly easy solution involves simply increasing the boom's draft length to compensate for the planing effect and help restore a larger containment pocket. Contrary to logic, extending the boom's draft has virtually no effect on the boom's ability to reduce carryunder in swift currents. OHMSETT tests show that increasing a boom's draft to even 36 inches does not appreciably limit carryunder. In fact, field trials in which the boom's draft was actually anchored to the stream bed and held immobile against current flow reveal that carryunder persists in currents exceeding 1.5 knots. Various mathematical models have been proposed to account for this phenomenon. To date, no satisfactory model has been formulated. The fact remains that conventional booming procedures fail due to carryunder in currents comparable to those in the Marcus Hook area.

-6-

Figure 15-2 (continued).

Methods of implementing boom containment

Figure 3 is introduced.

Traditional Boom Deployment Methods

As mentioned previously, containment booming is the only federally approved response tactic available for use by B. P. emergency response crews. Yet, past spills have demonstrated conclusively that conventional booms fail in swift currents. It appears that B. P. emergency teams are saddled with only one approved first line response tactic and that it is doomed to fail in the swift currents of the Delaware River.

Traditionally, spill containment booms are deployed in a straight line between 2 points (A&B) as illustrated in Figure 3.

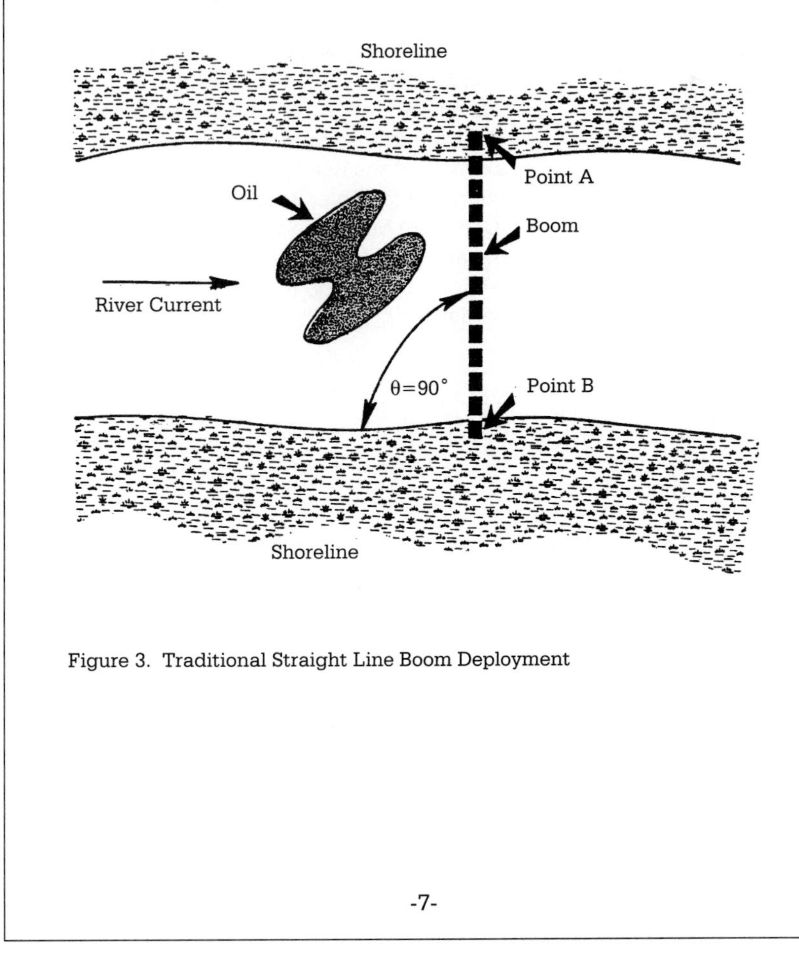

Figure 3. Traditional Straight Line Boom Deployment

-7-

Figure 15-2 (continued).

Figure 3 is discussed.

Note that in Figure 3 that θ (the angle between the boom and the shoreline) is a 90° angle. The rationale underlying this straight line deployment method is difficult to explain. It seems likely to have evolved as a result of untrained crews stringing the boom along the shortest distance between 2 points as a simple matter of convenience. Boom manufacturers, too, may have inadvertently contributed to this practice by demonstrating their products deployed in such a configuration. Finally, since some booms cost as much as $95 per linear foot, this straight line deployment requires a minimal amount of boom to seal off a waterway.

Adapting Containment Booms to Swift Current Applications

A site analysis of the Marcus Hook waterfront terminal was conducted by C. B. Strong and J. L. Payne during the week of November 1–7, 1990. An evaluation of current speeds, available containment equipment, and traditional response techniques reinforced the notion that traditional boom deployment methods were not feasible. Recalling equipment tests conducted by the Canadian Environmental Protection Service (EPS) on the St. Lawrence River in 1980, Strong and Payne considered the boom to shoreline angle (θ) to be the most easily manipulated deployment variable.

-8-

Figure 15-2 (continued).

Logic suggested that by varying θ, the boom could be angled directly into the current to reduce the current load on the boom. A reduction in current loading along the face of the boom, it was thought, might also restrict carryunder. String and Payne also suggested that the boom—even if some carryunder occurred—would improve containment by deflecting oil out of the middle of the stream where currents tend to be greatest towards the calmer waters along the shoreline (see Figure 4).

Figure 4 visualizes written recommendations.

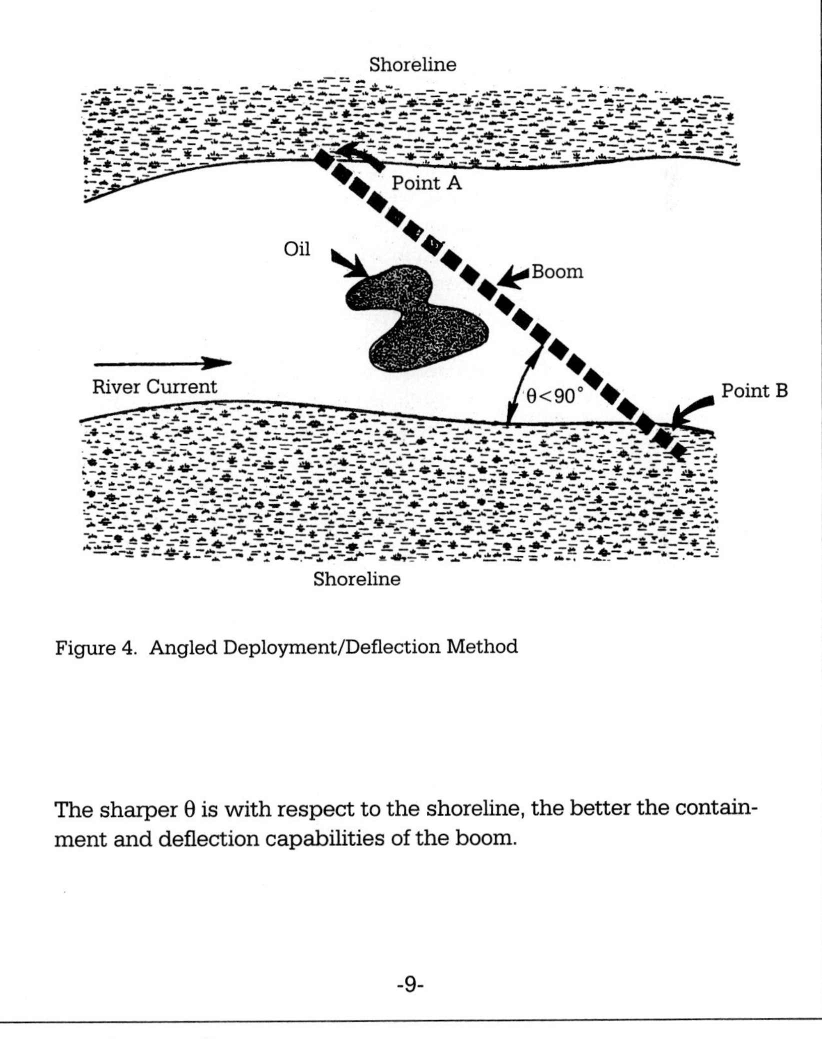

Figure 4. Angled Deployment/Deflection Method

The sharper θ is with respect to the shoreline, the better the containment and deflection capabilities of the boom.

-9-

Figure 15-2 (continued).

Research supporting the use of boom containment

FINDINGS

Field Trials

Field trials conducted at Marcus Hook on November 5, 1990, showed that such an angled deployment strategy effectively contained a test slick of dyed rapeseed oil (a biodegradable product approved by the EPA for field tests) in a current estimated at 2.0 knots. The boom was deployed at an angle of roughly 40 degrees to the shoreline as compared with the traditional 90 degree angle.

Containment boom used during the trials was SLICKBAR MARK IV harbor boom. A total of 1,500 feet (three 500 foot sections joined with coupling plates) was towed into place by a 21 foot MONARK aluminum tri-hull workboat powered by twin 50 h.p. Johnson outboard motors. Both the boom and the workboat were part of B. P.'s emergency response equipment available at Marcus Hook.

No difficulties were noted during the trials. The boom was easily towed into the 2.0 knot current by the workboat and performed well. Boom recovery operations were also routine.

While greater lengths of boom are required to implement this deployment strategy, there do not appear to be any major drawbacks. As noted, the tri-hull MONARK craft easily handled the boom in swift currents. However, attempting to deploy the boom from a single engine shallow draft boat is likely to create problems. Use of the facility's 17 foot Boston Whaler as a tracking or chase vessel revealed that this shallow draft vessel was subject to problems caused by wind advection. Such a tendency to drift would probably combine with river currents in an additive fashion making the boat unmanageable and dangerous to operate.

The SLICKBAR MARK IV round boom seemed well suited to the task. Towing characteristics were good and the boom deflected the test slick well. There is some speculation that a compactible boom may reduce current drag and be a bit easier to tow. However, since the vessel had no trouble towing the facility's present boom, this advantage is slight.

Further trials need to be conducted in order to compare the performance

-10-

Figure 15-2 (continued).

Narrative statement of recommendations that are briefly listed on p. 2.

of round booms to others in this deployment mode. It seems likely that round booms will perform better in swift currents than fence booms; however, this is mere speculation based on general characteristics of boom design.

Recommendations

While B. P.'s emergency response crews face a difficult situation in the event of an oil spill, the challenge can be dealt with effectively. Response teams must be trained promptly in the alternative method of swift water boom deployment outlined in this report. Traditional straight line boom deployment techniques should be discontinued.

In addition, response crews must have access to appropriate containment equipment. At Marcus Hook, an additional 3,500 feet of SLICKBAR MARK IV should be purchased to complement the 1,500 feet already available at the facility. The boom (a total of 5,000 feet) should be stockpiled in two strategic locations (2,500 feet at each site) for rapid access and deployment. The estimated cost for the additional boom at $30 per foot is $105,000.00.

Although an argument can be made for replacing the SLICKBAR MARK IV boom with one of the increasingly popular compactible booms, such a move does not seem economically feasible. Replacing the SLICKBAR would require the purchase of 5,000 feet of boom. Costs for compactible boom range in excess of $60 per foot. Hence, minimum estimated costs for the boom would be $300,000.00. Since the SLICKBAR boom has already demonstrated its effectiveness in field trials, there seems to be little rationale for replacing it.

One additional tow vessel comparable to B. P.'s 21 foot MONARK craft should be purchased and placed on standby to help deploy the boom. The estimated cost for a fully outfitted 21 foot MONARK vessel is $56,500.00. It is essential that 2 boats be available (1 vessel for each 2,500 foot stockpile) to simultaneously deploy the boom. Again, the 21 foot MONARK has already demonstrated its suitability in field trials. The 17 foot Boston Whaler currently used as a backup by dockside personnel is not sufficient as a deployment vessel. This vessel's shallow draft and

-11-

Figure 15-2 (continued).

limited power are not suited to swift currents and using it under such conditions would endanger its crew.

B. P. should provide adequate hands-on training for all of the emergency response team members at the facility. Swift water containment strategies should be emphasized and the response personnel should become familiar with their equipment and its limitations. Cooperative training efforts with the Del-Bay Oil Spill Cooperative under the direction of Mr. Paul Eckert are recommended.

Once the necessary equipment has been purchased and response crews have received proper training and practice, B. P.'s Marcus Hook group should be able to contend effectively with a major spill incident at the facility.

-12-

Figure 15-2 (continued).

Situation 2 (Figure 15-2)

Larry Payne, a consultant with R.J. Meyers Environmental Engineering and Consulting, writes a final analytical report with recommendations for British Petroleum. The task of the consulting organization was to determine whether oil spills could be controlled by using containment booming, as EPA guidelines would not permit use of chemical dispersants for controlling oil spills. Note that the report begins with a short introduction, as readers will be expecting the report, and then a detailed summary composed of four subheadings. The report explains how the recommended boom containment works and its results in field trials. Recommendations for the booms follow from the discussion of their use and the performance in field tests.

The strength of this analytical report is its narration of the problem, which is achieved by an almost conversational style. The author uses drawings to show how boom containment works. The conclusions, given in abbreviated form in the summary, are more fully discussed at the end of the report. How the conclusions evolved from the discussion of the problem and the results of the field tests is easy to see.

Recommendation Reports

A common type of analytical report is the recommendation report. The focus of this type of analytical report is the recommendations. You must analyze a problem or situation, present possible solutions, analyze each solution as it relates to the problem, and then recommend the one you think is best. Figure 15–3 illustrates a recommendation report that provides final recommendations for solving a problem.

Situation 3 (Figure 15-3)

Roberto Ramirez is a mechanical engineer whose group is designing a salt water disposal system for Chemco, which manufactures agricultural chemicals such as pesticides and fertilizers. Amelia Stouffer, Roberto's supervisor, has asked Roberto to explain the group's work thus far and to outline the options the group is considering for designing the salt water disposal plant. Because Amelia has requested this report, Roberto does not need a detailed introduction. He uses only a brief introduction (which includes only subject and scope) and then launches into the conclusions and then the recommendation of the best salt water disposal option studied thus far. In the discussion section, each option is presented and analyzed.

Because Roberto Ramirez, a technician, is writing to Amelia Stouffer, his supervisor, who is an engineer familiar with the topic, Roberto

The introduction is brief—contains only a statement of purpose—because the reader is familiar with the report problem.

The writer gives the results of his analysis before he presents his recommended course of action on page 2.

TO: Amelia Stouffer DATE: July 13, 1993

FROM: Roberto Ramirez

SUBJECT: Mimmstown Produced Salt Water Disposal System (SWD)

This report documents a study of past, present, and anticipated produced water from the WC 66 Field for disposal at the Mimmstown Separation Station. It also evaluates the adequacy of the present system and forecasts future needs.

CONCLUSIONS

1. The produced water disposal system at the Mimmstown Separation Station periodically receives surges from liquid dumps on FWKO and treaters above the rates 1700 BWPD capacity of the SWD system.

2. The skimmer tank and Wemco arrangement, as designed, limits the surge capacity to only what the Wemco can accommodate.

3. With workover operations still underway on the WC 66A platform, only a fraction of the expected yield of produced salt water is arriving at Mimmstown; therefore at present there are no surge problems. Workover rig is expected to move off WC 66A by 10/1/93. Thereafter, peak rates are again expected.

4. Installation of water polishing equipment on WC 66 Field platforms, produced water will be diverted to Mimmstown P/L for treatment and disposal at Mimmstown.

5. Upon upset of polishing equipment on the WC 66 Field platforms, produced water will be diverted into Mimmstown P/L for treatment and disposal at Mimmstown.

6. Present indication from USGS and EPS is that now and in the future, produced water offshore can be treated to specific ppm and disposed of overboard.[1] See Table A.

7. Modifications are required for the existing SWD system to accommodate the surges (which create operating problems) that will continue during the interim period until water polishers are placed offshore.

[1] Information provided by LAS.

Figure 15–3 Recommendation Report, Situation 3. (Note: Figures and tables mentioned in this report are not included in this figure.)

A. Stouffer
Page 2
July 13, 1993

RECOMMENDATION

Engineering recommends that modification to the existing SWD system include installation of a small centrifugal pump downstream of skim tank to pump into the Model 36 Wemco to provide a constant flow to the Wemco. By pumping from the skimmer tank, a lower operating level can be maintained in the tank to allow for as much as a 250 bbl instantaneous surge situation before creating operating problems.

RATIONALE FOR RECOMMENDATION

Analysis of the problem

At the Mimmstown Separation Station, produced water is separated from the product steam in the free water knock-out (FWKO) and the fired production heater treaters. The produced water then flows through a gas boot and into a 1000 bbl tank. The water gravity flows from the skimmer tank, through a water leg, into the Wemco model 36 water clarifier.

A control valve on the discharge of the Wemco regulates the flow of treated water into the 220 bbl injection pump suction tank for disposal into one of two wells. Each SWD injection pump is capable of delivering 4500 BWPD at 100 psig MAWP.

Since a relatively constant level is maintained in the Wemco it should not be expected to handle surges entering at a rate greater than 1700 B/D. The present water leg (6" diameter) and skimmer arrangement cannot retain the surge in the tank nor restrict the flow of water in the Wemco, as the skimmer contains no flow outlet regulating valve or pneumatic LLC device.

During adverse surge conditions, spill-over from the oil skim chamber of the Wemco flows into the oil holding tank and is pumped to the bad oil tank for recirculation to the fired production treaters.

To prevent overflow, the operator may manually by-pass the Wemco by throttling the block valve on the 6 inch by-pass line and, thus, divert a portion of the flow directly into injection pump suction tank.

Since it is undesirable to allow BS&W to enter the bad oil tank from the SWD system as well as to inject improperly treated water (in excess of 100 ppm oil content) into the disposal wells, modification to the existing system is warranted.[2]

[2]Information provided by SID office.

Figure 15-3 (continued).

A. Stouffer
Page 3
July 13, 1993

With the expected production forecast (See Table B) derived from well test data and information suppled by KWJ, the following calculations forecast the anticipated maximum daily average produced water yield for the existing and intended platforms in WC 66 Field (including WC 66A, B, C, & D platforms).

To use maximum case for expected water field

by 12/93	$\frac{3700\ \text{BOPD}}{2.3}$	=	1610 BWPD
Produced water off H.P. gas			1500 BWPD
	Total		3110 BWPD
by 6/94	$\frac{4200\ \text{BOPD}}{2.3}$	=	1825 NEPF
Produced water off H.P. gas			1500 BWPD
	Total		3325 BWPD

With these estimated yields of water production, sizing Mimmstown facility for ultimate maximum BWPD @ 3400 BWPD (max.) is adequate without placing water polishing equipment offshore. As indicated in Table B, maximum daily injections rate was 2700 bbl (3.79); the maximum average per day rate was 1875 BWPD (3/79).

Several methods were investigated to remedy the periodic surges that occur. For this investigation, three possible options may be available:

OPTION 1

Description of each option considered in solving the problem and the cost of employing each option

Modify the salt water disposal as shown in Figure 3 enabling the water clarification equipment to allow for surges, in addition to the maximum skim capacity of 1700 BWPD. Installation of water polishing equipment offshore in WC 66 Field should eliminate the need for any greater skim capacity than currently exists at Mimmstown.

This modification includes:

1. Modified skimmer tank:
   a. Add external cage
   b. Add sight glasses
   c. Add level trip for pumps

Figure 15-3 (continued).

A. Stouffer
Page 4
July 13, 1993

2. Dual pump package w/controls

3. Added piping, valves and fittings

Estimated cost of this installation is $18,500.

This option will allow the 1000 bbl tank to accept periodic surges by maintaining lower level in the tank (pumping down the level to retain surges totally in the tank and not causing spillover in the Wemco).

OPTION 2

Modify existing SWD as indicated in Figure 3 to allow for maximum average anticipated yield of 3400 BWPD to ensure adequate capacity of Mimmstown SWD facility in the event of complete upset of off-shore clarification equipment or changes in OCS requirements for overboard water disposal.

A synopsis of required items for this option includes:

1. Additional Wemco (1700 BWPD skim capacity)

2. Modified skimmer tank

   a. Add external cage
   b. Add sight glasses
   c. Add level-trip for pumps

3. Dual pump package w/control panel

4. Additional piping, valves, and fittings

The estimated cost of this equipment and installation is $48,650.

OPTION 3

Continue to operate with the existing system and continue to utilize bad oil system for surge situations, as the intended offshore installation of water clarification equipment will eliminate the need for additional SWD equipment.

Option 3 requires no additional equipment.

Figure 15-3 (continued).

Final argument for selecting Option 1

A. Stouffer
Page 5
July 13, 1993

FINAL OBSERVATION

Although installation of the recommended changes in the SWD system would require temporary shutdown and emptying of the SWD skimmer tank, the length of time required for welding on the tank itself would be minimal (estimate 2 days). The remainder of the pump, piping, and valves can be installed with SWD in service (estimate 2 weeks). In addition, installation of these SWD modifications should not interfere with the intended offset workover of SWD wells 1 and 2.

As noted in the rationale, the planned water polishing equipment in WC 66 Field will decrease the amounts of produced water arriving at Mimmstown. However, this modification would allow for capability of handling surges whenever such situations arose as previously explained and would solve for present needs and long term operating problems.

Prepared by: R. Ramirez
R. Ramirez

Approved: A. Stouffer
A. Stouffer

Figure 15-3 (continued).

knows that his technical language and acronyms will be understood by both.

The visual design of the report allows readers to find the sections they want: They can read only the conclusions and recommendations, or they can read the analysis (or parts of it) that leads to the detailed statement of the recommendations.

Feasibility Studies

Another major type of analytical report is the feasibility study, which is launched to determine whether an organization should take an action it is considering taking. The central question involves doing something or not doing it or determining which option among several options is the best choice. The feasibility study is similar to the recommendation report, but the feasibility study is usually much more involved. A number of issues need to be presented and addressed, factors that usually produce a rather long, complex report. The following situations call for a feasibility study.

- A university research group wants to know whether fire ants can be effectively controlled in open fields of 100 acres or less by using chemicals such as x, y, and z.
- The state legislature wants to know whether partially treated sewage can be deposited on land in a semi-arid part of the state to improve the soil without leaking harmful bacteria into the water table.
- A company wants to know whether land it currently owns near a major interstate highway should be developed into a shopping mall, a business park, or some combination of both.
- A major city wants to know whether building an additional airport on property it owns near a major waterway will cause flooding in residential areas downstream from the proposed airport.
- A university writing program wants to know whether personality measurement tests might be useful to writing teachers in helping them ascertain their own and their students' personality characteristics.

Usually, the feasibility study helps organizations to make decisions based on the value of a solution as well as its cost. The feasibility report presents the written analysis and the resulting decision. Like other analytical reports, the feasibility study and the resulting report will do the following:

- Set the purpose and scope of the study
- Gather and check information
- Analyze information and data
- Reach conclusions
- Arrive at a decision

As in all composition, a good deal of back and forth movement occurs in the investigative process.

For any report, information or analytical, you need to know how to gather, check, and study your information; how to develop a discussion, conclusions, and recommendations; and how to use principles of document design to present your study and its results. Because formulating purpose and scope are so critical to successful analytical reports, particularly feasibility studies, we discuss that at some length here.

Purpose

Just as the introduction to any report always contains a statement of purpose, the feasibility or analytical study should begin with a clear purpose, which will become part of the statement of purpose for the report. Because the feasibility study usually answers a question, the following purpose statement shows the direction the study will take:

> This study decides whether the establishment of an E-mail system among all departments on campus is feasible given the computer facilities available and current funding provided by the State for network expansion.

Although the need to articulate a clear purpose statement seems obvious, its importance to the success of the feasibility study cannot be underestimated. If investigators, as well as writers, do not have a clear direction for their study, the resulting information they gather will be incoherent.

Here are some additional examples of purpose statements:

- The purpose of this study is to determine the feasibility of using particle board to sheathe the interior of houseboats.
- The purpose of this investigation is to determine whether instructional materials are available for teaching school children about the effects of toxic pollution on water quality.
- Our primary objective is to decide what computing options will provide the best design programs for landscape architecture students, given the limited financial resources of the department.

Scope

Once the purpose of the feasibility study has been clearly decided and stated, the methods of accomplishing the purposes must be determined. These methods form the scope of the study, the actions to be taken, the range of data to be gathered, the bounds to be set for studying the problem, and the criteria against which possible solutions will be measured. For example, a company that is deciding whether to establish a new branch in New Town will need to answer the following questions:

Purpose

Should X Company establish a branch plant in New Town?

Scope

Does X Company now have, or can it anticipate, enough profit in New Town to justify a plant there?

Does New Town offer adequate physical facilities, utilities, transportation, and communication?

What is the price of land? Is existing office space available that would be suitable?

What is the tax structure in New Town? Are the tax code and building regulations favorable for a plant like X?

Can the required staff be obtained, whether by local hiring or moving personnel into the area?

What is the general quality of life in New Town?

What effect would opening a branch plant in New Town have upon overall company organization, operations, policy, and financial conditions?

From such an initial list of questions, the investigation can proceed, although each question will probably be broadened and rephrased as research warrants. Devising a scope statement that guides the investigation is crucial, but as an investigator you should not remain blindly committed to your initial scope statement. You should reexamine it from time to time in light of the information you gather. Look for holes, overlaps, irrelevant questions. Often, a person unacquainted with the study is in a far better position than you to spot shortcomings and illogical assumptions in the scope statement. Therefore, someone outside the study should be asked to review and react to the list of scope items you compile.

Once you have carefully researched and answered your questions stated in your scope statement, you will need to plan the design of your report. Your written feasibility report may include all or some of the following elements, discussed in Chapter 10:

- Letter of transmittal or preface
- Title page
- Table of contents
- List of illustrations
- Glossary of terms
- Executive summary
- Introduction
- Discussion
- Factual summary
- Conclusions
- Recommendations
- Appendixes
- References

How many of these elements you include will depend upon audience factors and the length and complexity of the report. For example, a long report aimed at a narrow audience of several people should have a letter of transmittal. A long report for a more general audience would have a preface instead. A short feasibility report of only several pages may be written as a memorandum or letter and essentially consist of only an executive summary, discussion, conclusions, and recommendations.

For our discussion of feasibility reports within the context of report development in general, we want to focus on the development of the key internal elements: the introduction, discussion, factual summary, conclusions, and recommendations—common elements in any feasibility report.

Introduction

Although we have already discussed the components of the introduction, in a feasibility study you may want to include several types of information in addition to the standard elements used in introductions:

- Subject, purpose, and scope
- Reasons for conducting the study
- Identification and characteristics of the person or company performing the study (if not given in a preface or letter of transmittal)
- Definition and historical background of the problem studied
- Any limitations imposed upon the study
- Procedures and methods employed in the study

- Acknowledgements to those who were instrumental in preparing the study (if not given in the transmittal letter or the preface)
- Topics that will be covered in the discussion

Discussion

The discussion of any report can be the most difficult task for any writer. How should the research be organized for effective presentation?

Using the feasibility report as an example, consider your purpose statement and your scope, both included in the introduction. The report will, of course, begin with a summary and an introduction. Then, the scope statement presented earlier suggests the main divisions in the analysis and then the report itself. To illustrate how the segments of the scope statement can lead to the plan of the discussion, let's return to the feasibility example we studied earlier.

The answer to the purpose statement—"Should X Company establish a branch plant in New Town?"—was deemed to depend on the answers to seven questions that comprised the scope statement. Taking each of these questions and the answers to each as the bulk of the discussion, we would have the following plan for the report:

Executive Summary: Should X Company establish a branch plant in New Town? Why or why not?

Introduction

Subject, purpose, and scope of the study
Reasons for conducting the study
Procedures used in conducting the study

Analysis of Factors Determining the Establishment of a Branch Plant in New Town

Estimated Profitability of a New Branch in New Town

Evaluation of Facilities

Existing office space
Utilities
Transportation
Communication
Land prices and availability
Local construction prices

Business Climate

Tax structure
Building codes
Business regulation
Economic health

Personnel

Local labor market
Personnel available for transfer

Quality of Life in New Town

Schools
Cost of living

Effect on X Company of Establishing a Branch Plant in New Town

Existing organization
Existing operations
Company policy
Financial resources

Factual Summary of Research

Conclusions

Recommendations

References
Appendices

Note that each of the main issues covered in the scope statement becomes a main segment of the discussion, which works to answer the controlling question.

Once the analysis is complete, the writer can begin to write the conclusion: Should X Company establish a branch plant in New Town? Why or why not? Based on the results of the research, the writer can make appropriate recommendations. For example, the analysis might suggest that a branch plant would be profitable but that X Company should wait until a new industrial park is completed in 1999 because leased space can be obtained at attractive preconstruction prices. Or, the analysis might suggest that establishing a new branch is feasible only if the company can negotiate tax abatements. Without these, initial profitability could be undercut so much that the company's financial structure could be weakened. Or, lack of adequate transportation facilities might make the establishment of a branch plant totally infeasible.

Factual Summary of Research

The term *factual summary* distinguishes this type of summary from those that contain conclusions and recommendations as well as essential facts. Creating a good factual summary is one of the hardest challenges you face as a writer. In meeting that challenge, keep these principles in mind:

- You can summarize only the information appearing in the discussion of the report. A factual summary must not introduce new information.

- Every statement in a factual summary should be an assertion of facts presented. Opinions should be omitted.
- Facts, as they appear in various sections of the report, should be collected for the reader.
- The facts should become a springboard for the conclusions and recommendations that will follow.

Conclusions

The conclusions section acts as the intermediate step between the facts in the factual summary and the recommendations. Conclusions are inferences and implications you draw from your data.

Conclusions can be presented in normal essay style or in a series of short, separate statements. However you present your conclusions, you should arrange them in a clear order that reveals your thinking process to the reader. If your readers can read your introduction, factual summary, conclusions, and recommendations and believe that you have justified your recommendations, then you have probably presented your study well. Your readers should not read your conclusions and wonder how they relate to your factual summary and recommendations: They should see the connections immediately.

Recommendations

In contrast to conclusions, a recommendation is an action statement. That is, it recommends that the report users take some proposed action or refrain from taking it. Always be sure that the first (or only) recommendation discharges the purpose set forth early in the report.

If the purpose is to determine whether X Company should establish a branch plant in New Town, then the recommendations should address this purpose:

> X Company should establish a branch office in New Town.

or

> X Company should delay establishing a branch plant in New Town until adequate tax abatements are offered by the City Council.

or

> X Company should not establish a branch plant in New Town because an adequate work force will not be available to staff a new office.

Further recommendations expand on the initial one. If the first recommendation favors establishment of a branch plant, subsequent recommendations may detail the necessary steps to carry out the plan. If the first recommendation opposed the establishment of a branch, the reasons for the negative recommendation should be succinctly restated. Subsequent recommendations may propose alternative plans.

References will include sources of the study. These sources may include documentation of interviews, state reports, census statistics, journal articles, even unpublished studies—whatever sources of information you used to gather information under each scope statement.

The **appendix** may include items such as drawings of proposed buildings, examples of building codes that would affect construction decisions, Chamber of Commerce reports on potential growth of the community, statement of tax structures secured from the city attorney and the local tax office—data used to support analysis and the resulting conclusions.

Figure 15-4 exemplifies a feasibility study. Examine the situation and the resulting report to see how each report element is used to determine the feasibility of the proposed personality test.

Situation 4 (Figure 15-4)

Nadine Miller is a graduate student who serves as a tutor in the university's writing laboratory. The writing lab serves students who are enrolled in courses in the university's writing program. In her research in theory of writing instruction, Nadine discovers that the Myers–Briggs Type Indicator (MBTI) is an effective way of helping teachers understand their own and their students' cognitive preferences. Nadine's initial goal was to study the MBTI as part of her research in educational psychology. However, her initial study led her to question how it might be used in teaching writing to help teachers understand how their perceptions differ from those of their students and how these differences might influence the presentation of writing instruction. Her initial purpose statement in studying the feasibility of using the MBTI test in teaching composition was as follows:

Purpose

Can the MBTI help teachers understand their own personality types and the personalities of their students to improve writing instruction?

Scope

What is the MBTI?
How does it work?
How has it been used?
How could it be used in teaching writing?
What research could be done in applying MBTI to writing instruction?
How much would implementing MBTI cost?
How could we implement it in The Writing Program?

Seminars for Teaching Technical Writing, Inc.

Myers–Briggs Type Indicator

A Feasibility Study for Writing Program Courses

Nadine Miller

Figure 15–4 Feasibility Report, Situation 4
Reprinted with permission of Nadine Miller.

Table of Contents

Figure 15-4 (continued).

Executive Summary

In Brief

The Writing Program course instructors have need of every possible advantage when it comes to teaching. This feasibility study examines the possibility of implementing the Myers–Briggs Type Indicator (MBTI) assessment as a way of providing information to teachers on the cognitive preferences of their students, as well as themselves. The study analyzes the practical benefits of such a program and weighs them against the potential cost of implementation. According to the most recent price information concerning the MBTI assessment materials, initial start-up costs would be approximately $2249, and following this, the only maintenance costs would be purchasing scoring forms, or approximately $1449 per semester. Alternatively, a personality typing instrument similar to the MBTI, and in the public domain is that published by Keirsey and Bates. Using such an instrument, with in-house development and duplication of forms, start-up cost would be approximately $559, with following semester cost of $279. Training costs in either case would be negligible, as it could be accomplished in a short workshop, such as are currently conducted at the beginning of each semester for GATs. Accordingly, I recommend that we implement a program using the Keirsey–Bates set of public domain questions.

Myers–Briggs Type Indicator **Executive Summary • ii**

Figure 15-4 (continued).

Introduction

Myers–Briggs Type Indicator

The Myers–Briggs Type Indicator (MBTI) is a personality assessment based upon Jungian personality type principles. These principles are useful in identifying the cognitive approaches and methods that people use when dealing with their environment and when solving problems; such information may be useful to teachers of composition. As my final project for the graduate seminar in teaching technical writing, I have studied the theory behind MBTI and research which has used the MBTI and evaluated the feasibility of using the instrument in Writing Program courses.

My study (1) examined the practical use of the MBTI for teachers, particularly of composition and technical writing; (2) examined the cost that would be incurred in using the assessment; (3) determined the possibility of useful research information resulting from use of the MBTI in Writing Program courses; and (4) on the basis of cost to practical value, determined the feasibility of the project.

This report summarizes the results of the study, draws conclusions, and recommends actions to be taken. The central portion of the report provides a detailed discussion of the results of the study.

Myers–Briggs Type Indicator **Introduction • 1**

Figure 15-4 (continued).

Overview: Myers–Briggs Type Indicator

Background of MBTI

The Myers–Briggs Type Indicator is a self-reporting instrument for the identification of personality type. Its personality type system is based on the personality types of Carl Jung, founder of analytical psychology. Jung postulated that individuals gravitate toward certain cognitive preferences. These cognitive preferences do not determine behavior, but rather, explain it. Thus, personality typing is a method by which individuals can understand their own cognitive styles, as well as become more aware of others' styles. Understanding styles helps individuals to recognize the differences in approach people have when dealing with the outer world, thinking processes, judgmental processes, and perceptual processes. Jung's, and by extension, Myers and Briggs's theory, in no way places value on any preference. The purpose of assessing personality and getting to know differing types is to learn to see the value in others' ways of thinking. By understanding others' processes, an individual is capable of using his or her own talents in combination with others' to be more effective. Additionally, understanding different personality types, as well as our own, allows us to realize what areas of our own cognitive processes need development. It also shows how we may favor our own processes so much as to be biased toward other processes—it points out our "blind spots."

Myers–Briggs Type Indicator **Overview: Myers–Briggs Type Indicator • 2**

Figure 15-4 (continued).

Personality Type Areas

The MBTI breaks down personality into four single areas, each having two possible preferences. When combined, the four cognitive preferences interact, resulting in sixteen possible combinations. The dichotomies are: Introverted/Extraverted[1] (I/E); Sensing/Intuitive (S/N); Feeling/Thinking (F/T); and Perceiving/Judging (P/J). Each of these categories refers to a particular area of cognitive processing.

Introverted/Extraverted (I/E)

First, consider the most obvious of the four areas: introverted versus extraverted. Jensen and DiTiberio refer to the introversion/extraversion dichotomy as the "basic attitude towards life" (8). Introverted personalities tend to reflect more and speak less unless prompted. Introverted personalities are also inward-directed. They look to themselves to find confirmation of their beliefs and feelings. Extraverts, on the other hand, are more out-going. Rather than reflecting about things at length, they talk about their concerns in life. They tend to be more physically active, generally, using body motion to aid them in their cognitive processes. When seeking to confirm their beliefs and feelings, they look to the outer world. Extraverts are not necessarily people oriented, either. They may prefer to interact more with the "things" of their world. Introverts may also be at ease in social settings, particularly when surrounded by a few close friends, but generally they have less endurance when dealing with large groups (Jensen and DiTiberio 9)[2].

In terms of the classroom experience, extraverts need activity. Their cognitive strategies are oriented towards talking and group work, with little planning, relying on trial and error systems (*PTC* 9). Introverts, on the other hand, work best when allowed their solitude, and the teacher may need to draw them out to get them to participate in classroom discussions (*PTC* 10). Introverts usually plan extensively and anticipate problems.

Sensing/Intuitive (S/N)

The sensing/intuitive dichotomy refers to the way individuals perceive the world. Sensing personality types rely on sensory data for their information about the way the world works, concentrating on the verifiable. They develop abstract ideas only after analyzing the details of experience (*PTC* 11). Intuitive types rely on the opposite set of ideals. They note the interrelationships of things and ideas, the gestalt, first, and note the details later (*PTC* 11). Intuitive types tend to take hunches seriously, and use them effectively.

In the classroom, sensing types like to learn concrete, practical skills which have applications. They want clear and precise directions from their teachers (*PTC* 11). The sensing philosophy of learning is "to learn a skill or procedure, perfect it,

[1]The spelling "extravert" is the preferred spelling of Jungian psychoanalysts.

[2]Hereafter referred to as *PTC*.

Myers–Briggs Type Indicator **Overview: Myers–Briggs Type Indicator • 3**

Figure 15-4 (continued).

and then practice it without much variation" (*PTC* 11). Intuitive types resist routine and mechanized learning processes. Their impressionistic approach leads them to also prefer open-ended assignments.

Additionally, note that "[c]ollege level English instructors tend to be intuitive types in an intuitive world" (*PTC* 11). In a small informal survey at the 1983 and 1984 Conference on College Composition and Communication, Jensen and DiTiberio found that about 90% of college composition teachers were intuitive types. This is significantly different from the relatively even distribution among high school teachers as found by Hoffman and Betkouski. Though the number of intuitive students in college is generally higher than the number of sensing students (McCaulley and Natter) Myers has suggested that it is because of teaching methodology that sensing students have a high drop-out rate and have more trouble in school. These findings suggest that in this area particularly, teachers need to be more aware of their own biases in evaluating student responses, and that teachers should develop the less preferred side of typology so that they can be more flexible in the classroom environment and in dealing with individual students.

Feeling/Thinking (F/T)

This aspect of the MBTI concerns the judgment process of a person. Thinking types follow a logical decision process, and attempt to maintain an impersonal, impartial position. Feeling types follow a process of valuing to make decisions, basing decisions on their personal, subjective value system (Lawrence 8). Thinking types use an ethical system as well, but they strive to determine criteria and principles which transcend people and the variabilities of life (*PTC* 13). The goal of thinking types is to apply these criteria consistently to all people; though they sometimes slight the "people issues," thinking types are usually consistent and truthful. Feeling types, on the other hand, "seek to clarify their own personal values" (*PTC* 13), to figure out what is most important to them. They are generally very concerned about how their decisions will affect others, and are interested in group harmony.

In the classroom, thinking types do better work when given clearly explained performance criteria. They also need a logical set of explanations which describe the value of the information they are working on; they want to know how and why what is being taught is useful to understanding the systematic way the world works (*PTC* 14). Feeling types, on the other hand, want to know the human relations value of what is being taught (*PTC* 14). They need personal motivation to do their best work, in the sense that they need to have their hearts in it to be most effective.

The communication style of thinking types and feeling types is dramatically different. Thinking types often explain their decisions by naming "reasons" and using cues which indicate "syllogistic thought: thus, therefore, in conclusion" (*PTC* 14). Kinneavy refers to this type of discourse as "thing-centered" (88). The communication of feeling types is filled with expressive cues such as "I feel," or "I believe" (*PTC* 14). Their communication is more "people-centered" (Kinneavy 88), and they may soften or adjust their message to fit their audience.

Myers–Briggs Type Indicator **Overview: Myers–Briggs Type Indicator • 4**

Figure 15-4 (continued).

Perceiving/Judging (P/J)

Perceiving and judging refer to an individual's attitude toward the outer world. Judging types "tend to adopt a position and adhere to it" while perceiving types "delay closure in order to explore a wide range of options" (*PTC* I5). Judging types like to plan, manage, and organize things that have to do with their dealings with the outer world (as opposed to their internal thought activities); structure and getting things done are their watchwords. Perceiving types resist structure, preferring to be adaptive, flexible, spontaneous, and inquisitive. They seek to engage the world in its "free, unaffected state" (*PTC* 15). Judging types usually come to decisions expediently, while perceiving types defer conclusions until all possibilities are explored. While judging types are usually productive, they can rush to decisions, or become rigid and controlling as deadlines approach. Perceiving types are usually very thorough, but can become bogged down, being indecisive and procrastinating.

In the classroom, judging types tend to be very organized, focussing on projects and meeting deadlines. Teachers often perceive them as overachievers, thinking them more motivated and organized than perceiving types (*PTC* 15). The tendency for perceiving types to spread out their focus on a multitude of projects, to overcommit themselves, and to wait until the last minute to make decisions leads teachers to believe that they are unmotivated (*PTC* 15). Teachers need to recognize that both methods can work to produce good results, as perceiving types often have more thorough research or make more thorough conceptualizations of projects.

Type Distribution

The distribution of the type dichotomies in the general population[3] of adults in the United States according to Jefferies (48-9) is as follows:

Extravert (E) 70%	Introvert (I) 30%
Sensory (S) 70%	Intuitive (N) 30%
Thinking Judger (TJ) 50%	Feeling Judger (FJ) 50%
Judger (J) 55%	Perceiver (P) 45%

Although it appears that TJs and FJs are equally split in the population, this trait is the only one that shows a male–female bias. Of the 60% reporting F, about 66% are female, and of the 60% reporting T, about 66% are male. This gender bias has stayed constant over the course of many years of testing.

[3]There is a caveat to these data, as the MBTI has been given mostly in situations where the majority of people are white and middle-class, and in secondary school, post-secondary school, or business situations. No demographic data (other than gender) are collected on the MBTI questionnaire.

Myers-Briggs Type Indicator **Overview: Myers–Briggs Type Indicator • 5**

Figure 15-4 (continued).

Keirsey estimates that the sixteen types break down into the following percentages:

ISTJ, ISFJ, ISTP, ISFP	6% each
INTJ, INFJ, INTP, INFP	1% each
ESTP, ESFP, ESTJ, ESFJ	13% each
ENFP, ENTP, ENFJ, ENTJ	5% each

From these two sets of figures it is clear that the previously noted percentage of 90% intuitives teaching college composition (*PTC* 165) is dramatically different from the population at large, where approximately 30% are Ns. More dramatically, the Jensen and DiTiberio survey showed that 55% were INs, which should account for only 4% of the population at large (167).

Generalized Descriptions of the Sixteen Types

The descriptions in the table on the following page are adapted from a table by Margaret K. Morgan in Lawrence (52-3), and are specifically oriented towards instructional strategies at primary and secondary school levels. Nonetheless, they are useful for generalizing about student behavior and personality type.

Figure 15-4 (continued).

ISTJ	ISFJ	INFJ	INTJ
• Linear learner with strong need for order (SJ) • Likes direct experience (S) • Likes audiovisuals (S) • Likes lectures (I) • Enjoys working alone (I) • Likes well-defined goals (ST) • Needs predictability (ISJ) • Prefers practical tests (S)	• Linear learner with strong need for order (SJ) • Likes direct experience (S) • Likes listening to lectures (I) • Likes audiovisuals (S) • Enjoys working alone (I) • Likes practical tests (S) • Needs predictability (ISJ)	• Can be global or linear (NJ) • Wants to consider theory first, then applications (N) • Enjoys working alone (I) • Prefers open-end instruction (N) • Needs **harmony** in group work (F)	• Can be global or linear (NJ) • Wants to consider theory first, then applications (N) • Enjoys working alone (I) • Prefers open-end instruction (N) • Good at paper-and-pencil tests (NT)
ISTP	**ISFP**	**INFP**	**INTP**
• Linear learner, but needs help in organizing (SP) • Likes direct experience (S) • Likes lectures (S) • Likes audiovisuals (S) • Enjoys working alone (I) • Wants logically structured, efficient materials (IT)	• Linear learner, but needs help in organizing (SP) • Likes direct experience (S) • Needs well-defined goals (S) • Needs **harmony** in group projects (F) • Likes audiovisuals (S) • Likes practical tests (S) • Enjoys working alone (I) • Needs sensitive instructor (IF)	• Global learner but may need help organizing (NP) • Likes reading (N) • Likes listening (N) • Wants to consider theory first, then applications (N) • Prefers open-end instruction (N) • Enjoys working alone (I) • Likes autonomy (NP)	• Global learner but needs help coming to closure (NP) • Likes reading (N) • Likes listening (N) • Wants to consider theory first, then applications (N) • Good at paper-and-pencil tests (NT) • Prefers open-end instruction (N) • Enjoys working alone (I) • Likes autonomy (NP)
ESTP	**ESFP**	**ENFP**	**ENTP**
• Linear learner, but needs help in organizing (SP) • Needs to know **why** before doing something (S) • Likes direct experience (S) • Likes group projects, class reports (E) • Likes team competitions (E) • Likes audiovisuals (S) • Likes practical tests (S) • May like lecture (T)	• Linear learner, but needs help in organizing (SP) • Likes direct experience (S) • Needs to know **why** before doing something (S) • Likes group projects, class reports (E) • Likes team competitions (E) • Likes audiovisuals (S) • Needs orderly, well-defined goals (S) • Likes to help others (EFP) • May need personal encouragement (ESF)	• Global learner, but needs choices and deadlines (NP) • Likes seminars (EN) • Likes reading if can settle down long enough (ENF) • Likes **harmonious** group projects (EF) • Likes team competition (E) • Likes class reports (E) • Likes autonomy (NP) • Needs help with organizing (NP)	• Global learner, but needs choices and deadlines (NP) • Likes seminars (EN) • Likes reading (NT) • Likes listening (N) • Wants to consider theory, then applications (N) • Good at paper-and-pencil tests (NT) • Prefers open-end instruction (N)
ESTJ	**ESFJ**	**ENFJ**	**ENTJ**
• Linear learner with strong need for structure (SJ) • Needs to know **why** before doing something (S) • Likes direct experience (S) • Likes group projects, class reports (E) • Likes team competition (E) • Likes audiovisuals (S) • Likes practical tests (S) • May like lecture (T)	• Linear learner with strong need for structure (SJ) • Needs to know **why** before doing something (S) • Needs well-defined goals • Values **harmonious** group projects (EF) • Likes class reports, team competition (E) • Likes audiovisuals (S) • Likes practical tests (S) • Likes direct experience (S)	• Can be global or linear learner (NJ) • Likes seminars (EN) • Likes reading if can settle down long enough (ENF) • Values **harmonious** group projects (EF) • Likes class reports, team competition (E) • Likes listening (N) • Prefers open-end instruction (N) • Wants to consider theory, then applications	• Can be global or linear learner (NJ) • Likes seminars (EN) • Likes reading (NT) • Likes group projects (E) • Likes class reports, team competition (E) • Likes listening (N) • Likes paper-and-pencil tests (NT) • Prefers open-end instruction (N) • Wants to consider theory, then applications (N)

Myers–Briggs Type Indicator   Overview: Myers–Briggs Type Indicator • 7

Figure 15-4 (continued).

Usefulness of MBTI in the Classroom

Applications

As has been noted under each of four dichotomies, and in the table on page 7, the implications of personality types in the classroom are multi-faceted. Students' as well as teachers' personality types can manifest themselves in a variety of ways, ranging from the need for structured learning environments, to the need for personal attention. Sometimes, simply changing word choice in communication, or tone, can aid students in their perceptions of assignments (and teachers).

Additionally, at the college level, it is clear from the data noted in the overview of MBTI section that the majority of college composition instructors are of a personality type that is in the minority: only 30% of the general population are Ns, but up to 90% of college composition instructors may be Ns (Jefferies 48, *PTC* 165). Instructors need to be aware of their own type so that they can avoid a biased learning situation which might hamper students of differing personality types. The structured environment of the school system itself reflects this bias: favoring those students who can work quietly and alone; stressing abstract systems of communication (reading and writing) rather than hands-on systems (talking); teaching the abstractions first (math rules and concepts) and the applications later, perhaps with some chance for the students to use the rules (Lawrence 40–41).

Clearly, teachers cannot make the classroom a place where every student gets all instruction directed specifically at his or her cognitive approach. However, by balancing teaching strategies, teachers can develop students' less preferred cognitive processes, while giving them a chance to excel at their preferred strategy as well. Lawrence, and to some extent Jensen and DiTiberio as well, advocate active MBTI use in the classroom. Lawrence especially advocates teaching

Figure 15-4 (continued).

students about MBTI, for he feels that it can only enhance their understanding of their own cognitive methods, as well as showing them how they can work with others to get the best of alternative cognitive processes.

Jensen and DiTiberio extensively discuss strategies that instructors can use to coach their student writers. They boil their experience down into a set of descriptions of the differing strengths and weaknesses of writers and coaching strategies that teacher can employ. An adaption of these tables appears in Appendix A.

Extant Research

Current research in personality typing and education covers a wide range of topics. Schurr and fellow authors examine the interaction between personality type and instruction from a statistical standpoint to attempt to determine if personality type affects the accuracy of grades in an introductory writing course. Baden has discussed the teaching of pre-writing strategies that address the needs of feeling students as well as thinking students. Ching examines the type of college faculty members and the effects that it has on their writing. Hoffman and Betkouski survey the extant research on personality type and education in their 1981 article, "A Summary of Myers–Briggs Type Indicator Research Applications in Education," as do Provost and Anchors in their 1987 book *Applications of the Myers–Briggs Type Indicator in Higher Education*.

Several articles concerning the usefulness of type to composition education are of particular importance. Walters examines the personality types of students, instructors, and the resulting grades. In her research, which covered English, journalism, and business communication courses, she found that students who were similar in type to their instructors, and also closest to the modal type of their discipline, received the highest grades. In *PTC* Jensen and DiTiberio mention research by Gowen on the effect of rater's personality type and level of experience on their scoring of student essays (140). Her research indicated that experience level of the rater was not important to the consistency of rating, but that personality type was (*PTC* 140). These two researchers' work indicates that instructors need to clarify their reasoning in evaluating students' work, as well as re-evaluating researcher's biases. The findings on personality type biases are reminiscent of Gilligan's criticism of male theorists in their rating of female developmental traits as consistently lower, or less mature, than male developmental traits (33).

Possible Research Topics

The possible research topics available for use with the MBTI are at least as varied as the types themselves. Jensen and DiTiberio have made extensive use of case study examples for coaching students with writing problems. Case studies of coaching using personality types in other fields would be equally useful. Also, most of Jensen and DiTiberio's examples are taken from the ranks of basic writers that they have coached. Case studies concerning more advanced composition

Figure 15-4 (continued).

students, technical writers, and fiction writers would be useful. Equally useful would be further studies on the success of students in given courses, and the influence that their personality types have on their grades. Walter's work begins in this area, but more research would add weight to her findings concerning instructors' biases. A potential research topic which would be particularly interesting in this area would be an analysis of a wide population of students in business communication classes, as compared to the same set of instructors' evaluation of experienced business writers.

One of the difficulties of any examination of personality type is the almost impossible task of getting a balanced sample in the population. Because of the imbalance of types in the population at large, and the compounding fact that every vocation, college major, and elective class in college has certain personality types which gravitate toward them, it is difficult to report accurate assessments of what constitutes evaluative bias. Distinguishing between bias on the part of the instructor being examined, or the researcher, and what bias is a result of the skew in the population being considered becomes troublesome. Research attempting to overcome this difficulty would be very useful to the research pool on MBTI.

Myers–Briggs Type Indicator **Usefulness of MBTI in the Classroom • 10**

Figure 15-4 (continued).

Costs of Implementation

MBTI Materials

There are two possible ways of going about MBTI assessment. Consulting Psychologists Press makes the instrument available for a reasonable price in small quantities. For example in the table below, it is clear that on a small scale, administering the MBTI is quite feasible. However, on a larger scale things become more problematical.

Estimated Costs for full MBTI

MBTI Assessment Materials Purchased from CPP	Cost/No. of Items*	Total Cost/No. of Items†
Question Booklets	$10 / 25	$400 / 1000
Response Sheets	$7 / 25	$1449 / 5175
Manual	$20 / 1	$200 / 10
Total for Written Materials		$2049
Optional Software (MS-DOS)	$200 / 1	$200
Total with Software		$2249

*Prices are approximate as I was unable to contact Consulting Psychologists Press for more recent pricing information.

† Based upon section estimates discussed in the text below.

Obviously, the $600 for question booklets and the manuals is a one-time cost, as would be the optional $200 software. However, the semesterly output of $1449 for response sheets is not feasible in the current fiscal environment. One possible

Figure 15-4 (continued).

option would be the development of our own response sheet, using desktop publishing software available in the Writing Center. Such a step could significantly reduce costs. Costs could be reduced by approximately 80% to about $290. However, it is unknown at this time if implementing such an answer sheet would be in violation of CPP's copyright.

The number of question booklets needed is based upon rotating use of booklets as there are never more than twelve sections of Writing Program classes in session at the same time. The figure of one thousand booklets would allow for two sets of twelve sections (with forty students per section) to be administered. The total number of answer sheets, 5,175, is based upon the total number of sections of English 104, 210, and 301 from the fall of 1992, with forty students per section. There were 72 sections of English 104, 20 sections of English 210, and 37 sections of English 301.

Alternative Materials

Another possibility would be to use a similar instrument to the MBTI which is in the public domain. Keirsey and Bates, in *Please Understand Me* (1978), a book which examines their personal theories concerning type, have a self-reporting inventory similar to the Myers–Briggs, which they have opened up for use in any non-profit situation, given that they are credited. While this assessment has not undergone the rigorous testing that the MBTI has, it could serve as an effectual way to implement personality type concepts without a large outlay of funds. In this case, the resultant data could not be used for publishable research, but would still serve to aid teachers in their relations with students. It would still be advisable to purchase at least 3–5 Myers–Briggs manuals for teacher reference purposes, and perhaps a few question booklets for teacher use, so that teachers could compare the Keirsey–Bates to the full MBTI. In this case, total cost for implementation would be more manageable.

Estimated Costs for Alternative

Keirsey–Bates Materials	Cost*	Total Cost/No. of Items†
Question Forms printed front and back, 2 pages (one-time cost)	$00.20 / 2pp	$200.00 / 1000
Response Sheets	$00.05 / pg	$258.75 / 5175
MBTI Manual from CPP	$20.00 / 1	$60.00 / 3
Training Handouts for Teachers printed front and back, 4 pages	$00.40 / 4 pp	$40.00 / 100
Total		$558.75

*Based on common reproduction costs.

† Based on section estimates discussed in conjunction with full MBTI.

Myers–Briggs Type Indicator **Costs of Implementation • 12**

Figure 15-4 (continued).

These costs are much more compatible with the current state of the department's budget. Question forms would be a one-time cost, as would be the MBTI manuals for reference. Training handouts would probably cost only 1/2 of the first semester cost in following semesters, as it would only be necessary to provide handouts to new teachers. Based on these factors, second and following semester costs would be approximately $279, or about one-half the cost of the first semester's start-up costs. The full cost for the first year (2 full semesters, 2 summer sessions) would be approximately $978. This estimate is based on the stated start-up semester cost of $558.75, second semester cost of $279, and summer sessions costs of $140 (one-half the number of students in a regular semester). Second year costs would be approximately $700.

I would develop the question and response forms, as well as the teacher handout, at no cost to the English department.

Training Costs

Regardless of which test system is chosen, the training costs (outside of materials) for implementing such a program are negligible. The advantage of both the MBTI and the Keirsey–Bates system is their ease of use. In both cases, the basic concepts of typology and testing methods can be taught in a very short period of time. A three hour workshop would amply cover the topics necessary. Such a workshop could be integrated into the beginning of the semester workshops for new teachers. Interested graduate students or faculty members could present the workshop. Having a colleague of the teachers present the workshop would have the added advantage of giving credibility to the usefulness of personality typing concepts. The one potential problem with this system is assuring that there will be a graduate student or faculty member motivated and knowledgable enough to present the workshop. Solving this problem may be done by making sure that graduate students and faculty members in the Writing Program get enthusiastic support for understanding and using personality typing. By getting teachers involved in active and innovated problem-solving, they will be motivated to learn more and willing to pass their knowledge on to others.

Myers–Briggs Type Indicator **Costs of Implementation • 13**

Figure 15-4 (continued).

Conclusions and Recommendations

Factual Summary

The value of the MBTI in the workplace has been acknowledged in the number of seminars given to businesses annually on the subject. Similarly, Jensen and DiTiberio have advocated the usefulness of MBTI as a constructive aid for understanding the cognitive processes of students of writing, giving examples of coaching and intervention strategies for students. The Myers–Briggs typing instrument is uncomplicated, its use can easily be taught, and it has been widely tested under a variety of circumstances. In 1982, Lawrence reported figures from the Center for Applications of Psychological Type, indicating over ten thousand teachers and administrators from all levels had taken the assessment (21).

The MBTI instrument and materials necessary for administration are available from Consulting Psychologists Press. Based upon 1987 prices, the cost of using the full MBTI instrument would be approximately $2249 in the start-up semester, and about $1449 in the semesters thereafter. Alternatively, an instrument in the public domain, developed by Keirsey and Bates, could be used by the Writing Program at far less expense. First semester costs would be approximately $559, with following semesters costing about $279. This instrument would have more in-house development associated with it, as the instrument and associated materials would have to be assembled. I am willing to develop the materials associated with the use of the Keirsey–Bates question set at no cost to the department.

Training could be done at negligible cost via pre-semester workshops such as those currently used for newly assigned GATs. A three hour workshop would suffice for introducing the teachers to the personality typing concepts and the administration of the question set. Such workshops would only require a volunteer graduate or faculty member to present the materials.

Myers–Briggs Type Indicator **Conclusions and Recommendations • 14**

Figure 15-4 (continued).

Problems associated with the implementation of such a program lie mainly in the area of training. Assuring an enthusiastic workshop leader would be the most difficult portion. The most viable solution to this problem is presenting the personality typing system in a way which clearly shows its value, and by enthusiastic support of personality typing as a method of active and innovative problem solving by the workshop instructor and other permanent Writing Program staff.

Conclusions

Several conclusions come out of this feasibility study.

- Teachers can utilize personality typing to better understand their own cognitive approaches to writing and teaching.
- Personality typing is useful for coaching basic writing students such as those students found in English 104.
- Personality typing is also useful for coaching advanced writing students, particularly those who are having difficulties.
- Many research opportunities exist for using personality typing for the analysis of writing, cognitive strategies, and teaching strategies.
- Question forms for 5,175 students per regular semester would cost $1449. Implementing the full MBTI instrument is therefore cost prohibitive for all Writing Program classes due to the number of students taking these classes.
- Implementation of a similar personality typing instrument (Keirsey–Bates) in the public domain, would allow the benefits of learning students' personality types, but would not allow research opportunities. The cost of this alternative would be only $559 in the first semester, and $279 thereafter, compared to the full MBTI costs of $2249 and $1449 respectively.
- Training costs would be negligible, as volunteer led pre-semester workshops would cover all necessary information.
- The problem of getting volunteers to lead the workshops would be forestalled by an enthusiastic implementation of the program, and by providing useful applications of personality typing.
- The alternative plan of developing the Keirsey–Bates question set is feasible. The first steps are the approval of the preliminary project steps by the Writing Programs staff and the development of the question and answer forms.

Myers–Briggs Type Indicator **Conclusions and Recommendations • 15**

Figure 15-4 (continued).

Recommendations

The English department and the Writing Program staff should take the necessary steps for implementing the Keirsey–Bates personality typing system in Writing Programs classes. The steps are as follows:

- Approve the plan so that I can develop the written materials necessary to administer the instrument and lead the pre-semester workshop on Myers–Briggs personality typing.
- Propose the project to the English department so that it can receive consideration for the '93–'94 fiscal year.
- If approved, reproduce the materials necessary for the administration of the instrument to students, and the materials for the workshop. Schedule the workshop so that GATs can plan and be prepared for it well in advance.

Myers–Briggs Type Indicator **Conclusions and Recommendations • 16**

Figure 15-4 (continued).

Appendix A

Qualities of Writers and Coaching Strategies

The table on the following two pages is adapted from tables in *PTC* (98–99 113–4), and summarizes the strengths and weaknesses of the different personality variables in terms of writing, as well as giving coaching strategies for teachers to use with students.

Figure 15-4 (continued).

Extraversion (E)	Introversion (I)	Sensory Perception (S)	Intuitive Perception (N)
Physical proximity to topic and audience.	Physical distance from topic and audience.	Focus on concrete, sensory data.	Focus on generalities, implications, inferences, and possibilities.
Strengths: Es excel at writing from experience. Their prose is likely to be vital and reflect a clear connection between experience and thought. They tend to excel at writing dialogues from outer speech and usually have a clear sense of voice. At their best, they will write a vital, informal prose that reflects an immediacy of experience.	**Strengths:** Is will tend to write more intensely about a more limited range of ideas or topics (especially if also J). They will usually reflect on their topic enough to make abstractions from it and clearly perceive the audience as someone with values different than their own. At their best, they will condense ideas into a naturally formal style.	**Strengths:** Ss excel at following directions closely, at attending to concrete observations and accurately presenting data. At their best, they will write accurate descriptions and sound technical reports; they may also be better at dealing with complex data sets than Ns.	**Strengths:** Ns excel at developing unique approaches to a topic. At their best, they will write imaginatively and originally about sound concepts and theories.
Weaknesses: Es may write more fragmentarily, touching superficially on a broad range of topics (especially if also P). They may not adequately reflect on their topic or fail to differentiate their values from those of their audience (especially if also T). At their worst, they present undigested information in an inappropriate and converstaional style.	**Weaknesses:** Is' writing may be so distant from experience that it lacks vitality or fails to reflect clearly the connection between experience and thought. They may be reluctant to express ideas and feelings, even on paper. At their worst, they will produce a lifeless and needlessly formal prose (especially if also T).	**Weaknesses:** Ss may fail to present the ideas and concepts behind their concrete data. They may fail to see the unique demands of the rhetorical situation and fail to adjust their writing to meet those demands. At their worst, they will present mere facts in a hackneyed formula.	**Weaknesses:** Ns may leap into the middle of the piece or narrative and fail to provide background information. They may fail to include support (examples, facts, etc.) for their ideas, and they may find it difficult to follow directions. At their worst, their writing is incomprehensibly abstract and complex, based on flighty, inaccurate hunches.
Making Contact: The best way to make contact with Es is through their preferred channel of communication: talk. They usually respond better to oral than to written feedback. Since they value active experience, they respond well to compliments about the vitality and energy of their writing.	**Making Contact:** Is respond better to any situation when given advance notice and when not expected to "think on their feet." Before talking to Is, it is best to provide them with an agenda, and then follow up later. This will allow them time to think about what they may want to say. Is often respond well to teachers who acknowledge that they have more to say (or write) than they have offered. Is will talk (or write) more when they trust their teacher and when they are not forced to share their ideas.	**Making Contact:** Ss attend better to communications that begin with the concrete. Facts, concrete examples, and practical solutions appeal to them. They respond well to compliments about being accurate, reliable, and precise.	**Making Contact:** Ns attend better to communications that begin with concepts, theories, or inferences. They like sentences that begin with 'What if...?' They value being innovative, original, and theoretical.
Words of Praise that Appeal to Es: experience, vitality, lively, action, doing, initiative	**Words of Praise that Appeal to Is:** thoughtful, serious, sincere, deep	**Words of Praise that Appeal to Ss:** practical, realistic, solid, concrete, sensible, here-and-now, responsible, careful	**Words of Praise that Appeal to Ns:** innovation, possibilities, hunches, inspiration, fantasy, dream, imagination

Myers–Briggs Type Indicator

Appendix A • 18

Figure 15-4 (continued).

Thinking Judgment (T)	Feeling Judgment (F)	Judging (J)	Perceiving (P)
Emotional distance from topic and audience.	Emotional proximity to topic and audience.	Exclusion of data and ideas to enhance decisiveness.	Inclusion of data and ideas to enhance thoroughness.
Strengths: Ts excel at the logical development of essays. They tend to write objectively and analytically. At their best, they will present content clearly and develop sound and consistent organizational patterns.	**Strengths:** Fs excel at using personal examples to develop their essays and at writing expressively. They also excel at making contact with the audience and at conveying the deep personal conviction behind their beliefs. At their best, they produce stylish, interesting, and personal prose.	**Strengths:** Js tend to be decisive and can usually reach conclusions quickly and thus forcefully assert a proposition. They tend to write quickly, meet deadlines, and complete more projects than Ps. At their best, they write expediently and emphatically.	**Strengths:** Ps tend to investigate their topic thoroughly and present carefully considered ideas. Their writing will tend to be fully developed, and their ideas well supported. At their best, they are thorough and present well-qualified conclusions.
Weaknesses: Ts may regard their beliefs as universally held (especially if also E) and thus write abrasively or dogmatically. They may objectify ideas and examples until they lose personal appeal. Their organizational patterns may be too structured. For example, they may make a narrative read like a technical report. At their worst, they may, by writing abrasively or dryly, fail to connect with their audience.	**Weaknesses:** Fs' writing may be gushy or overly sentimental. They often dislike analyses or making critical remarks and may, out of concern for others' feelings, soften criticism until it is unclear or lacks force. Their organization may be so free flowing that their thoughts seem contradictory (especially if also E). At their worst, they write sentimentally and unclearly.	**Weaknesses:** Js, in their haste to meet deadlines, are in danger of reaching premature closure. Their conclusions may be ill-considered or arbitrary. At their worst, their writing is opinionated, unambitious, and underdeveloped; their statements unqualified.	**Weaknesses:** Ps may write on topics that are too broad, and in general, be over-inclusive. Their writing may be so thorough that it is tediously long, or it may be rife with digressions. They may be reluctant to assume a position in persuasive essays, open-endedly discussing both sides of the issue. At their worst, their writing seems to ramble endlessly without clear focus.
Making Contact: Ts value being logical and objective. They react most favorably to a logical rationale and analytical thought. They prefer criticism that is to the point, rather than criticism that is softened or indirect. Praise should, at least in part, relate to the content of their writing.	**Making Contact:** Fs tend to respond well to communication after personal contact has been made. They prefer to chat informally before getting down to business, especially if also E. They respond well to any acknowledgment of them as individuals: learning their name, asking about their interests and values, a smile, a pat on the back, etc. They often respond well to an instructor saying that he or she enjoyed reading their essays. Praise should, at least in part, relate to the process of their writing, e.g., the way that they worded their ideas	**Making Contact:** Js are particularly concerned about wise use of time, and usually respond well to compliments about how efficient, expedient or punctual they are. Js also value being decisive. They tend to view their work, once it is submitted, as finished, and prefer to hear comments about how they can improve the next essay rather than about how to improve this essay.	**Making Contact:** Ps are usually most concerned about being inclusive, so they tend to respond well to compliments about the breadth of their research. They tend to view their work as ongoing. They prefer to hear about how to improve this essay rather than about how to improve the next essay. They also like to see how one project can relate to the next, even though both are unfinished.
Words of Praise that Appeal to Ts: objective, analytical, logical, valid, systematic	**Words of Praise that Appeal to Fs:** beliefs, values, personal, heartfelt, touching, interesting, us, we, together, invested	**Words of Praise that Appeal to Js:** complete, finished, hard working, punctual, decisive	**Words of Praise that Appeal to Ps:** thorough, extensive, in-progress, ongoing, adaptable, flexible, open-minded, questioning

Myers–Briggs Type Indicator — Appendix A • 19

Figure 15-4 (continued).

References

Works Cited

Baden, R. "Pre-writing: The Relationship Between Thinking and Feeling." *College Composition and Communication* 26: 368–70.

Ching, M. K. L. "The Effect of Psychological Type Upon the Composing Process of University Faculty." *Bulletin of Psychological Type* 8 (Spring 1986): 34–7.

Gilligan, Carol. *In a Different Voice: Psychological Theory and Women's Development*. Cambridge, MA: Harvard UP, 1984.

Hall, Calvin S. and Gardner Lindzey. *Theories of Personality*. 2nd ed. New York: John Wiley & Sons, Inc., [1970].

Hoffman, J. L. and M. Betkouski. "A Summary of Myers–Briggs Type Indicator Research Applications in Education." *Research in Psychological Type* 3: 3–41.

Jensen, George H. and John K. DiTiberio. *Personality and the Teaching of Composition*. Writing Research: Multidisciplinary Inquiries into the Nature of Writing. Norwood, NJ: Ablex Publishing Corporation, [1989].

Keirsey, D. and M. Bates. *Please Understand Me: An Essay on Temperament Styles*. Del Mar, CA: Prometheus Nemesis Books, 1978.

Kinneavy, J. L. *A Theory of Discourse*. Englewood Cliffs, NJ: Prentice-Hall, 1971.

Lawrence, Gordon. *People Types and Tiger Stripes: A Practical Guide to Learning Styles*. 2nd ed. Gainesville, FL: Center For Applications of Psychological Type, Inc., [1982].

Figure 15-4 (continued).

McCaulley, Mary H. and Frank L. Natter. *Psychological Type Differences in Education*. Gainesville, FL: Center for Applications of Psychological Type, 1974.

Myers, I. B. *Gifts Differing*. Palo Alto, CA: Consulting Psychologists Press, 1980.

Pervin, Lawrence A., ed. *Handbook of Personality Theory and Research*. New York: The Guilford Press, [1990].

Provost, J. A. and S. Anchors. *Applications of the Myers–Briggs Type Indicator in Higher Education*. Palo Alto, CA: Consulting Psychologists Press, 1987.

Schurr, K. Terry, Forrest Houlette, and Arthur Ellen. “The Effects of Instructors and Student Myers–Briggs Type Indicator Characteristics on the Accuracy of Grades Predicted for an Introductory English Composition Course.” *Educational and Psychological Measurement* 46.4 (Win 1986): 989–1000.

Walter, K. M. *Writing in Three Disciplines Correlated with Jungian Cognitive Styles*. PhD Thesis, University of Texas at Austin, 1984.

Wiggins, Jerry S. “Myers–Briggs Type Indicator.” *The Tenth Mental Measurements Yearbook*. Eds. Jane Close Conoley and Jack J. Kramer. Lincoln, NE: Buros Institute of Mental Measurements of the University of Nebraska-Lincoln, 1989.

Myers–Briggs Type Indicator **References • 21**

Figure 15-4 (continued).

From accumulating research to answer each of these questions, Nadine developed the report shown in Figure 15–4. Note that each section of the report, shown in the Table of Contents, reflects the answers she found to the questions that formed the scope of her project. Because the report will be read by audiences other than the program director who authorized the report, Nadine submitted the report as a formal, rather than an informal document.

Note that Nadine chose a visually effective format—three type sizes with second-level headings delineated by a line placed over the heading. Main section heads are centered and printed in 36-point type. Second-level headings are printed in 24-point type, and third-level headings are printed in 18-point type. The supporting text is typed in 12-point type.

Empirical Research Reports

As a student and later a technical specialist, you may be assigned the task of designing some device, determining its reliability, or testing the validity of some idea. This kind of analytical report is called the empirical research report because it analyzes a problem based on extant knowledge, proposes a solution, tests that solution, and then concludes whether the solution is workable. For example, the student report, segments of which will be presented below, reports the results of a study to determine whether a link exists between canine and human osteosarcoma, a deadly form of bone cancer. If such a link can be shown to exist, dogs can be used as research models to test possible treatment for this type of bone cancer.

Other examples of empirical research studies might generate reports such as the following:

- A report on research conducted by a student who was attempting to design a monitor for use with infants who may be prone to Sudden Infant Death Syndrome. Like many research reports, this one reports the progress of the research up to the point at which the report had to be submitted. Thus, its conclusions are not definitive, but suggestive of what also needs to be done to pursue this research further.

This empirical study opens with the statement of the problem (introduction), the known causes of the problem, description of the problem, best known intervention method (monitoring), the design the student has chosen for the monitor, the theory by which it operates, and the conclusion about the possibility of this device at the time the report was submitted.

- A report summarizing research to integrate a digital computer into the control loop of an M-B shake test machine to develop a random vibration testing apparatus, using the computer to generate and control the vibration spectrum.

Following the introduction, this study begins with a discussion of the theory behind computer integration, then moves to presentation of the model studied for integration, then to results of the model studied, discussion of the results, costs of implementation, and suggestions for further improvement of the model.

The empirical research report generally includes the following elements in some form:

- Introduction and review of current knowledge
- Materials and methods for solving the problem
- Results
- Discussion
- Conclusion

To illustrate how this type of report is developed, we will show the major segments for the report described in Situation 5:

Situation 5

Ann Underwood, a senior biochemistry major, develops a research project to study the retinoblastoma (RB) gene in normal canine tissues and in canine osteosarcoma (OS). Her goal is to see whether a relationship exists between the canine RB gene and OS. This relationship is present in human osteosarcoma. Thus, if her research can establish a firm relationship between the canine RB gene and osteosarcoma, canines may be used as research models to study ways of monitoring or slowing OS in humans. The empirical research report that follows reports the results of her research and provides a useful example for studying how empirical research is reported.

Introduction and Literature Review

Like all introductions, the empirical research report introduction gives the subject, scope, significance, and objectives of the research. The literature review explains what is known about the problem, as this knowledge has been reported in published articles and reports. The writer may also use the literature review to explain a choice of method or to show the rationale for his or her investigation.

The introduction and literature review may be separate sections, or they may be integrated into one section. Examine the introduction and literature review of the retinoblastoma research report.

Background and Significance

Cancer cells exhibit many abnormal traits. The most obvious trait is loss of contact inhibition. Cancer cells continue dividing even when crowded: They stack on top of each other to form an aggregated mass, eventually forming a tumor. They have a different shape than noncancerous cells, as they rely to a greater extent on anaerobic metabolism. Their outer membrane displays special tumor antigens, which confer distinct immunological properties on the cell (Weinberg, 1983). Cancer cells, unlike normal cells, will grow in semi-solid media and will form tumors when implanted into nude mice.

Oncogenes and Anti-Oncogenes—Their Relevance

A relatively small number of genes, termed oncogenes or proto-oncogenes, may profoundly influence transformation of a cellular phenotype from normal to neoplastic (Weinberg, 1983). Oncogenes are converted from normal cellular genes, the proto-oncogenes. Proto-oncogenes may be converted to active oncogenes by a number of somatic and/or hereditary events (Deion, 1984). Oncogene mutations as discrete as a single base pair change, resulting in a single amino acid substitution have been shown, in some cases, to promote a transformed phenotype. Various oncoproteins may promote cellular growth, regular cellular signaling and/or regular transcriptional rates (Hunter, 1984). The mechanisms by which oncoproteins work are diverse and clearly involve fundamental cellular processes and changes in regulation of these processes (Goades, 1989).

More recently, a new mechanism for the development of cancer has been studied. In this mechanism, certain proteins are in normal cells not to promote proliferation but to suppress it (Weinberg, 1988). Thus, the loss of growth-suppressor proteins causes unregulated cell growth by removing a normally present control. Genes encoding such growth-suppressor proteins have been termed anti-oncogenes, or recessive oncogenes. The essential difference between a dominant and a recessive oncogene is that for the former it is the presence of the product, whereas for the latter it is the absence of the product that leads to transformation (Green, 1988).

Retinoblastoma and Osteosarcoma—Their Relationship

Retinoblastoma (RB) serves as a prototype for this recessive class of cancer genes. The normal function of the RB gene is to suppress growth. Loss of function is thus associated with the appearance of malignancy. Specific changes resulting in homozygosity or hemizygosity for the mutant or inactive allele appear to be a key mechanism leading to tumor formation (Murphree, 1984).

Retinoblastoma is a rare cancer found in the retinoblasts of children. Retinoblasts are the precursors of retinal cells. After differentiation into a specialized retinal cell, a retinoblast stops dividing and no longer serves as a target for tumorigenesis. Retinoblastoma occurs with a frequency of l in 10,000 (Benedict, 1988) to 1 in 20,000 live births (McFall, 1987), where the age of onset is anywhere from 0 to 7 years. No discernable geographic, racial or sex-specific clustering was apparent in the cases studied (Cavenee, 1986). Affected individuals fall into two categories: a) those having multifocal tumors that develop at an early age and b) those with unilateral tumors that appear at a later age. Individuals affected with multifocal tumors tend to be affected with the hereditary form of retinoblastoma, while those with unilateral tumors are usually affected with the sporadic form.

In 1971, Knudson hypothesized that retinoblastoma results from a mutation in both alleles of chromosome 13. Two events are necessary in the development of both the hereditary and the sporadic forms of retinoblastoma. The first event is a germinal mutation in the hereditary form and a somatic mutation in the sporadic form. The second event is somatic in both forms (Knudson, 1978). In hereditary cases, the mutant allele is usually inherited from the father (Bookstein, 1990) and is present in all cells of the body. As a result only one additional mutation must occur in the retinoblast in order for cancer to develop. With the sporadic form, both mutations must occur somatically.

The heritable form of RB is transmitted as a dominant cancer susceptibility trait (Lee, 1988). Each of the offspring of the carrier parent has a 50% chance of inheriting the trait, and, of these, 90% will develop RB. While the predisposing mutation is inherited in a dominant fashion, it is recessive at the cellular level. The gene represents the prototype for a recessive class of human tumor cancer genes. The inherited mutation is not sufficient to cause cancer. It merely predisposes the cell to the development of retinoblastoma. For cancer to develop, the second allele must also undergo mutation.

The method by which the second mutation occurs is not known. Possibilities include nondisjunction, nondisjunction with reduplication of the mutated allele and mitotic recombination. Any of these events results in homozygosity or hemizygosity for the abnormal chromosome 13 (Figure 1). The mutation apparently can be caused by either a microscopic or submicroscopic change (Knudson, 1978). Whether or not these tumors have identifiable structural changes within the RB gene, they either have an absence or abnormal expression of the RB transcript (Benedict, 1988). Through reverse transcription, it has been shown that 470 nucleotides are lost near the 5' end of the gene. As the size of the deletion is only about 50kb, it is often too small to be cytogenetically visible (Lee, 1988).

In contrast, the normal protein produced by the RB gene is 110 kD in size (Lee, 1988). RB tumors lack this protein.

Experiments have shown that in normal cells, the protein is located within the nucleus, where it is associated with DNA binding activity. This association supports the proposed role of the protein in regulating other genes. It is these genes which, when unregulated, cause the uncontrolled proliferation seen in cancer cells.

Studies have shown that individuals affected with the hereditary form of retinoblastoma show a greatly increased risk of developing independent secondary neoplasms (Herd, 1987). The most common of these secondary neoplasms is osteosarcoma (OS). Studies show that cells in some osteosarcoma tumors exhibit a deletion at the Rb1 locus. It is therefore thought that some OS arises by the same mechanism as retinoblastoma (Cavenee, 1986). Other etiologic possibilities have been suggested for OS, such as radiation effects (Baack, 1990), but other studies have also revealed cases in which OS developed outside the field of radiation in the absence of such treatment (Hansen, 1985).

Comparative Features of Canine OS and Human OS

Osteosarcoma in dogs is the most commonly occurring skeletal neoplasm. It affects 7 in 100,000 dogs annually. Dogs diagnosed with OS have a very poor prognosis, with only 10–20% surviving one year following amputation. As in humans, OS readily metastasizes, with the most common site being the lungs.

Many other similarities exist between OS in man and OS in dogs. (See Table 1.) OS primarily affects the appendicular skeleton of both species. The bones most often affected are the proximal humerus, the distal femur, the proximal tibia, and the distal radius. A genetic correlation is seen according to size: breeds of large dogs (over 80 lbs.) and tall humans. Familiar trends are also evident. A high occurrence of OS is found among families of St. Bernards and rottweilers as well as in several pairs of human siblings (Haines, 1989).

Treatment of Osteosarcoma

Current treatment for human osteosarcoma involves removal of the tumor followed by radiation therapy and chemotherapy. In vitro research is currently underway to test the possibility of gene therapy as a mode of treatment. Huang et al. (1988) demonstrated that by inserting a functional RB protein into cells containing inactivated endogenous RB genes, the neoplastic phenotype was suppressed. Cells were infected with a retrovirus carrying the functional RB gene. Infected cells in culture subsequently expressed a RB protein indistinguishable from the native RB in terms of molecular weight, cellular location and phosphorylation. Infected cells also became enlarged and exhibit severely inhibited growth. When injected into nude mice, the infected cells lacked the ability to form tumors.

Furthermore, Bookstein et al. (1990) successfully restored normal RB expression in human prostate cancer cell line DU145, which was shown to possess a RB mRNA transcription that lacked 105 nucleotides by exon 21. Normal RB expression was restored in DU145 cells using retrovirus mediated gene transfer. The transformed cells, expressing exogenous RB protein, subsequently lost their ability to form tumors in nude mice.

The results obtained by Huang and Bookstein illustrate the possibility of gene therapy as a new form of treatment of malignant cancers. Gene therapy would be beneficial in that it is based on the permanent correction of an underlying defect in tumor cells. Unlike conventional cytotoxic cancer therapies, gene therapy should not be harmful to normal cells and therefore need not be specifically targeted.

Research Goals

My research focuses on OS in canines with the hope that such a study will lead to discovery of a correlation between canines and humans. If OS in dogs is shown to share molecular features with human RB and OS, dogs with hereditary diseases may then be considered as research models for the benefit of both species.

My research will thus have three specific aims:

1. Using Northern analysis, to characterize RB gene transcription in normal canine tissues.
2. Using Northern analysis, to characterize RB gene transcription in canine osteosarcoma.
3. To determine if the RB transcript from canine osteosarcoma differs from the transcript of normal tissues, and then to determine the frequency of RB mRNA changes in canine osteosarcoma.

Note that the student uses parenthetical references to show what is known about retinoblastoma and osteosarcoma, as these findings relate to the proposed research. Extant research shows the rationale for the proposed research on canines.

The objectives of the research are sometimes stated as a hypothesis. They can often be stated as questions. The important point to note is that the study has a rationale, based on the existing literature.

In much research reporting, passive voice and past tense are used to report what is known from previous research. Present tense is used to indicate current knowledge or status of the problem. You may wish to examine the use of tense and voice in the student report.

Materials and Methods

The researcher's goal in this section is to allow other experienced researchers to duplicate the research. This section usually contains the following parts:

- Design of the investigation—what you planned to do in your research, perhaps one sentence
- Materials—objects and equipment you used for the research
- Procedures—how you conducted the research
- Methods for observation and interpretation of what you found

However, each part may not be distinctly titled. In the retinoblastoma research report, the investigation involved seven steps, as indicated by the seven subheadings under Materials and Methods. Within each step, the writer states what materials were used for that step, the procedure used (with explanation added when she deems that to be necessary), with the final step devoted to analysis of the data collected.

Materials and Methods

The research required seven major phases. Crucial analysis occurred during RNA Extractions, Nucleic Acid Electrophoresis, and Northern Hybridization.

Tissue Sample and Cell Lines Utilization

Normal tissues (n = 6) were collected from terminal teaching dogs immediately after euthanasia. No history was available on these animals.

Canine osteosarcoma tissue was collected from animals presented for tumor biopsy or removal at Colorado State University Veterinary Teaching Hospital (n = 24) and the Texas A&M Veterinary Teaching Hospital (n = 6). All tissue and tumor samples were stored at -80° C until use.

Two human cells were also utilized. Human retinoblastoma cell line, Y-79, is known to express a truncated RB message of 4.0kb. A human fibroblast cell line, CDD-45, is expected to express a normal RB transcript of 4.7kb. Both cell lines served as homologous controls for hybridization with the human RB cDNA probe. Y-79 cells were grown in RPMI with 15% fetal calf serum; CDD-45 cells were grown in MEM with 10% FCS. Both cell lines were grown at 37° C in 5% CO_2 and 95% O_2.

Description of each research phase

RNA Extractions

Approximately 0.5g of frozen tissue and tumor samples was ground under liquid N_2 using a mortar and pestle. Following grinding, the samples were denatured and solubilized in guanidinium thiocyanate solution (4M guanidinium thiocyanate, 2-mercaptoethanol). The tissue was passed through a 16G needle followed by a 20G needle to reduce viscosity. Sodium acetate was added to lower pH. The samples were extracted with phenol followed by chloroform:isoamyl alcohol (24:1). The aqueous phase was precipitated with equal volume isopropanol at -20° C for 1.5 hrs. The pellet was resuspended in guanidinium solution and reextracted with phenol and chloroform:isoamyl alcohol. The aqueous phase was precipitated again under the same conditions. The pellet was resuspended and precipitated with 2.5 vol. ethanol. The new pellet was washed in 70% ethanol, dried, and resuspended in 100mcl 0.5% SDS.

Quantification of RNA

Extracted RNA was diluted 1:20 in 0.5% SDS and quantitated using a Beckman DU-70 spectrophotometer. The spectrophotometer was used on the dual wavelength program and set to compare O.D.260 and O.D.280. The 260/280 ratio should be close to 2.0 and not less than 1.6 for acceptable RNA purification. Calculations used were as follows:

mcg RNA = (O.D.260) (constant) (dilution factor) (ml. sample)

constant = 40mcg/ml

dil. factor = 20 (5mcl/100mcl = 1/20)
ml. sample = 100mcl-5mcl = 0.095

Nucleic Acid Electrophoresis

10mcg of sample were dried and resuspended in 10mcl sample buffer [37% formaldehyde (17.5% vol), 2x di-formaldehyde (50%), 5x MOPS running buffer (10%), TE pH8.0 (22.5%)]. 5mcl of 6x tracking dye was added and the samples were loaded into the wells of an 0.5% agarose horizontal gel; 4mcl ethidium bromide were added to the gel to stain the nucleic acids during electrophoresis. The gel was run in 1x MOPS running buffer at 50V for 3–4 hrs. A UV-illuminated photograph was taken of each gel for comparison of rRNA band size. The nucleic acids were then transferred overnight from the gel to a Nytran membrane using 1x sodium transfer buffer (pH 6.5).

Probe Labeling

3.8R and 0.9R human cDNA probes were prepared; 3.8R hybridizes to the 3' end and 0.9R to the 5' end of the Rb transcript. The probes were labeled using the BRL Random Primers DNA Labeling System. Fifty ng of DNA was added to 23mcl 1xTE and denatured. Six mcl dNTPs, 15mcl random primer buffer solution, and 5mcl ^{32}P dCTP were added on ice and mixed; 1mcl Klenow fragment was added and the mixture briefly centrifuged. The probe was allowed to incubate at 37° C for 30 min. Equal volume stop buffer was added and the resulting mixture was separated through a column of Sephadex G50 with TE. The labeled probe was collected. 1mcl was spotted on a glass filter (Whatman GF/C) and counted using a LKB 1211 Rack-Beta Liquid Scintillation Counter.

Northern Hybridization

Membranes were hybridized using a rapid hybridization protocol. Northern blots were prehybridized with shaking at 65° C for 15 min in rapid hybridization buffer (pre-warmed to 65° C). The probe was denatured at 95–100° C for 2–5 min. The probe was added to the hybridization buffer at a concentration of 1¥10^{6} cpm/mcl. The membranes were allowed to hybridize with shaking at 65° C overnight (minimum of 2 hrs. recommended in protocol). The membranes were washed once at room temperature for 15 min. in 1% SDS and 1xSSC, then twice at 65° C for 30 min. in 0.5% SDS and 0.5xSSC. The washed membranes were placed in a seal-a-meal bag and autoradiographed with 2 intensifying screens at -80° C for 48–120 hrs.

Analysis

The developed autoradiographs were analyzed by comparison to known molecular weight markers to determine the size of the RB transcript in normal tissues and in canine osteosarcoma.

Results

You may begin your results section with an overview of what you have learned. In a results section, the supporting details, if at all possible, are presented in tables and graphs. (We have omitted those included in the student report.) You do not need to restate these details. But you may want to refer to key data, both to emphasize their significance and to help your readers comprehend your tables and graphs.

Note that in the Results section of the retinoblastoma report, the student specifically states her results from the three crucial analysis phases she mentions at the beginning of Materials and Methods:

Results

The main results occurred in the three crucial phases.

The results of the three crucial analytical phases are as follows:

RNA Extraction

Total RNA was extracted from normal canine tissues, control cell lines, and canine osteosarcomas. RNA was quantitated and evaluated for intactness (Tables 2 and 3).

Since nucleic acids absorb at a wavelength of 260 nm and proteins absorb wavelengths of 280 nm, the desired range for the 260/280 ratio is 1.6 to 2.0. Most samples yielded a ratio in the desired range, while some (103573, 105989, 106463, 106685, 106899-B, OS2.7) had lower ratios, indicating the presence of excess protein. Only 3 samples did not yield at least 10mcg of RNA (091091, 099580, 107393). The entire quantity of RNA available for these three samples was loaded per lane during electrophoresis.

Nucleic Acid Electrophoresis

CCD-45 is a normal human fibroblast cell line and as such is expected to express a typical RB mRNA of 4.7kb. Y-79, a cell line derived from human retinoblastoma, expresses a shortened mRNA transcript (4.0kb). RNA from both of these cell lines were included as controls.

The integrity of RNA from all samples was evaluated. Following electrophoresis for 3–4 hours, the gels were examined under UV light for bands in the regions of the 18S and 28S subunits of rRNA, indicating the presence of intact RNA. These bands were seen for all normal tissues (Figs. 4 and 5), 3 lanes of CCD (Figs. 4–9), one lane of Y-79 (Fig. 7), and 14 lanes of tumors (Figs. 6–9). Other lanes indicated the presence of degraded RNA. (See Figs. 4–9 and Tables 2 and 3.)

Northern Hybridization

Nucleic acids were transferred from gels to Nytran membranes. Following transfer, the membrane was probed using 3.8R and 0.9R. At no time were acceptable hybridization signals apparent with the 3.8R probe. Using 0.9R, signals were obtained for CCD45 and 5 normal tissues (lymph node, liver, testes, kidney, uterus) (Fig. 10). However, no signal was seen with any of the later membranes. Possible signals were obtained on some tumors, but were so faint as to be inconclusive.

Discussion

The discussion interprets the results. It answers questions such as these:

- Do the results really answer the questions raised?
- Are there any problems with the results? If so, why?
- Were the research objectives met?
- How do the results compare with results from previous research? Are there disagreements? Can these disagreements be explained?

Although the discussion section may cover a lot of ground, keep it tightly organized around the answers to the questions that need to be asked. In the retinoblastoma report, the writer directs the discussion to the three main research steps and explains the problems she encountered and the possible explanation:

Discussion

Problems occurred within all three analytical phases. These problems affected the results of the research.

RNA Extraction

Total cellular RNA was extracted from six normal canine tissues, 30 osteosarcoma tumors, and two human cell lines. The quantity and quality of RNA obtained were calculated and are shown in Tables 2 and 3. The majority of samples yielded high amounts of RNA, with the exception of 091091, 099580, and 107393. Ratios obtained for 260/280 were primarily between 1.6 and 2.0. Ratios less than that indicate the presence of unextracted protein contaminating the DNA.

Nucleic Acid Electrophoresis

For each of the crucial phases in the research, the writer explains the negative results, the possible reasons for the results, and ways that better results could be obtained.

All electrophoresis was accomplished using an RNA ladder for determination of transcript size and RNA from CCD-45 cells for comparison to typical human RB transcript. RNA from Y-79 cells was included on three of the gels, until the available quantity ran out. Y-79 RB transcripts are known to be truncated and thus represent an important positive control for RB gene mutation. Following electrophoresis for 3–4 hours, the gels were examined for the presence of intact rRNA. Bands localized at the 18S and 28S subunits of rRNA suggest the presence of undegraded RNA. In some cases where the bands were not seen, clumps of degraded RNA were observed at the end of the lane. Other lanes showed no evidence of nucleic acids (Figs. 4–9).

The rRNA bands were seen for all normal tissues, CCD-45, Y-79, and for 14 out of 30 tumors. There are many possible explanations for the lack of intact RNA. 24 of the tumors were sent to Texas A&M from Colorado State University. It is not known how the tumors were handled before their arrival. Perhaps they were allowed to sit at room temperature for a period of time before freezing. The manner in which the samples were shipped could also have been less than ideal. If at any time the samples were allowed to thaw, it would allow time for endogenous RNAses to degrade the RNA.

The error may also lie within the extraction procedure. I extracted the tumors in 3 batches of 10. From the first extraction, 5 tumors indicated intact RNA; from the second, only 2 had intact RNA; and from the third, 7. One might assume that mishandling occurred, especially during the second batch. However, mistakes made during the extraction cannot account for the degradation seen in all the samples as some from each batch, as well as all normal tissues and cells, had intact RNA.

Northern Hybridization

The writer is clear and specific about the results.

The success with which the probe bound to the canine RNA was minimal. All membranes were hybridized first with 3.8R then with 0.9R. It was hoped that the size of the mRNA could be determined by analyzing the location of the bands. However, as none of the mRNA was successfully labeled with the radioactive probe, no bands were found.

When using the 0.9R probe, which binds to the 5' end of the transcript, hybridization was seen with one membrane. Signals were detected with CCD-45 and 5 normal tissues (lymph node, liver, testes, kidney, and uterus) (Fig. 10). The approximate transcript size of mRNA from normal tissues and CCD-45 cells was 2.4kb. Another band around 6.5kb was found for the CCD-45 cells and for LN-4, U-5, and L-2. One tumor (107393) did hybridize to the 0.9R probe. The approximate transcript size was 2.5kb, with another band at 7.5kb (Fig. 11).

I am unsure as to exactly why the hybridizations were unsuccessful. Lack of homology between canine RNA and a human DNA probe was one possibility. However, the human probe should have labeled the RNA from the CCD and Y-79 cells, both of which were derived from human samples. Also, the 0.9R probe hybridized with normal canine tissues at an earlier date. It would therefore be reasonable to expect a signal from the other normals probed with 0.9R at a later date. In addition, if homology exists between human probe and normal canine tissue RNA, the same homology should exist with canine tumor RNA.

Since lack of homology can essentially be ruled out, other reasons must be explored. A problem may have existed with the probe itself. Perhaps the ^{32}P dCTP was not fresh or did not incorporate adequately into the DNA. However, on the basis of the specific activity, it is thought that the probe should have been adequate.

The probe DNA may not have been allowed to denature long enough before being added to the membrane. Repetition of this technical error on all blots would have been unlikely.

Another possible cause lies in the transfer of the nucleic acids from the gel to the Nytran membrane. All transfers were allowed to proceed overnight, and all were set up in the manner detailed in the protocol. When dismantling the transfer, in each case it was noted that the tracking dye had transferred from the gel to the membrane, indicating that the nucleic acid should also have transferred.

Steps that can be taken in the future to help increase hybridization include a) cloning the canine RB gene to create a homologous cDNA probe, b) preparing oligonucleotide probes to conserved areas of human RB gene that might be expected to hybridize more efficiently with canine transcripts and/or c) using a reverse polymerase chain reaction (rPCR) to amplify the canine RB message. These methods increase the possibility of documenting a hybridization for the canine RB transcript.

Cloning of the canine RB gene would require the production of a canine cDNA library from canine mRNA. Screening using nucleic acid probes would require production of a canine cDNA library from canine mRNA. Screening using nucleic acid probes to RB sequences or antibodies to expressed RB fusion proteins should allow selection of positive clones with subsequent sub-cloning and RB cDNA purification. This procedure would eventually allow for a homologous Northern screening system of canine osteosarcomas using a canine RB cDNA probe. The investment of time and energy for this approach would be substantial.

A more practical method would be to prepare oligonucleotide probes for conserved segments of the human RB gene. Since the sequence of the human RB gene is known and functional regions of the gene delineated, a specific sequence from a conserved region of the gene can be synthesized. Conserved gene sequences in the human gene would presumably be represented in canines, allowing more efficient hybridization.

Another possibility would be to use reverse polymerase chain reaction to amplify canine RB mRNA. Canine mRNA would be reverse transcribed to DNA, which would then be used in the PCR reactions. Two oligonucleotide primers flanking the desired segment of RB DNA and oriented with their 3' ends facing each other would be designed so that DNA synthesis extended across the DNA segment between them. The template DNA would be denatured in the presence of the oligo primers and the 4 dNTPs. The mixture would be cooled, allowing the oligos to anneal to the target segment. Annealed primers would then be extended by DNA polymerase. After repeating the cycle several times (~30x), an amplification level of 10^6 of the desired segment should have occurred. This procedure would yield a much greater amount of canine RB mRNA available for Northern hybridizations (23).

Conclusion

A conclusion, if necessary, allows you to add any final evaluating remarks about the project in general. In the retinoblastoma report, the writer explains the partial success (as well as the partial failure) of the research and what might be done in subsequent research. The conclusion can state the value of the research in this project.

Conclusion

With the data collected, the RB gene can be characterized in the canine species and research into osteosarcoma and possible treatments, including gene therapy, can continue. Using the data already collected, it has been possible to characterize RB transcripts as found in normal canine tissue as compared to normal human fibroblast cell line CCD-45.

While many explanations exist for lack of adequate hybridization with the RB probe, none provides an adequate explanation in itself. The cause may be a factor of all those reasons detailed or may be something I have not considered. However, while no signal was obtained for the tumors, signal was obtained for some normal tissues. Also, the RNA from the tumors remaining after electrophoresis has been quantitated and stored at -80° C. Thus, it is feasible that given more time, the errors and drawbacks encountered could be worked out and sufficient data collected from the tumors to allow analysis of RB transcripts.

A Final Word

In this section we have given you general advice about reporting empirical research. If you are to become a professional in any field that requires such reporting, doing it well will be of vital importance to you. Therefore, we strongly urge you to examine representative journals and student theses in your discipline. Observe closely their format and style. Most journals have a section labeled something like "Information for Contributors." This section will inform you about manuscript preparation and style.

Many disciplines in engineering and science require senior students to pursue an empirical research project. If you are assigned such a project, you may well find that your empirical research can be the topic of a formal report assignment in your technical writing class. Professors in your disciplinary courses will guide you in the development of your research, and the above guidelines should guide you in planning how to report your research. If you decide to develop an empirical research report, examine the example segments in this chapter and the student report in the appendix. By studying each, you should have a general idea about how to go about developing similar sections of your own report.

Planning and Revision Checklists

The following questions are a summary of the key points in this chapter, and they provide a checklist when you are planning and revising any document for your readers.

PLANNING

- What is the purpose of your report? Have you stated it in one sentence?
- What is the scope of your report?
- Who is your reader? What is your reader's technical level?
- What will your readers do with the information?
- What information will you need to write the report?
- How long should the report be?
- What format should you use for the report?
- What report elements will you need?
- What elements do you need to include in your introduction?
- What arrangement will you use in presenting in your report?
- What graphics will you need to present information or data?

REVISION

- Have you stated your purpose clearly and specifically?
- Has the information in your discussion fulfilled your purpose?
- Is your information correct?
- Is your information clearly stated?
- Does your introduction adequately prepare your reader for what will come in the report itself?
- Do your conclusions follow logically from your discussion?
- Do your recommendations follow logically from your conclusions?
- Is the presentation of your report logical?
- Do your graphics immediately show what they are designed to show?
- Have you adequately documented all information sources?

Exercises

1. Locate a journal in your field that reports empirical studies of major issues in your field. Survey issues during the last two years. Write an informal information report to your instructor describing the kinds of topics covered and the approach required by the journal for reporting the research.

2. Choose one empirical research article. Write an information report describing how the writer develops each part: introduction and literature review, materials and methods, results, and discussion. Add a final section summarizing the instructions to authors for papers submitted to this journal.
3. Reports for the situations that follow may be written either individually or collaboratively. Your instructor may divide the class into work groups to develop a report responding to each issue. If you have such a collaborative assignment, be sure that your groups work separately from the other groups. After completing individual or group reports, you may want to compare reports written by other individuals or other groups, raising these questions:
    - What design differences do you see?
    - Are the differences significant as they affect the overall effectiveness of each report?
    - Based on how writers perceive the report situation, how do these differences reflect content and structure?

Report Planning

As you begin the planning process for each report situation, answer the following questions:

- Who will be the readers of this report? What are the differences among these readers in terms of their knowledge about what you will say?
- What is your purpose in writing this report?
- What main ideas do you want to include?
- Should this be an informal report or a formal report?
- What main sections do you want to include?
- How should these sections be arranged?
- Where will you place the ideas you listed in question 3?
- Will any visual presentations, such as tables or graphs, be important in presenting your information?
- What kind of format considerations do you want to use?

Report Situations

A. Your professional society has just received a letter from the dean of your college explaining that a wide-ranging study is underway to evaluate the required courses in each department. As part of this evaluation, he asks your organization to submit a report explaining what you found to be the worst required courses you have taken. He also asks you to evaluate the best courses you have taken. The dean states that a committee composed of faculty and students from each department, from outside each department, and from the dean's office will be examining all required courses in each department. He definitely wants input from students with majors in each department, as these evaluations are deemed important in the college's examination of its degree programs.

B. By April 15 of every academic year, your professional society is required to submit a budget and an action plan for the coming year to the college. In this budget report, you must describe what activities the society will pursue the next school year, the expenditures planned for these activities, and any extra funding the society will need for special projects or for other activities. The dean of students of the college must approve the action plan and the budget by the first meeting in September.

C. The student professional societies on campus have been asked to write reports listing and describing the major problems students confront on the campus and suggest any recommendations. These reports will be sent to the Office of Student Affairs, who will compile the reports from student groups and submit one report to the board of regents.

D. You are a student assistant in the main office of your department. Your department director has decided that the department needs to develop a brochure telling new students how to survive in college. You have been asked to work with other majors in the department to develop such a document, which will be given to new students by the department's advisors. The department director asks you to keep the document to under 10 pages, fewer if at all possible, as the department plans to make several hundred copies.

E. Write a report that defines and then describes your major field or discipline. This report, which should be four pages or less, will be available at the university's College Night. Students like you will be available to answer questions about your department and the majors available in the department during College Night. This report will be available for high school seniors who are thinking about what discipline they would like to study in college. As you plan such a document, you might want to discuss the importance of this discipline and its subfields, job opportunities available for new college graduates in this discipline, coursework required for a major in this discipline, etc. In short, what would a high school student be interested in knowing about your major field of specialization?

F. The chairperson of your department has asked you to write a report recommending the best calculator for students in your major. Write a report to your department's advisor explaining the kind(s) of calculators you believe students should purchase. Explain the rationale for your views.

Oral Exercises

Before you attempt either oral exercise, please consult the guidelines in Chapter 18, "Oral Reports." These reports are based on B and C above.

- Report Situation B. Assume that as an officer in your professional society, you are asked to give an oral version of your budget report to the dean of students and her staff. Your presentation should be 10 minutes maximum. Be sure to include visual aids to explain your proposed expenditures.
- Report Situation C. The Office of Student Affairs has invited your professional organization to give an oral briefing describing your concerns about campus issues to the university's academic council. You will be allowed 10 minutes to give an oral summary of your written report. Use visual aids if they will enhance your points.

CHAPTER 16

INSTRUCTIONS

Instructing others to follow some procedure is a common task on the job. Sometimes the instructions are given orally. When the procedure is done by many people or is done repeatedly, however, written instructions are a better choice. Instructions may be quite simple, as in Figure 16-1, or exceedingly complex, requiring a bookshelf full of manuals. They may be highly technical, dealing with operating machinery or programming computers, for example. Or they may be executive- or business-oriented, such as explaining how to complete a form or how to route memorandums through a company. The task is not to be taken lightly. A Shakespearean scholar who had also served in the British Army wrote the following:

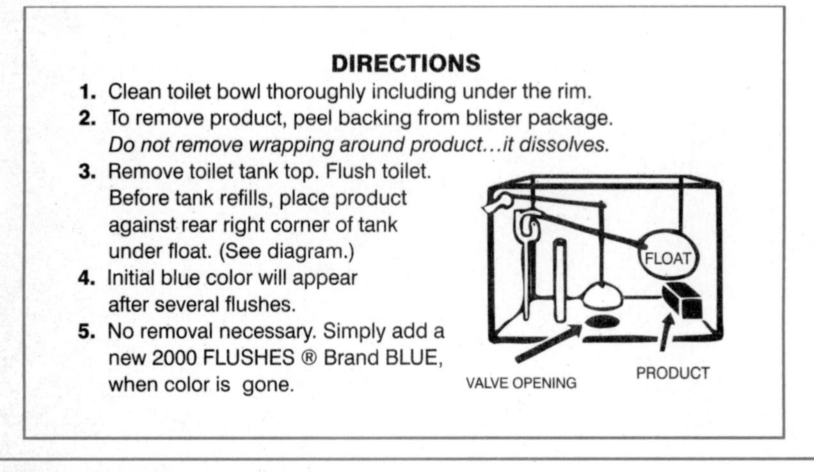

DIRECTIONS

1. Clean toilet bowl thoroughly including under the rim.
2. To remove product, peel backing from blister package. *Do not remove wrapping around product...it dissolves.*
3. Remove toilet tank top. Flush toilet. Before tank refills, place product against rear right corner of tank under float. (See diagram.)
4. Initial blue color will appear after several flushes.
5. No removal necessary. Simply add a new 2000 FLUSHES ® Brand BLUE, when color is gone.

Figure 16–1 Simple Instructions
Source: Reprinted with permission of Block Drug Company, Inc. ©Block Drug Company, Inc., 1990.

> The most effective elementary training [in writing] I ever received was not from masters at school but in composing daily orders and instructions as staff captain in charge of the administration of seventy-two miscellaneous military units. It is far easier to discuss Hamlet's complexes than to write orders which ensure that five working parties from five different units arrive at the right place at the right time equipped with the proper tools for the job. One soon learns that the most seemingly simple statement can bear two meanings and that when instructions are misunderstood the fault usually lies with the original order.[1]

To help you write instructions, we discuss situational analysis for instructions, the possible components of instructions, creating an accessible format, and checking with your readers.

Situational Analysis for Instructions

In preparing to write instructions, follow the situational analysis we describe on pages 17–21 in Chapter 2, "Composing." In addition, pay particular attention to the answers to these questions.[2]

What Is the Purpose of My Instructions?

Be quite specific about the purpose of your instructions. Keep your purpose in mind because it will guide you in choosing your content and in arranging and formatting that content. State your purpose in writing, like this:

To instruct the plant managers, the corporate treasurer, and the plant accountant in the steps they need to follow to establish a petty cash fund.

What Is My Reader's Point of View?

Don't be satisfied with a general description of a reader, such as "the average consumer" or "a typical car owner." You'll gain more accurate insights if you put yourself in the place of someone you know who fits that general description. For example, if I were my mother, what would be my point of view if I had to work with these instructions on how to complete this form? What questions and problems might I have? In what order might these questions and problems arise? Are there terms and concepts involved that I might not understand? What information do I really need? What information would be irrelevant? And so forth.

How and Where Will My Readers Use These Instructions?

Will your readers read your instructions carefully from beginning to end? Evidence indicates that they will not.[3] Readers most often scan instructions and then enter them at those points where they need clarification. In other words, they usually read them as a reference work rather than as an essay or a novel. Where will my readers use these instructions? In a comfortable, well-lighted workshop, well stocked with tools? In a cold, drafty, poorly lighted garage with only those tools they thought to bring from the workshop? In the cockpit of a boat under emergency conditions, reading by a flashlight? Standing in line in a government office? The answers to such questions will help you organize and format your instructions.

What Content Does My Reader Really Need and Want?

Understanding your purpose and your reader's point of view is essential for answering this question. You can include many things in a set of instructions: theory, descriptions of mechanisms, troubleshooting advice, and so forth. We discuss such things in the next section, "Possible Components of Instructions." You should include everything that is really relevant and nothing that is not relevant. If your reader has a need or a desire for theory, then furnish it. If theory is not needed or desired, furnishing it would be wasted effort for all concerned. Unneeded material is worse than irrelevant. It may obscure the relevant information to the point where the reader has difficulty finding it.

How Should I Arrange My Content?

Answers to all the previous questions help you to make decisions about arranging the content. If a good deal of theory is important and needed, your arrangement should probably include a separate section for it. If only brief explanations of theory are needed for the reader to understand a few steps in the instructions, place the explanations with the steps. For example, you might put the whys and wherefores of using a carpenter's level at the point in the instructions where the reader needs to use a level.

For arranging the actual instructions on how to perform the process, you must understand the process fully. If you can perform the process, taking notes as you go, do so. If that is not possible or convenient, analyze the process in your mind. Break it into its major steps and substeps. Be alert for potential trouble spots for your reader.

Possible Components of Instructions

Sets of instructions may contain as many as eight components:

- Introduction
- Theory or principles of operation
- List of equipment and materials needed
- Description of the mechanism
- Warnings
- How-to instructions
- Tips and troubleshooting procedures
- Glossary

We do not present this list as a rigid format. For example, you may find that you do not need a theory section, or you may include it as part of your introduction. You may want to vary the order of the sections. You may want to describe or list equipment as it is needed while performing the process rather than in a separate section. Often, nothing more is needed than the how-to instructions. We present the components primarily as a guide to your discovery of the material you will need.

Introduction

Normally, introductions to instructions state the purpose of the instructions and preview the contents. Frequently, they provide motivation for reading and following the instructions. They may also directly or indirectly indicate who the intended readers are. The following from a publication concerning cholesterol does all these things:

> High blood cholesterol is a serious problem. Along with high blood pressure and cigarette smoking, it is one of the three major modifiable risk factors for coronary

Intended audience

heart disease. Approximately 25 percent of the adult population 20 years of age and older has "high" blood cholesterol levels—levels that are high enough to need intensive medical attention. More than half of all adult Americans have a blood cholesterol level that is higher than "desirable."

Motivation

Purpose of instructions

Because high blood cholesterol is a risk to your health, you need to take steps to lower your blood cholesterol level. The best way to do this is to make sure you eat foods that are low in saturated fat and cholesterol. The purpose of this brochure is to help you learn how to choose these foods. The brochure will also introduce you to key concepts about blood cholesterol and its relationship to your diet. For example, it includes basic (but very important) information about saturated fat—the dietary component most responsible for raising blood cholesterol—and about dietary cholesterol—the cholesterol contained in food.

Preview of contents

This brochure is divided in three parts. The first part of the brochure gives background information about high blood cholesterol and its relationship to heart disease. The second part introduces key points on diet changes and better food choices to lower blood cholesterol levels.

Finally, in the third part more specific instructions are given for modifying eating patterns to lower your blood cholesterol, choosing low-saturated fat and low-cholesterol foods, and preparing low-fat dishes.

Reference to glossary

The "Glossary" provides easy definitions of new or unfamiliar terms. The appendices that follow the Glossary list the saturated fat and cholesterol content of a variety of foods.[4]

This introduction begins with motivation, stating that high blood cholesterol is a serious problem and providing support for that statement. The audience for the brochure, adult Americans, is indicated in an indirect manner. The purpose is clearly stated: "The purpose of this brochure is to help you learn how to choose these foods [that are low in saturated fat and dietary cholesterol]." Following the motivation and purpose steps, the introduction previews what is to come in the rest of the brochure and refers to the glossary.

Introductions to instructions are often not much different from the introductions we describe for you in Chapter 10, "Design Elements of Reports." However, short sets of instructions may have very abbreviated introductions or, in some cases, no introduction at all. On the other hand, when introductions are longer than the one we have shown you, it is usually because the writers have chosen to include theory or principles of operation in the introduction. This is an accepted practice, and we tell you how to give such information in our next section.

Theory or Principles of Operation

Many sets of instructions contain a section that deals with the theory or principles of operation that underlie the procedures explained. Sometimes historical background is also included. These sections may be called "Theory" or "Principles of Operation," or they may have substantive titles such as "Color Dos and Don'ts," "Purpose and Use of Conditioners," and "Basic Forage Blower Operation." Information about theory may be presented for several reasons. Some people have a natural curiosity about the principles behind a procedure. Others may need to know the purpose and use of the procedure. The good TV repair technician wants to know why turning the vertical control knob steadies the picture. Understanding the purposes behind simple adjustments enables the technician to investigate complex problems. What if nothing happens when the vertical control knob is turned? With a background in theory, the technician will know more readily where to look in the TV set to find a malfunction.

Such sections can be quite simple. In the following excerpt labeled "Color Dos and Don'ts," some basic color design theory is presented in easy-to-understand language:

> **Do** use light colors in a small room to make it seem larger.
> **Do** aim for a continuing color flow through your home—from room to room using harmonious colors in adjoining rooms.
> **Do** paint the ceiling of a room in a deeper color than the walls, if you want it to appear lower; paint it in a lighter shade for the opposite effect.
> **Do** study color swatches in both daylight and indoor light. Colors often change in artificial light.
> **Don't** paint woodwork and trim of a small room in a color that is different from the background color, or the room will appear cluttered and smaller.
> **Don't** paint radiators, pipes, and similar projections in a color that contrasts with walls, or they will be emphasized.
> **Don't** choose neutral or negative colors just because they are safe, or the result will be dull and uninteresting.
> **Don't** use glossy paints on walls or ceilings of living areas since the shining surface creates glare.[5]

As simple as this excerpt is, it presents color principles. Readers are told not only to use light colors in a small room, but why—"to make it seem larger."

Theory sections can be more complex. Figure 16–2 presents a portion of the theory section from the cholesterol brochure. It describes the relationship between cholesterol and atherosclerosis (hardening of the arteries). Understanding the theory helps the readers to understand the guidelines for cholesterol levels and motivates them to follow these guidelines. The entire section is written on a very personal level: What does this theory mean for the reader? Through the use of format, graphics, questions, and simple language, the writers of the brochure make the theory quite accessible for the intended audience.

What You Need to Know About High Blood Cholesterol

Why Should You Know Your Blood Cholesterol Level?

There are important reasons for you to be concerned about your blood cholesterol level. Over time, cholesterol, fat, and other substances can build up in the walls of your arteries (a process called *atherosclerosis*) and can slow or block the flow of blood to your heart. Among many things, blood carries a constant supply of oxygen to the heart. Without oxygen, heart muscle weakens, resulting in chest pain, heart attacks, or even death. However, for many people there are no warning symptoms or signs until late in the disease process.

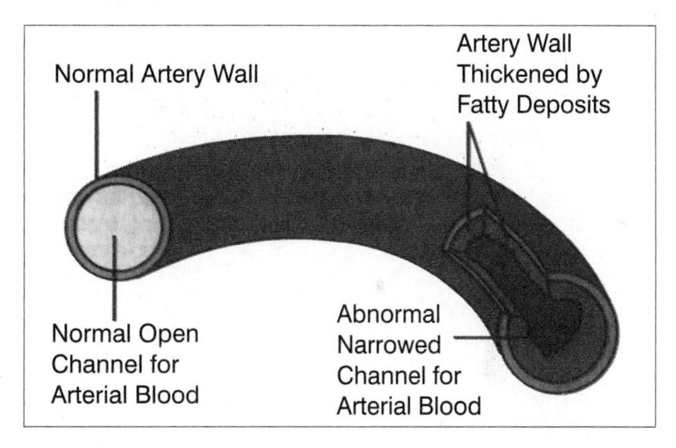

Heart disease is the leading cause of death in this country. Scientists have known for a long time that high blood cholesterol, high blood pressure, and smoking all increase the risk of heart disease.

Research now shows that the risk of developing atherosclerosis or coronary heart disease also increases as the blood cholesterol level increases. And it has now been proven that lowering high blood cholesterol, like controlling high blood pressure and avoiding smoking, will reduce this risk.

How High Is Your Blood Cholesterol Level?

The medical community recently set guidelines for classifying blood cholesterol levels. They advise that a total cholesterol level less than 200 mg/dl is **"desirable"** for adults–above 200 mg/dl the risk of coronary heart disease steadily increases. The classifications of total blood cholesterol in the following chart are related to the risk of developing heart disease.

Figure 16–2 Theory Section

Source: U.S. Department of Health and Human Services, *Eating to Lower Your Blood Cholesterol* (Washington, DC: GPO, 1989) 2–3.

Does Your Total Blood Cholesterol Level Increase Your Risk for Developing Coronary Heart Disease?

Desirable Blood Cholesterol	Borderline-High Blood Cholesterol	High Blood Cholesterol
Less than 200 mg/dl	200-239 mg/dl	240 mg/dl and above

If your total cholesterol level is in the range of 200-239 mg/dl, you are classified as having **"borderline high"** blood cholesterol and are at increased risk for coronary heart disease compared to those with lower levels. However, if you have no other factors that increase your risk for coronary heart disease,* you should not need intensive medical attention. But you should make dietary changes to lower your level and thus reduce your risk of coronary heart disease.

On the other hand, if you have borderline-high blood cholesterol and have coronary heart disease or two other risk factors for coronary heart disease, you need special medical attention. In fact, you should be treated in the same way as people with **"high"** blood cholesterol–240 mg/dl or greater–who could be at high risk for developing coronary heart disease and warrant more detailed evaluation and medical treatment.

Additional evaluation helps your physician determine more accurately your risk of coronary heart disease and make decisions about your treatment. Specifically, your doctor will probably want to measure your low density lipoprotein (LDL) cholesterol level–since LDL-cholesterol more accurately reflects your risk for coronary heart disease than a total cholesterol level alone. LDL-cholesterol levels of 130 mg/dl or greater increase your risk for developing coronary heart disease. After evaluating your LDL-cholesterol level and other risk factors for coronary heart disease, your physician will determine your treatment program.

Remember: *As your cholesterol level rises, your risk of developing coronary heart disease increases.*

*Risk factors for coronary heart disease include high blood pressure, cigarette smoking, family history of coronary heart disease before the age of 55, diabetes, vascular disease, obesity, and being male.

Figure 16-2 (continued).

The section uses some unfamiliar terms, such as *atherosclerosis,* which are defined in the glossary mentioned in the last paragraph of the introduction. However, the authors would have served their readers better by mentioning the glossary again the first time it is needed and giving its page numbers. Remember our advice on page 65 about directing your readers. Locating a glossary for them is a good example of such direction.

As our two excerpts illustrate, many diverse items of information can be placed in a theory or principles section. Remember, however, that the main purpose of the section is to emphasize the principles that underlie the actions later described in the how-to instructions. In this section, you're telling your readers *why.* Later, you'll tell them *how.* Theory is important, but don't get carried away with it. Experts in a process sometimes develop this section at too great length, burying their readers under information they don't need and obscuring more important information that they do need. Make this section, if you have it at all, only as full and as complex as your analysis of purpose and readers demands.

List of Equipment and Materials Needed

In a list of equipment and materials, you tell your readers what they will need to accomplish the process. A simple example would be the list of cooking utensils and ingredients that precedes a recipe. Sometimes in straightforward processes, or with knowledgeable audiences, the list of equipment is not used. Instead, the instructions tell the readers what equipment they need as they need it: "Take a rubber mallet and tap the hubcap to be sure it's secure."

When a list is used, often each item is simply listed by name. Sometimes, however, your audience analysis may indicate that more information is needed. You may want, for instance, to define and describe the tools and equipment needed, as is done in Figure 16–3. If you think your readers are really unfamiliar with the tools or equipment being used, you may even give instruction in their use, as in Figure 16–4. If the equipment cannot be easily obtained, you'll do your readers a service by telling them where they can find the hard-to-get items. As always, your audience analysis determines the amount and kind of information you present.

Description of the Mechanism

Instructions devoted to the operation and maintenance of a specific mechanism usually include a section describing the mechanism. Also, when it is central to some process, the mechanism is frequently described. In such sections, follow the principles for technical

Basic Tools

You'll need a few basic tools for most home maintenance jobs, and some special tools for special jobs. Some are expensive, and are not needed very often. Is there a place where you can borrow or rent those?

Here are some basic tools and materials you may need for doing simple repairs on the outside of your house.

Nail Set

A *nail set* is a small metal device used to sink the heads of nails slightly below the surface you are driving them into (fig. 1).

Squares

The *framing square* is a handy measuring tool for lining up materials evenly and making square corners. It is usually metal (fig. 2).

The *try square* is smaller and is also used for lining up and squaring material. One side is made of wood and is not marked to measure with (fig. 3).

Miter Box

With a *miter box,* you can saw off a piece of board at an exact angle. It may be of wood, to use with a separate saw (fig. 4). Or it may be steel, with the saw set in the steel box (fig. 5).

Masonry Trowels and Jointer

The *trowel* is used to build or repair masonry walls, sidewalks, etc. It has a flat, thin, steel blade set into a handle. The "brick trowel" is the larger and is used for mixing, placing, and spreading mortar. The smaller "pointing trowel" is used to fill holes and repair mortar joints (fig. 6). This process is called "pointing."

The *jointer* is another masonry tool, used to finish joints after the wall is laid (fig. 7). Finish joints are made on the outside of a masonry wall to make it more waterproof and to improve appearance. The "V" and "concave" joints are the most weather tight. A different type of jointer is needed for each type of joint used.

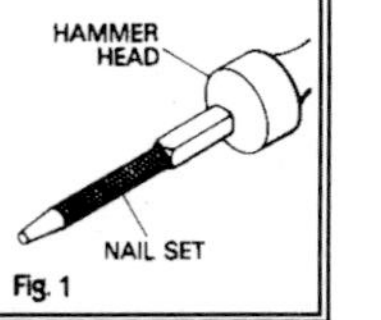

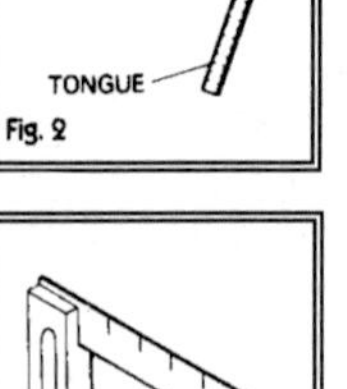

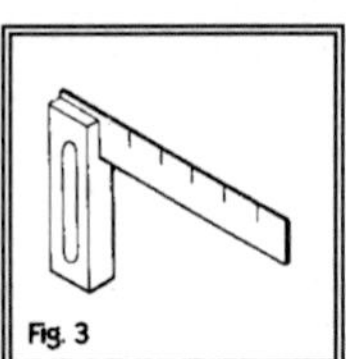

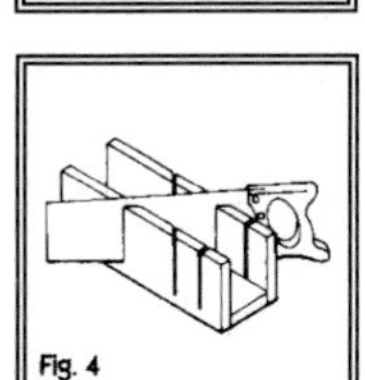

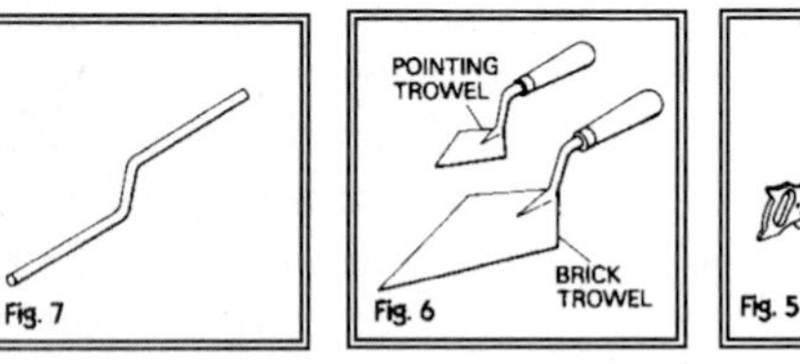

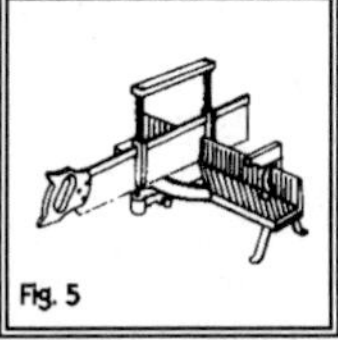

Figure 16–3 List of Tools

Source: U.S. Department of Agriculture, *Simple Home Repairs: Outside* (Washington, DC: GPO, 1986) 4.

description given in Chapter 7. Break the mechanism into its component parts and describe how they function.

Figure 16–5 shows you the mechanism description for the Macintosh mouse, the device that allows a user to move a pointer around the

NAILS, SCREWS, AND BOLTS

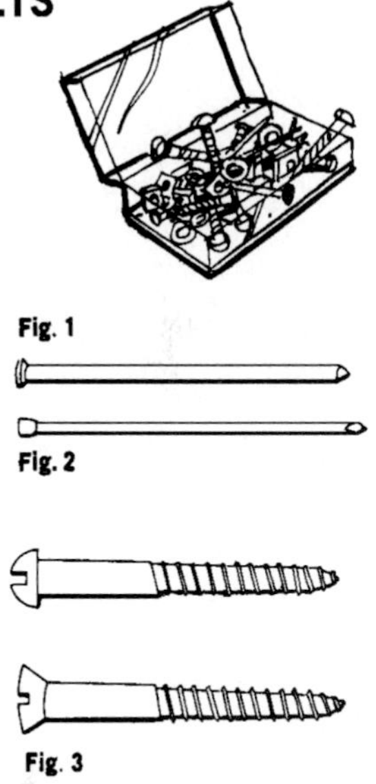

Nails, screws, and bolts each have special uses. Keep them on hand for household repairs.

NAILS come in two shapes.

Box nails have large heads. Use them for rough work when appearance doesn't matter. (Fig. 1.)

Finishing nails have only very small heads. You can drive them below the surface with a nail set or another nail, and cover them. Use them where looks is important, as in putting up panelling or building shelves. (Fig. 2)

SCREWS are best where holding strength is important. (Fig. 3.) Use them to install towel bars, curtain rods, to repair drawers, or to mount hinges. Where screws work loose, you can refill the holes with matchsticks or wood putty and replace them.

Use **molly screws** or **toggle bolts** on a plastered wall where strength is needed to hold heavy pictures, mirrors, towel bars, etc.

Molly screws have two parts (Fig. 4). To install, first make a small hole in the plaster and drive the casing in even with the wall surface. Tighten screw to spread casing in the back. Remove screw and put it through the item you are hanging, into casing, and tighten.

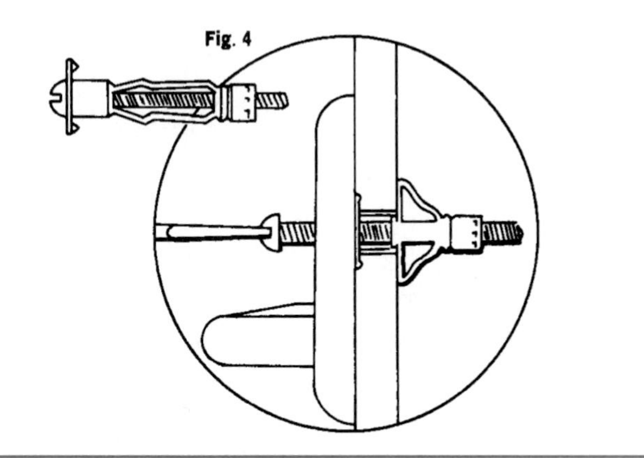

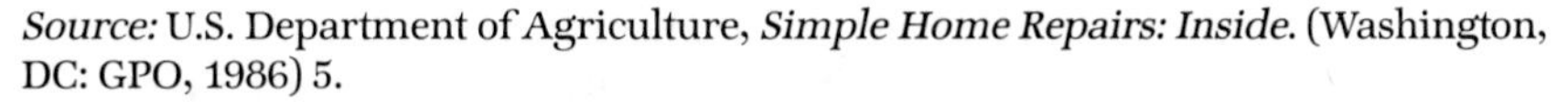

Figure 16–4 Instruction in Equipment Use

Source: U.S. Department of Agriculture, *Simple Home Repairs: Inside.* (Washington, DC: GPO, 1986) 5.

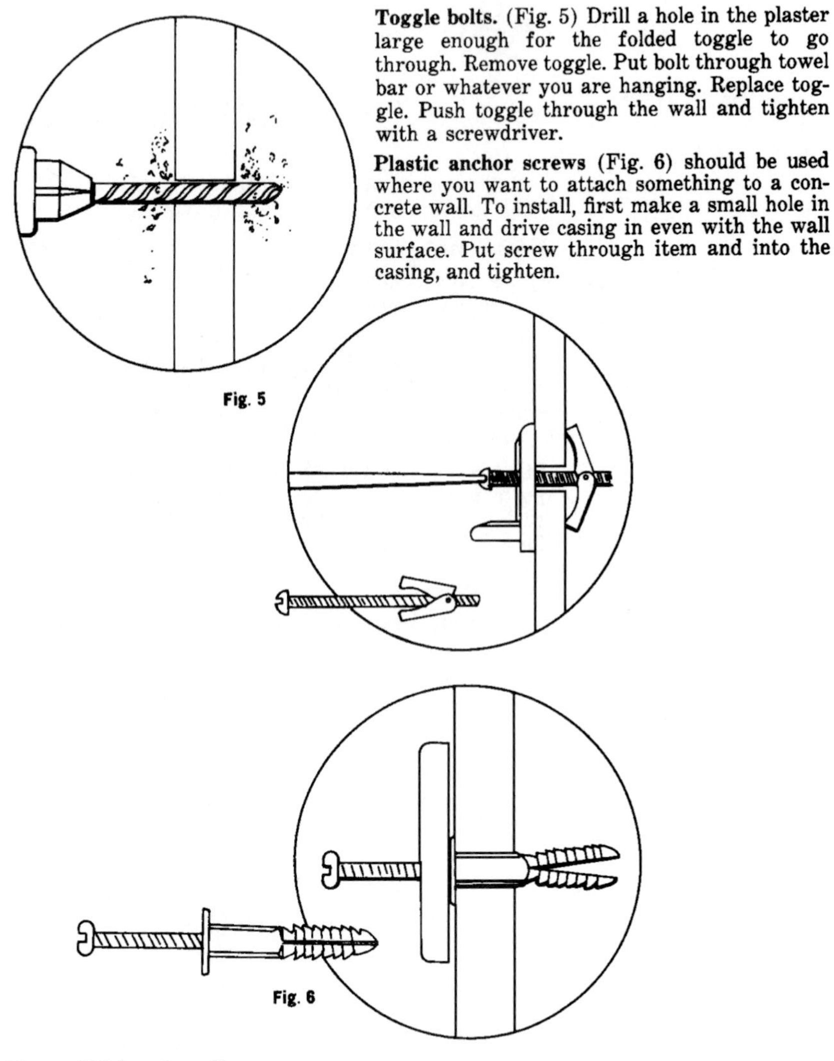

Toggle bolts. (Fig. 5) Drill a hole in the plaster large enough for the folded toggle to go through. Remove toggle. Put bolt through towel bar or whatever you are hanging. Replace toggle. Push toggle through the wall and tighten with a screwdriver.

Plastic anchor screws (Fig. 6) should be used where you want to attach something to a concrete wall. To install, first make a small hole in the wall and drive casing in even with the wall surface. Put screw through item and into the casing, and tighten.

Figure 16-4 (continued).

Macintosh screen. The description describes the function of the mouse and explains how it fulfills that function. With the help of a graphic, the description breaks the mouse into its component parts and describes the function of each part.

Mechanism descriptions are generally accompanied by numerous illustrations like those in Figures 16–5, 16–6, and 16–7. Such illustrations show only necessary detail, and to be effective they normally have to be well annotated. Some, like Figure 16–7, are exploded views. We hasten to add in this context, *exploded* means that the mechanism is

The mouse

The Macintosh mouse is a hand-operated device that lets you easily control the location of the pointer on your screen and make selections and choices with the mouse button. Coupled with the graphic elements of the Macintosh User Interface—icons, windows, pull-down menus, and so on—the mouse makes ordinary operation of the system almost effortless: you view your work on the screen and interact with it merely by pointing with the mouse and clicking the mouse button.

The Macintosh mouse contains a rubber-coated steel ball that rests on the surface of your working area. When the mouse is rolled over that surface, the ball turns two rotating axles inside the mouse, each connected to an interrupter wheel. These wheels contain precisely spaced slots through which beams of light are aimed at detectors. The wheels track vertical and horizontal motions. As the axles turn the wheels, the light beams shining through their slots are interrupted; the detectors register the changing optical values, and a small integrated circuit inside the mouse interprets them and signals the operating system to move the pointer on your screen accordingly. (There's also a third axle that helps balance the ball and keep it rolling smoothly.)

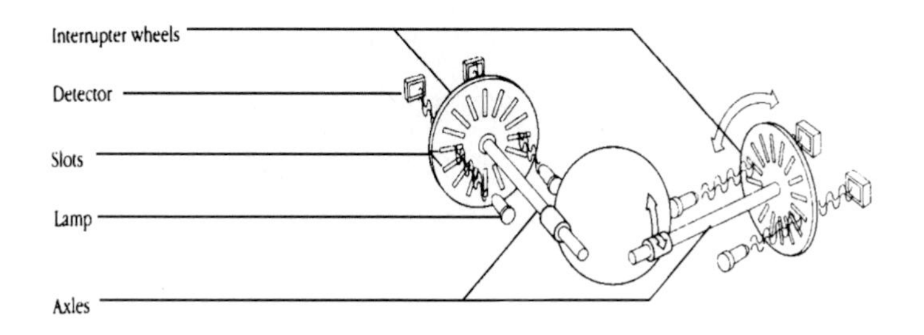

The mouse registers relative movement only; the operating system can tell how far the mouse has moved and in which direction, but not the mouse's absolute location. That's why you can pick up the mouse and move it to another place on your table or desk surface and the pointer will not move. You can adjust the speed with which the pointer on the screen responds to the mouse's movements by using the Control Panel desk accessory in the Apple menu.

Figure 16–5 Description of the Macintosh Mouse
Source: Apple Computer, Inc., *Macintosh SE Owner's Guide* (Cupertino, CA: 1988) 37–38. Copyright © 1988 Apple Computer, Inc. Reprinted by permission.

drawn in such a way that its component parts are separated and thus easier to identify. Figure 16–7 makes the concept clear.

Warnings

We live in an age of litigation. People who hurt themselves or damage their equipment when following instructions in the use of that

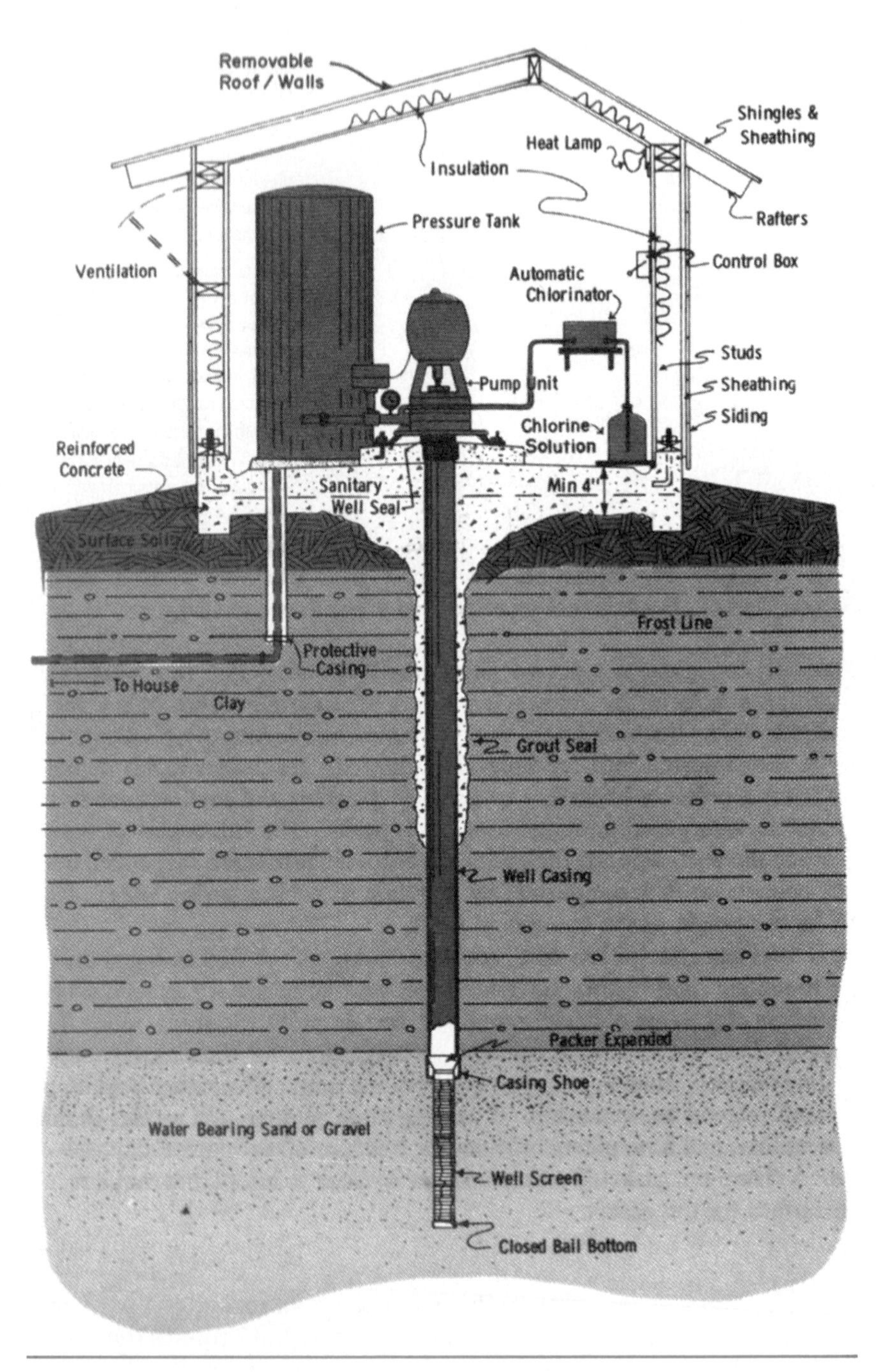

Figure 16–6 Pumphouse

Source: United States Environmental Protection Agency, *Manual of Individual and Non-Public Water Supply Systems* (Washington, DC: GPO, 1991) 121.

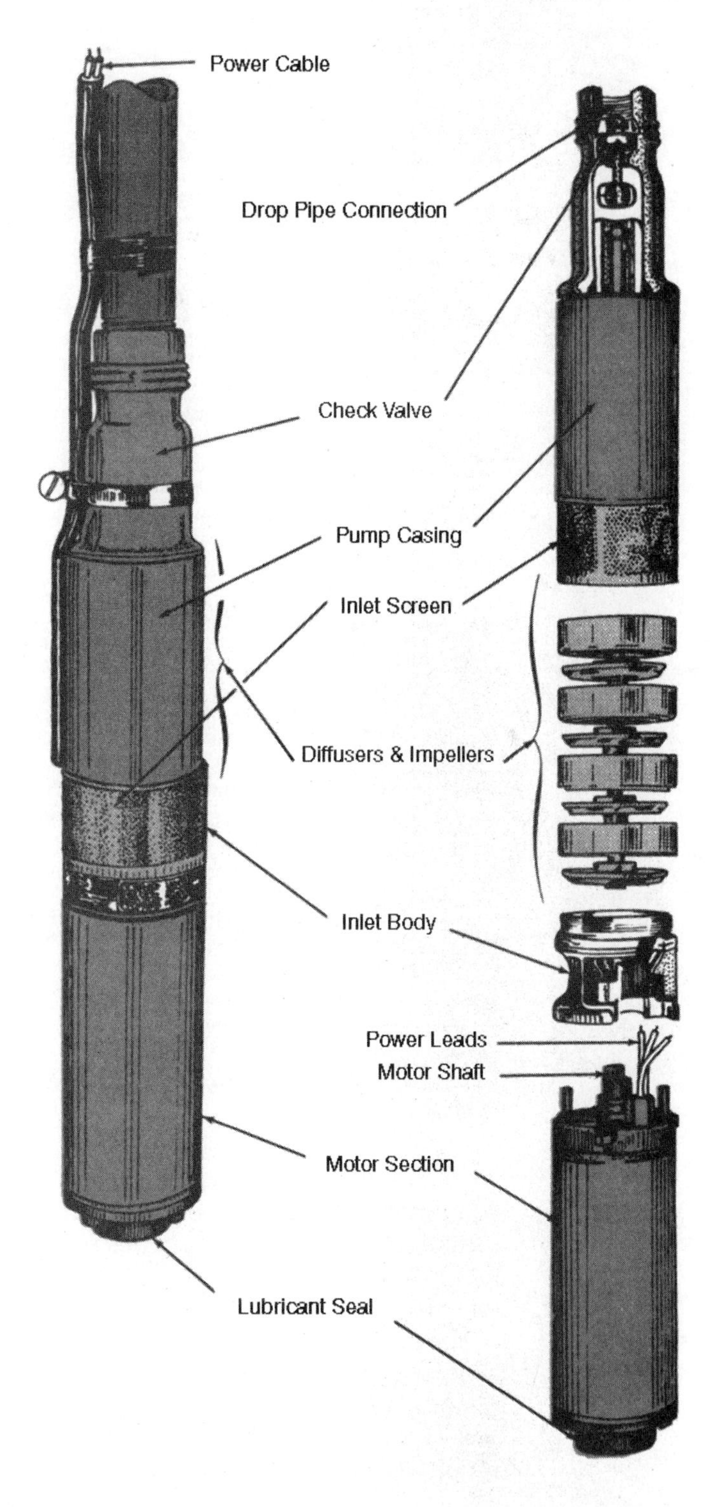

Figure 16–7 Exploded View of Submersible Pump

Source: United States Environmental Protection Agency, *Manual of Individual and Non-Public Water Supply Systems* (Washington, DC: GPO, 1991) 101.

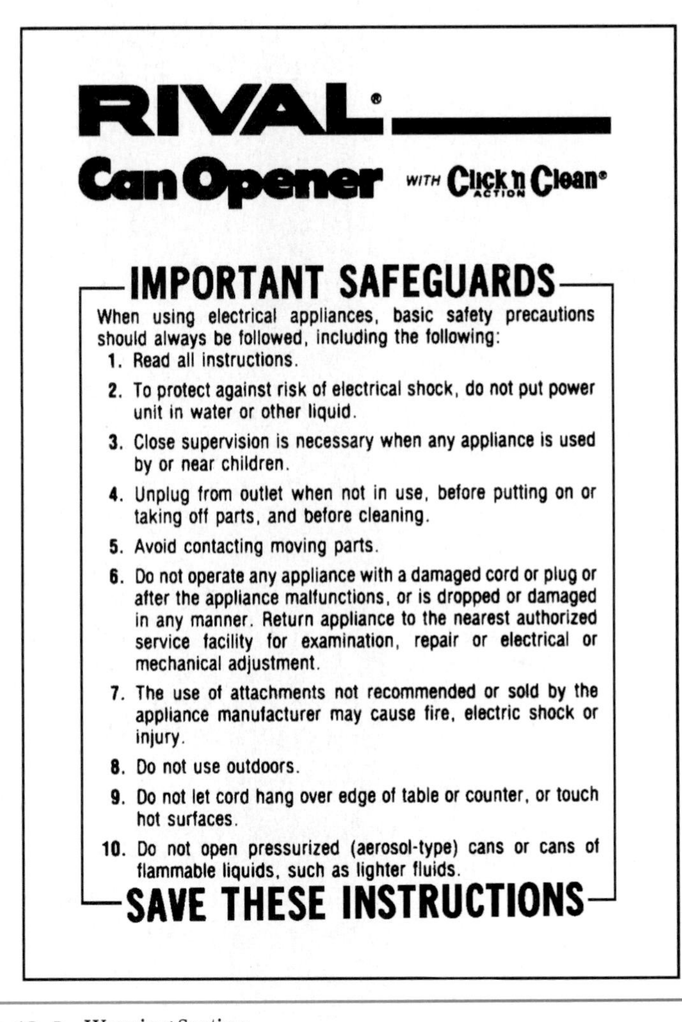

Figure 16–8 Warning Section

Source: Reprinted by permission of Underwriters Laboratories, Northbrook, Illinois.

equipment often sue for damages. If they can prove to a court's satisfaction that they were not sufficiently warned of the dangers involved, they will collect large sums of money. For this reason, warnings have increasingly become an important part of instructions.

How seriously do corporations take this need to warn people of possible dangers? We recently saw a shoe box that contained boating shoes. The box was decorated with an oceanographic chart. On the side of the box was a warning stating, "This chart is not intended to be used as a navigational aid and is not reliable for that purpose."[6] Figure

Check Valve Test

- Place the mouthpiece shut-off valve in the **Diving** position.
- Place the mouthpiece in your mouth, squeeze the inhalation hose closed, and attempt to inhale through the mouthpiece. If it is possible to inhale with the inhalation hose closed off, the check valve is missing or defective.

CAUTION

If the mouthpiece shut-off valve is in the **Open** position, the test will incorrectly indicate a defective or missing check valve.

Figure 16–9 A Caution Message

16–8 shows that something as simple to operate as an electric can opener comes with a set of warnings.

If they are extensive enough, the warnings may be put into a separate section, as they are in Figure 16–8. But often they are embedded in the how-to instructions. In either case, be sure they are prominently displayed in some manner that makes them obvious to the reader. You may surround them with boxes, print them in type different from and larger than the surrounding text, print them in a striking color, or mark them with a symbol of some sort. Often, you will use some combination of these devices.

Not only must you make the warnings stand out typographically, you must use language and, when appropriate, graphics that are absolutely clear about the nature, severity, and consequences of the hazards involved. You must clearly state how to avoid the hazard. Any lack of clarity can result in a preventable accident, almost certainly followed by a costly law suit against your employer.[7]

No terminology is completely agreed upon for warnings. However, three levels of warning have been widely accepted, designated by the words *caution, warning,* and *danger.*[8]

Caution Use **caution** to alert the reader that not following the instructions exactly may lead to a wrong or inappropriate result. In a caution, no danger to people or equipment is involved. Figure 16–9 shows how a caution might be used to advise a technician to follow the steps of a procedure in proper order. Sometimes, **note** is used for this level of warning.

Warning Use **warning** to alert the reader to faulty procedures that may cause minor or moderate personal injury or damage equipment,

⚠ **WARNING** ⚠

- Do not use force to open or close the disk tray. Force may result in a damaged tray.
- Place nothing but a compact disk in the tray. Inserting objects other than disks in the tray may result in a damaged tray.

Figure 16–10 A Warning Message

as in the warning from a compact disc player manual shown in Figure 16–10. The exclamation point inside the triangle in Figure 16–10 is a commonly accepted symbol to attract the reader's attention and to stress the importance of the message. You will see it used on all three levels of warnings.

Danger Use **danger** for the highest level of warning: a warning to prevent major personal injury or death. Obviously, you must make danger messages stand out typographically and write them with utter clarity. Figure 16–11 presents a good example.

⚠ **DANGER** ⚠

- **Use no oil.**
- Oil coming in contact with a high-pressure connection in diving equipment may result in an **explosion.**
- To prevent serious injury or death, **use no oil.**

Figure 16–11 A Danger Message

How-to Instructions

The actual instructions on how to perform the process lie at the heart of any set of instructions. For sample sets of how-to instructions, see Figures 16–12 through 16–15. The same general principles apply to all how-to instructions. What we now tell you about writing them is well illustrated by these samples.

Style When you are writing how-to instructions, one of your major concerns is to use a clear, understandable style. To achieve this, write your instructions in the active voice and imperative mood: *Turn the mouse upside down and rotate the plastic dial counterclockwise as far as it will go.* The imperative mood is normal and acceptable in instructions. It's clear and precise and will not offend the reader. By using the format shown in Figure 16–12, you can use the imperative mood even when several people with distinct tasks have to carry out the procedure.

DELUXE CHECK PRINTERS, INC. **STANDARD OPERATING PROCEDURES**

Procedure C-9
Establishing, Changing, or Eliminating the Petty Cash Fund
Accounts Payable and Purchasing Manual - C

SUMMARY: The petty cash fund is a fixed cash fund reserved for minor expenditures of $50 or less. This procedure explains how to establish, change, or eliminate the petty cash fund.

NOTE: When a petty cash fund is established, the plant manager should assign responsibility to no more than two cash drawer custodians, with one individual having primary responsibility. The accounts payable clerk must not be a custodian of the petty cash fund.

See Procedure C-17 to disburse petty cash. See Procedure C-18 to replenish the petty cash fund. See Appendix J for petty cash fund controls.

RESPONSIBILITY	ACTION
Plant manager	1. Request authorization from corporate treasurer for one of the following: • establish petty cash fund • change amount of existing petty cash fund • eliminate existing petty cash fund
Corporate treasurer	2. Review request and approve or disapprove and notify plant accountant of decision.
Plant accountant	3. Notify plant manager of decision.
	4. If establishing fund or increasing existing fund, have check prepared from Account 1030 (Regular Cash Account) for authorized amount, payable to cash, debiting Account 1010 (Petty Cash Account) on check voucher.
	4a. Place check in check cashing fund box and withdraw authorized amount of cash.
	4b. Place cash in petty cash fund, and notify plant manager that petty cash fund is established or increased.
	-OTHERWISE-
	5. If decreasing or eliminating existing fund, use Daily Report of Cash, form A-30-Q (Exhibit 24), to credit Account 1010.

Rewritten by: Kathy Huebsch

Figure 16–12 Standard Operating Procedures
Source: Reprinted by permission of Deluxe Check Printers, Inc.

In the format shown, the headings to the left identify the responsible actor, allowing you to use the imperative mood in the right action column. It's an efficient system. (For more on the active voice and imperative mood, see Chapters 5 and 7.)

Sample Calculations

The first house plan discussed in Chapter 8 (Figure 8.3) is analyzed. Orlando weather data are used. The calculations follow.

Form for Calculating Window Areas in Naturally Ventilated Houses

Project ______________________

Analyst ______________________

1. House conditioned floor area = 1334 ft^2 (1)

2. Average ceiling height = 8 ft (2)

3. House volume = 1334 (Step 1) x 8 (Step 2) = 10,672 ft^3 (3)

4. Design air change rate/hr = (recommended value is 30) 30 ACH (4)

5. Required airflow rate, cfm = 10,672 (Step 3) x 30 (Step 4) ÷ 60 = 5336 cfm (5)

6. Design month = may (6)

   (Recommended months = May for Florida and the Gulf Coast
   = June for more northern southeast cities)

7. Nearest city location with weather data = Orlando (7)

8. From weather data in Appendix B, determine windspeed (WS) and wind direction (WD) for design month

   8a. WS = 8.8 mph (8a)

   8b. WD = SE (8b)

9. From prevailing wind direction and building orientation, determine incidence angle on windward wall. Incidence angle = about 10 degrees (9)

10. From Table A1, determine inlet-to-site 10 meter windspeed ratio = 0.35 (10)

Figure 16–13 Sample Calculations

Source: Subrato Chandra, Philip W. Fairey III, and Michael M. Houston, *Cooling with Ventilation* (Washington, DC: GPO, 1986) 64–65.

Notice that the sets of how-to instructions we've shown you all use a list format. Each numbered step usually contains only one instruction and at the most two or three closely related instructions. The list format keeps each step in a series clear and distinct from every other step. Listing has several other advantages as well:

11. Determine windspeed correction factors

11a. For house location and ventilation strategy, determine terrain correction factor from Table A2 = 0.67 (11a)

11b. For neighboring buildings, determine neighborhood convection factor from Table A3 = 0.77 (11b)

Assume h = 8 ft, =
g = 24 ft

11c. If sizing windows for the second floor or for house on stilts, use a height multiplication factor of 1.15. Otherwise, use 1.0. Selected value = 1.0 (11c)

12. Determine overall windspeed correction factor

= 0.67 (Step 11a) x 0.77 (Step 11b) x 1.0 (Step 11c) = 0.52 (12)

13. Determine site windspeed in ft/min

= 8.8 (Step 8a) x 0.52 (Step 12) x 88 = 403 (13) ft/min

14. Determine window inlet airspeed

= 403 (Step 13) x 0.35 (Step 10) = 141 (14) ft/min

15. Determine net aperture inlet area

= 5336 (Step 5) ÷ 141 (Step 14) = 37.8 (15) ft^2

16. Determine total inlet + outlet area, insect screened. Assumes fiberglass screening with a porosity of 0.6

= 2 x 37.8 (Step 15) x 1.67 = 126 (16) ft^2

17. Since typical window or door framing is about 20% of the gross area, determine gross total operable area required as

= 1.25 x 126 (Step 16) = 156 (17) ft^2

18. Determine gross operable area as a % of floor area

= 158 (Step 17) ÷ 1334 (Step 1) x 100 = 11.8 (18) %

Note: This gross operable area requirement can be met the by same area of windows if the windows are 100% operable (awning, casement, hopper, etc.). The window area required will be twice this value if single-hung or sliding windows are used which have only 50% of the area as operable.

Figure 16-13 (continued).

- It makes it obvious how many steps there are.
- It makes it easy for readers to find their place on the page.
- It allows the reader to use the how-to instructions as a checklist.

Use familiar, direct language and avoid jargon. Tell your readers to *check* things or to *look them over*. Don't tell them to *conduct an investigation*. Tell your readers to *use* a wrench, not to *utilize* one. Fill your instructions with readily recognized verbs such as *adjust, attach, bend, cap, center, close, drain, install, lock, replace, spin, turn*, and *wrap*. For more on good style, see Chapter 5, "Achieving a Readable Style."

The mouse

Be careful not to drop the mouse or let it hang from a table by its cable. Use common sense in treating it as carefully as you can.

The surface your mouse moves on should be as smooth, clean, and dust-free as possible. And give the mouse itself an occasional cleaning.

Here's how to clean the mouse:

1. **Turn the mouse upside down and rotate the plastic dial counterclockwise as far as it will go.**

2. **Holding one hand over the ball and dial to catch them, turn the mouse back right side up. The dial and the ball will drop into your hand.**

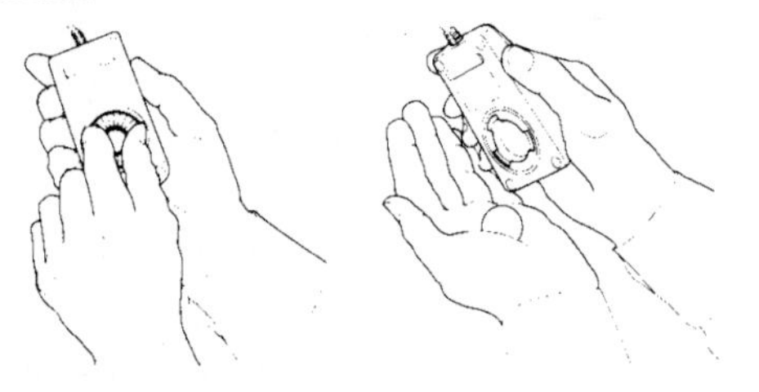

3. **Inside the case are three plastic rollers, similar to those on a tape recorder. Using a cotton swab moistened with alcohol or tape head cleaner, gently wipe off any oil or dust that has collected on the rollers, rotating them to reach all surfaces.**

4. **Wipe the ball with a soft, clean, dry cloth. (Don't use tissue or anything that may leave lint, and don't use a cleaning liquid.)**

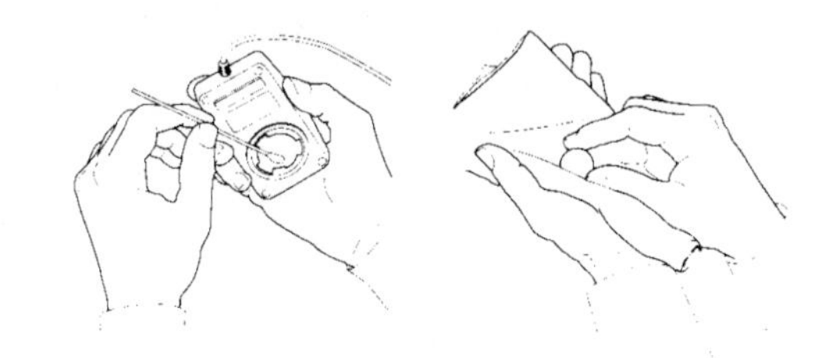

5. **Blow gently into the case to remove any dust that has collected there.**

6. **Put the ball back into its case and, lining up the indicator on the dial with the O on the back of the case, reinsert the dial and turn it clockwise as far as it will go.**

Figure 16–14 How-to Instructions: Mouse
Source: Apple Computer, Inc., *Macintosh SE Owner's Guide* (Cupertino, CA: 1988) 64–65. Copyright © 1988 Apple Computer, Inc. Reprinted by permission.

If your how-to instructions call for calculations, include sample calculations to clarify them for the reader. See Figure 16–13 for an example of such calculations.

Graphics Be generous with graphics. Word descriptions and graphics often complement each other. The words tell *what* action is to be done. The graphics show *where* it is to be done and often *how* to do it. Our samples demonstrate well the relationship between words and graphics. Note that graphics are often annotated to allow for easy reference to them. As Figures 16–14 and 16–15 demonstrate, graphics can be used to show the worker, or at least the worker's hands, actually performing the job.

Arrangement When writing performance instructions, arrange the process being described into as many major routines and subroutines as needed. For example, a set of instructions for the overhaul and repair of a piece of machinery might be broken down as follows:

- Disassembly of major components
- Disassembly of components
- Cleaning
- Inspection
- Lubrication
- Repair
- Reassembly of components
- Testing of components
- Reassembly of major components

ELECTRIC PLUGS—Repair or Replace

YOUR PROBLEM

- Lamps or appliances do not work right.
- A damaged plug is dangerous.
- It's hard to hire help for small repairs.

WHAT YOU NEED

- New plug—if your old one cannot be used. (Buy one with a UL label)
- Screwdriver
- Knife

HOW-TO

1. Cut the cord off at the damaged part. (Fig. 1.)
2. Slip the plug back on the cord. (Fig. 2.)
3. Clip and separate the cord. (Fig. 3)
4. Tie Underwriters' knot. (Fig. 4)
5. Remove a half-inch of the insulation from the end of the wires. **Do not cut any of the small wires.** (Fig. 5)

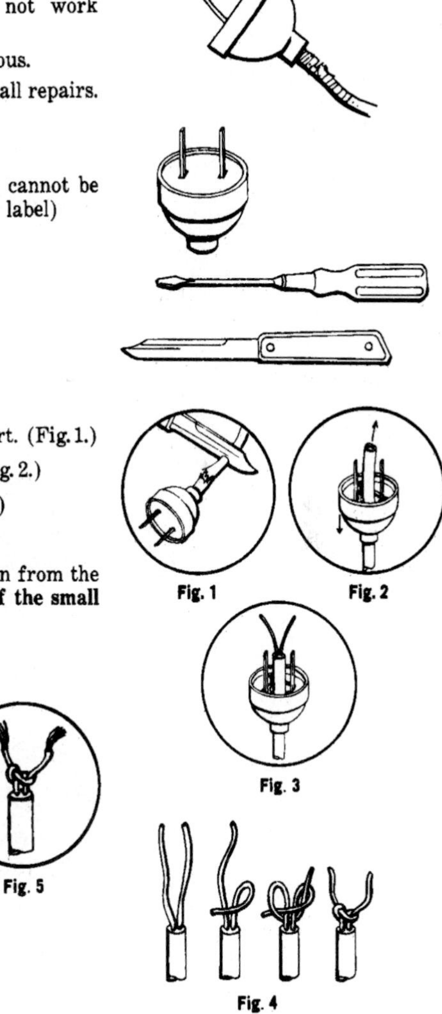

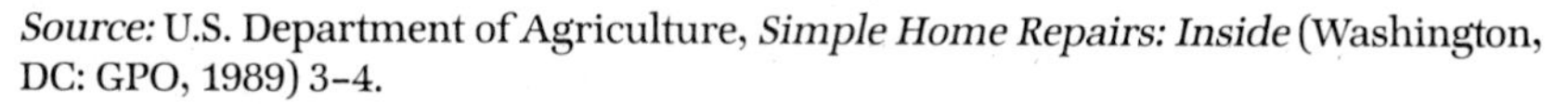

Figure 16–15 How-to Instructions: Home Repair

Source: U.S. Department of Agriculture, *Simple Home Repairs: Inside* (Washington, DC: GPO, 1989) 3–4.

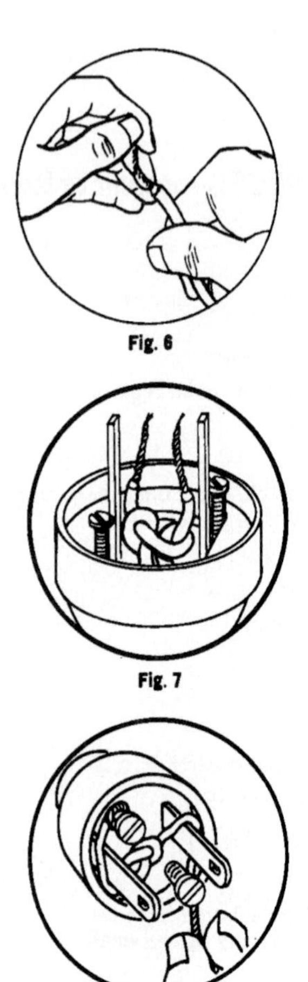

6. Twist small wires together, clockwise. (Fig. 6)
7. Pull knot down firmly in the plug. (Fig. 7)
8. Pull one wire around each terminal to the screw. (Fig. 8)
9. Wrap the wire around the screw, clockwise. (Fig. 9)
10. Tighten the screw. Insulation should come to the screw but not under it. (Fig. 10)
11. Place insulation cover back over the plug. (Fig. 11)

YOUR REWARD

- The appliance or lamp is back in working condition.
- You have eliminated a possible cause of a fire or shock.
- You have saved money by doing the repair yourself.

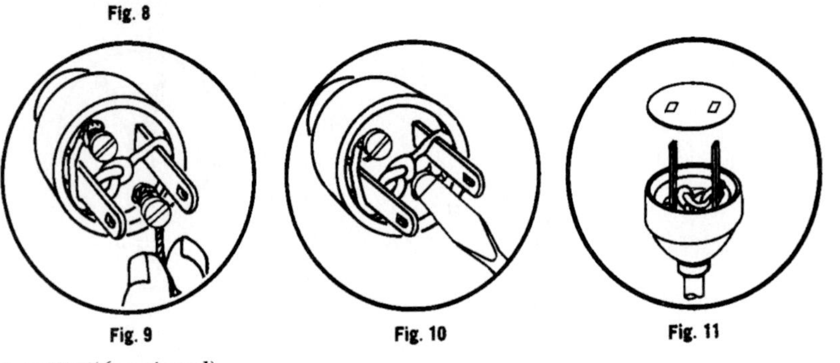

Figure 16-15 (continued).

Notice that in this case the steps are in chronological order. Our samples also demonstrate chronological order.

If there are steps that are repeated, it is sometimes a legitimate practice to tell the reader to "Repeat steps 2, 3, and 4." But whether you do so depends upon your analysis of the reader's situation. Visualize your reader. Maybe he or she will be perched atop a shaky ladder, your instructions in one hand, a tool in the other. Under such circumstances, the reader will not want to be flipping pages around to find the instructions that need to be repeated. You will be wiser and kinder to print once again all the instructions of the sequence. If the reader will be working in a comfortable place with both feet on the ground, you will probably be safe enough saying, "Repeat steps...."

Such reader and situation analysis can help you make many similar decisions. Suppose that your readers are not expert technicians, and the process calls for them to use simple test equipment. In such a situation, you should include the instructions for operating the test equipment as part of the routine you are describing. On the other hand, suppose your readers are experienced technicians following your instructions at a comfortable workbench with a well-stocked library of manuals nearby. You can assume that they know how to operate any needed test equipment, or you can refer them to another manual that describes how to operate the test equipment.

Only the instructions in Figure 16–15 have a conclusion of sorts, labeled in this case "Your Reward." For the most part, instructions have no conclusions. They simply end with the last instruction. On occasion, particularly when writing for lay people, you might wish to close with a summary of the chief steps of the process or, perhaps, a graceful close. However, such endings are not general practice.

Tips and Troubleshooting Procedures

Many sets of instructions contain sections that give the reader helpful tips on how to do a better job or that provide guidance when trouble occurs.

Tips You may present tips in a separate section, as illustrated in Figure 16–16. Or you may incorporate them into the how-to instructions, as in this excerpt on setting flexible tile. In the excerpt, the last sentence in instructions 1, 2, 3, and 5 gives the reader a tip that should make the task go more easily:

1. Remove loose or damaged tile. A warm iron will help soften the adhesive.
2. Scrape off the old adhesive from the floor or wall. Also from the tile if you're to use it again.
3. Fit tiles carefully. Some tile can be cut with a knife or shears, others with a saw. Tile is less apt to break if it's warm.

New Ways to Make Sauces and Soups

Sauces, including gravies and homemade pasta sauces, and many soups often can be prepared with much less fat. Before thickening a sauce or serving soup, let the stock or liquid cool – preferably in the refrigerator. The fat will rise to the top and it can easily be skimmed off. Treat canned broth-type soups the same way.

For sauces that call for sour cream, substitute plain low-fat yogurt. To prevent the yogurt from separating, mix 1 tablespoon of cornstarch with 1 tablespoon of yogurt and mix that into the rest of the yogurt. Stir over medium heat just until the yogurt thickens. Serve immediately. Also, whenever you make creamed soup or white sauces, use skim or 1% milk instead of 2% or whole milk.

New Ways to Use Old Recipes

There are dozens of cookbooks and recipe booklets that will help you with low-fat cooking. But there is no reason to stop using your own favorite cookbook. The following list summarizes many of the tips. Using them, you can change tried and true recipes to low-saturated fat, low-cholesterol recipes. In some cases, especially with baked products, the quality or texture may change. For example, using vegetable oil instead of shortening in cakes that require creaming will affect the result. Use margarine instead; oil is best used only in recipes calling for **melted** butter. Substituting yogurt for sour cream sometimes affects the taste of the product. Experiment! Find the recipes that work best with these substitutions.

Instead of	Use
1 tablespoon butter	1 tablespoon margarine or ¾ tablespoons oil
1 cup shortening	⅔ cup vegetable oil
1 whole egg	2 egg whites
1 cup sour cream	1 cup yogurt (plus 1 tablespoon cornstarch for some recipes)
1 cup whole milk	1 cup skim milk

Low-Fat Cooking Tips

Your kitchen is now stocked with great tasting, low-saturated fat, low-cholesterol foods. But you may still be faced with the temptation to fix your favorite higher fat meats, rich soups, and baked breads and cookies. The suggestions below will help you to reduce the amount of total and saturated fats in these foods.

New Ways to Prepare Meat, Poultry, Fish, and Shellfish

When you prepare meats, poultry, and fish, remove as much saturated fat as possible. Trim the visible fat from meat. Remove the skin and fat from the chicken, turkey, and other poultry. And, if you buy tuna or other fish that is packed in oil, rinse it in a strainer before making tuna salad or a casserole, or buy it packed in water.

Changes in your cooking style can also help you remove fat. Rather than frying meats, poultry, fish, and shellfish, try broiling, roasting, poaching, or baking. Broiling browns meats without adding fat. When you roast, place the meat on a rack so that the fat can drip away.

Finally, if you baste your roast, use fat-free ingredients such as wine, tomato juice, or lemon juice instead of the fatty drippings. If you baste turkeys and chickens with fat use vegetable oil or margarine instead of the traditional butter or lard. Self-basting turkeys can be high in saturated fat–read the label!

Figure 16–16 Tips

Source: U.S. Department of Health and Human Services, *Eating to Lower Your Cholesterol* (Washington, DC: GPO, 1989) 23–24.

4. Spread adhesive on the floor or wall with a paint brush or putty knife.
5. Wait until adhesive begins to set before placing the tile. Press tile on firmly. A rolling pin works well.[9]

Troubleshooting Procedures You may incorporate troubleshooting procedures into your how-to instructions, as in this excerpt:

> Tighten screws in the hinges. If screws are not holding, replace them one at a time with a longer screw. Or insert a matchstick in the hole and put the old screw back.[10]

Perhaps more often, troubleshooting procedures will have a section of their own, as in Figures 16–17 and 16–18. Both figures illustrate a typical format, a three-column chart with headings such as *Problem, Possible Cause,* and *Possible Remedy.* Notice that the chart in Figure 16–17 uses graphics to illustrate the problem, an excellent technique you should use where possible. Notice also that the remedies are given as instructions in the active voice, imperative mood.

The example in Figure 16–18 gives page references when appropriate to guide the reader to additional information, also an excellent idea.

TROUBLE-SHOOTING

Most baler operating problems are caused by improper adjustment or delayed service. This chart is designed to help you when a problem develops, by suggesting a probable cause and the recommended solution.

Apply these suggested remedies carefully. Make certain the source of the trouble is not some place other than where the problem exists. A thorough understanding of the baler is a must if operating problems are to be corrected satisfactorily. Refer to the operator's manual for detailed repair procedures.

TROUBLE-SHOOTING CHART

Knotter Difficulties — Twine Baler

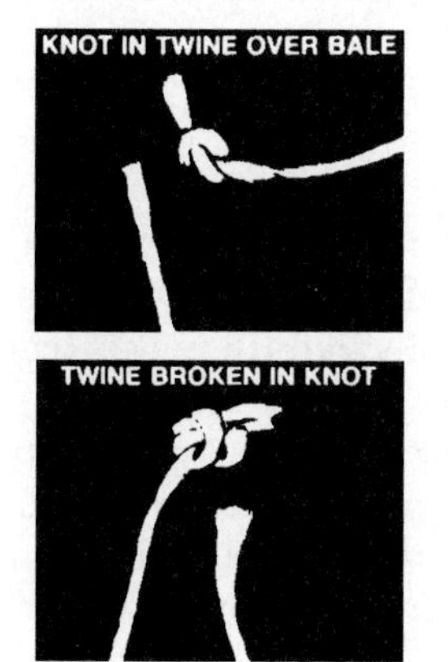

PROBLEM	POSSIBLE CAUSE	POSSIBLE REMEDY
KNOT IN TWINE OVER BALE	Tucker fingers did not pick up needle twine or move it into tying position properly.	Adjust tucker fingers. Adjust needles or twine disk. Check twine disk and twine-box-tension. Install plungerhead extensions.
	Hay dogs do not hold end of bale.	Free frozen hay dogs. Replace broken hay-dog springs. Reduce feeding rate. Install plungerhead extensions.
TWINE BROKEN IN KNOT	Extreme tension on twine around billhook during tying cycle causes twine to shear or pull apart.	Loosen twine-disk-holder spring. Smooth off all rough surfaces and edges on billhook.

Figure 16–17 Troubleshooting Chart
Source: Reprinted by permission of Deere and Company.

ELECTRICAL SYSTEM

PROBLEM	POSSIBLE CAUSE	POSSIBLE REMEDY	PAGE REFERENCE
Battery will not charge	Loose or corroded connections	Clean and tighten connections	66
	Sulfated or worn-out battery	Check electrolyte level and specific gravity	67
	Loose or defective alternator belt	Adjust belt tension or replace belt	52
"CHG" indicator glows with engine running	Low engine speed	Increase speed	
	Defective battery	Check electrolyte level and specific gravity	67
	Defective alternator	Have your John Deere dealer check alternator	

Figure 16–18 Troubleshooting Chart
Source: Reprinted by permission of Deere and Company.

Glossary

If your audience analysis tells you that your reader will not comprehend all the terminology you plan to use in your instructions, you'll need to provide definitions. If you need only a few definitions, you can define terms as you use them. You can even provide graphic definitions, as in the definition of the "underwriters' knot" in Figure 16–15 on page 523.

If you must provide many definitions, you'll probably want to provide a glossary as a separate section. Figure 16–19 illustrates a typical layout and style. Arrangement is alphabetical and the first sentence in each definition is a fragment. The rest of the definition is in complete sentences. See also Chapter 7, where we discuss definitions, and Chapter 10, where we discuss glossaries.

Accessible Format

Your major goal in setting up your format in instructions should be to make the information accessible for your readers. Our Part III, "Document Design," is especially helpful in this regard.

The theory section shown in Figure 16–2 on pages 507–508 demonstrates excellent accessibility. The type is large and readable, and the format is especially helpful for readers who scan the document. The headings standing apart to the left of the print allow the reader to scan quickly, looking for points of interest. Also, headings phrased as questions are more likely to arrest the attention of scanning readers and attract them to read the text. Curiosity is put to work. They may want to know the answers to the questions.

Glossary

1. **Atherosclerosis** - A type of "hardening of the arteries" in which cholesterol, fat, and other blood components build up on the inner lining of arteries. As atherosclerosis progresses, the arteries to the heart may narrow so that oxygen-rich blood and nutrients have difficulty reaching the heart.

2. **Carbohydrate** - One of the three nutrients that supply calories (energy) to the body. Carbohydrate provides 4 calories per gram—the same number of calories as pure protein and less than half the calories of fat. Carbohydrate is essential for normal body function. There are two basic kinds of carbohydrate—simple carbohydrate (or sugars) and complex carbohydrate (starches and fiber). In nature, both the simple sugars and the complex starches come packaged in foods like oranges, apples, corn, wheat, and milk. Refined or processed carbohydrates are found in cookies, cakes, and pies.

   - **Complex carbohydrate** - Starch and fiber. Complex carbohydrate comes from plants. When complex carbohydrate is substituted for saturated fat, the saturated fat reduction helps lower blood cholesterol. Foods high in starch include breads, cereals, pasta, rice, dried beans and peas, corn, and lima beans.

   - **Fiber** - A nondigestible type of complex carbohydrate. High-fiber foods are usually low in calories. Foods high in fiber include whole grain breads and cereals, whole fruits, and dried beans. The type of fiber found in foods such as oat and barley bran, some fruits like apples and oranges, and some dried beans may help reduce blood cholesterol.

3. **Cholesterol** - A soft, waxy substance. It is made in sufficient quantity by the body for normal body function, including the manufacture of hormones, bile acid, and vitamin D. It is present in all parts of the body, including the nervous system, muscle, skin, liver, intestines, heart, etc.

   - **Blood cholesterol** - Cholesterol that is manufactured in the liver and absorbed from the food you eat and is carried in the blood for use by all parts of the body. A high level of blood cholesterol leads to atherosclerosis and coronary heart disease.

Figure 16–19 Glossary
Source: U.S. Department of Health and Human Services, *Eating to Lower Your Blood Cholesterol* (Washington, DC: GPO, 1989) 27.

The graphic of the narrowed artery and the table showing cholesterol levels highlight the two key points in the section. The scanning reader who stops only long enough to absorb the information in the two graphics will at least know the principal danger of high cholesterol and what a desirable cholesterol level is.

Look now at Figure 16–20, a government document instructing its readers how to file a form to establish their relationship with alien

U.S. Department of Justice
Immigration and Naturalization Service

PETITION TO CLASSIFY STATUS OF ALIEN RELATIVE
FOR ISSUANCE OF IMMIGRANT VISA

READ INSTRUCTIONS CAREFULLY. FEE WILL NOT BE REFUNDED.

Not all of these instructions relate to the type of case which concerns you. Please read carefully those which do relate. Failure to follow instructions may require return of your petition and delay final action.

1. **Eligibility.** A petition may be filed by a citizen or a lawful permanent resident of the United States to classify the status of alien relatives as follows:

   a. *By citizen of the United States:* Except as noted in paragraph 2, a citizen of the United States may submit a petition on behalf of a spouse or sons and daughters (regardless of age or marital status). A United States citizen at least 21 years of age may submit a petition for a parent, brother, or sister. If the petition is for a son or daughter who is married or at least 21 years of age, or both, or for a brother or sister, do not submit petitions for the beneficiary's spouse or unmarried children under 21 years of age. If the petition is approved, the beneficiary's spouse and unmarried children under 21 years of age, if accompanying or following to join him/her, will automatically be eligible for the same preference status.

   b. *By a lawful permanent resident alien:* Except as noted in paragraph 2, an alien lawfully admitted to the United States for permanent residence may submit a petition on behalf of a spouse or an unmarried child regardless of age. However, if a lawful permanent resident alien is married to a citizen and wishes to petition for an unmarried child, such alien should consult the nearest office of the Immigration and Naturalization Service for advice as to whether it would be preferable, or *necessary*, for the United States citizen spouse to submit the petition instead. If the petition is for an unmarried son or daugther, do not submit patitions for the beneficiary's unmarried children under 21 years of age. If the petition is approved, the beneficiary's unmarried children under 21 years of age, if accompanying or following to join him/her, will automatically be eligible for the same preference status.

2. **Petitions which cannot be approved.** Approval cannot be given to a petition on behalf of—

   a. A parent, brother, or sister, unless the petitioner is a United States citizen and at least 21 years of age.

   b. An adoptive parent, unless the relationship to the United States citizen petitioner exists by virtue of an adoption which took place while the child was under the age of 16, and the child has thereafter been in the legal custody of, and has resided with, the adopting parent or parents for at least 2 years. While the legal custody must be after the adoption, residence occurring prior to the adoption can satisfy the residence requirement.

   c. A stepparent, unless the marriage creating the status of stepparent occurred before the citizen stepchild reached the age of 18 years.

   d. An adopted child, unless the child was adopted while under the age of 16 and has thereafter been in the legal custody of, and has resided with, the adopting parent or parents for at least 2 years. While the legal custody may be after the adoption, residence occurring prior to the adoption can satisfy the residence requirement.

   e. A stepchild, unless the child was under the age of 18 years at the time the marriage creating the status of stepchild occurred.

   f. A wife or husband by reason of any marriage ceremony where the contracting parties thereto were not physically present in the presence of each other, unless the marriage shall have been consummated.

   g. A grandparent, grandchild, nephew, niece, uncle, aunt, cousin, or in-law.

3. **Supporting documents.** The following documents must be submitted with the petition:

   a. *To prove United States citizenship of petitioner* (where petition is for relative of a citizen).

      (1) If you are a citizen by reason of birth in the United States, submit your birth certificate. If your birth certificate is unobtainable, see "Secondary Evidence" below for submission of document in place of birth certificate.

      (2) If you were born outside the United States and became a citizen through the naturalization or citizenship of a parent or husband, and have not been issued a certificate of citizenship in your own name, submit evidence of the citizenship and marriage of such parent or husband, as well as termination of any prior marriages. Also, if you claim citizenship through a parent, submit your birth certificate and a separate statement showing the date, port, and means of all your arrivals and departures into and out of the United States. (Do not make or submit a photostat of a certificate of citizenship.)

      (3) If your naturalization occurred within 90 days immediately preceding the filing of this petition, of if it occurred prior to September 27, 1906, the naturalization certificate must accompany the petition. Do not make or submit a photostat of such certificate.

   b. *To prove family relationship between petitioner and beneficiary.*

      (1) If petition is submitted on behalf of a wife or husband, it must be accompanied by a certificate of marriage to the beneficiary and proof of legal termination of all previous marriages of both wife and husband.

      (2) If a petition is submitted by a mother on behalf of a child (regardless of age), the birth certificate of the child, showing the name of the mother, must accompany the petition. If the petition is submitted by a father or stepparent on behalf of a child (regardless of age), certificate of marriage of the parents, proof of termination of their prior marriages, and birth certificate of the child showing the names of the parents thereon, must accompany the petition.

      (3) If petition is submitted on behalf of a brother or sister, your own birth certificate and the birth certificate of the beneficiary, showing a common mother, must accompany the petition. If the petition is on behalf of a brother or sister having a common father and different mothers, marriage certificate of your parents, and proof of termination of their prior marriages must accompany the petition.

      (4) If petition is submitted on behalf of a mother, your own birth certificate, showing the name of your mother, must accompany the petition. If petition is submitted on behalf of a father or stepparent, your own birth certificate, showing the names of the parents thereon, and marriage certificate of your parents must accompany the petition, as well as proof of termination of prior marriages of your parents.

      (5) If either the petitioner or the beneficiary is a married woman, marriage certifcate(s) must accompany the petition. However, when the relationship between the petitioner and beneficiary is that of a mother and child (regardless of age), the mother's marriage certificate need not be submitted if the mother's present married name appears on the birth certificate of the child.

      (6) If the petitioner and the beneficiary are related to each other by adoption, a certified copy of the adoption decree must accompany the petition.

   c. *Secondary evidence.*

      If it is not possible to obtain any one of the required documents or records shown above, the following may be submitted for consideration:

      (1) Baptismal certificate.—A certificate under the seal of the church where the baptism occurred within two months

Figure 16–20 Government Instructions before Revision

relatives who wish to immigrate to the United States. The document is an example of inaccessible format. The headings and the print are small. The page is cluttered and intimidating.

The headings are not worded in a way to lead readers to the information they seek. Terms such as "Eligibility," "Documents previously submitted," and "Documents in general" are probably meaningful to the person who wrote them but not to the typical reader of these instructions. The format violates most of the principles discussed in Chapter 9, "Document Design." Furthermore, the style of the

...er birth, showing date and place of the child's birth, date of baptism, and the names of the child's parents.

(2) School record.—A letter from the school authorities having jurisdiction over school attended (preferably the first school), showing the date of admission to the school, child's date of birth or age at that time, place of birth, and the names and places of birth of parents, if shown in the school records.

(3) Census Record.—State or federal census record showing the name(s) and place(s) of birth, and date(s) of birth or age(s) of the person(s) listed.

(4) Affidavits.—Written statements sworn to or affirmed by two persons who were living at the time, and who have personal knowledge, of the event you are trying to prove—for example, the date and place of birth, marriage, or death. The persons making the affidavits may be relatives and need not be citizens of the United States. Each affidavit should contain the following information regarding the person making the affidavit: his/her full name and address; date and place of birth; relationship to you, if any; full information concerning the event; and complete details concerning how he/she acquired knowledge of the event.

d. Documents and secondary evidence unavailable.

If you are unable to submit required evidence of birth, death, marriage, divorce or adoption because the event took place in a foreign country which does not record such events, and secondary evidence is unavailable, attach a statement to this effect, setting forth the date and place of each of your entries into the United States. Also attach any letters, photographs, remittances or similar documents which tend to support the claimed relationship and three passport type photographs of yourself.

e. Documents previously submitted.

If your birth abroad, or the birth abroad of any person through whom citizenship is claimed by you, was registered with an American consul, submit with this petition any registration form that was issued. If any required documents were submitted to and retained by the American consul in connection with such registration, or in connection with the issuance of a United States passport or in any other official matter, and you wish to use such documents in support of this petition instead of submitting duplicate copies, merely list such documents in an attachment to this petition and show the location of the consulate. If you wish to make similar use of required documents contained in any Immigration and Naturalization Service file, list them in an attachment to this petition and identify the file by name and number. Otherwise, the documents required in support of this petition must be submitted.

f. Documents in general.

All supporting documents must be submitted in the original. If you desire to have the original returned to you, and if copies are by law permitted to be made, you may submit photostatic or typewritten copies. Photostatic copies unaccompanied by the original may be accepted if the copy bears a certification by an immigration or consular officer that the copy was compared with the original and found to be identical. Any document in a foreign language must be accompanied by a translation in English. The translator must certify that he is competent to translate and that the translation is accurate. (Do not make a copy of a certificate of naturalization or citizenship.)

4. **Preparation of petition.** A separate petition for each beneficiary must be typewritten or printed legibly with pen and ink.

   (If you need more space to answer fully any questions on this form, use a separate sheet(s), identify each answer with the number of the corresponding question, and date and sign each sheet.) **Be sure this petition and attached Form I-130A are legible.**

5. **Submission of petition.** If you are residing in the United States, send the completed petition to the office of the Immigration and Naturalization Service having jurisdiction over your place of residence. If you are residing outside the United States consult the nearest American consulate as to the consular office or foreign officer of the Service designated to act on your petition. If you are a United States citizen petitioning for an immediate relative classification in behalf of your unmarried child, the petition must be submitted in sufficient time for action to be completed on the petition and for the child to obtain a visa and reach the United States before the date on which he/she will be 21 years of age.

6. **Approval of petition.** Upon approval of a petition filed by a United States citizen for his/her alien spouse, unmarried minor child, or parent, an immigrant visa may be issued to the alien without regard to the annual limitation on immigrant visa issuance. In the cases of all other aliens for whom immigrant visa petitions are approved, an immigrant visa number will be required. Availability of an immigrant visa number depends on the volume of demand by aliens in the same visa classification who have an earlier priority date on the visa waiting list.

7. **Fee.** A fee of thirty-five dollars ($35) must be paid for filing this petition. It cannot be refunded regardless of the action taken on the petition. DO NOT MAIL CASH. ALL FEES MUST BE SUBMITTED IN THE EXACT AMOUNT. Payment by check or money order must be drawn on a bank or other institution located in the United States and be payable in United States currency. If petitioner resides in Guam, check or money order must be payable to the "Treasurer, Guam." If petitioner resides in the Virgin Islands, check or money order must be payable to the "Commissioner of Finance of the Virgin Islands." All other petitioners must make the check or money order payable to the "Immigration and Naturalization Service." When check is drawn on an account of a person other than the petitioner, the name of the petitioner must be entered on the face of the check. If petition is submitted from outside the United States, remittance may be made by bank international money order or foreign draft drawn on a financial institution in the United States and payable to the "Immigration and Naturalization Service" in United States currency. Personal checks are accepted subject to collectibility. An uncollectible check will render the petition and any document issued pursuant thereto invalid. A charge of $5.00 will be imposed if a check in payment of a fee is not honored by the bank on which it is drawn.

8. **Penalties.** Severe penalties are provided by law for knowingly and willfully falsifying or concealing a material fact or using any false document in the submission of this petition.

9. **Authority.** The authority for collecting the information requested on this form is contained in 8 U.S.C. 1154(a). Submission of the information solicited is voluntary. The principal purpose for which the information is solicited is to determine the eligibility of the beneficiary for the benefits sought. The information solicited may also, as a matter of routine use, be disclosed to other federal, state, local, and foreign law enforcement and regulatory agencies, the Department of Defense including any component thereof (if either the beneficiary or petitioner has served, or is serving in the Armed Forces of the United States), the Department of State, Central Intelligence Agency, Interpol, and individuals and organizations, during the course of investigation to elicit further information required by this Service to carry out its functions. Failure to provide any or all of the solicited information may result in the denial of the petition.

Figure 16-20 (continued).

instructions violates most of the principles discussed in Chapter 5, "Achieving a Readable Style."

Now look at Figure 16–21, which is the same document after it has been revised and given a new format to make it accessible. Certain things are immediately obvious. The print is bigger and there is more white space. The headings are more meaningful and informative. They are phrased from the reader's point of view. They have become the kinds of questions that someone approaching this process might reasonably ask: "Who can file?" and "For whom can you file?" have replaced "Eligibility." Such new headings lead and inform the readers rather than confuse them. The format and style of the instructions is now a readable style, showing a knowledge and application of the principles discussed in Chapters 5 and 9. The result is a readable document.

Finally, when a set of instructions runs more than several pages, you should furnish a table of contents (TOC) to help your readers find their

U.S. Department of Justice
Immigration and Naturalization Service (INS)

Petition for Alien Relative

Instructions

Read the instructions carefully. If you do not follow the instructions, we may have to return your petition, which may delay final action.

1. Who can file?

A citizen or lawful permanent resident of the United States can file this form to establish the relationship of certain alien relatives who may wish to immigrate to the United States. You must file a separate form for each eligible relative.

2. For whom can you file?

A. If you are a citizen, you may file this form for:

1) your husband, wife, or unmarried child under 21 years old
2) your unmarried child over 21, or married child of any age
3) your brother or sister if you are at least 21 years old
4) your parent if you are at least 21 years old.

B. If you are a lawful permanent resident you may file this form for:

1) your husband or wife
2) your unmarried child

NOTE: If your relative qualifies under instruction A(2) or A(3) above, separate petitions are not required for his or her husband or wife or unmarried children under 21 years old. If your relative qualifies under instruction B(2) above, separate petitions are not required for his or her unmarried children under 21 years old. These persons will be able to apply for the same type of immigrant visa as your relative.

3. For whom can you *not* file?

You cannot file for people in these four categories:

A. An adoptive parent or adopted child, if the adoption took place after the child became 16 years old, or if the child has not been in the legal custody of the parent(s) for at least two years after the date of the adoption, or has not lived with the parent(s) for at least two years, either before or after the adoption.
B. A stepparent or stepchild, if the marriage that created this relationship took place after the child became 18 years old.
C. A husband or wife, if you were not both physically present at the marriage ceremony, and the marriage was not consummated.
D. A grandparent, grandchild, nephew, niece, uncle, aunt, cousin, or in-law.

4. What documents do you need?

You must give INS certain documents with this form to show you are eligible to file. You must also give INS certain documents to prove the family relationship between you and your relative.

A. For each document needed, give INS the original and one copy. However, because it is against the law to copy a Certificate of Naturalization, a Certificate of Citizenship or an Alien Registration Receipt Card (Form I-151 or I-551), give INS the original only. **Originals will be returned to you.**

B. If you do not wish to give INS the original document, you may give INS a copy. The copy must be certified by:

1) an INS or U.S. consular officer, or
2) an attorney admitted to practice law in the United States, or
3) an INS accredited representative
(INS still may require originals).

C. Documents in a foreign language must be accompanied by a complete English translation. The translator must certify that the translation is accurate and that he or she is competent to translate.

5. What documents do you need to show you are a United States citizen?

A. If you were born in the United States, give INS your birth certificate.
B. If you were naturalized, give INS your original Certificate of Naturalization.
C. If you were born outside the United States, and you are a U.S. citizen through your parents, give INS:
1) your original Certificate of Citizenship, or
2) your Form FS-240 (Report of Birth Abroad of a United States Citizen).
D. In place of any of the above, you may give INS your valid unexpired U.S. passport that was initially issued for at least 5 years.
E. If you do not have any of the above and were born in the United States, see the instructions under 8, below, *"What if a document is not available?"*

6. What documents do you need to show you are a permanent resident?

You must give INS your alien registration receipt card (Form I-151 or I-551). Do not give INS a photocopy of the card.

7. What documents do you need to prove family relationship?

You have to prove that there is a family relationship between your relative and yourself.

In any case where a marriage certificate is required, if either the husband or wife was married before, you must give INS documents to show that all previous marriages were legally ended. In cases where the names shown on the supporting documents have changed, give INS legal documents to show how the name change occurred (for example, a marriage certificate, adoption decree, court order, etc.).

Find the paragraph in the following list that applies to the relative you are filing for.

If you are filing for your:

A. **husband or wife,** give INS:

1) your marriage certificate
2) a color photo of you and one of your husband or wife, taken within 30 days of the date of this petition

Figure 16–21 Government Instructions after Revision

way and to provide an overview of the instructions. The headings in the TOC should duplicate those in the instructions. Figure 16–22 shows a useful TOC in which the headings are meaningful and informative to the readers.

These photos must have a white background. They must be glossy, un-retouched, and not mounted. The dimension of the facial image should be about 1 inch from chin to top of hair in 3/4 frontal view, showing the right side of the face with the right ear visible. Using pencil or felt pen, lightly print name (and Alien Registration Number, if known) on the back of each photograph.

3) a completed and signed Form G-325A (Biographic Information) for you and one for your husband or wife. Except for name and signature, you do not have to repeat on the G-325A the information given on your I-130 petition.

B. **child** and you are the **mother,** give the child's birth certificate showing your name and the name of your child.
C. **child** and you are the **father or stepparent,** give the child's birth certificate showing both parents' names and your marriage certificate.
D. **brother or sister,** give your birth certificate and the birth certificate of your brother or sister showing both parents' names. If you do not have the same mother, you must also give the marriage certificates of your father to both mothers.
E. **mother,** give your birth certificate showing your name and the name of your mother.
F. **father,** give your birth certificate showing the names of both parents and your parents' marriage certificate.
G. **stepparent,** give your birth certificate showing the names of both natural parents and the marriage certificate of your parent to your stepparent.
H. **adoptive parent or adopted child,** give a certified copy of the adoption decree and a statement showing the dates and places you have lived together.

8. What if a document is not available?

If the documents needed above are not available, you can give INS the following instead. (INS may require a statement from the appropriate civil authority certifying that the needed document is not available.)

A. Church record: A certificate under the seal of the church where the baptism, dedication, or comparable rite occurred within two months after birth, showing the date and place of child's birth, date of the religious ceremony, and the names of the child's parents.
B. School record: A letter from the authorities of the school attended (preferably the first school), showing the date of admission to the school, child's date and place of birth, and the names and places of birth of parents, if shown in the school records.
C. Census record: State or federal census record showing the name, place of birth, and date of birth or the age of the person listed.
D. Affidavits: Written statements sworn to or affirmed by two persons who were living at the time and who have personal knowledge of the event you are trying to prove; for example, the date and place of birth, marriage, or death. The persons making the affidavits need not be citizens of the United States. Each affidavit should contain the following information regarding the person making the affidavit: his or her full name, address, date and place of birth, and his or her relationship to you, if any; full information concerning the event; and complete details concerning how the person acquired knowledge of the event.

9. How should you prepare this form?

A. Type or print legibly in ink.
B. If you need extra space to complete any item, attach a continuation sheet, indicate the item number, and date and sign each sheet.
C. Answer all questions fully and accurately. If any item does not apply, please write "N/A".

10. Where should you file this form?

A. If you live in the United States, send or take the form to the INS office that has jurisdiction over where you live.
B. If you live outside the United States, contact the nearest American Consulate to find out where to send or take the completed form.

11. What is the fee?

You must pay $35.00 to file this form. **The fee will not be refunded, whether the petition is approved or not.** DO NOT MAIL CASH. All checks or money orders, whether U.S. or foreign, must be payable in U.S. currency at a financial institution in the United States. When a check is drawn on the account of a person other than yourself, write your name on the face of the check. If the check is not honored, INS will charge you $5.00.

Pay by check or money order in the exact amount. Make the check or money order payable to "Immigration and Naturalization Service". However,

A. if you live in Guam: Make the check or money order payable to "Treasurer, Guam", or
B. if you live in the U.S. Virgin Islands: Make the check or money order payable to "Commissioner of Finance of the Virgin Islands".

12. When will a visa become available?

When a petition is approved for the husband, wife, parent, or unmarried minor child of a United States citizen, these relatives do not have to wait for a visa number, as they are not subject to the immigrant visa limit. However, for a child to qualify for this category, all processing must be completed and the child must enter the United States before his or her 21st birthday.

For all other alien relatives there are only a limited number of immigrant visas each year. The visas are given out in the order in which INS receives properly filed petitions. To be considered properly filed, a petition must be completed accurately and signed, the required documents must be attached, and the fee must be paid.

For a monthly update on dates for which immigrant visas are available, you may call (202) 632-2919.

13. What are the penalties for submitting false information?

Title 18, United States Code, Section 1001 states that whoever willfully and knowingly falsifies a material fact, makes a false statement, or makes use of a false document will be fined up to $10,000 or imprisoned up to five years, or both.

14. What is our authority for collecting this information?

We request the information on this form to carry out the immigration laws contained in Title 8, United States Code, Section 1154(a). We need this information to determine whether a person is eligible for immigration benefits. The information you provide may also be disclosed to other federal, state, local, and foreign law enforcement and regulatory agencies during the course of the investigation required by this Service. You do not have to give this information. However, if you refuse to give some or all of it, your petition may be denied.

It is not possible to cover all the conditions for eligibility or to give instructions for every situation. If you have carefully read all the instructions and still have questions, please contact your nearest INS office.

Figure 16-21 (continued).

Reader Checks

When you are writing instructions, check frequently with the people who are going to use them. Bring them a sample of your theory section. Discuss it with them. See if they understand it. Does it contain too much

Table of Contents

Figure 16–22 Table of Contents
Source: U.S. Department of Health and Human Services, *Eating to Lower Your Blood Cholesterol* (Washington, DC: GPO, 1989) i.

theory or too little? Submit your how-to instructions to the acid test. Let members of the audience for whom the instructions are intended—but who are not familiar with the process—attempt to perform the process following your instructions. Encourage them to tell you where your instructions are confusing. A procedure called **protocol analysis** can be a help at this point. In this procedure, you ask the person following your instructions to speak into a tape recorder, giving his or her observations about the instructions while attempting to follow them. Here is an excerpt from a set of such observations made by someone trying to use a computer manual and on-line help to aid him in a word processing exercise:

> Somehow I've got the caps locked in here. I can't get to the lowercase. OK, I'm struggling with trying to come off those capitals. I'm not having any luck. So, what do I need to do. I could press help. See if that gets me anything. Using the keyboard. I'll try that. 2.0. I can't do that because it's in this mode. I'm getting uppercase on the numbers so I can't type in the help numbers. So I'll reset to get rid of

that. Big problem. Try reset. Merging text, formatting, setting margins, fixing problems. I can't enter a section number because I can't get this thing off lock. Escape. Nothing helps. Well, I'm having trouble here.[11]

Such information pinpoints troublesome areas in instructions. If you were writing instructions that were to be used by many people, protocol analysis would be a worthwhile investment in time and money. In any case, whether you use protocol analysis or not, if your readers can't follow your instructions, don't blame them. Rather, examine your instructions to see where you have failed. Often, you will find you have left out some vital link in the process or assumed knowledge on the part of your readers that they do not possess.

Planning and Revision Checklists

You will find the planning and revision checklists that follow Chapter 2, "Composing" (pages 38-39 and inside the front cover), and Chapter 4, "Writing for Your Readers" (pages 81-82), valuable in planning and revising any presentation of technical information. The following questions specifically apply to instructions. They summarize the key points in this chapter and provide a checklist for planning and revising.

PLANNING

- What is the purpose of your instructions?
- What is your reader's point of view?
- How and where will your readers use these instructions?
- What content does your reader really need and want?
- How should you arrange your content? Which of the following components should you include as a separate section? Which should you omit or include in another component, such as theory in the introduction?
    - Introduction
    - Theory or principles of operation: How much theory do your readers really need or want?
    - List of equipment and materials needed: Are your readers familiar with all the needed equipment and material? Do they need additional information?
    - Description of the mechanism: Does some mechanism play a significant role in these instructions?
    - Warnings: Are there expected outcomes that will be affected by improper procedure? Are there places in the instructions where improper procedure will cause damage to equipment or injury or death to people?
    - How-to instructions: Can your instructions be divided into routines and subroutines?

What is the proper sequence of events for your how-to instructions?

Tips and troubleshooting procedures: Are there helpful hints you can pass on to the reader? What troubles may come up? How can they be corrected?

Glossary: Do you need to define enough terms to justify a glossary?

- What graphics will help your instructions? Do you have them available or can you produce them?

REVISION

- Have you made the purpose of your paper clear to your readers?
- Can your readers scan your instructions easily, finding what they need?
- Do you have sufficient headings?
  - Do your headings stand out?
  - Are they meaningful to your readers?
  - Would it help to cast some as questions?
- Is all terminology unfamiliar to the reader defined somewhere?
- Is your print size large enough for your readers and their location?
- Is all your content relevant? Is it needed or desired by your readers? Have you made it easy for your readers to scan and to skip parts not relevant to them?
- Have you covered any needed theory adequately?
- Do your readers know what equipment and material they will need?
  - Do they know how to use the equipment needed?
  - If not, have you provided necessary explanations?
- Have you provided any necessary mechanism description?
- Are your caution, warning, and danger messages easy to see and clear in their meaning? Are you sure you have alerted your readers to every situation in which they might injure themselves or damage their equipment?
- Have you broken your how-to instructions into as many routines and subroutines as necessary?
- Are your steps in chronological order, with no steps out of sequence?
- Are your how-to instructions written in the active voice and imperative mood?
- Have you used a list format with short entries for each step of the instructions?
- Have you used simple, direct language and avoided jargon?
- If needed, have you provided sample calculations?
- Have you used graphics whenever they would be helpful? Are they sufficiently annotated?
- Have you provided helpful tips that may help your readers to do the task more efficiently?
- Have you anticipated trouble and provided troubleshooting procedures?
- If troubleshooting is covered in a separate section, is the section laid out in a way to clearly distinguish among problem, cause, and remedy?
- Do you have enough definitions to warrant a glossary?

- Are your instructions long enough to warrant a table of contents?
- Have you checked with your readers? Have you allowed a typical reader to attempt the procedure using your instructions? Have you corrected any difficulties such a check revealed?
- Have you checked thoroughly for all misspellings and mechanical errors?

Exercises

1. Writing instructions offers a wide range of possible papers. Short papers might consist of nothing more than an introduction and a set of how-to instructions. Examples—good and bad—of such short instructions can be found in hobby kits and accompanying such things as toys, tents, and furniture that must be assembled. Textbook laboratory procedures frequently are examples of a short set of instructions. Using the Planning and Revision Checklists for this chapter, write a short set of instructions. Here are some ideas:

   Developing film
   Drawing a blood sample
   Applying fertilizer
   Setting a bicycle gear
   Completing a form
   Accomplishing some do-it-yourself task around a house
   Replacing a part in an automobile or some other mechanism
   Cleaning a carpet
   Balancing a checkbook

2. Using the Planning and Revision Checklists for this chapter, write a set of instructions that includes at least six of the eight possible components listed on page 504. The components do not have to be in separate sections, but they must be clearly recognizable for what they are. Here are some suggested topics:

   Testing electronic equipment
   Writing (or following) a computer program
   Setting up an accounting procedure for a small business
   Conducting an agronomy field test
   Checking blood pressure
   Painting an automobile

3. Figure 16–23 is a set of instructions for ridding a house of termites. It is usable in its current form, but it could be greatly improved. In a collaborative group, examine and discuss the instructions. Using the

Termites/Search and Destroy

Your Problem
- Wood materials are threatened by termites.
- Wood materials have deteriorated.
- The house needs to be checked for the presence of termites.
- Preventive action against termites is required.

What You Need
- Flashlight
- Penknife or icepick
- Pick and shovel
- Chemical solution (consult an exterminator to determine the type of chemicals or treatment to use).

How-To: Checking for Termites

Wood decay and damage by insects are threats to the upkeep of the home.

The insects most destructive to wood in buildings are termites. There are two varieties: The "drywood" termite and the "subterranean" or "ground-nesting termite." Both thrive on wood for food.

"Drywood" termites can live without moisture, so that protection against them is very difficult. However, there are not many "drywood" termites in this country.

"Subterranean" or *"ground-nesting" termites* are a serious problem in the southern States. Subterranean termites live in colonies in the ground and require moisture to survive. The worker termites attack damp wood which is in contact with the ground. They may build earthen tunnels from the ground up to the wood. They will sometimes completely eat away the inside of a piece of wood while leaving the outside surface intact.

1. Check for termites at least twice a year.

2. During the spring and summer (termite mating season), call an exterminator to identify large numbers of flying insects that you cannot identify.

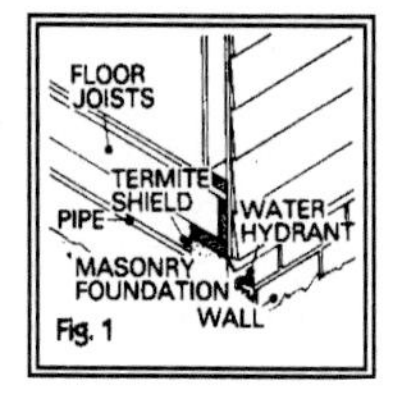

3. Look for earthen tunnels in the following locations:
- Along masonry foundation and basement walls
- Around openings where pipes enter walls
- Along the surface of metal pipes (fig. 1).

4. Examine all cracks in slabs, and loose mortar in masonry walls. Check all joints where wood meets with concrete or masonry, at walls, slabs, piers, etc.

5. Inspect all wood and wood structures that are near the ground. Pay special attention to any that touch the house, such as fences, wood trellises, carports, etc. Examine crawl spaces that provide moist conditions.

6. Check windowsills, door thresholds, porches and the underside of stairs. Be on the lookout for peeling and blistering paint.

Figure 16–23 Termite Protection Instructions

Source: U.S. Department of Agriculture, *Simple Home Repairs: Outside* (Washington, DC: GPO, 1986) 38–40.

7. If you suspect that wood has termite damage, probe with a sharp point, such as an icepick or penknife (fig. 2). If the point penetrates the wood to a depth of ½ inch, when you use only hand pressure, it's a good indication of wood damage by termites.

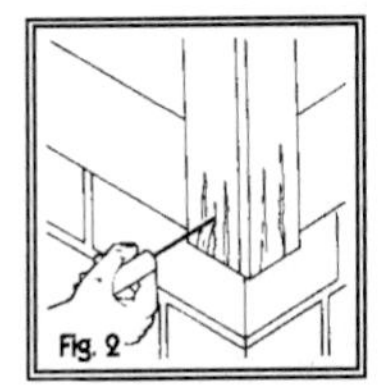

How-To: Protecting Against Termites

Chemicals needed to control termites are toxic to animals and plant life. There is also danger of contaminating the water supply. The chemicals should be applied with extreme caution and preferably by an experienced person.

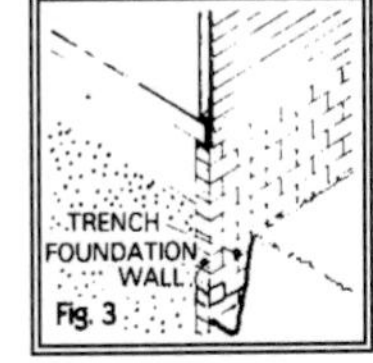

1. The following procedure should be followed when chemical treatments are necessary for an existing building:

(a) Dig a trench, approximately 1 foot wide and 3 feet deep, adjacent to the foundation wall (fig. 3).

(b) Prepare a solution of the insecticide. (Consult your County Extension Office or local exterminator regarding the recommended type and mixing instructions.)

(c) Pour the insecticide against the exposed wall surface and into the trench as it is backfilled. The solution should be applied to all other locations where wood and masonry meet at a joint. It should also be applied to other areas that have earth floors.

(d) Use extreme caution with these chemicals since they will also be poisonous to humans and pets. If a chemical is used inside the house, the room or space must be well ventilated and vacated for a period of time.

2. All surface water should be directed away from the building, allowing no water to accumulate at the foundations.

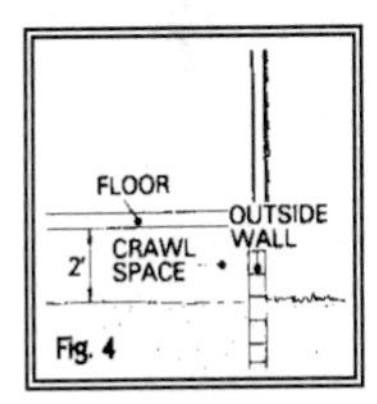

3. Cover the earth of unpaved basements with plastic film 4 mil or heavier.

4. Keep crawl spaces well ventilated. A house of 1,000 square feet should have at least 8 vents, 16 inches x 8 inches, open at all times. Crawl spaces should be at least 2 feet in height (fig. 4). Keep the space clear of wood scraps.

5. Untreated wood should not come closer than 6 inches to the ground.

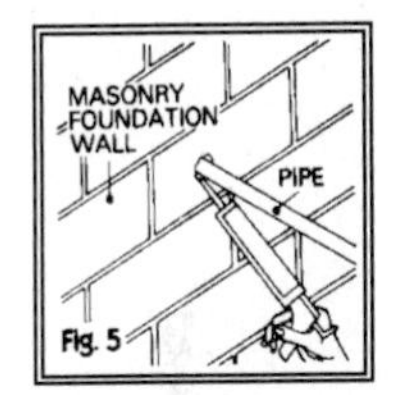

6. Using caulking compound, seal all openings where pipes pass through foundation walls or other walls of the house (fig. 5). Also, seal any cracks or points of loose mortar in masonry walls.

7. If there is a termite shield around the foundation it should be straightened and turned down (at least 2 inches) at approximately a 45-degree angle (fig. 6).

8. Make sure all scraps of lumber or stumps are removed when a building project is complete.

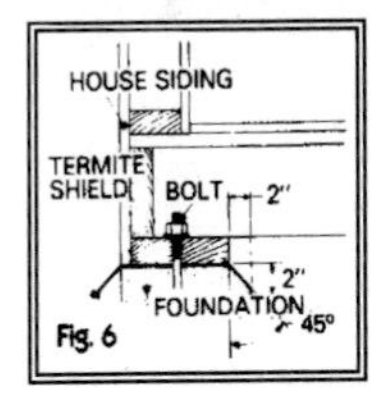

Your Benefits

- Controlling and preventing wood deterioration in your home.
- Preventing costly repairs later.
- Assuring an attractive appearance of the wood in your home.

Figure 16-23 (continued).

Revision Checklists for this chapter, decide on ways to improve them. When the discussion is finished, each member of the group should individually prepare a revision. If you have access to word processing, use some typographical variation to make the instructions more accessible.

CHAPTER 17 PROPOSALS AND PROGRESS REPORTS

The Relationship Between Proposals and Progress Reports

Many times as an employee in an organization, you will generate a variety of documents relating to one particular problem or situation. You may send several E-mail memoranda to colleagues within the organization; you may write letters to individuals outside the organization concerning the problem or situation; you may write memo reports "To File" that document your activities on the problem or situation; you may also write a detailed formal report, such as a formal feasibility study, discussed in Chapter 15, at the conclusion of your work on the situation. In short, you will write various documents to different audiences about one project, problem, or topic.

Proposals and progress reports are two additional documents that are often written in response to one project. The proposal, as its name implies, describes the work that will be done, the reasons it should be done, and the methods that will be used to accomplish the work. The progress report, as its name implies, describes and evaluates the project as the work is being done. Thus, if an individual or an organization decides to begin a work or research project, particularly one that requires several months or even several years to complete, the individual or organization will usually need to *propose* the project and then *report the progress* on that project at intervals agreed upon while the proposal is accepted and the resulting agreement or contract is being negotiated. When the topic of the progress report emanates from the project that is proposed, the content and organization of the progress report are often directed by the content and organization of the written proposal.

However, in other instances, employees may need to report progress on a variety of projects or problems on which they are working. In situations like these, the employee writes a progress report (or status report, as it may be called) to inform supervisors or other individuals about what has been accomplished in completing a job or solving a problem. By keeping these individuals up-to-date on work activities, the employee uses the status report to document what has been accomplished and by whom. The progress or status report thus becomes an official and even a legal record of work.

To help you understand how to design and write progress reports and proposals, we will divide this chapter into two parts. In the first part, we will discuss progress reports in general, as these discuss the status of progress or work on a project. To illustrate progress reports, we will present two example situations in which a student and an employee must report the status of a project: The first example shows a report designed to describe and evaluate the work completed on a student research project. The second example shows a report designed to describe, evaluate, and document the status of work on a routine project to which the employee has been assigned.

In the second part, to discuss the development of proposals and to show how they may relate to progress reports, we will use two additional examples: Our first example will include a student's research project proposal and her progress report on her research. The second situation discusses a project in progress and a status report written by one of the project leaders to share concerns he was having about the way the project was going. In this second situation, we will show you the letter proposal that initiated the project—a writing-instruction workshop in an agricultural chemical company—as well as the instructor's informal status report to the president of Write, Inc. who had submitted the proposal.

After you study this chapter, you should be able to develop the typical sections included in the proposal as well as in progress reports or status reports. As the following examples will show you, progress reports and proposals—in fact any kind of document we discuss in this book—can be submitted in a memo or letter format, as discussed in Chapters 12 and 14, or as a longer, formal document, as discussed in Chapter 15. The length of each document as well as the audience and the context in which the document is generated and received will determine which format you use.

Progress Reports

In this section, we discuss the nature and structure of progress reports.

The Nature of Progress Reports

When a soliciting organization requests a proposal, it often states that a specific number of progress reports will be required, particularly if the project covers a long time period. As their name suggests, progress reports, sometimes known as status reports, tell readers how work is progressing on a project. They are usually submitted at specific intervals agreed upon at the beginning of a project that requires several months or even years to complete. Their immediate purpose is to inform the authorizing person of the activities completed on a project, but their long-range purpose should be to show the proposing organization's or the individual's competence in pursuing a task and completing it.

As we mentioned at the beginning of this chapter, an employee may write progress reports routinely to report the status of projects on which the employee is working. These progress reports explain what the employee has done so that others interested in the progress are kept informed. They also help the employee or work group provide evidence of their activities. Whether a progress report is written to describe work done on a proposed project or whether the report is written to report activity on any job, progress reports have three main purposes that provide *documentation* of work accomplished:

- They explain to the reader what has been accomplished and by whom, the status of the work performed, and problems that may have arisen that need attention.
- They explain to your client how time and money have been spent, what work remains to be done, and how any problems encountered are being handled.
- They enable the organization or individual conducting the proposed work to assess the work and plan future work.

The Structure of Progress Reports

Several different strategies are used in designing progress reports. They should begin with an introduction and a project description to familiarize the reader with the project. A summary of work completed follows. The middle section then explains what has been accomplished on specific tasks as well as what work remains, followed by a statement of work planned for the next progress report period. The final section assesses the work done thus far. Any problems that are encountered are also presented, along with methods of addressing those problems in the form of conclusions and recommendations. Cost can be dealt with in either the middle or final section.

Structure by Work Performed The structure of a progress report might follow one of the two basic plans portrayed below. In the left-hand column, the middle is organized around work completed and work remaining. In the right-hand column, the middle is organized around tasks. Figure 17-1, based on Situation 1, illustrates the task model.

Beginning

- Introduction/project description
- Summary

Middle

• Work completed Task 1 Task 2, etc.	• Task 1 Work completed Work remaining
• Work remaining Task 3 Task 4	• Task 2 Work completed Work remaining
• Cost	• Cost

End

- Overall appraisal of progress to date
- Conclusions and recommendations

In this general plan, the writer emphasizes what has been done and what remains to be done and supplies enough introduction to be sure that the reader knows what project is being discussed.

Situation 1 (Figure 17–1)

Karim Tedesco, a senior engineering technology student, is developing his semester research project. He is surveying current industrial methods of controlling sulfate-producing bacteria in oil wells to explain current control methods.

Response to Situation 1

Karim divides his research into three main areas and reports his progress on his research in each area. Because he is one of 35 students, he includes a detailed introduction to refresh his instructor's memory about his project—his purpose and his justification for pursuing this topic:

Beginning

- Introduction
- Summary

Middle

Progress on work by phases

- Phase 1
- Phase 2
- Phase 3

- Phase 1
  (Work completed)

- Phase 2
  Work completed
  Work remaining

- Phase 3
  etc.

- Cost
- Final outline

End

- Evaluation

Situation 2 illustrates a simpler version of this standard progress report.

Situation 2 (Figure 17–2)

Dean Smith, a training manager for a software development company, is responsible for developing training sessions for new sales personnel. His main responsibility is to provide any training to employees, to decide what training should be conducted, and then to develop the training courses. Dean routinely writes progress reports to the Director of Personnel, Sharon Sanchez, to keep her informed of his activities—training he thinks will need to be offered, training programs he is currently developing or planning to develop, and training programs he and other training staff are currently teaching or directing.

Response to Situation 2

Dean writes these reports to Sharon about once a month (or whenever he has something he wants her to know about). Because she is familiar with what Dean does, he does not need to include an elaborate introduction. He begins with a concise summary and then proceeds to describe pertinent activities. In this particular report, he is asking for increased funding for a training program. Thus he includes an action required statement in the heading.

For this routine progress report, Dean modifies the general plan as follows:

Beginning

- Introduction

  Purpose of the report—to report the status of a project
  Purpose of the work being performed
  Summary of current status

Middle

- Tasks completed on the project

  Task A
  Task B
  Task C

End

- Conclusions and perhaps recommendations

In Situation 3, pages 564–571, Laurie Tinker divides her research into three tasks: research, analysis, and writing and editing. In her progress report, she explains what she has done and not done in her research and then in her analysis of her research. At this point, she has not yet begun to write the draft of her final report. (See Figure 17-5.)

For progress reports that cover more than one period, the basic design can be expanded as follows:

Beginning

- Introduction
- Project description
- Summary of work to date
- Summary of work in this period

Middle

- Work accomplished by tasks (this period)
- Work remaining on specific tasks
- Work planned for the next reporting period
- Work planned for periods thereafter
- Cost to date
- Cost in this period

End

- Overall appraisal of work to date
- Conclusions and recommendations concerning problems

Structure by Chronological Order If your project or research is set up by time periods, your progress report can be structured to emphasize the periods.

Beginning

- Introduction/project description
- Summary of work completed

Middle

- Work completed
- Period 1 (beginning and ending dates)
  Description
  Cost
- Period 2 (beginning and ending dates)
  Description
  Cost
- Work remaining
- Period 3 (or remaining periods)
  Description of work to be done
  Expected cost

End

- Evaluation of work in this period
- Conclusions and recommendations

Situation 4, pages 571–583, illustrates a progress report structured by periods.

Structure by Main Project Goals Many research projects are pursued by grouping specific tasks into major groups. Then, when progress is reported, the writer describes progress according to work done in each major group and perhaps the amount of time spent on that group of tasks. Alternatively, a researcher may decide to present a project by research goals—what will be accomplished during the project. Thus, progress reports will explain activities performed relevant to the achievement of those goals. In the plans below, the left-hand column is organized by work completed and remaining and the right-hand column by goals.

Beginning

- Introduction/project description
- Summary of progress to date

Middle

- Work completed
  Goal 1
  Goal 2
  Goal 3, etc.
- Work remaining
  Goal 1
  Goal 2
  Goal 3, etc.
- Cost

- Goal 1
  Work completed
  Work remaining
  Cost
- Goal 2
  Work completed
  Work remaining
  Cost

End

- Evaluation of work to date
- Conclusions and recommendations

Proposals

All projects have to begin somewhere and with someone. In universities, in business, and in research organizations, the starting point is often a proposal. In simplest terms, a proposal is an offer to provide a

DATE; APRIL 10, 1989

TO: ELIZABETH TEBEAUX

FROM: KARIM TEDESCO

SUBJECT: PROGRESS ON FORMAL REPORT

INTRODUCTION

Subject and Purpose

I am completing work on the study and development of a growing problem in the production and recovery of oil and natural gas: bacterial control. As the major outcome of the proposed investigation, I will recommend from among the options the method most feasible for identifying and controlling oilfield bacteria.

Background

Petroleum microbiology is a relatively new field of interest. Because research is in its infant stages, conclusive data are somewhat scarce. Currently, however, petroleum production and recovery companies are financing research on a large-scale basis. Oilfield engineers, in conjunction with microbiologists, are directing research toward the accurate identification of petroleum bacteria, oil and natural gas recovery problems, and methods of eliminating and/or controlling bacterial infestation.

What is the significance of bacteria in petroleum production? Oilfield bacteria corrode pumping machinery, storage tanks, and other installations. They contaminate natural gas associated with oil deposits; they grow in injection waters and may clog the system; and they may have roles in the formation, release, and transformation of oil hydrocarbons.

SUMMARY

My investigation concerning feasible methods of identifying and controlling oilfield bacteria has progressed slowly, attributable primarily to limited sources, highly technical information, and conflicting data. I have, however, obtained invaluable information from Tretolite Chemicals, a corrosion inhibiting company. Additionally, I have acquired needed supplementary data from the University of Texas Library in Austin.

PROGRESS ON WORK BY PHASES

I have constructed what I consider to be an effective procedure for conducting the research, which has fallen into three separate phases:

Figure 17–1 Routine Progress Report—Response to Situation 1

Elizabeth Tebeaux -2- April 10, 1989

Phase 1: Description of Various Bacterial Groups

Phase 2: Recognition of Bacterial Problems

Phase 3: Alleviation of Bacterial Problems

Various Bacterial Groups

I have completed the general description of the three bacterial groups proven most detrimental to the oilfield industry. My report will focus primarily on the sulfate-reducing bacteria. These organisms represent the bulk of oilfield related problems. Additionally, two less prevalent bacterial species, the iron-oxidizing bacteria and the slime-forming bacteria, will be presented in this report. My two primary sources used in compiling this information are Zobell's Ecology of Sulfate Reducing Bacteria and Anderson's Petroleum Microbiology.

Recognition of Bacterial Problems

Work Completed

My analysis of the various methods of identifying bacterial problems is partially complete. I have successfully compiled a list of techniques for identifying bacterial problems. Tretolite Chemicals' "Training Manual for Sales Engineers" has proven extremely beneficial.

Work Remaining

I have not started my analysis concerning the feasibility of the various problem identification techniques. I plan to use Tretolite's training manual in conjunction with relevant data found in Producers Monthly.

Alleviation of Problems

Work Completed

I have compiled sufficient information pertaining to methods of eliminating and/or controlling oilfield bacteria. Producers Monthly and Tretolite Chemicals' "Recognition and Chemical Treatment of Bacterial Problems in the Oilfield" represent the bulk of my source.

Work Remaining

The work remaining in regard to this aspect of my investigation consists of a discussion of the feasibility of each method according to three criteria: effectiveness (desired results), financial considerations, and ease of application.

Figure 17-1 (continued).

Elizabeth Tebeaux -3- April 10, 1989

COST ANALYSIS

My estimated costs proposed on March 12 and my costs to date are outlined below.

Item	Estimated Cost	Cost to Date
Travel	$ 43.20	$ 35.00
Phone calls	6.00	5.00
Typing	60.00	20.00
Salaries	1000.00	700.00
	$ 1109.20	$ 760.00

FINAL OUTLINE

I. Introduction

II. Summary of Findings

III. Type of Organisms

A. Sulfate-Reducing Bacteria

1. What are SRB?
2. Where are they found?
3. When is their presence a problem?
4. Why do SRB cause problems?

B. Slime-Forming Bacteria

C. Iron-Oxidizing Bacteria

IV. Recognition of Bacterial Problems

A. Primary Indicators

1. Microscopic observations
2. Bacterial enumeration methods
3. Advantages and disadvantages of each
   a. Economic feasibility
   b. Ease of application
   c. Desired results

B. Secondary Indicators

1. Corrosion
2. Plugging
3. Sulfide build

Figure 17-1 (continued).

Elizabeth Tebeaux -4- April 10, 1989

V. Alleviation of Problems

A. Physical and Mechanical Changes

B. Chemical Treatment

1. Initiate with microbiocide
2. Criteria for microbiocide treatment
3. Compounds known to possess microbiocidal properties
4. Mode of action

C. Maintaining Control and Monitoring

VI. Conclusions

EVALUATION OF RESEARCH

The number of sources I will use may be less than I anticipated. Industry journals are difficult to retrieve. However, the research I have gathered is extensive and detailed. My research shows that industry research has made rapid strides in identifying bacteria and finding cost-efficient ways of dealing with these bacteria. Costs of implementing these control methods are also reasonable.

Figure 17-1 (continued).

DATE; November 19, 1993

TO: Sharon Sanchez

FROM: Dean Smith

SUBJECT: TQI Workshop Plans for Spring 1994

ACTION REQUIRED BY 12/10/93

Summary of Plans

Planning for our TQI workshops, scheduled for April and May, are nearly complete. Our TQI facilitators have produced detailed training schedules for technical service and customer relations. I have approved these, and materials are being ordered and prepared. A TQI program package, developed by HI-TOP Sales Materials, specifically for computer sales personnel, can be purchased for $3,600. The format would be an excellent follow-up for all employees. As you and I had already discussed, quality off-the-shelf TQI packages should be considered.

TQI Workshop Activities as of 11/19/93

TQI Preparations for Technical Service Personnel

Johnette Darden and her group have been preparing materials for the technical service employees. The workshop will help the staff to look at software purchases from the customer's perspective and to begin to look for ways of decreasing the time required to resolve customer complaints. The second part of the workshop will encourage TSP to examine faulty products and find ways of eliminating the problems by working with design. We will probably recommend a quality control unit be formed that uses people from both technical service and design.

TQI Preparations for Customer Relations

Responding to customer needs and filling orders rapidly will be the main focus of the TQI workshop for CR employees. Robert Newmann will be working with the group (1) to reassess what we are doing in our current customer relations efforts, and then (2) to improve and even eliminate processes that do not help us serve customers quicker and better.

Figure 17–2 Routine Progress Report—Response to Situation 2

Sharon Sanchez
Page 2
11/19/93

HI-TOP Sales Program Available for Purchase by 12/15

HI-TOP developed a superb TQI program for sales personnel about two years ago. The program was so effective that HI-TOP is now selling the program, which has received rave reviews from a half-dozen companies that sell hardware and software. I reviewed the program two days ago and think it would be an excellent follow-up for any TQI work we do. I would like to purchase the program and use it with both Technical Services and Customer Relations. We can continue to use the package with other TQI training.

An outline of the program is attached. As you can see, it addresses the major issues we believe our own TQI programs need to emphasize.

Please Note

HI-TOP will sell us the program package priced at $4,500 for $3,600 if we act before December 10. They are currently redesigning the program to include extensive teaching aids, which are nice but unnecessary for our purposes. The new package will be available 1/1 for $6,000. **The original package will not be available after 12/15**.

Please give me a call so that we can discuss.

Figure 17-2 (continued).

service or a product to someone in exchange for money. Usually, when the organization—frequently a federal, state, or city government or a business enterprise— decides to have some sort of work done, it naturally wants the best job for the best price. To announce its interest, the soliciting organization may advertise the work it wants done and invite interested individuals or organizations to contact the organization. In a university setting, the research and grants office may notify departments that money is available for research projects in a specific area. Faculty members are invited to submit project proposals that explain how much time they will need to complete the project; any financial resources required for equipment, salaries, and release time from regular teaching duties; and the goals and benefits of the research to the individual researcher and the university. Thus, the proposal process usually begins with an organization that is interested in work or research being done in response to a specific need or problem. The proposal is the written document that launches a proposed solution to this need or problem by individuals or groups qualified to deal with the matter.

When an organization disseminates a description of the work it wants done, this document is usually called a request for a proposal (RFP) or a statement of work (SOW). The soliciting organization may send selected companies an RFP that includes complete specifications of the work desired, or it may describe the needed work in general terms and invite interested firms to submit their qualifications. This type of request is usually called a request for qualifications (RFQ). The responding organization explains its past accomplishments, giving the names of companies for which the work was performed, describing the work it did, and giving references who can substantiate the organization's claims. Based on the qualifications of the responding firms, the soliciting organization will send full descriptions of the work to the groups it believes to be best qualified.

Alternatively, the soliciting organization may describe the kind of work it wants done and invite interested companies to describe briefly what they offer, how they would do the work, and approximately how much it will cost. This kind of request is called a request for quote (also referred to as an RFQ). Responding firms that give a price that best approximates what the soliciting organization wants to pay will be sent a full description of the work needed and will be invited to submit full proposals.

To understand some of the many ways that proposals initiate projects, consider the following examples.

1. Professor X of the university's sociology department notes in the *Federal Register* that the U.S. Department of Health and Human Services (HHS) is soliciting studies of educational problems experienced by school-age children of single-parent

families. Because Professor X has established a research record in this field and is looking for new projects, she decides to request a copy of the RFP. After studying it carefully, she decides to submit a proposal. In her proposal she describes her planned research and explains its benefits. She states her qualifications to conduct the research and details the costs of the project.

2. A county in Texas decides that it wants to repave a heavily used rural road and extend the paving another five miles beyond the existing pavement. The county public works office runs an advertisement in several county and state newspapers briefly describing the work. Public works officials also send copies of the advertisement to road construction firms that have reputations for doing quality work at a fair price. The construction companies interested in submitting bids will notify the county officials and will be invited to attend a bidder's conference in which requirements of the job are discussed further. Public works officials may take potential contractors on a tour of the area. Those who decide to bid on the paving project will have four weeks to submit bids that meet the minimum specifications given in the published RFP as well as in the bidder's conference.
3. Alvin Cranston, a manager for a local telephone company, is charged with redesigning the operator service facilities for the company. Alvin knows that he will need to consider a number of issues (lighting, furniture, computers, as well as building layout), so he decides to publish a request for qualifications in telephone trade publications. He also asks the company's marketing department to help him locate a list of companies that specialize in ergonomic design. With a list of these companies, he writes each one and explains, in general terms, what his company wants to do and invites ergonomic design firms to submit their qualifications for performing such work.
4. Biotech Corporation is considering the development of a new organic dispersant for combating major oil spills in fresh water industrial lakes. They want to know how much containers for transporting this new dispersant would cost, what kinds of containers are currently available to transport the dispersant by rail or air freight, and whether chemical transport container companies would be interested in providing the containers and shipping it to purchasers.

In short, each aspect of the solicitation process, the RFP, the RFQ, or the SOW, has an appropriate use, but one or more of them is necessary to initiate action on a project.

The Context of Proposal Development

Because proposals are time consuming to write—most require substantial research and analysis on the part of the proposing organization—individuals and organizations wishing to respond to an RFP study it carefully. They do not want to submit a proposal that is not likely to be accepted. Thus, the proposer—whether a university professor seeking research funds or a highway construction firm seeking to win a contract from a county to repave its rural roads—will approach the decision to prepare a proposal carefully.

The responding company must decide whether to respond to the proposal. This decision is based on careful study of the RFP or the RFQ with a number of questions in mind: Can we do the work requested? Can we show that we can do this work, based on what we have already done? Can we do it within the time limit given in the RFP? Businesses responding to RFPs are also interested in economic issues: How much will our proposed approach cost? How much money can we make? Who else will be submitting proposals? What price will they be quoting for the same work? Will we be competitive? What other projects are we currently involved in? Could problems arise that would make us unable to complete the job on time and at the price we quote? Do we have the personnel qualified for this project?

Many business entities requesting proposals will hold a bidder's conference where companies interested in submitting a proposal can ask questions about the project or seek clarification of the needs described in the RFP. Most RFPs require that proposals be submitted by a deadline and contain specific information. Proposals that do not contain the information requested may be omitted from consideration. Therefore, once an organization decides to submit a proposal, the RFP is carefully studied and the information requirements are separated. Each information requirement is given to an individual or a group who will be responsible for furnishing necessary material and data.

Some proposals, such as university research proposals, may be written by one person. In complex proposals, however, different sections may be written by individuals in different areas of the organization. An editor or proposal writer will then compile the final document. This writer/editor may be assisted by readers, who help check the accuracy of the developing proposal to be sure that all requested information is included and that the information is correct. Once the proposal is written and submitted, it becomes a legally binding document. The proposing company or individual is legally committed to do what it says it will do at the cost stated and within the time limit stated. For that reason, the proposing organization carefully checks all information for accuracy. Figure 17-3 will help you visualize the proposal process.

In requests for proposals in which a large number of bidders submit proposals, the soliciting organization may select several finalists and allow each finalist to give an oral version of the proposal. During this oral presentation, the soliciting group asks questions; the proposing group has one more opportunity to argue for the value of what they are proposing, the merits of their organization to do the work, and the justification for the cost attached to the proposed work.

Effective Argument in Proposal Development

All writing is persuasive in that it must convince the reader that the writer has credibility and that the writer's ideas have merit. However, the success of the proposal rests totally on the effectiveness of the argument—how convincingly the writer argues for a plan, an idea, a product, or a service to be rendered and how well the writer convinces the reader that the proposing organization is the best one to do the work or research needed. In planning the content of the proposal, the proposer must harmonize the soliciting company's needs with the proposer's capabilities. The writer must be acutely sensitive to what readers will be looking for but not propose action that cannot be done because it is outside the capability of the proposing individual or organization. The proposing individual or organization has an ethical responsibility to explain accurately and specifically what work can be done and not done so that there is no intent to deceive readers with promises that cannot be fulfilled.

The following questions are useful in analyzing the effectiveness of the argument, whether in a written or oral proposal:

What does the soliciting organization really want?
What is the problem that needs to be solved?
What approaches to the solution will be viewed most favorably?
What approaches will be viewed unfavorably?
What objections will our plan elicit?
Can we accomplish the goals we propose?

To answer these questions, the proposer may be required to do research on the organization, its problems, its corporate culture and the perspective and attitudes stemming from its corporate culture, its current financial status, goals, and problems. As each part of the proposal is developed, you should examine it from the intended readers' perspective.

What are the weaknesses of the plan, as we—the writer(s)—perceive them?
How can we counter any weaknesses and readers' potential objections?
How can we make our plan appealing?
How can we show that we understand their needs?

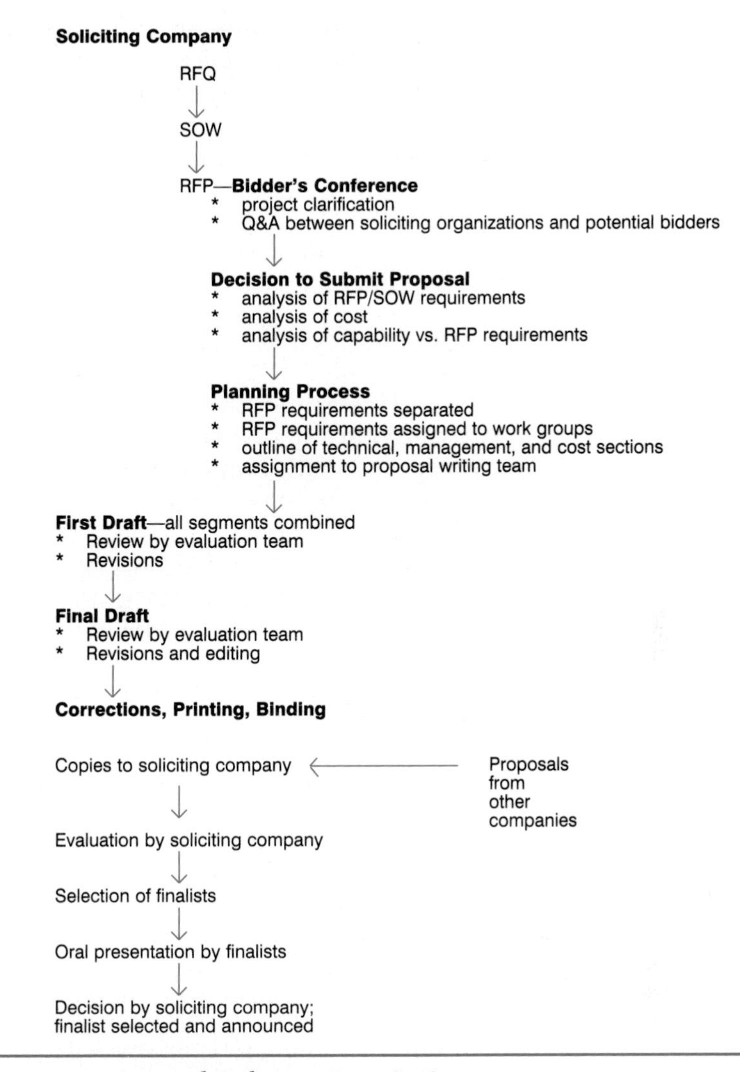

Figure 17–3 Proposal Cycle in an Organization

How can we best present our capability to do this project?
What are our strengths?
From our own knowledge of our organization, what are our weaknesses—in personnel, in overall capability to complete this project as proposed?
Do we need to modify our proposed plan to avoid misleading readers about our ability to perform certain tasks on time, as proposed, and at cost?
Can we sell our idea without compromising the accuracy of what we can actually do?

As you consider each question, you should determine what evidence you will need to support the merits of your idea and the arguments needed to refute any objections. Every sentence should argue for the merits of your plan and your or your organization's ability to complete it. Although the proposal is designed to be a sales document, the writer is still ethically obligated to present a plan that meets the soliciting organization's needs and requirements. In considering the ethical issues that confront proposal writers, you will want to review Chapter 2 on ethical considerations in technical writing.

Standard Sections of Proposals

Proposals generally include three main divisions: a summary, main body, and attachments. Major proposals are submitted in complete report format, which requires a letter of transmittal, a title page, a submission page (perhaps), a table of contents, and a summary. The main body focuses on the three main parts of the proposal: what the proposal's objectives are (technical proposal), how the objectives will be achieved (management proposal), and how much the project will cost (cost proposal). You may find it helpful to visualize the structure in this way:

- Project summary
- Project description (technical proposal)
  - Introduction
  - Rationale and significance
  - Plan of the work
  - Facilities and equipment
- Personnel (management proposal)
- Budget (cost proposal)
- Appendices

Shorter proposals may be written in a memo or letter format. Whatever the format, the main elements will be required, although how they appear will vary with each proposal. In most RFPs, the soliciting organization will explain what should be included in the proposal, either specific information to be included or major elements, as shown in the Department of Agriculture RFP in Figure 17–8 at the end of this chapter. Often, RFPs indicate the maximum number of pages allowed in a proposal. Writers are well advised to follow these instructions carefully to ensure that the proposal is not rejected during the initial screening process because it fails to follow preparation guidelines stipulated by the RFP.

Summary The summary is by far the most important section of the proposal. Many proposal consultants believe that project approval can be won or lost according to the effectiveness of the summary, which is your readers' first introduction to what you are proposing. The summary should concisely describe the project, particularly how your work meets the requirements of the soliciting organization, your plan for doing the work, and your or your company's main qualifications. The summary should be a concise version of the detailed plan, but it should be written to convince readers that you understand what the soliciting firm needs and wants; that what you are proposing can be done as you describe; and that your approach is solid because you have the required knowledge and expertise. From reading the summary, readers should want to read more of your proposal.

Project Description (Technical Proposal)

Introduction The proposal introduction should explain what you are proposing, why you are proposing this idea, and what you plan to accomplish. The introduction contains the same elements as any introduction. In short proposals, the summary and introduction can be combined.

Rationale and Significance Much of your success in convincing readers that you should be granted permission to do the work you propose rests on your success in convincing them that you understand the project. In this section, you need to make it clear that you understand their needs—as stated in the summary or introduction—and that you have designed your goals by analyzing and defining their needs. Although you will clearly be selling your idea, you should recognize and answer any questions your readers may have as you argue the merits of your project. Convincing your readers that you fully understand what they are looking for is critical in establishing your credibility. In short,

- You may want to define the problem, to show that you understand it.
- You may want to explain the background of the problem, how it evolved, by providing a historical review of the problem.
- If you are proposing a research project, you may want to explain why your research needs to be done and what results can be expected from your research.
- You may want to describe your solution and the benefits of your proposed solution.

Of greatest importance, however, is the *feasibility* of the work you propose. Is your proposed work doable? Is it suitable, appropriate, economical, and practicable? Have you given your readers an accurate view of what you can and will do?

Plan of the Work This section is also critical, particularly to expert readers who will attempt to determine whether you understand the breadth of the work you are proposing. In this section, you will describe how you will go about achieving the goals you have stated. You will specify what you will do in what order, explaining and perhaps justifying your approach as you believe necessary. A realistic approach is crucial in that a knowledgeable reader will sense immediately if your plan omits major steps. A flawed work plan can destroy your credibility as well as the merits of the goals or the solution you are proposing.

Scope The work plan section may need to describe the scope of the proposed work. What will you do and not do? What topics will your study or your work cover and not cover? What are the limits of what you are proposing? What topics will be outside the scope of your project? Readers need to know, from an ethical and a legal perspective, the limits of your responsibility.

Methods A work plan may also require a statement of the methods you will use. If you are going to do on-site research, how will you do this research? If you plan to collect data, how will you analyze it? How will you guarantee the validity of the analysis? If you are going to conduct surveys, how will you develop them? If you plan to do historical research or a literature review of a topic, how will you approach such a review to ensure that your findings are representative of what is currently known about a subject area? A precise, carefully detailed description of your work methods can add to your credibility as one who is competent to perform the proposed work.

Task Breakdown Almost all proposals require you to divide your work into specific tasks and to state the amount of time allotted to each task. This information may be given in a milestone chart, as illustrated in the student research report (Figure 17-4). The task breakdown further subdivides the work plan to show how much time you plan to devote to each task. A realistic time schedule also becomes an effective argument. It suggests to readers that you understand how much time your project will take and that you are not promising miracles just to win approval of your proposal or business plan.

If a project must be completed by a deadline, the task breakdown and work schedule need to indicate exactly how you plan to fit every job into the allotted time. However, do not make time commitments that will be impossible for you to meet. Readers who sense that your work plan is artificial will immediately question your credibility. Remember, too, that a proposal is a binding commitment. If you cannot do what you propose, what the soliciting organization requires within the required time, you can destroy your professional credibility and leave yourself open to litigation.

Problem Analysis Few projects can be completed without problems. If you have carefully analyzed the problem or work you intend to do, you should anticipate where difficulties could arise. Problems that may be encountered can often be discussed in the rationale sections. However, if you discover major obstacles that you believe will occur during the course of the project, you may wish to isolate and discuss these in a separate section. Many organizations that request work or solicit research proposals are aware of problems that may arise. Reviewers in these organizations look carefully at the problem analysis section, wherever it occurs, to see whether the proposer has anticipated these problems and explained the course of action he or she will use to deal with them. Anticipating and designing solutions to problems can further build your credibility with readers, who will not be impressed if you fail to diagnose points in your work plan that could be troublesome and even hinder your completion of the project as proposed.

Facilities The facilities section of the proposal is important if you need to convince the reader that the proposing company has the equipment, plant, and physical capability to do the proposed work. Facilities descriptions are particularly crucial if hardware is to be built at a plant site owned by the proposing organization. Even in study proposals, your readers may want to know what research resources you will use. Sometimes existing facilities are not adequate to do a particular job and the company must purchase specific equipment. The facilities section enables the proposer to explain this purchase and how it will be included in the cost proposal.

In study proposals, researchers may need to visit special libraries or research sites that require travel. The amount of money needed for this travel will be part of the cost proposal. Thus, the nature of any extra research support, its importance, and its cost to the project will need to be explained here.

Personnel (Management Proposal) Any technical proposal or project is only as good as the management strategy that directs it. The management proposal should explain how you plan to manage the project: who will be in charge and what qualifications they have for this kind of work. Management procedures should harmonize with the methods of pursuing the work described in the technical proposal.

Descriptions of the proposer's management philosophy and hierarchy may need to be related to the company's management philosophy and culture. Readers should see the same kind of management applied to the proposed work as to the company and other projects it manages. Any testimony to or evidence of the effectiveness of the management approach will lend credibility to the technical proposal. Proposal reviewers must be convinced that you and the organization have a sound approach supported by good management of that approach.

In research proposals, the researcher who is soliciting funds will want to explain his or her expertise in the subject area proposed. This support may include educational background, previous projects successfully undertaken, published research on the topic, and general experience.

Cost (Cost Proposal) The cost proposal is usually the final item in the body of the proposal, even though cost may ultimately be the most crucial factor in industrial proposals. Cost is usually given last and appears as a budget for the length of the proposal period. The technical and management proposals, with their descriptions of methods, tasks, facilities, required travel, and personnel, should help justify the cost. They should have already explained the rationale for items that will produce the greatest cost. However, any items not previously discussed in the technical and management proposals, such as administrative expenses, additional insurance benefit costs, and unexpected legal costs, should be explained. An itemized budget is often submitted as a separate document. It includes items such as the proposing organization's liability for not meeting project deadlines, for cost overruns, and for unforeseen strikes and work stoppages. Many budget sections include standard statements, such as descriptions of union contracts with labor costs, insurance benefits costs, non-strike costs, and statements of existing corporate liability for other projects—any existing arrangements that affect the cost of the proposed contract. Clearly, the goal is to explain exactly how much the project will cost and where the cost is determined. How extensive the budget will be depends on the magnitude of the project.

Conclusion Like any report, the proposal includes a final section that reiterates what the proposal offers the potential client or the soliciting agency, why the proposer should be selected to perform the work, and the benefits that the project, when completed, will yield for the client. The conclusion presents the final restatement of the central argument.

Appendixes Again, as in any report, the appendix section will include supporting materials for information given in the main body of the proposal—in the technical, management, or cost proposal. For example, the appendix might include résumés of principal investigators, managers, or researchers. These résumés should highlight their qualifications as they pertain to a specific project.

To help you understand how proposals can be written by students, please study the following situation that led to a proposal and the student's research proposal written in response to this situation (Figure 17-4):

Situation 3

Laurie Tinker is a mathematics major whose senior research project requires her to explore an area that applies mathematics to the solution of a problem in science. Laurie chooses an oceanography topic. Her research shows how a mathematical method called matched field processing can be used in underwater acoustics to find the range, depth, and bearing of an acoustic source. She provides a general mathematical description of the method and then explains how the method works when it is used to track an acoustic signal. See Figure 17-4.

Physical Appearance of Proposals and Progress Reports

The importance of the appearance of any proposal or progress report cannot be overestimated. Competence of the proposing organization or individual is suggested by a report that is neat and effectively formatted. Proposals and progress reports that exceed letter or memorandum length should have a protective cover. The title page should be tasteful. The type or print quality should be high quality. Colored paper and covers should convey a professional attitude of the individuals or the organization. A professional outward appearance is the first method of arguing for the merits of the proposal.

Style and Tone of Proposals and Progress Reports

The proposal and its related report documents are sales documents, but writers have an ethical commitment to present information about a project in a clear but accurate manner. Proposals, once accepted, become legally binding documents. Because contracts are based on proposals, organizations must be prepared to stand behind their proposals. Thus, the style should be positive and suggest the competence of the proposer. The style should be vigorous, firm, and authoritative. Generalizations must be bolstered by detailed factual accomplishments. Problems should be honestly discussed, but positive solutions to problems should be stressed. Neither the proposal nor the progress report should resort to vague, obfuscatory language.

The student progress report in Figure 17-5 shows how progress by main tasks was used by Laurie Tinker to report the status of her report on Matched Field Processing in Underwater Acoustics.

Situation 3 (continued)

During the course of the senior projects, mathematics students are required to submit one written progress report whenever the student's committee requests one. The committee wants to know whether the proposed project is on schedule to ensure that it is completed within one semester, the time allowed for the senior research project.

DATE; March 10, 1990

TO: Elizabeth Tebeaux
Coordinator of Technical Writing

FROM: Laurie Tinker

SUBJECT: Proposal for a Research Report on Matched Field Underwater Acoustics

PROJECT SUMMARY

Matched field processing is a technique used in underwater acoustics to locate a sound source. This technique uses parameter estimation to find the range, depth, and bearing of a source from the signal field propagating in an acoustic waveguide. Extensive research has been done and is still being done in this area of underwater acoustics.

I plan to research one of the many methods of matched field processing: The maximum likelihood method, commonly referred to as MLM. Upon completion of my research I will be able to determine advantages and disadvantages of this method. I will also speculate on possibilities of what can be done to overcome the drawbacks.

This research will aid me to a great extent in the career I plan to pursue upon graduation, a position as an analyst in underwater acoustics. It will provide me with specialized background, helping me to better understand the type of mathematical models I shall encounter in my work.

The manner in which I plan to approach this research project requires mostly library research. A great number of specialized journal articles have been published in the past ten years that discuss matched field processing methods. If I encounter any problems finding information, I will contact my supervisor at the Naval Research Laboratory or a professor in either the Department of Physics or the Department of Oceanography at Texas A&M.

Costs for this project are minimal. I will have costs for photocopies of articles and research studies, transportation to and from Rice University, and at least two long distance calls to ask my supervisor for advice.

PROJECT DESCRIPTION

In the field of underwater acoustics, research is always in progress to develop new methods and to improve existing methods of source detection. One of the methods currently being researched is matched field processing. In this project, I plan first to give a description of matched field processing as background. Following the description, I will present the parameters which govern this mathematical model. Through the description of the parameters affecting this model, its advantages and disadvantages should become apparent. I will describe the advantages and disadvantages of the method. The final section of my research will attempt to determine the disadvantages and advantages of the maximum likelihood method and ways of overcoming drawbacks for using the method. I may not find any conclusive results because my training in this area is limited. However, this lack of knowledge is my

Figure 17–4 Proposal Memorandum Response to Situation 3
Reprinted with permission of Laurie Tinker.

Dr. Elizabeth Tebeaux -2- March 10, 1990

primary rationale for conducting my research. Most of my research will be done in the library where I will use recent journal articles and government documents.

PROBLEM JUSTIFICATION

New developments are constantly occurring in matched field processing: methods are repeatedly revised to improve the quality of signal processing. As a cooperative education student at the Naval Research Laboratory, I became familiar with a small part of the field of matched field processing as it is applied in underwater acoustics. Since the Naval Research Laboratory is a research facility, all the work explored is relatively new or in beginning stages. Considering that I am interested in pursuing a career at the Naval Research Laboratory in underwater acoustics, I would like to learn more about the recent developments in matched field processing. My initial assignment will likely include assisting research in perfecting matched field processing.

I was hired as a mathematics student trainee, and I had not yet taken any physics courses. While working at NRL, I discovered that although I understood the mathematics and the computer programming involved, I was many times unable to grasp the overall concepts involved with my work because of my lack of training in physics, particularly underwater acoustics. I believe that pursuing a research project that expands my knowledge of underwater acoustics as it utilizes matched field processing will increase my skills.

PERSONNEL QUALIFICATIONS

I am a senior student in applied mathematical sciences at Texas A&M. My overall grade point average is 3.5. My ten courses in applied mathematics provide me the theoretical background to study MFP. During the fall semester of 1989, I worked as a co-op student at the Naval Research Laboratory in Washington, D.C. I became familiar with the basics of matched field processing at the lab by modifying and running programs to evaluate the effects of various parameters which control the quality of matched field processing. This co-op experience in addition to my educational background, which now includes two courses in physics, provides me with the background to extend my knowledge of matched field processing methods.

TASK BREAKDOWN

I anticipate my research will require three main phases: research, analysis of research, and compilation of findings for the written report. In a recent conversation with my supervisor, Dr. Michael Hansen, I have set up a time on April 10 and April 14 to discuss my research. Because of requirements in my other classes, I am compelled to follow the following work schedule in order to complete the report by the due date, April 28.

Research	Feb. 1–March 20
journal articles	
technical reports	
Analysis	
mathematical models	March 21–April 3
consultation with NLR supervisor	April 10

Figure 17-4 (continued).

Dr. Elizabeth Tebeaux -3- March 10, 1990

Compilation/Writing April 11–April 19
- rough draft of discussion
- first draft of complete report
- compilation with NLR supervisor — April 14
- final draft written and typed
- final draft submitted — April 28—report due date

The specific milestone chart of my proposed schedule is as follows:

Matched Field Processing Research Report

Project Tasks	**Work Schedule by Weeks** Feb. 1–4	March 1	March 2	March 3	March 4	April 1	April 2	April 3	April 4
Research journal articles	—	—	—	—	→				
Research technical reports	—	—	—	—	→				
Research mathematical models	—	—	—	—	—	—	→		
Consult with supervisor or professor							4/10		
Write rough draft of body							—	→	
Complete first draft of whole report							—	—	→
Consult with supervisor or professor								4/14	
Write final draft								—	→
Type final draft								—	→
Turn in final draft									4/28

PROBLEMS

The only problems I foresee are difficulties in understanding the information I am planning to examine. My two courses in physics, taken since I completed my co-op work in Washington, should alleviate some of the problems I originally encountered. However, Dr. Hansen has offered to provide additional information if I need more background. I also have available research assistants in both the Department of Physics and the Department of Oceanography if I encounter difficulties in deciphering technical reports that utilize extensive acoustical physics principles.

Figure 17-4 (continued).

Dr. Elizabeth Tebeaux -4- March 10, 1990

TENTATIVE OUTLINE

I. Introduction—What is Matched Field Processing?
   A. Definition
   B. Uses of matched field processing
   C. Method of matched field processing to be researched: Maximum Likelihood

II. Maximum Likelihood Method
   A. Concepts defining MLM
   B. Mathematical model of MLM
   C. Definition of parameters
   D. Problems encountered using MLM
      1. Noise interference
      2. Shallow water
      3. Other problems
   E. Solutions already found to some problems
   F. Advantages and disadvantages of MLM
   G. Comparison of MLM to other methods of matched field processing

III. Conclusion
   A. Summary of defining MLM concepts
   B. Summary of advantages and disadvantages of MLM
   C. Suggestions of possible approaches to solving existing problems

COST

Photocopies of journal articles and other information	$ 20.00
Phone calls to Washington, D.C.	25.00
Travel expense to Houston	10.00
Research time—35 hrs. @ $10.00/hr.	350.00
Compilation/Writing/Wordprocessing/Binding—15 hrs.	150.00
plus supplies	10.00
Total expected cost	$565.00

INITIAL BIBLIOGRAPHY

Alexandrou. 1987. *Boundary reverberation rejection via constrained adaptive beamforming*. The Journal of the Acoustical Society of America, October, 82: 1274–90.

Baggeroer, Kuperman, and Schmidt. 1988. *Matched field processing: Source localization in correlated noise as an optimum parameter estimation problem*. The Journal of the Acoustical Society of America, February, 83: 571–87.

Baggeroer and Wagstaff, eds. 1983. *Full field ambiguity function processing in a complex, shallow water environment*. High resolution spatial processing in underwater acoustics, NORDA proceedings, NORDA.

Figure 17-4 (continued).

Dr. Elizabeth Tebeaux -5- March 10, 1990

Bouvet and Schwartz. 1988. *Underwater noises: Statistical modeling, detection, and normalization.* The Journal of the Acoustical Society of America, March, 83: 1023–33.

Del Balzo, Feuillade, and Rowe. 1988. *Effects of water-depth mismatch on matched-field localization in shallow water.* The Journal of the Acoustical Society of America, June, 83: 2180–5.

NATO Advanced Study Institute on Adaptive Methods in Underwater Acoustics; North Atlantic Treaty Organization/Scientific Affairs Division. 1987. *Adaptive methods in underwater acoustics.* The Journal of the Acoustical Society of America, June, 81: 1999–2000.

Robust Methods for Array Signal Processing (Final Report), Lowell University, MA. Dept. of Mathematics, 1987.

Tamg, Yang, and McDaniel. 1988. *Acoustic beam propagation in a turbulent ocean.* The Journal of the Acoustical Society of America, November, 84: 1808–12.

Underwater acoustic signal processing: Special issue. 1987. IEEE Journal of Oceanic Engineering, January, 12: 2–278.

Ziomek. 1987. *Underwater acoustics: A linear systems theory approach.* Physics Today, December, 40: 91–2.

If you need to discuss this proposal with me further, please contact me at 845-6098.

Figure 17-4 (continued).

Beginning

1. She begins with a project description that states the purpose of the research and then gives the original list of tasks.
2. She next divides her tasks into research and analysis and provides a brief outline of these tasks. Note that the actual writing is not scheduled to begin until after this progress report.

Middle

3. She lists each major task—research and then analysis—and explains what work has been completed and what work remains on each of the first parts of the project. She then provides an outline of the final report.

End

4. She provides an overall appraisal of her work and explains her costs to date.
5. She proposes a table of contents for her report and then concludes.

Other Forms of Proposals and Progress Reports

Proposals and progress reports can be prepared in a variety of formats: as memo reports, as formal reports, and as letters. Yet, no matter what the format, proposals and progress reports will incorporate the same elements described above and illustrated in the reports written by the senior mechanical engineering student. To see how proposals and progress reports might operate in another format and another context, examine Situation 4 next, which illustrates a letter proposal and a memo status report on the project:

Situation 4

Write, Inc. specializes in helping companies improve employee writing. The company will design workshops for business organizations depending on each organization's particular writing problems. However, Write, Inc. will also rewrite or edit manuals, such as computer manuals, technical manuals, and policy and procedure manuals.

The company depends on referrals, satisfied client organizations who then recommend Write Inc. to other companies. However, repeat business is also important. Write, Inc. attempts to provide continuing teaching or writing services to its clients, and the company seeks to make its services appealing through reasonable fees and instruction that achieves results,that is, employees whose writing shows improvement because of the instruction the company provides.

Jon Sanchez, president of Write, Inc. recently received a call from Cindy Easton, Personnel Vice-President for Lambert Chemical Company in Minnesota. Cindy tells Jon that she has heard good things about the workshops conducted by Jon's instructors and is interested in having Write, Inc. do a workshop for all the technical sales personnel on how to design effective correspondence. Cindy says that

DATE; April 10, 1990

TO: Elizabeth Tebeaux
Coordinator of Technical Writing

FROM: Laurie Tinker

SUBJECT: Progress on Research Report on Matched Field Processing (Maximum Likelihood Method)

PROJECT DESCRIPTION

The purpose of my report on Matched Field Processing is to research the Maximum Likelihood Method of source detection. I plan to summarize the advantages and disadvantages of using this method in underwater acoustics and to suggest possibilities for overcoming problems posed by this method.

Original Task Breakdown

As stated in my proposal to you of March 10, my project has three main phases: research, analysis, and compilation/writing.

Research	Feb. 1–March 20
journal articles	
technical reports	
Analysis	
mathematical models	March 21–April 3
consultation with NLR supervisor	
Compilation/Writing	April 11–April 19
rough draft of discussion	
first draft of complete report	
consultation with NLR supervisor	April 14
final draft written and typed	
final draft submitted	April 28—report due date

SUMMARY OF WORK TO DATE

I have not yet begun to write the report, but my research and analysis phases have involved the following tasks:

Library research
- journal articles
- books
- technical reports

Analysis of research
- analysis of maximum likelihood mathematical model
- consultation with faculty for help in model analysis
- consultation with faculty for help with technical language

Figure 17–5 Progress Report Memorandum to Situation 3
Reprinted with permission of Laurie Tinker.

Dr. Elizabeth Tebeaux -2- April 10, 1990

My library research has taken more time than I anticipated. I have had difficulty with the model analysis, but consultation with Professor Jergenson, scheduled for April 12, should enable me to complete this aspect of the research.

LIBRARY RESEARCH

Work Completed

The journal articles I have consulted have been the most helpful. My main source article is "Source localization in correlated noise as an optimum parameter estimation problem," published by The Journal of the Acoustical Society of America. From studies referenced in this article, I have found three additional studies that have broadened my knowledge of MFP theory and application.

During our first phone consultation, Dr. Andersen of the Naval Research Laboratory recommended four books on acoustical theory that have helped me broaden my understanding of the physics underlying the MFP method. These books have also provided definitions to strengthen my understanding of the technical terminology. The special issue of IEEE Journal of Oceanic Engineering devoted to underwater acoustic signal processing and Lowell University's Robust Methods for Array Signal Processing (Final Report) have been the most useful supplements to the study by Baggeroer, Kupeman, and Schmidt on matched field processing. Many studies referenced by sources I listed in my initial bibliography have been too technical to be useful to me at this point in my education. However, I have been able to read and absorb enough information to complete the study that I outlined in my proposal.

Work Remaining

Dr. Andersen is sending me copies of research reports taken from his personal files. When these arrive, I will study them and call him if I have any questions. Mr. Hazad Mezziz, a graduate student in oceanography, has offered to help me with analysis of any reports I need in beginning the written draft.

ANALYSIS OF RESEARCH

Work Completed

Webber and Schmidt's book-length study of Acoustic Beam Propagation in Turbulent Oceans and Vizeck and Howetz's Extensive Report on Problems in Matched Field Processing—two books recommended by Dr. Andersen—have been the most useful sources for analyzing the mathematical model and showing the disadvantages and advantages of MFP methods. Consultations with Dr. Biggs of the Department of Oceanography and Dr. Robert Randall of Ocean Engineering have enabled me to complete analysis of the theory underlying the Maximum Likelihood Method (Section II of the Table of Contents).

Figure 17-5 (continued).

Dr. Elizabeth Tebeaux -3- April 10, 1990

Work Remaining

I will review four studies published in The Journal of the Acoustical Society of America before writing the draft of my discussion. Dr. Biggs has offered to read this draft and make suggestions.

EVALUATION OF WORK TO DATE

Although I am continuing to develop my knowledge of the concepts involved, I will not be able to do as intensive an analysis of the mathematical model as I originally proposed. I originally planned to have completed all my research by March 20, and the analysis of the mathematical model by April 3. However, the complexity of the math has caused me to work at a slower pace. Once I have the articles from Dr. Andersen and am able to study them, I can begin my draft. A more realistic plan is to have both the draft of the discussion and the analysis of the model completed by April 16.

COST TO DATE

To date, I have spent 32 hours analyzing research and an additional three hours in consultations with faculty. I have spent $10.00 in photocopies, less than projected, and I have not had to drive to Rice for consultation with Mr. Leo Bahrer, an engineering graduate student who is currently pursuing an engineering project in MFP. My phone costs have exceeded my estimates by $12.00 because of the extended questions I have needed to ask Dr. Andersen. Basically, however, my project is on budget, although the research and analysis time will exceed my original estimate by about 12–15 hours. The final cost of the project will exceed the original budget by approximately $90.00 as a result of time required for consultation with faculty, additional research, and analysis of this new material.

PROPOSED TABLE OF CONTENTS

Based on my initial outline, which has guided my research and analysis, the table of contents of my report will be as follows. I have deviated only slightly from my original outline:

I. Introduction—What is Matched Field Processing?
   A. Definition
   B. Uses of matched field processing
   C. Methods of matched field processing to be researched: Maximum Likelihood

II. Maximum Likelihood Method
   A. Basic concepts defining MLM
   B. A simplified mathematical model of MLM
   C. Definition of parameters
   D. Problems encountered using MLM
      1. Noise interference
      2. Shallow water
      3. Other problems

Figure 17-5 (continued).

Dr. Elizabeth Tebeaux -4- April 10, 1990

E. Advantages and disadvantages of MLM
G. Why MLM is superior to other methods of matched field processing

III. Conclusion
A. Summary of MLM theory
B. Summary of advantages and disadvantages of MLM
C. Problems with using MLM

CONCLUSION

My report will be completed and submitted on schedule. The magnitude of the project is greater than I anticipated, but I have been able to acquire more knowledge of matched field processing and the physics of underwater acoustics. While the depth of my research is not as extensive as I had hoped it would be because of the technical depth of the mathematics as well as the physics involved, the project is achieving its goal of broadening my knowledge of matched likelihood methods. I have certainly read more technical studies than I originally thought would be necessary, but the discussions with Drs. Andersen, Biggs, and Randall has been invaluable.

If you have suggestions, please contact me.

Figure 17-5 (continued).

about eleven salespeople have shown a definite interest in the workshop because of several customer problems that have surfaced from Lambert correspondence that was interpreted as rude or misleading by customers. Jon and Mark Clayton, one of Write, Inc.'s instructors, make an appointment with Cindy, examine the types of technical sales letters that Lambert has to write, and see two letters that Cindy describes as "problematic," as well as several internal memoranda written by various individuals. Cindy tells Jon that she would like the workshops held early in the mornings, but that she is open to the number of workshop sessions that will be necessary to correct the sales correspondence problems. Cindy suggests that Jon drop her a letter that includes details— course content, time schedule, and cost—which she will then present to the Lambert president for approval.

Response to Situation 4: Proposal

Jon asks Mark to write the letter proposal, which he will examine and possibly revise before sending it on the Cindy. Even though Mark writes the letter, it will bear Jon's signature. See Figure 17–6 for Mark's proposal. Note that this proposal is developed by specific tasks.

Situation 4: Status Report

Cindy Easton accepts Write, Inc.'s proposal. Mark spends three weeks developing the cases and instructional materials before the teaching sessions begin. After the third week of the writing workshop, Mark faxes the status report shown in Figure 17–7 to Jon, who is currently on vacation but who plans to visit with Cindy Easton during week 4 to offer another proposal to write the safety policy manual for Lambert. Jon is particularly interested in the participants' responses to the workshop as he plans his meeting with Cindy.

Note that the progress report specifically deals with each proposed task—those accomplished, those remaining, and the problems that have occurred in planning the workshop.

W R I T E, Inc. → Instruction
4 W. 11th Place Writing
Minneapolis, MN 55842 Editing
(612) 846-2921

November 2, 1993

Ms. Cindy Easton, Vice-President for Personnel
Lambert Agricultural Chemical Co.
Minneapolis, MN 55841

Dear Ms. Easton:

We enjoyed visiting with you last Thursday and discussing how our instructional programs can improve the quality of your technical correspondence at Lambert. After assessing your description of the kind of instruction you would like to make available for all sales personnel, and after again reviewing copies of several of the problem letters you sent us for further examination, I would like to submit our proposal.

Proposed Workshop for Lambert Agricultural Chemical Co.

Write, Inc. will present a correspondence workshop designed specifically for Lambert. Because of the schedule of your officers, the workshop will be presented from 8–10 a.m. for five Tuesdays. In addition to presentation of basic strategies for the design of all correspondence, the workshop will require participants to develop three letters to two case problems that simulate actual company situations that require letter responses. We will evaluate each response, make transparencies of these responses, and share each participant's response with the group to allow group analysis of each response. A fourth and final assignment will divide participants into groups of three or four to collaboratively develop and then present a single shared response to a third case. Each group will critique these responses and suggest revisions.

In addition to in-class instruction, practice, and assignments, each participant will receive a notebook of readings and instructional materials that will be useful during the workshop and afterward as a reference.

Workshop Schedule

The workshop will be organized as follows:

Figure 17–6 Proposal Letter Response to Situation 4

Ms. Cindy Easton -2- November 2, 1993

First Tuesday

Principles of Effectiveness for all Writing: Analysis of readers, definition of purpose, relationship of writer and reader, attitude and tone in writing. Assignment 1: Reading and practice in notebook.

Second Tuesday

Discussion of reading and practice assignment.

Organization of business writing. Organization of business correspondence. Analysis of cases and letter responses in notebook.

Assignment 2: Develop a letter responding to the situation in Lambert Case #1.

We will collect all assignments during the week and evaluate them before the third class. Each participant will receive an evaluation of his/her response.

Third Tuesday

Peer review and critique of each assignment. Analysis of cases and letter responses in notebook, Section 2. Review of tone, organization, audience needs and perception.

Assignment 3: Develop two letters responding to the situations described in Lambert Case #2.

Again, all assignments will be collected and evaluated prior to the fourth class.

Fourth Tuesday

Peer review and critique of each assignment. Practice on tone and clarity of sentences and paragraphs.

Assignment 4: Collaboration exercise. Participants will be divided into groups of three or four and will develop a shared response to Lambert Case #3. Groups will have at least one hour to develop their response and submit it. We will type all responses and have these ready for distribution at the final meeting.

Fifth Tuesday

Presentation of each group's response. Explanation of rationale for content, organization, tone, and style.
Peer review of each response.

Figure 17-6 (continued).

Ms. Cindy Easton -3- November 2, 1993

Wrap-Up. Final review of principles of effective design of business correspondence. Evaluation of workshop. Write, Inc. will provide each participant an evaluation form to allow everyone to respond anonymously to the workshop. These will be collected and given immediately to you or your representative. After you have seen them, we would appreciate your sharing them with us.

Cost

As we discussed, the workshop will have 15–16 sales people. Because notebooks will be developed specifically for this workshop, much of the fee is allocated to case preparation, notebook preparation, instructional time, and assignment evaluation. Please note, however, that the fee covers all costs.

Preparation Fee Case development Assignment development Notebook preparation	$2,000.00
Instructional time five 2-hr. sessions at $150/hr.	1,500.00
Assignment Evaluation Complete analysis of two assignments for each participant; analysis of collaborative assignment	300.00
Total	$3,800.00

Course Materials

Each participant in a Write, Inc. workshop receives a notebook of teaching materials and practice exercises. Material in the notebook will be the focus of class discussion and will have space for participants to take notes. All exercises will be done in the notebook, which also features cases and responses for each participant's analysis. After the workshop is complete, each participant has an instructional guide that can later be used for reference and review of correspondence design strategies.

Each participant will also receive a complimentary copy of Strunk and White, The Elements of Style, a handy guide to correctness. This guide will be placed in the front pocket of each folder.

Rationale for Write. Inc.'s Approach

Individuals learn to write by writing, not just by listening. Unlike many writing instruction services, we require participants to practice the concepts we present. We evaluate each person's

Figure 17-6 (continued).

Ms. Cindy Easton -4- November 2, 1993

work, not once, but three times, to be sure participants understand the strategies we present as well as the logic that underlies these strategies.

Qualifications of Writing. Inc.

Materials for the Lambert notebook will be developed by the instructional materials groups of Write, Inc. Each member of this group holds an M.A. in English and has had at least three years' experience teaching professional writing. Your workshops will be conducted in their entirety by Mr. Mark Clayton, who has been with our organization for four years. Mark holds a B.A./M.B.A. in English—Writing Emphasis and Finance and an M.A. in English from Texas A&M University, where he taught professional writing for three years. Mark has taught 19 workshops for business organizations throughout the State.

Conclusion

Ms. Easton, if you accept our proposal, I need to meet with you or someone in Lambert to begin developing cases for the letter exercises. We believe that writing assignments that mirror common situations within an organization provide the most meaningful contexts in which employees can improve their writing skills. Development of cases that will help your employees address common customer questions and problems will require several days, as we will need opportunity to work with Lambert managers to refine the cases. We have used this method for five years and found that workshop participants are pleased with this approach.

If you have further questions about the approach I have proposed, or if you would like to see example materials we have developed for use by other organizations, please give me a call. As we discussed, the day and times can be arranged at whatever day and hour you choose.

Sincerely,

Jon Sanchez

Jon Sanchez
President, Write, Inc.

Figure 17-6 (continued).

TO: Jon Sanchez DATE: January 24, 1994

FROM: Mark Clayton

SUBJ: Status Report on Workshop at Lambert Chemical

SUMMARY

I have currently completed instruction for Week 3 of the Lambert workshop. While I am confident that the workshop will be deemed successful for the 21 participants, you need to know about two issues that might influence your discussion with Ms. Easton about our redoing her policy manual. Employee attitude about communications is typical of what I have seen in companies that have communications problems: underestimation of what is required to write any technical report or letter effectively. With this lack of understanding comes problems in explaining cost for our services.

WORK COMPLETED

First Tuesday

Principles of Effectiveness for all Writing: Analysis of readers, definition of purpose, relationship of writer and reader, attitude and tone in writing. Assignment 1: Reading and practice in notebook.

Second Tuesday

Discussion of reading and practice assignment. Organization of business writing. Organization of business correspondence. Analysis of cases and letter responses in notebook.
Assignment 2: Develop a letter responding to the situation in Lambert Case #1.

These cases were evaluated and returned. The responses exhibited the usual problems that writers have when they do not think in terms of their customer's perspective.

Third Tuesday

Peer review and critique of each assignment. Analysis of cases and letter responses in notebook, Section 2. Review of tone, organization, audience needs and perception.

Participants had difficulty during the first hour understanding the concept of tone and audience perception. However, by the end of the session, about half the class were

Figure 17–7 Status Report Response to Situation 4

Jon Sanchez—2 January 24, 1994

catching on to the problems in some of the letters I showed them.

Assignment 3: Develop two letters responding to the situations described in Lambert Case #2.

I have collected and evaluated both letters on this case. The improvement is extraordinary.

WORK REMAINING

Fourth Tuesday

Peer review and critique of each assignment. Practice on tone and clarity of sentences and paragraphs.
Assignment 4: Collaboration exercise.

Fifth Tuesday

Presentation of each group's response. Explanation of rationale for content, organization, tone, and style.
Peer review of each response.

Wrap-Up. Evaluation of workshop.

MAJOR CONCERNS

Excessive Enrollment

Lambert representatives were not amenable to my request for two workshops of ten people each. With 21 people enrolled, I am having to redesign the collaborative response for Week 4 to divide the group into seven groups of three each. We will discuss as many of the responses as possible. I will provide detailed written critiques for each group to ensure that the response of each group is evaluated.

The size of this workshop, to my mind, reduces its effectiveness. Participants do not feel comfortable discussing examples or asking questions. We are using one of Lambert's instructional rooms, which contains 25 seats arranged in theater style. As you well know, we usually want our workshops to be conducted in seminar room arrangement to improve communication.

Perspective on Writing

Lambert, like many other technical firms, seems to have few people who are what I would call good writers. Developing the

Figure 17-7 (continued).

Jon Sanchez—3 January 24, 1994

cases was extremely time consuming and required about 11 hours more than I had expected to spend. The senior environmental engineers seem to think that good writing is simply a matter of correct mechanics and were not enthusiastic about the need for cases, once we began working with them on cases.

Participants, however, seem interested in what I have to say, and I am fairly confident that they will view the workshop as a success. Their writing, and their attitudes, on what is required to communicate effectively, are definitely changing for the better.

Cost Overruns

Because of the time overruns for case development and the large class, we will lose money on this workshop. I will have to expend about 15 hours extra for case development and grading.

RECOMMENDATIONS

In your forthcoming conversations with Ms. Easton, you may want to emphasize step-by-step what will be required to redesign their safety manual. Before we submit a proposal, we need to know if they will comply with our requests for information, if they have available the OSHA standards we need, and if they will supply the company liaison that we will need. In short, we need to be better prepared to estimate how much time we will need to do the policy manual so that our cost proposal is fair to us and to Lambert.

I believe that the evaluations will show that participants have had difficulties in writing sales letters because they have underestimated the writing process. If this fact comes through in the evaluations, you should be able to use this information in your discussion with Easton.

You can reach me at 2125. I have my voice mail activated.

Figure 17-7 (continued).

Planning and Revision Checklists

The following questions are a summary of the key points in this chapter, and they provide a checklist when you are planning and revising any document for your readers.

PROPOSALS

Planning

- Have you studied the RFP carefully?
- Have you made a list of all requirements given in the RFP?
- Who are your readers? Do they have technical competence in the field of the proposal? Is it a mixed audience, some technically educated, some not?
- What problem is the proposed work designed to remedy? What is the immediate background of the problem? Why does the problem need to be solved?
- What is your proposed solution to the problem? What benefits will come from the solution? Is the solution feasible (both practical and applicable)?
- How will you carry out the work proposed? Scope? Methods to be used? Task breakdown? Time and work schedule?
- Do you want to make statements concerning the likelihood of success or failure and the products of the project?
- What facilities and equipment will you need to carry out the project?
- Who will do the work? What are their qualifications for doing the work? Can you obtain references for past work accomplished?
- How much will the work cost? Consider such things as materials, labor, test equipment, travel, administrative expenses, and fees. Who will pay for what?
- Will you need to include an appendix? Consider such things as biographical sketches, descriptions of earlier projects, and employment practices.
- Will the proposal be better presented in a report format or a letter or memo format?
- Do you have a student report to propose? Consider including the following in your proposal:
  - Subject, purpose, and scope of report
  - Task and time breakdown
  - Resources available
  - Your qualifications for doing the report

Revision

- Does your proposal have a good design and layout? Does its appearance suggest the high quality of the work you propose to do?
- Does the project summary succinctly state the objectives and plan of the proposed work? Does it show how the proposed work is relevant to the readers' interests?
- Does the introduction make the subject and the purpose of the work clear? Does it briefly point out the so-whats of the proposed work?
- Have you defined the problem thoroughly?
- Is your solution well described? Have you made its benefits and feasibility clear?

- Will your readers be able to follow your plan of work easily? Have you protected yourself by making clear what you will do and what you will not do? Have you been careful not to promise more results than you can deliver?
- Have you carefully considered all the facilities and equipment you will need?
- Have you presented the qualifications of project personnel in an attractive but honest way? Have you asked permission from everyone you plan to use as a reference?
- Is your budget realistic? Will it be easy for the readers to follow and understand?
- Do all the items in the appendix lend credibility to the proposal?
- Have you included a few sentences somewhere that urge the readers to take favorable action on the proposal?
- Have you satisfied the needs of your readers? Will they be able to comprehend your proposal? Do they have all the information they need to make a decision?

PROGRESS REPORTS

Planning

- Do you have a clear description of your project available, perhaps in your proposal?
- Do you have all the project tasks clearly defined? Do all the tasks run in sequence or do some run concurrently? In general, are the tasks going well or badly?
- What items need to be highlighted in your summary and appraisal?
- Are there any problems to be discussed?
- Can you suggest solutions for the problems?
- Is your work ahead of or behind schedule?
- Are costs running as expected?
- Do you have some unexpected good news you can report?

Revision

- Does your report present an attractive appearance?
- Does the plan you have chosen show off your progress to its best advantage?
- Is your tone authoritative with an accent on the positive?
- Have you supported your generalizations with facts?
- Does your approach seem fresh or tired?
- Do you have a good balance between work accomplished and work to be done?
- Can your summary and appraisal stand alone? Would they satisfy an executive reader?

Exercises

1. Examine the RFP from the Department of Agriculture in Figure 17–8 and answer these questions:

27910 Federal Register / Vol. 58, No. 89 / Tuesday, May 11, 1993 / Notices

DEPARTMENT OF AGRICULTURE

Extension Service (ES)

Rural Technology and Cooperative Development Grants Program; Fiscal Year 1993; Request for Proposals; Application Guidelines

AGENCY: Extension Service, USDA.
ACTION: Notice.

FOR FURTHER INFORMATION CONTACT: Dr. Ted Maker, 202–720–7185.

Program Description

(a) Purpose

Proposals are requested from qualified non-profit institutions for the purpose of awarding competitive grants for fiscal year 1993 to establish and operate centers for rural technology or for cooperative development. The authority for this program is contained in section 2347 of the Food, Agriculture, Conservation, and Trade Act of 1990, Public Law 101–624, 7 U.S.C. 1932(f). The Program is Administered by the Extension Service (ES) of the U.S. Department of Agriculture (USDA). Under the program, ES will award competitive grants to non-profit institutions for the purpose of enabling such institutions to establish and operate centers for rural technology or cooperative development.

Available Funding

For fiscal year 1993, $1 million is available for the Program. Individual grants may be awarded in amounts ranging from $50,000 to a maximum of $960,000. Grants under this program may defray up to 75 percent of the administrative costs incurred by awardees to carry out projects for which grants are made. For the purpose of determining the non-Federal share of such costs, consideration will be given to contributions in cash and in kind, fairly evaluated, including but not limited to premises, equipment, and services.

(c) Eligibility

Proposals are invited from non-profit institutions as defined below.

In addition to the above, an applicant must qualify as a responsible applicant in order to be eligible for a grant award under the Program. To qualify as responsible, an applicant must meet the following standards:

(1) Adequate financial resources for performance, the necessary experience, organizational and technical qualifications, and facilities, or a firm commitment, arrangement, or ability to obtain same (including any to be obtained through subagreements (s));

(2) Ability to comply with the proposed or required completion schedule for the project;

(3) Adequate financial management system and audit procedures that provide efficient and effective accountability and control of all funds, property and other assets;

(4) Satisfactory record of integrity, judgment, and performance, including, in particular, any prior performance under grants and contracts from the Federal government; and

(5) Otherwise be qualified and eligible to receive a grant under the applicable laws and regulations.

(d) Definitions

For the purpose of awarding grants under this Program, the following definitions are applicable.

(1) "Administrative costs" means the total of direct and indirect costs, as defined in the Departmental Assistance Regulations, related to the establishment and operation of a center under this program;

(2) "Awarding official" means the Administrator of the Extension Service.

(3) "Grant" means the award by the Administrator to a grantee for the purpose of enabling nonprofit institutions to establish and operate centers for rural technology or cooperative development;

(4) "Grantee" means the entity designated in the grant award document as the responsible legal entity to whom a grant is awarded;

(5) "Nonprofit institution" means any organization or institution, including an accredited institution of higher education, no part of the net earnings of which inures or lawfully inure to the benefit of any private shareholder or individual;

(6) "Peer review panel" means the appropriate employees of the U.S. Department of Agriculture;

(7) "Project" means the particular activity within the scope of the Program as identified herein;

(8) "Project director" means an individual who is responsible for the technical direction of the project, as designated by the grantee in the grant proposal and approved by the Administrator;

(9) "Project period" means the total time approved by the Administrator for conducting the proposed project as outlined in an approved grant proposal or approved portions thereof;

(10) "United States" means the several states, the District of Columbia, the Commonwealth of Puerto Rico, the Virgin Islands, Guam, American Samoa, and other territories and possessions of the United States.

Proposal Preparation

(a) Proposal Cover Page

Title of Proposal.

(1) The title of the proposal must be brief (80-character maximum) yet represent the major thrust of the project.

(2) Other information.

Also include the following information on the proposal cover page:

(A) Name, address, telephone and fax numbers of applicant and Project Director.

(B) Signatures and date.

The cover page must contain the original signatures of the Project Director and the Authorized Organizational Representative who possesses the necessary authority to commit the entity's time and other relevant resources.

(b) Project Summary

Each proposal must contain a project summary which may not exceed 2 pages in length. The project summary should contain the following:

(1) Overall project goal(s) and supporting objectives;

(2) A brief description of the plans to accomplish project goal(s); and

(3) Relevance or significance of the project to rural economic development.

(c) Project Description

The specific aims of the project must be included in all proposals. The text of the project description may not exceed 15 pages and must contain the following components:

(1) Introduction.

A clear statement of the goal(s) and supporting objectives of the proposed project should preface the project description.

(2) Background and Existing Situation.

Provide a detailed description giving rise to the need for the effort in the proposed project.

(3) Project Plan.

Applicants shall submit a plan for the establishment and operation by the applicant of a center for rural technology or cooperative development.

Such a plan shall contain the following elements:

(A) A provision substantiating that the center will effectively serve rural areas in the United States by creating job opportunities, strengthening existing rural business, and demonstrating innovative methods to deliver services.

(B) A provision substantiating that the primary objective of such center will be to improve the economic condition of rural areas by promoting the development (through technological innovation, cooperative development,

Figure 17–8 Example Request for Proposal

- What work does the RFP want done or what product? What problem does the RFP present to be solved?
- Does it specify a length for the proposal?

Federal Register / Vol. 58, No. 89 / Tuesday, May 11, 1993 / Notices 27911

and adaptation of existing technology) and commercialization of:

(i) New services and products that can be produced or provided in rural areas;

(ii) New processes that can be utilized in the production of products in rural areas; and

(iii) New enterprises that can add value to on-farm production through processing and marketing.

(C) A description of the activities that the center will carry out to accomplish the objective of improving rural economic conditions in rural areas to be served by the center, including, but not limited to the following:

(i) Programs for technology research, investigations, and basic feasibility studies in any field or discipline for the purpose of generating principles, facts, technical knowledge, new technology, or other information that may be useful in rural industries, cooperatives, agribusinesses, and other persons or entities in the development and commercialization of new products, processes, or services;

(ii) Programs for the collection, interpretation, and dissemination of principles, facts, technical knowledge, new technology or other information that may be useful to rural industries, cooperatives, agribusinesses, and other persons in the development and commercialization of new products, processes, or services;

(iii) Programs providing training and instruction for individuals with respect to the development (through technological innovation, cooperative development, and adaption of existing technology) and commercialization of new products, processes, or services;

(iv) Programs providing loans and grants to individuals, small businesses and cooperatives for purposes of generating evaluating, developing, and commercializing new products, processes, or services;

(v) Programs providing technical assistance and advisory services to individuals, small businesses, cooperatives, and individuals for purposes of developing and commercializing new products, processes, or services; and

(vi) Programs providing research and support to individuals, small businesses, cooperatives and industries for the purposes of developing new agricultural enterprises to add value to on-farm production through processing or marketing.

(D) A description of the contributions that the activities described above are likely to make the improvement of the economic conditions of the rural areas for which such a center will provide services.

(E) Provisions substantiating that the center, in carrying out the activities described above, will seek, where appropriate, the advice, participation, expertise, and assistance of representatives of business, industry, educational institutions, and the Federal, State and local governments.

(F) Provisions ensuring the center

(i) Will consult with any college or university administering any program under title V of the Rural Development Act of 1972 in the State in which such center is located; and

(ii) Will cooperate with such college or university in the coordination of such activities and such programs.

(G) Provisions that the center will strive for self-sufficiency by taking all practicable steps to develop continuing sources of financial support for such center, particularly from sources in private sector.

(H) Provisions for

(i) Evaluating and monitoring of center activities by the institution operating the center; and

(ii) Accounting for money received by the institution under this section.

(I) Provisions ensuring that the center will pursue the optimal application of technology and cooperative development in rural areas, especially those areas affected by adverse agricultural economic conditions through the establishment of demonstration projects and subcenters for

(i) Rural technology development where the technology can be implemented by communities, community colleges, businesses, cooperatives, and other institutions; or

(ii) Cooperative development where such development can be implemented by cooperatives to improve local economic conditions.

(4) Evaluation. Give specific evaluation objectives including impact factors and indicators of effectiveness and efficiency in accomplishing objectives.

(d) Collaborative Arrangements

If the nature of the proposed project requires collaboration or subcontractural arrangements with other entities, the applicant must identify the collaborator/subcontractor and provide a full explanation of the nature of the relationship.

(e) Project Management and Personnel

Provide a management plan for the project, and describe the skills, qualifications and experience of key project personnel who will be involved.

(f) Budget

A budget and a detailed narrative in support of the budget is required. Show all funding sources, including Federal funds and non-Federal matching funds, and itemize costs by the following line items: Personnel costs, equipment, material and supplies, travel and all other costs.

Funds may be requested under any of the line items listed above provided that the item or service for which support is requested is identified as necessary for successful conduct of the project, is allowable under the authorizing legislation, the applicable Federal cost principles, and is not prohibited under any applicable Federal statute. Salaries of project personnel who will be working on the project may be requested in proportion to the effort that they will devote to the project.

Proposal Submission

(a) What to Submit

An original and two copies of the proposal must be submitted. Each copy of each proposal must be stapled securely in the upper lefthand corner (Do Not Bind). All copies of the proposal must be submitted in one package.

(b) Where and When to Submit

Proposals submitted through regular mail must be postmarked by June 21, 1993, and sent to the address below. Hand-delivered proposals must be submitted by June 21, 1993 to an express mail or courier service or brought to the following address: Extension Service—USDA, Cooperative Funds Division, Cotton Annex, 2nd Floor Mez., 300 12th Street SW., Washington, DC 20250–0900.

Proposal Review, Evaluation, and Disposition

(a) Proposed Review

All proposals received will be acknowledged. Prior to technical examination, a preliminary review will be made for responsiveness to this solicitation. Proposals that do not fall within the solicitation guidelines will be eliminated from competition. All accepted proposals will be reviewed by a peer review panel and recognized specialists in the areas covered by the proposals received. The peer review panel and specialists will be selected and organized to provide maximum expertise and objective judgment in the evaluation of proposals and shall be comprised entirely of Department employees. Proposals will be ranked and support levels will be recommended by the panel within the

Figure 17-8 (continued).

- Does the RFP make clear the information the proposal must contain?
- Does the RFP furnish an outline to follow? If so, what does the outline require?
- Does the RFP require a specific format for the proposal? What is it?
- Does the RFP make clear the criteria by which submitted proposals will be evaluated and who will do the evaluation?

27912 **Federal Register** / Vol. 58, No. 89 / Tuesday, May 11, 1993 / Notices

limitation of total funding available in fiscal year 1993.

(b) Evaluation Criteria

(1) In evaluating proposals, the peer review panel and the awarding official will take into account the degree to which the proposal:

(A) Demonstrates the capability to transfer, for practical application in rural areas, the technology generated at such centers and the ability to commercialize products, processes, services, and enterprises in such rural areas;

(B) Will effectively serve in rural areas that have—

(i) Few rural industries and agribusinesses;

(ii) High levels of unemployment or underemployment;

(iii) High rates of outmigration of people, businesses, and industries; and

(iv) Low levels of per capita income; and

(C) Will contribute the most to the improvement of economic conditions of rural areas.

(2) Additional criteria for selecting proposals are as follows:

(A) The extent to which the proposal is responsive to issues associated with identified national needs and priorities, including those that may be interdisciplinary in nature;

(B) The extent to which the proposal has the potential for improving effectiveness, efficiency, or appropriateness of educational programs;

(C) The degree to which the proposal demonstrates originality, practicality, and creativity in developing and testing innovative, effective solutions to existing or anticipated issues or problems of targeted audiences;

(D) The degree to which the proposal is adequate, sound, and appropriate to rural economic development;

(E) The expected results of the project, including plans for demonstrating and sharing information with governmental and nongovernmental entities;

(F) The extent to which the applicant proposes cooperation with the programs of other entities;

(G) The experience, qualifications, competence, and availability of personnel to direct and carry out the project;

(H) The extent to which the applicant is committing resources to the project.

(c) Proposal Disposition

When the peer review panel has completed its deliberations, the USDA program staff, based on the recommendations of the peer review panel, will recommend to the Awarding Official that the project be (a) approved for support from currently available funds or (b) declined due to insufficient funds or unfavorable review. USDA reserves the right to negotiate with the Project Director and/or the submitting entity regarding project revisions (e.g., reductions in scope of work), funding level, or period of support prior to recommending any project for funding.

A proposal may be withdrawn at any time before a final funding decision is made. One copy of each proposal that is not selected for funding (including those that are withdrawn) will be retained by USDA for one year, and remaining copies will be destroyed.

SUPPLEMENTARY INFORMATION:

(a) Programmatic Contact

For additional information on the program, please contact: Dr. Ted Maher, U.S. Department of Agriculture, Extension Service, 14th & Independence Avenue, SW., Room 3901–South Building, Washington, DC 20250–0900. (202) 720–7185

(b) Paperwork Reduction

The Office of Management and Budget has approved the information collection requirements of this notice, in accordance with the Paperwork Reduction Act of 1980 and 5 CFR part 1320. Public reporting burden for the information collections contained in this notice is estimated to be 4 hours per response, including the time for reviewing instructions, searching existing data sources, gathering and maintaining the data needed, and completing and reviewing the collection of information. Send any comments on this notice to the Department of Agriculture, Clearance Officer, OIRM, rm. 404–W, Washington, DC 20250; and to the Office of Management and Budget, Paperwork Reduction Project (OMB document No. 05270011), Washington, DC 20503.

(c) Grant Awards

Within the limit of funds available for such purpose, the awarding official shall make grants to those responsible, eligible applicants whose proposals are judged most meritorious under the evaluation criteria and procedures set forth in this solicitation and application guidelines.

The date specified by the awarding official as the beginning of the project period shall be not later than September 30, 1993.

All funds granted under the Program shall be expended solely for the purpose for which the funds are granted in accordance with the approved application and budget, the terms and conditions of any resulting award, the applicable Federal cost principles, and the Department's Federal assistance regulations.

(d) Obligation of the Federal Government

Neither the approval of any application nor the award of any grant commits or obligates the United States in any way to provide further support of a project or any portion thereof.

(e) Other Applicable Federal Statutes and Regulations that Apply

Several other Federal statutes and regulations apply to grant proposals considered for review or grants awarded under the Program. These include, but are not limited to the following:

7 CFR part 1B—USDA Implementation of the National Environmental Policy Act;

7 CFR part 3—USDA implementation of OMB Circular A–129 regarding debt collection;

7 CFR part 1.1—USDA implementation of the Freedom of Information Act;

7 CFR part 15, Subpart A—USDA implementation of title VI of the Civil Rights Act of 1964;

7 CFR part 3015—USDA Uniform Federal Assistance Regulations, implementing OMB directives (i.e., Circular Nos. A–110, A–21, and A–122) and incorporating provisions of 31 U.S.C. 6301–6308 (formerly, the Federal Grant and Cooperative Agreement Act of 1977, Pub. L. 95–224), as well as general policy requirements applicable to recipients of Departmental financial assistance;

7 CFR part 3016—USDA Uniform Administrative Requirements for Grants and Cooperative Agreements to State and Local Governments;

7 CFR part 3017, as amended—USDA implementation of Governmentwide Debarment and Suspension (nonprocurement) and Governmentwide Requirements for Drug-Free Workplace (Grants);

7 CFR part 3018—USDA implementation of New Restrictions on Lobbying. Imposes prohibitions and requirements for disclosure and certification related to lobbying on recipients of Federal contracts, grants, cooperative agreements, and loans;

29 U.S.C. 794, section 504—Rehabilitation Act of 1973, and 7 CFR part 15B (USDA implementation of the statute), prohibiting discrimination based upon physical or mental handicap in Federally assisted programs; and

35 U.S.C. 200 et seq.—Bayh-Dole Act, controlling allocation of rights to inventions made by employees of . . . .

Figure 17-8 (continued).

2. Write an information report based on Figure 17-8 to your professor summarizing what content items the proposal should emphasize and what criteria will be used to evaluate the proposals.
3. You will probably be required to write a complete technical report as part of the requirements for your course in technical writing.
   - Choose two or three potential topics you would consider to be suitable for a one-semester project.

- Write a feasibility report to your instructor examining each topic in terms of availability of information, suitability of the topic for the amount of time available during the semester, and the significance of each topic to your discipline or to your career goals. Decide which topic seems most feasible.
- Once your instructor has approved your choice of topic, write a proposal to your instructor, using memo format. In your proposal, include all elements commonly found in proposals.
- Write a progress report to explain the status on your semester report project. Design the progress report to reflect the tasks or project goals you have used in developing your project report.

4. The following progress report is poorly organized and formatted. Reorganize it and rewrite it. Use a letter format, but furnish a subject line and the headings readers need to find their way through the report.

Forestry Research Associates
222 University Avenue
Madison, Wisconsin 53707
June 30, 1991
Mr. Lawrence Campbell, Director
Council for Peatlands Development
420 Duluth Street
Grand Forks, ND 58201
Dear Mr. Campbell:

Well, we have our Peatland Water-Table Depth Research Project underway. This is our first progress report. As you know, by ditching peatlands, foresters can control water-table depths for optimum growth of trees on those peatlands. Foresters, however, don't have good data on which water-table depths will encourage optimum tree growth. This study is an attempt to find out what those depths might be. We have to do several things to obtain the needed information. First, we have to measure tree growth on plots at varying distances from existing ditches on peatland. Then we have to establish what the average water-table depth is on the plots during the growing season of June, July, August, and September. To get meaningful growth and water-table depth figures, we have to gather these data for three growing seasons. Finally, we have to correlate average water-table depth with tree growth. Knowing that, foresters can recommend appropriate average water-table depths.

When the snow and ice went out in May, we were able to establish 14 tree plots on Northern Minnesota peatlands. Each plot is one-fortieth of a hectare. The distances of the plots from a drainage ditch vary from 1 to 100 meters. The plots have mixed stands of black spruce and tamarack. During June, we measured height and diameter at breast height (DBH) of a random selection of trees on each plot. We marked the measured trees so that we can return to them for future measurements. We will measure them again in September of this year and in June and September of the next two years.

While we were measuring the trees, we began placing two wells on each plot. The wells consist of perforated plastic pipes driven eight feet into the mineral soil that underlies the plot. We should have all the wells in by the end of next week. We'll measure water-table depths once a month in July, August, and September. We will also measure water-table depths any time there is a rainfall of one inch or more on the plots.

We have our research well underway, and we're right on schedule. We have made all our initial tree measurements and will soon obtain our first water-table depth readings.

By the way, the entire test area seems to be composed of raw peat to a depth of about twenty cm with a layer of well-decomposed peat about a meter thick beneath that. However, to be sure there are no soil differences that would introduce an unaccounted-for variable into our calculations, we'll do a soil analysis on each of the 14 plots next summer. We will do this additional work at no extra cost to you. During the cold-weather months, between growing seasons, we'll prepare water-table profiles that will cover each plot for each month of measurement. At the completion of our measurements in the third year, we'll correlate these profiles with the growth measurements on the plots. This correlation should enable us to recommend a water-table depth for optimum growth of black spruce and tamarack. We have promised two progress reports per growing season and one each December. Therefore, we will submit our next report on September 30.

Sincerely,
Robert Weaver
Principal Investigator

Oral Exercises

Before doing these exercises, see Chapter 18, "Oral Reports."

5. Prepare an oral version of your proposal, which you will deliver to your class. You will be allowed 8 minutes maximum. Enhance your presentation with graphics (computer graphics or overhead transparencies) to show anticipated costs, your project schedule, and any visuals that will help explain the significance of your project or the methods you will use.

6. Prepare an oral progress report, which you will deliver to your class. You will be allowed 5 minutes maximum. Enhance your presentation with graphics to show work completed, work remaining, project costs to date, and status of your project.

CHAPTER 18 ORAL REPORTS

Oral reports are a major application of reporting technical information. You will have to report committee work, laboratory experiments, and research projects. You will give reports at business or scholarly meetings. You will instruct, if not in a teacher–student relationship, perhaps in a supervisor–subordinate relationship. You may have to persuade a group that a new process your section has devised is better than the present process. You may have to brief your boss about what your department does to justify its existence. In this chapter we discuss preparing and presenting your oral report, with a heavy emphasis on the ways in which you can provide visual support.

Preparation

In many ways, preparing an oral report is much like composing a written report. The situational analysis is virtually identical to the situational analysis described in Chapter 2, "Composing." You have to consider your purpose and audience, discover your material, and arrange your material. You will not in most cases write out your material beyond an outline stage. However, you should rehearse its delivery several times, which is akin to writing and revising. In addition to these tasks, you will have others peculiar to the speech situation.

Find out as much as you can about the conditions under which you will speak. Inquire about the size of the room you will speak in, the time allotted for the speech, and the size of the audience. If you have to speak in a large area to a large group, will a public address system be available to you? Find out if you will have a lectern for your notes. If you plan to use visual aids, inquire about the equipment. Does the sponsoring group have projectors to show 35 mm slides or transparencies? Many speakers have arrived at a hall and found all of their vital visual aids worthless because projection equipment was not available. Find out if there will be someone to introduce you. If not, you may have to work your credentials as a speaker into your talk. Consider the time of day and day of the week. An audience listening to you at 3:30 on Friday afternoon will not be nearly as attentive as an audience earlier in the day or earlier in the week. Feel free to ask the sponsoring group any of these questions. The more you know beforehand, the better prepared and therefore the more comfortable you will be.

Delivery Techniques

There are four basic **delivery techniques,** but you really need to think about only two of them. The four are (1) impromptu, (2) speaking from memory, (3) extemporaneous, and (4) reading from a manuscript.

Impromptu speaking involves speaking "off the cuff." Such a method is too risky for a technical report, where accuracy is so vital. In speaking from memory, you write out a speech, commit it to memory, and then deliver it. This gives you a carefully planned speech, but we cannot recommend it as a good technique. The drawbacks are (1) your plan becomes inflexible; (2) you may have a memory lapse in one place that will unsettle you for the whole speech; (3) you think of words rather than thoughts, which makes you more artificial and less vital; and (4) your voice and body actions become stylized and lack the vital spark of spontaneity.

We consider the best methods to be extemporaneous speech and the speech read from a manuscript, and we will discuss these in more detail.

The Extemporaneous Speech

Unlike the impromptu speech, with which it is sometimes confused, the extemporaneous speech is carefully planned and practiced. In preparing for an extemporaneous speech, you go through the planning and arranging steps described in Chapter 2. But you stop when you complete the outline stage. You do not write out the speech. Therefore, you do not commit yourself to any definite phraseology. In your outline, however, include any vital facts and figures that you must present accurately. You will want no lapses of memory to make you inaccurate in presenting a technical report.

Before you give the speech, practice it, working from your outline. Give it several times, before a live audience preferably, perhaps a roommate or a friend. As you practice, fit words to your outlined thoughts. Make no attempt to memorize the words you choose at any practice session, but keep practicing until your delivery is smooth. When you can go through the speech without faltering, you are ready to present it. When you practice a speech, pay particular attention to timing. Depending upon your style and the occasion, plan on a delivery rate of 120-180 words per minute. Nothing, *but nothing*, will annoy program planners or an audience more than to have a speaker scheduled for 30 minutes go for 40 minutes or an hour. The long-winded speaker probably cheats some other speaker out of his or her allotted time. Speakers who go beyond their scheduled time can depend upon not being invited back.

We recommend that you type your outline. Use capitals, spacing, and underlining generously to break out the important divisions. Use boldface type if you have word processing capability. But don't do the entire outline in capitals. That would make it hard to read. As a final refinement, place your outline in a looseleaf ring binder. By so doing you can be sure that it will not become scattered or disorganized.

There are several real advantages of the extemporaneous speech over the speech read from manuscript. With the extemporaneous speech you will find it easier to maintain eye contact with your audience. You need only glance occasionally at your outline to keep yourself on course. For the rest of the time you can concentrate on looking at your audience.

You have greater flexibility with an extemporaneous speech. You are committed to blocks of thought but not words. If by looking at your audience you see that they have not understood some portion of your talk, you are free to rephrase the thought in a new way for better understanding. If you are really well prepared in your subject, you can bring in further examples to clarify your point. Also, if you see you are running overtime, you can condense a block by leaving out some of your less vital examples or facts.

Finally, because you are not committed to words, you retain conversational spontaneity. You are not faltering or groping for words, but neither are you running by your audience like a well-oiled machine.

The Manuscript Speech

Most speech experts recommend the extemporaneous speech above reading from a manuscript. We agree in general. However, speaking in a technical situation often requires the manuscript speech. Papers delivered to scientific societies are frequently written and then read to the group. Often, the society will later publish your paper. Often, technical reports contain complex technical information or extensive statistical material. Such reports do not conform well to the extemporaneous speech form, and you should plan to read them from a manuscript.

Planning and writing a speech are little different from writing a paper. However, in writing your speech try to achieve a conversational tone. Certainly, in speaking you will want to use the first person and active voice. Remember that speaking is more personal than writing. Include phrases like "it seems to me," "I'm reminded of," "Just the other evening, I," and so forth. Such phrases are common in conversation and give your talk extemporaneous overtones. Certainly, prefer short sentences to long ones.

Type the final draft of your speech. Just as you did for the extemporaneous speech outline, be generous with capitals, spacing, and underlining. Plan on about three typed pages per five minutes of speech. Put your pages in order and place them in a looseleaf binder.

When you carry your written speech to the lectern with you, you are in no danger of forgetting anything. Nevertheless, you must practice it, again preferably aloud to a live audience. As you practice, remember that because you are tied to the lectern, your movements are restricted. You will need to depend even more than usual on facial expression, gestures, and voice variation to maintain audience interest. Do not let yourself fall into a sing-song monotone as you read the set phrases of your written speech.

Practice until you know your speech well enough to look up from it for long periods of good eye contact. Plan an occasional departure from your manuscript to speak extemporaneously. This will aid you to regain the direct contact with the audience that you so often lose while reading.

Arranging Content

For the most part you will arrange your speech as you do your written work. However, the speech situation does call for some differences in arrangement and even content, and we will concentrate on these differences. We will discuss the arrangement in terms of introduction, body, and conclusion.

Introduction

A speech introduction should accomplish three tasks: (1) create a friendly atmosphere for you to speak in, (2) interest the audience in your subject, and (3) announce the subject, purpose, scope, and plan of development of your talk.

Be alert before you speak. If you can, mingle and talk with members of the group to whom you are going to speak. Listen politely to their conversation. You may pick up some tidbit that will help you to a favorable start. Look for bits of local color or another means to establish a common ground between you and the audience. When you begin to speak, mention some member of the audience or perhaps a previous speaker. If you can do it sincerely, compliment the audience. If you have been introduced, remember to acknowledge and thank the speaker. Unless it is a very formal occasion, begin rather informally. If there is a chairperson and a somewhat formal atmosphere, we recommend no heavier beginning than, "Mr. Chairman (or Madam Chairwoman), ladies and gentlemen."

Gain attention for your subject by mentioning some particularly interesting fact or bit of illustrative material. Anecdotes are good if they truly tie in with the subject. But take care with humor. Avoid jokes that really don't tie in with the subject or the occasion. Forget about risqué stories.

Be careful also about what you draw attention to. Do not draw attention to shortcomings in yourself, your speech, or the physical surroundings. Do not begin speeches with apologies.

Announce your subject, purpose, scope, and plan of development in a speech just as you do in writing. (See pages 251–256.) If anything, giving your plan of development is more important in a speech than in an essay. Listeners cannot go back in a speech to check on your arrangement the way that a reader can in an essay. So, the more guideposts you give an audience, the better. No one has ever disputed this old truism: (1) Tell the audience what you are going to tell them. (2) Tell them. (3) Tell them what you just told them. In instructional situations, some speakers provide their audiences with a printed outline of their talk.

Body

When you arrange the body of a speech, you must remember one thing: A listener's attention span is very limited. Analyze honestly your own attention span—be aware of your own tendency to let your mind wander. You listen to the speaker for a moment, and then perhaps you think of lunch, of some problem, or an approaching appointment. Then you return to the speaker. When you become the speaker, remember that people do not hang on your every word.

What can you do about the problem of the listener's limited attention span? In part, you solve it by your delivery techniques. We will discuss these in the next section of this chapter. It also helps to plan your speech around intelligent and interesting repetition.

Begin by cutting the ground you intend to cover in your speech to the minimum. Build a five-minute speech around one point, a 15-minute speech around two. Even an hour-long talk probably should not cover more than three or four points.

Beginning speakers are always dubious about this advice. They think, "I've got to be up there for 15 minutes. How can I keep talking if I have only two points to cover? I'll never make it." Because of this fear they load their speeches with five or six major points. As a result, they lull their audience into a state of somnolence with a string of generalizations.

In speaking, even more than in writing, **your main content should be masses of concrete information—examples, illustrations, little narratives, analogies, and so forth—supporting just a few generalizations.** As you give your supporting information, repeat your generalization from time to time. Vary its statement, but cover the same ground. The listener who was out to lunch the first time you said it may hear it the second time or the third. You use much the same technique in writing, but you intensify it even more in speaking.

We have been using the same technique here in this chapter. We began this section on the speech body by warning you that a listener's attention span is short. We reminded you that your listening span is short: same topic but a new variation. We asked you what you can do about a listener's limited span: same topic with only a slight shift. In the next paragraph we told you not to make more than two points in a 15-minute speech. We nailed this point down in the next paragraph by having a dubious speaker say, "I've got to be up there for 15 minutes. How can I keep talking if I have only two points to cover?" In the paragraph just preceding this one we told you to repeat intelligently so that the reader "who was out to lunch the first time you said it may hear it the second time or the third." Here we were slightly changing an earlier statement that "You listen to the speaker for a moment, and then perhaps you think of lunch...." In other words, we are aware that the reader's attention sometimes wanders. When you are paying attention we want to catch you. Try the same technique in speaking, because the listener's attention span is even more limited than the reader's.

Creating suspense as you talk is another way to generate interest in your audience. Try organizing a speech around the inductive method. That is, give your facts first and gradually build up to the generalization that they support. If you do this skillfully, using good material, your audience hangs on, wondering what your point will be. If you do not do it skillfully or use dull material, your audience will tune you out and tune into their private worlds.

Another interest-getting technique is to **relate the subject matter to some vital interest of the audience.** If you are talking about water pollution, for example, remind the audience that the dirtier their rivers get, the more tax dollars it will eventually take to clean them up.

Visual aids often increase audience interest. Remember to keep your graphics big and simple. No one is going to see typewritten captions from more than three or four feet away. Stick to big pie and bar graphs. If you have tables, print them in letters from two to three inches high. If you are speaking to a large group, put your graphic materials on transparencies and project them on a screen. Prepare your transparencies with care. Don't just photocopy typed or printed pages or graphics from books. No one behind the first row will see them. To work, letters and numbers on transparencies should be at least twice normal size. Large-type typewriters are available. Word processing makes large type fonts available. If you need an assistant to help you project visual aids, get one or bring one with you.

Do not display a visual aid until you want the audience to see it. While the aid is up, call your listeners' attention to everything you want them to see. Take the aid away as soon as you are through with it. If using a projector, turn it off whenever it is not in use. Be sure to key every visual aid into your speaker's script. Otherwise, you may slide right by it. (See also the section, "Visual Aids," pages 604–615.)

Conclusion

In ending your speech, as in your written reports, you have your choice of several closes. You can close with a summary, or a list of recommendations including a call for some sort of action, or what amounts to "Good-bye, it's been good to talk to you." As in the introduction, you can use an anecdote in closing to reinforce a major point. In speaking, never suggest that you are drawing to a close unless you really mean it. When you suggest that you are closing, your listeners perk up and perhaps give a happy sigh. If you then proceed to drag on, they will hate you.

Second, remember that audience interest is usually highest at the beginning and close of a speech. Therefore, you will be wise to provide a summary of your key points at the end of any speech. Give your listeners something to carry home with them.

Presentation

After you have prepared your speech you must present it. For many people giving a speech is a pretty terrifying business. Before speaking they grow tense, have hot flashes and cold chills, and experience the familiar butterflies in the stomach. Some people tremble before and even during a speech. Try to remember that these are normal reactions, for both beginning and experienced speakers. Most people can overcome them, however, and it is even possible to turn this nervous energy to your advantage.

If your stage fright is extreme, or if you are the one person in a hundred who stutters, or if you have some other speech impediment, seek clinical help. The ability to communicate ideas through speech is one of humanity's greatest gifts. Do not let yourself be cheated. Some of the finest speakers we have ever had in class were stutterers who admitted their problem and worked at it with professional guidance. Remember, whether your problems are large or small, the audience is on your side. They want you to succeed.

Physical Aspects of Speaking

What are the physical characteristics of good speakers? They stand firmly but comfortably. They move and gesture naturally and emphatically but avoid fidgety, jerky movements and foot shuffling. They look directly into the eyes of people in the audience, not merely in their general direction. They project enthusiasm into their voices. They do not mumble or speak flatly. We will examine these characteristics in detail—first movement and then voice.

Movement A century ago a speaker's movements were far more florid and exaggerated than they are today. Today we prefer a more natural mode of speaking, closer to conversation than oratory. To some extent, electronic devices such as amplifying systems radio, and television have brought about this change. However, you do not want to appear like a stick of wood. Even when speaking to a small group or on television (or, oddly enough, on the radio) you will want to move and gesture. If you are speaking in a large auditorium, you will want to broaden your movements and gestures. From the back row of a 2,500-seat auditorium, you look about three inches tall.

Movement during a speech is important for several reasons. First, it puts that nervous energy we spoke of to work. The inhibited speaker stands rigid and trembles. The relaxed speaker takes that same energy and puts it into purposeful movement.

Second, movement attracts attention. It is a good idea to emphasize an idea with a pointing finger or a clenched fist; and a speaker who comes out from behind the lectern occasionally and walks across the stage or toward the audience awakens audience interest. The speaker who passively utters ideas deadens the audience.

Third, movement makes you feel more forceful and confident. It keeps you, as well as your audience, awake. This is why good speakers while speaking over the radio will gesture just as emphatically as though the audience could see them.

What sorts of movements are appropriate? To begin with, **movement should closely relate to your content.** Jerky or shuffling motions that occur haphazardly distract an audience. But a pointing finger combined with an emphatic statement reinforces a point for an audience. A sideward step at a moment of transition draws attention to the shift in thought. Take a step backward and you indicate a conclusion. Step forward and you indicate the beginning of a new point. Use also the normal descriptive gestures that all of us use in conversation: gestures to indicate length, height, speed, roundness, and so forth.

For most people, **gesturing is fairly natural.** They make appropriate movements without too much thought. Some beginning speakers, however, are body-inhibited. If you are in this category, you may have to cultivate movement. In your practice sessions and in your classroom speeches, risk artificiality by making gestures that seem too broad to you. Oddly enough, often at the very point where your gestures seem artificial and forced to you, they will seem the most natural to your audience.

Allow natural gestures to replace nervous mannerisms. Some speakers develop startling mannerisms and remain completely oblivious of them until some brave but kind soul points them out. Some that we have observed include putting eyeglasses off and on; knocking a heavy ring over and over on the lectern; fiddling with a pen, pointer, chalk, cigar, microphone cord, ear, mustache, nose, you name it; shifting from foot to foot in time to some strange inner rhythm; and pointing with the elbows while the hands remain in the pants pockets. Mannerisms may also be vocal. Such things as little coughs or repeating comments such as "OK" or "You know" to indicate transitions may become mannerisms.

Listeners are distracted by such habits. Often they will concentrate on the mannerisms to the exclusion of everything else. They may know that a speaker put her eyeglasses on and off 22 times but not have the faintest notion of what she said. If someone points out such mannerisms in your speaking habits, do not feel hurt. Instead, work to remove the mannerisms.

Movement includes facial movement. Do not be a deadpan. Your basic expression should be a relaxed, friendly look. But do not hesitate to smile, laugh, frown, or scowl when such expressions are called for. A scowl at a moment of disapproval makes the disapproval that much more emphatic. Whatever you do, do not freeze into one expression, whether it be the stern look of the man of iron or the vapid smile of a smoker in a magazine ad.

Voice **Your voice should sound relaxed, free of tension and fear.** In a man, people consider a deep voice to be a sign of strength and authority. Most people prefer a woman's voice to be low rather than shrill. If you do not have these attributes, you can develop them to some extent. Here we must refer you to some of the good speech books in our bibliography (Appendix B), where you will find various speech exercises described. If, despite hard work, your voice remains unsatisfactory in comparison with the conventional stereotypes, do not despair. Many successful speakers have had somewhat unpleasant voices and through force of character or intellect directed their audiences to their ideas and not their voices.

Many beginning speakers speak too fast, probably because they are anxious to be done and sit down. A normal rate of speech falls between 120 and 180 words a minute. This is actually fairly slow. Generally, you will want a fairly slow delivery rate. When you are speaking slowly, your voice will be deeper and more impressive. Also, listeners have trouble following complex ideas delivered at breakneck speed. Slow up and give your audience time to absorb your ideas.

Of course, **you should not speak at a constant rate, slow or fast. Vary your rate.** If you normally speak somewhat rapidly, slowing up will emphasize ideas. If you are speaking slowly, suddenly speeding up will suggest excitement and enthusiasm. As you speak, change the volume and pitch of your voice. Any **change in volume,** whether from low to loud or the reverse will draw your listener's attention and thus emphasize a point. The same is true of a **change in pitch.** If your voice remains a flat monotone and your words come at a constant rate, you deprive yourself of a major tool of emphasis.

Many people worry about their accent. Normally, our advice is *don't.* If you speak the dialect of the educated people of the region where you were raised, you have little to worry about. Some New Englanders, for example, put *r*'s where they are not found in other regional dialects and omit them where they are commonly found. Part of America's richness lies in its diversity. Accents vary in most countries from one region to another, but certainly not enough to hinder communication.

If, however, your accent is slovenly—"Ya wanna cuppa coffee?"—or uneducated, do something about it. Work with your teacher or seek other professional help. Listen to educated speakers and imitate them. **Whatever your accent, there is no excuse for mispronouncing words.** Before you speak, look up any words you know you must use and about which you are uncertain of the pronunciation. Speakers on technical subjects have this problem perhaps more than other speakers. Many technical terms are jawbreakers. Find their correct pronunciation and practice them until you can say them easily.

Audience Interaction

One thing speakers must learn early in their careers is that they cannot count on the audience's hanging on every word. Some years ago an intelligent, educated audience was asked to record its introspections while listening to a speaker. The speaker was an excellent one. Despite his excellence and the high level of the audience, the introspections revealed that the audience was paying something less than full attention. Here are some of the recorded introspections:

> God, I'd hate to be speaking to this group....I like Ben—he has the courage to pick up after the comments....Did the experiment backfire a bit? Ben seems unsettled by the introspective report....I see Ben as one of us because he is under the same judgment....He folds his hands as if he was about to pray....What's he got in his pocket he keeps wriggling around....I get the feeling Ben is playing a role....It is interesting to hear the words that are emphasized....This is a hard spot for a speaker. He really must believe in this research....
>
> Ben used the word "para-social." I don't know what that means. Maybe I should have copied the diagram on the board....Do not get the points clearly...cannot interrupt...feel mad...More words....I'm sick of pedagogical and sociological terms....Slightly annoyed by pipe smoke in my face....An umbrella dropped....I hear a funny rumbling noise....I wish I had a drink....Wish I could quit yawning ...Don't know whether I can put up with these hard seats for another week and a half or not....My head itches....My feet are cold. I could put on my shoes, but they are so heavy....My feet itch....I have a piece of coconut in my teeth....My eyes are tired. If I close them the speaker will think I'm asleep....
>
> Backside hurts...I'm lost because I'm introspecting....The conflict between introspection and listening is killing me. Wish I didn't take a set so easily....If he really wants me to introspect, he must realize himself he is wasting his time lecturing....This is better than the two hour wrestling match this afternoon....This is the worst planned, worst directed, worst informed meeting I have ever attended....I feel confirmation, so far, in my feelings that lectures are only 5% or less effective....I hadn't thought much about coming to this meeting but now that I am here it is going to be O.K....Don't know why I am here....I wish I had gone to the circus....Wish I could have this time for work I should be doing....Why doesn't he shut up and let us react....The end of the speech. Now he is making sense....It's more than 30 seconds now. He should stop. Wish he'd stop. Way over time. Shut up....He's over. What will happen now?...[1]

As some of the comments reveal, perhaps being asked to record vagrant thoughts as they appeared made some members of the audience less attentive than they normally would have been. But most of us know that we have very similar thoughts and lapses of attention while we attend classes and speeches.

Reasons for audience inattention are many. Some are under the speaker's control; some are not. The speaker cannot do much about such physical problems as hard seats, crowded conditions, bad air, and physical inactivity. The speaker can do something about psychological problems such as the listeners' passivity and their sense of anonymity, their feeling of not participating in the speech.

Audience Analysis Even before they begin to speak, good speakers have taken audience problems into account. They have analyzed the audience's education and experience level. They have planned to keep their points few and to repeat major points through carefully planned variations. They plan interesting examples. While speaking they attempt to interest the audience through movement and by varying the speech rate, pitch, and volume.

But good speakers go beyond these steps and analyze their audience and its reactions as they go along. In an extemporaneous speech and even to some extent in a written speech, you can make adjustments based on this audience analysis.

To analyze your audience, you must have good eye contact. You must be looking at Ben, Bob, and Irma. You must not merely be looking in the general direction of the massed audience. Look for such things as smiles, scowls, fidgets, puzzled looks, bored expressions, interested expressions, sleepy eyes, heads nodded in agreement, heads nodded in sleep, heads shaken in disagreement. You will not be 100% correct in interpreting these signs. Many students have learned to smile and nod in all the proper places without ever hearing the instructor. But, generally, such physical actions are excellent clues as to how well you are getting through to your audience.

Reacting to Audience If your audience seems happy and interested, you can proceed with your speech as prepared. If, however, you see signs of boredom, discontent, or a lack of understanding, you must make some adjustments. Exactly what you do depends to some extent on whether you are in a formal or informal speaking situation. We will look at the formal situation first.

In the *formal* situation you are somewhat limited. If your audience seems bored, you can quickly change your manner of speaking. Any change will, at least momentarily, attract attention. You can move or gesture more. With the audience's attention gained, you can supply some interesting anecdotes or other illustrative material to better support your abstractions and generalizations. If your audience seems puzzled, you know you must supply further definitions and explanations and probably more concrete examples. If your audience seems hostile, you must find some way to soften your argument while at the same time preserving its integrity. Perhaps you can find some mutual ground upon which you and the audience can agree and move on from there.

Obviously, such flexibility during the speech requires some experience. Also it requires that the speaker have a full knowledge of the subject. If every bit of material the speaker knows about the subject is in the speech already, the speaker has little flexibility. But do not be afraid to adjust a speech in midstream. Even the inexperienced speaker can do it to some extent.

Many of the speaker's problems are caused by the speech situation's being a one-way street. The listeners sit passively. Their normal desires to react, to talk back to the speaker, are frustrated. The problem suggests the solution, particularly when you are in a more informal speech situation, such as a classroom or a small meeting

In the more *informal* situation, you can stop when a listener seems puzzled. Politely ask him where you have confused him and attempt to clarify the situation. If a listener seems uninterested, give him an opportunity to react. Perhaps you can treat him as a puzzled listener. Or, you can ask him what you can do to interest him more. Do not be unpleasant. Put the blame for the lack of interest on yourself, even if you feel it does not belong there. Sometimes you may be displeased or shocked at the immediate feedback you receive, but do not avoid it on these grounds. And do not react unpleasantly to it. You will move more slowly when you make speaking a two-way street, but the final result will probably be better. Immediate feedback reveals areas of misunderstanding or even mistrust of what is being said.

In large meetings where such informality is difficult, you can build in some audience reaction through the use of informal subgroups. Before you talk, divide your audience into small subgroups, commonly called **buzz groups.** Use seating proximity as the basis for your division if you have no better one. Explain that after your talk the groups will have a period of time in which to discuss your speech. They will be expected to come up with questions or comments. People do not like to seem unprepared, even in informal groups. As a result, they will be more likely to pay attention to your speech in order to participate well in their buzz groups.

Whether you have buzz groups or not, often you will be expected to handle questions following a speech. If you have a chairperson, he or she will field the questions and repeat them, and then you will answer them. If you have no chairperson, you will perform this chore for yourself. Be sure everyone understands the question. Be sure you understand the question. If you do not, ask the questioner to repeat it and perhaps to rephrase it.

Keep your answers brief, but answer the questions fully and honestly. When you do not have the answer, say so. Do not be afraid of conflict with the audience. But keep it on an objective basis; talk about the conflict situation, not personalities. If someone reveals through his question that he is becoming personally hostile, handle him courteously. Answer his question as quickly and objectively as you can and move on to another questioner. Sometimes the bulk of your audience will grow restless while a few questioners hang on. When this occurs release your audience and, if you have time, invite the questioners up to the platform to continue the discussion. Above all, during a question period be courteous. Resist any temptation to have fun at a questioner's expense.

Visual Aids

Today, good speakers increasingly use visual aids during their talks.

Purpose

You will use a visual aid (1) *to support* and *expand* the content of your message and (2) *to focus the audience's attention* on a critical aspect of your presentation.[2]

Support The first purpose of any visual material is *to support your message*—to enlarge on the main ideas and give substance and credibility to what you are saying. Obviously, the material must be relevant to the idea being supported. Too often a speaker gives in to the urge to show a visually attractive or technically interesting piece of information that has little or no bearing on the subject.

Suppose, for the purposes of our analysis, that you were asked to meet with government people to present a case for your company's participation in a major federal contract. Your visual support would probably include information about the company's past performances with projects similar to the one being considered. You would show charts reflecting the ingenious methods used by the company's development people to keep costs down; performance statistics to indicate your high-quality standards; and your best conception-to-production times to show the audience how adept you are at meeting target dates.

In such a presentation, before an audience of tough-minded officials, you wouldn't want to spend much of your time showing them aerial views of the company's modern facilities or photographs of smiling employees, antiseptic production lines, and the company's expensive air fleet. Such material would hardly support and expand your arguments that the company is used to working and producing on a Spartan budget.

Focus Your second reason for using visual aids is *focus of attention*. A good visual can arrest the wandering thoughts of your audience and bring their attention right down to a specific detail of the message. It forces their mental participation in the subject.

When you are dealing with very complex material, as you often will be, you can use a simple illustration to show your audience a single, critical concept within your subject.

Visual Aids Criteria

What about the visual aids themselves? What makes one better than another for a specific kind of presentation? Before we consider individual visual aids, let's look at the qualities that make a visual aid effective for the technical speaker.

Visibility First, a visual aid must be visible. If that seems so obvious that it hardly need be mentioned at all, it may be because you haven't experienced the frustration of being shown something the speaker feels is important—and not being able to read it, or even make out detail. To be effective, your visual support material should be clearly visible from the most distant seat in the house. If you have any doubts, sit in that seat and look. Remember this when designing visual material: **Anything worth showing the audience is worth making large enough for the audience to see.**

Clarity The second criterion for a good visual is *clarity*. The audience decides this. If they're able to determine immediately what they are seeing, the visual is clear enough. Otherwise, it probably calls for further simplification and condensation. Such obvious mistakes as pictorial material out of focus, or close-ups of a complex device that will confuse the audience, are easy to understand. But what about the chart that shows a relationship between two factors on *x*- and *y*-axes when the axes are not clearly designated or when pertinent information is unclear or missing? **Visual material should be immediately clear to the audience, understandable at a glance, without specific help from the speaker.**

Simplicity The third criterion for good visual support is *simplicity*. No matter how complex the subject, the visual itself should include no more information than absolutely necessary to support the speaker's message. If it is not carrying the burden of the message, it need not carry every detail. Limit yourself to *one* idea per visual—mixing ideas will totally confuse an audience, causing them to turn you off midsentence.

When using words and phrases on a visual, limit the material to key words that act as visual cues for you and the audience. If a visual communications expert wished to present the criteria for a good visual he might *think* something like this:

A good visual must be visible.
A good visual must be clear.
A good visual must be simple.
A good visual must be easy for the speaker to control.

What would he show the audience? If he knows his field as well as he should, he'll offer the visual shown in Figure 18–1.

The same information is there. The visual is being used appropriately to provide emphasis while the speaker supplies the ideas and the extra words. The very simplicity of the visual has impact and is likely to be remembered by the audience.

Control The fourth quality a visual aid needs is *control*, speaker control. You should be able to add information or delete it, to move forward

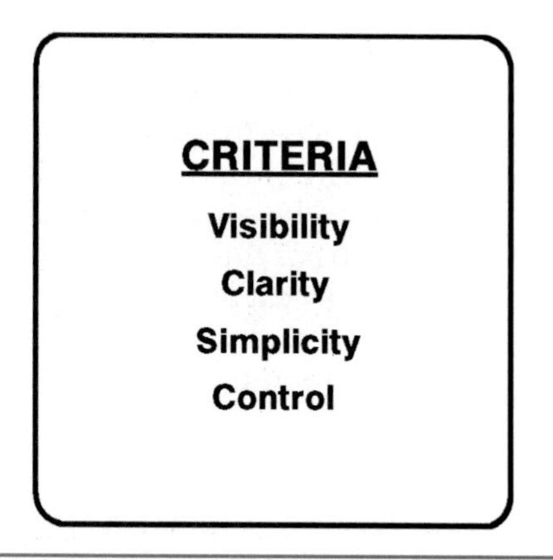

Figure 18–1 Criteria for Visual Aids

or backward to review, and, finally, to *take it away* from the audience to bring their attention back to you.

Some very good visual aids can meet the other criteria and prove almost worthless to a speaker because they cannot be easily controlled. The speaker, who must maintain a flow of information and some kind of rapport with the audience, can't afford to let visual material interfere with this task. Remember, visual material is meant to *support* you as a speaker, not to replace you.

Visual Content

So far we have discussed the *why* of visual support material. The remaining two questions of concern to you are: What do I use? How do I use it? Let's consider them in that order, applying the criteria already established as we go. Types of visual material can be roughly classified in six categories: (1) graphs, (2) tables, (3) representational art, (4) words and phrases, (5) cartoons, and (6) hardware. Graphs, tables, and representational art are discussed in detail in Chapter 11, "Graphical Elements." You'll want to apply the suggestions made there to the visual materials you use when speaking. In the next few pages, we'll discuss using words and phrases, cartoons, and hardware.

Words and Phrases There will always be circumstances in which you will want to emphasize key words or phrases by visual support, as in Figure 18–1. This type of visual can be effectively used in making the audience aware of major divisions or subdivisions of a topic, for instance.

There is danger, however, in the overuse of words—too many with too much detail. Some speakers tend to use visuals as a "shared" set of notes for their presentation, a self-limiting practice. Audiences who are involved in reading a long, detailed piece of information won't recall what the speaker is saying

With technical presentations, there is still another problem with the use of words. Too often, because they may be parts of a specialized vocabulary, they do more to confuse the audience than increase their understanding. Such terms should be reserved for audiences whose technical comprehension is equal to the task of translating them into meaningful thoughts.

Cartoons Cartooning is no more than illustrating people, processes, and concepts with exaggerated, imaginative figures—showing them in whatever roles are necessary to your purpose. (See Figure 18–2.) Not only does it heighten audience interest, but cartooning can be as specific as you want it to be in terms of action or position.

Some situations in which you might choose to use cartooned visual material are these:

- When dealing with subjects that are sensitive for the audience.
- When showing people-oriented action in a stationary medium (any visual aid outside the realm of motion pictures or video).

The resourceful speaker will use cartoons to help give additional meaning to other forms of visual support. The use of cartoons as elements in a block diagram tends to increase viewer interest.

Cartooning, like any other technique, can be overdone. There are circumstances in which the gravity of the situation would suggest that you consider only the most formal kinds of visual support material. On

Figure 18–2 Cartoon Used to Introduce Radon Reduction Methodologies
Source: U.S. Environmental Protection Agency, *Consumer's Guide to Radon Reduction: How to Reduce Radon Levels in Your Home* (Washington, DC: GPO, 1992) 7.

other occasions, cartooning may distract the audience or call too much attention to itself. Your purpose is not to entertain but to communicate.

Hardware After all this analysis of visual support material, you may wonder if it wouldn't be somewhat easier to show the real thing instead.

Certainly, there will be times when the best visual support you can have is the actual object. Notably, the introduction of a new piece of equipment will be more effective if it is physically present to give the audience an idea of its size and bulk. If it is capable of some unique and important function, it should definitely be seen by the audience. (The greatest difficulty with the use of actual hardware is control. The device that is small enough for you to carry conveniently may be too small to be seen from the audience.)

Even when the physical presence of a piece of equipment is possible, it is important to back it with supplementary visual materials. Chances are the audience will not be able to determine what is happening inside the machine, even if they understand explicitly the principle involved. With this in mind, you will want to add information with appropriate diagrams, graphs, and scale drawings.

In this discussion of visual support, the points of visibility, clarity, simplicity, and control have been stressed over and over. The reason for this is their importance to the selection and use of visual support by the technical communicator. In the end, it is you who can best decide which visual support form is required by your message and your audience.

You are also faced with the choice of visual tools for presenting your visual material. The next section will deal with popular visual tools, their advantages, disadvantages, and adaptability to the materials we've already discussed.

Visual Presentation Tools

The major visual tools are these:

- Chalkboards
- Charts
- Slides
- Movies/videos
- Overhead projection
- Computer presentations

In the next few pages we'll discuss them individually, with an eye on the advantages each one offers the technical speaker.

The Chalkboard Anyone who has attended school in the past half-century is familiar with the chalkboard. It has been a standard source of visual support for much longer than that and often the only means of presenting visual information available to the classroom teacher.

As a visual aid, it leaves something to be desired. In the first place, preparing information on a chalkboard, especially technical information in which every sliding scrawl can have significance, takes time. And after the material is in place, it cannot be removed and replaced quickly.

Second, the task of writing on a surface that faces the audience requires that you turn your back toward them while you write. And people don't respond well to backs. They want you to face them while you're talking to them.

Add to these problems the difficulties of moving a heavy, semipermanent chalkboard around, and you begin to wonder why anyone bothers.

Low cost and simplicity are the reasons. The initial cost of a chalkboard is higher than you may think; but the cost of erasers and chalk is minimal. In spite of its drawbacks, a chalkboard is also easy to use. It may take time, but there's nothing very complicated about writing a piece of information on a chalkboard. This simplicity, of course, gives it a certain flexibility, making it essentially a spontaneous visual aid on which speakers can create their visual material as they go.

There are specific techniques for using a chalkboard that make it a more effective visual tool and help overcome its disadvantages. Let's consider them one at a time.

- Plan ahead. Unless there is a clear reason for creating the material as you go, prepare your visual material before the presentation. Then cover it. Later, you can expose the information for the audience at the appropriate point in your speech.
- Be neat and keep the information simple and to the point. If your material is complex, find another way of presenting it.
- Prime the audience. Before showing your information, tell them what they are going to see and why they are going to see it.

This last point is especially important when you are creating your visual support as you go. Priming your listeners will allow you to maintain the flow of information and, at the same time, prepare them mentally to understand and accept your information.

Charts Charts take a couple of forms. The first is the individual **hardboard chart**, rigid enough to stand by itself and large enough to be seen by the audience from wherever they might be seated in the room. It is always prepared before the presentation, sometimes at considerable cost.

The second chart form is the **flip chart,** a giant-sized note pad that may be prepared before or during the presentation. When you have completed your discussion of one visual, you simply flip the sheet containing it over the top of the pad as you would the pages of a tablet.

The two types of charts have a common advantage. Unlike the chalkboard, they allow for reshowing a piece of information when necessary—an important aid to speaker control.

The following techniques will help you use charts more effectively during your presentation. They're really rules of usage, to be followed each time you choose this visual form for support.

- Keep it simple. Avoid complex, detailed illustrations on charts. A three-by-five-foot chart is seldom large enough for detailed visibility.
- Ask for help. Whenever possible, have an assistant on one side of your charts to remove each one in its turn. This avoids creating a break in your rapport with the audience while you wrestle with a large cardboard chart or a flimsy flip sheet.
- Predraw your visuals with a very light-colored crayon or chalk. During the presentation, you can simply draw over the original lines in darker crayon or ink. This allows you to create an accurate illustration a step at a time for clarity.
- Prime the audience. Tell them what they are going to see and why before you show each visual, for the same reason you would do it with the chalkboard.

Slides The 35 mm slide, with its realistic color and photographic accuracy, has always been a popular visual tool for certain types of technical presentations. Where true reproduction is essential, no better tool is available.

Modern projectors have two notable advantages over their predecessors. First, the introduction of slide magazines has made it possible for you to organize your presentation and keep it intact. Second, remote controls allow you to operate the projector—even to reverse the order of your material—from the front of the room.

To use slide projectors effectively, however, you must turn off the lights in the room. Any time you keep your audience in the dark, you risk damaging the direct speaker–listener relationship on which communication hinges. In a sense, it takes the control of the presentation out of your hands. Long sequences of slides tend to develop a will and a pace of their own. They tire an audience and invite mental absenteeism.

There are ways to handle the built-in problems of a slide presentation, simple techniques that can greatly increase audience attention and the effectiveness of your presentation.

- When using slides in a darkened room, light yourself. A disembodied voice in the dark is little better than a tape recorder; it destroys rapport and allows the audience to exit into their own thoughts. To minimize this effect, arrange your equipment so you may stay in the front of the room and use a lectern light or some other soft, nonglaring light to make yourself visible to the audience.

- Break the presentation into short segments of no more than five or six slides.
- Always tell the audience what they're going to see, and what they should look for.

Everything considered, slides are an effective means of presenting visual material. But like any visual tool, they require control and preparation on your part. The important thing to remember is that they are there only to support your message—not to replace you.

Movies and Videos Whenever motion and sound are important to the presentation, movies and videos are the visual forms available to the speaker that can accomplish the effect. Like slides, they also provide an exactness of detail and color that can be critical to certain subjects. There is really no other way an engineer could illustrate the tremendous impact aircraft tires receive during landings, for instance. The audience would understand the subject only if they were able to view, through the eye of the camera, the distortion of the rubber when the plane touches down.

But movies and videos *are* the presentation. They cannot be considered visual support material in the sense of the term developed in this book. They simply replace the speaker as the source of information, at least for their duration. If they become the major part of the presentation, the speaker is reduced to an announcer with little more to do than introduce and summarize their content.

This makes movies and videos the most difficult visual forms to control. Yet they can be controlled and, if they are to perform the support functions we've outlined, they must be. Some effective techniques are given here.

- Prepare the audience. Explain the significance of what you are about to show them.
- If a film or video is to be used, it should make up only a small part of the total presentation.
- Whenever possible, break the film or video into short three- or four-minute segments. Between segments you can reestablish rapport with the audience by summarizing what they have seen and refocusing attention on the important points in the next segment.

Overhead Projection Throughout this discussion of visual support, we've stressed the importance of maintaining a good speaker–audience relationship. It's an essential in the communication process. And it's fragile. Any time you turn your back to the audience, or darken the room, or halt the flow of ideas for some other reason, this relationship is damaged.

The overhead projector effectively eliminates all of these rapport-dissolving problems. The image it projects is bright enough and clear enough to be used in a normally lighted room without noticeable loss of visibility. And just as important from your point of view, it allows you to remain in the front of the room *facing* your audience throughout your presentation. The projector itself is a simple tool, and like all simple tools it may be used without calling attention to itself.

Visual material for the overhead projector is prepared on transparent sheets the size of typing paper. The methods for preparing these transparencies have become so simplified and made so inexpensive that the overhead has become a universally accepted visual tool in both the classroom and industry.

Perhaps the most important advantage of the overhead projector is the total speaker control it affords. With it, you may add information or delete it in a variety of ways, move forward or backward to review at will, and *turn it off* without altering the communicative situation in any way. It is the last capability that makes the overhead projector unique among visual tools. By flipping a switch, you can literally "remove" the visual material from the audience's consideration, bringing their attention back to you and what you are saying. Because the projector is used in a lighted room, this on-and-off process seldom distracts the audience or has any effect on the speaker–audience relationship.

There are three ways to add information to a visual while the audience looks on—an important consideration when you want your listeners to receive information in an orderly fashion. In order of their discussion, they are (1) overlays, (2) revelation, and (3) writing on the visual itself.

The **overlay technique** (Figure 18–3) combines the best features of preparing your visuals in advance and creating them at the moment of their need. It is the simple process of beginning with a single positive transparency and adding information with additional transparencies by "overlaying" them, placing each one over the first so they may be viewed by the audience as a single, composite illustration. Ordinarily, no more than two additional transparencies should be used this way, but it's possible to include as many as four or five.

The technical person, who must usually present more complex concepts a step at a time to ensure communication, can immediately see the applications of such a technique.

The technique of **revelation** is simpler. (See Figure 18–4.) It is the process of masking off the parts of the visual you don't want the audience to see. A plain sheet of paper will work. By laying it over the

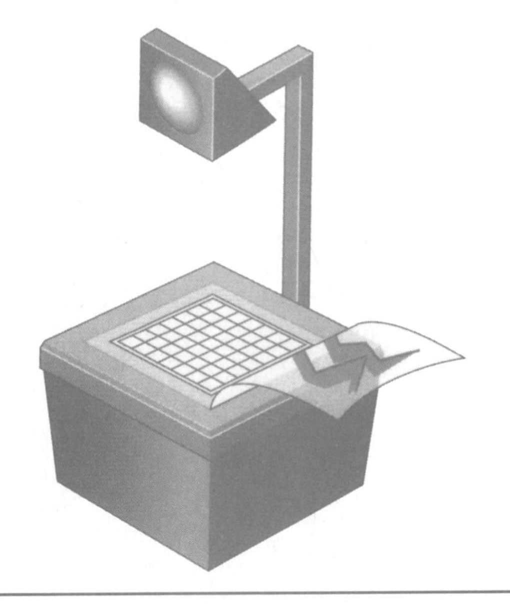

Figure 18–3 Using Overlays with an Overhead Projector

information you want to conceal for the moment, you can block out selected pieces of the visual. Then when you're ready to discuss this hidden information, you simply remove the paper. The advantage is

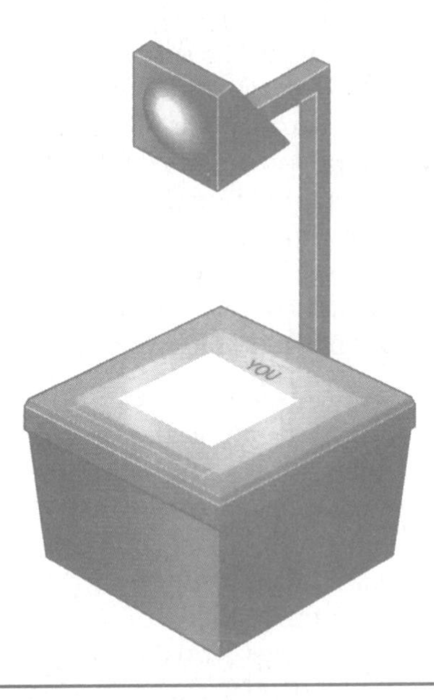

Figure 18–4 The Technique for Revelation

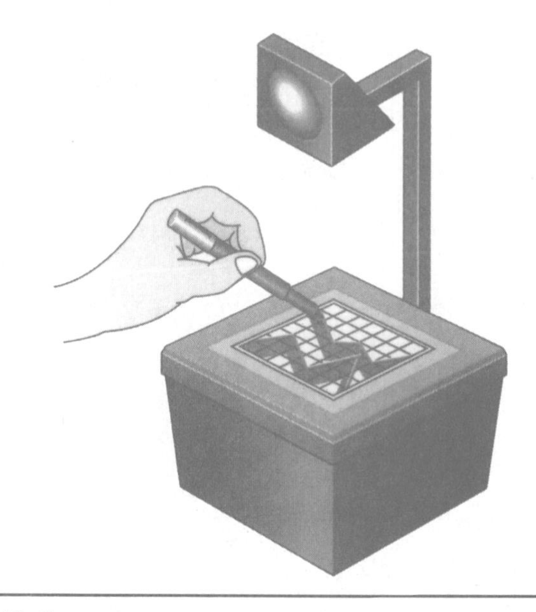

Figure 18–5 Writing on Transparencies

clear enough. If you don't want the audience to read the bottom line on the page while you're discussing the top line, this is the way to control their attention.

Writing information on an overhead-projection transparency (Figure 18–5) is nearly as easy as writing on a sheet of paper at your desk. Several felt-tipped pens available for this purpose may be used to create visual material in front of the audience. Often you can achieve your purposes by simply underlining or circling important parts of your visual—a means of focusing audience attention on the important aspects of your message.

A final way of directing audience attention with overhead-projection transparencies is simply to **use your pencil as a pointer.** The "profile" shadow of the pencil will be seen on the screen, directing the audience's attention to the proper place.

Computer Presentations As is to be expected, microcomputers play an increasingly important role in preparing and presenting graphics during talks. To begin with, computer equipment and programs are available to create transparencies and 35 mm slides of any graphic you can produce on your computer screen. You, therefore, can create for your talks computer graphics of the kind we discuss in Chapter 11, "Graphical Elements."

However, you can obtain even more versatility in your speech graphics by using computer projectors that display whatever is on your computer screen on a projection screen. The computer projector allows you to use your computer during your talk to create and display graphics and data as you need them. In essence, the **computer projector** is an overhead projector linked with a computer's capability for creating graphics.

Computer presentations allow you great control of your visual material. However, don't be misled by the great capability of the computer into creating graphics that are too complex to be readily readable. As with all graphics, keep your computer-generated graphics visible, clear, and simple. Use them not to impress your listeners but to focus their attention and to support and clarify your message.

Planning and Revision Checklists

You will find the planning and revision checklists that follow Chapter 2, "Composing" (pages 38–39 and inside the front cover), and Chapter 4, "Writing for Your Readers" (pages 81–82), valuable in planning and revising any presentation of technical information. The following questions specifically apply to oral reports. They summarize the key points in this chapter and provide a checklist for planning and revising.

PLANNING

- What are the conditions under which you will speak?
- What equipment is available to you?
- Which delivery technique will be more appropriate? Extemporaneous? Manuscript?
- If you are speaking extemporaneously, have you prepared a speech outline to guide you?
- If reading from a manuscript, have you introduced a conversational tone into your talk? Is your typed manuscript easy to read from?
- Do you have a good opening that will interest your audience and create a friendly atmosphere?
- Have you limited your major points to fit within your allotted time?
- Does your talk contain sufficient examples, analogies, narratives, and data to support your generalizations? Have you repeated key points?
- Can you relate your subject matter to some vital interest of your audience?
- Which visual aids do you plan to use?
  - Graphs?
  - Tables?
  - Representational art?
  - Words and phrases?
  - Cartoons?
  - Hardware?
- Which tools of presentation will you use?
  - Chalkboard?
  - Charts?
  - Slides?
  - Movies?
  - Overhead projection?
  - Computer presentations?
- Have you prepared your graphics? Do they successfully focus the listener's

attention and augment and clarify your message? Do they meet the four criteria that govern good graphics?

Visibility
Clarity
Simplicity
Control

- Do you have a good ending ready, perhaps a summary of key points or an anecdote that supports your purpose?
- Have you rehearsed your talk several times?

REVISION

In one sense, you can't revise a talk you have already given, unless, of course, you will have an opportunity to repeat it somewhere. But you can use revision techniques in your practice sessions. Most of the questions asked under "Planning" lend themselves to that. Also, you can critique your speeches, looking for ways to improve your delivery techniques in future speeches. These next questions lend themselves to that. It helps if you have someone in the audience to give you friendly but honest answers to all the questions asked under planning an revision.

- Did your gestures support your speech? Did they seem normal and relaxed? Did you avoid nervous mannerisms?
- Was your speech rate appropriate? Did you vary rate, pitch, and volume occasionally? Could everyone hear you?
- Did you pronounce all your words correctly?
- Did you have good interaction with your audience? Were they attentive or fidgety?
- Did your talk fit comfortably into the time allotted for it?
- Did the questions that followed your talk indicate a good understanding of it? Did the questions indicate friendliness or hostility to your key points?
- Were you sufficiently informed to answer the questions raised?
- Were there any indications that members of your audience could not see or readily comprehend any of your graphics?

Exercises

1. Deliver a speech in one of the following situations:

a. You are an instructor at your college. Prepare a short extemporaneous lecture on a technical subject. Your audience is a class of about 20 juniors and seniors.

b. You are the head of a team that has developed a new product or process. Your job is to persuade a group of senior managers from your own firm to accept the process or product for company use. Assume these managers have a layperson's knowledge about your subject. Speak extemporaneously.

c. You are a known expert on your subject. You have been invited to speak about your subject at the annual meeting of a well-known scientific association. You are expected to write out and read your speech. You are to inform the audience, which is made up of knowledgeable research scientists and college professors from a diversity of scientific disciplines, about your subject or to persuade them to accept a decision you have reached.

2. Change one of your written reports into an oral report. Deliver the report extemporaneously. Prepare several visual aids to support major points.

PART V
HANDBOOK

Any living language is a growing, flexible instrument with rules that are constantly changing to reflect the way it is used by its live, independent speakers and writers. Only the rules of a dead language are unalterably fixed.

Nevertheless, at any point in a language's development, certain conventions of usage are in force. Certain constructions are considered errors and mark the person who uses them as uneducated. It is with these conventions and errors that this handbook primarily deals. We also include sections on outlining and sexist language. To make the handbook easy to use as a reference, we have arranged it in alphabetical order. Each convention and error dealt with has an abbreviated reference tag. The tags are reproduced on the back endpapers, along with some of the more important proofreading symbols. If you are in a college writing course, your instructor may use some combination of these tags and symbols to indicate revisions needed in your reports.

- Abbreviations *(Ab)*
- Acronyms *(Acro)*
- Apostrophe *(Apos)*
  - Possessives
  - Missing Letters or Numbers
  - Plural Forms
- Brackets *(Brackets)*
- Capitalization *(Cap)*
  - Proper Nouns
  - Literary Titles
  - Rank, Position, Family Relationships
- Colon *(Colon)*
  - Introduction
  - Between Clauses
  - Styling Conventions
- Comma *(C)*
  - Main Clauses
  - Clarification
  - Nonrestrictive Modifiers
  - Nonrestrictive Appositives
  - Series
  - Other Conventional Uses
- Dangling Modifier *(DM)*
- Dash *(Dash)*
  - Typewriter
  - Typeset
- Diction *(D)*
- Ellipsis *(Ell)*
- Exclamation Point *(Exc)*
- Fragmentary Sentence *(Frag)*
- Hyphen *(Hyphen)*
  - Compound Numbers
  - Common Compound Words
  - Compound Words as Modifiers
  - Word Division
- Italicization *(Ital)*
  - Foreign Words
  - Words, Letters, and Numbers Used as Such
  - Titles
- Misplaced Modifier *(MM)*
- Numbers *(Num)*
  - Numbers as Words
  - Numbers as Sentence Openers
  - Compound Number Adjectives
  - Hyphens
  - Numbers as Figures
  - Fractions
- Outlining
- Parallelism *(Paral)*
- Parentheses *(Paren)*
  - Lists
  - Punctuation of Parentheses in Sentences
- Period *(Per)*
  - End Stop
  - Abbreviations
  - Decimal Point
- Pronoun-Antecedent Agreement *(P/ag)*
- Pronoun Form *(Pron)*
- Question Mark *(Ques)*
- Quotation Marks *(Quot)*
  - Short Quotations
  - Titles
  - Single Quotes
  - Punctuation Conventions
- Run-on Sentence *(Run-on)*
- Semicolon *(Semi)*
  - Independent Clauses
  - Series
- Sexist Usage *(Sexist)*
- Spelling Error *(Sp)*
- Verb Form *(Vb)*
- Verb-Subject Agreement *(V/ag)*
  - Words that Take Singular Verbs
  - Compound Subject Joined by *Or* or *Nor*
  - Parenthetical Expressions
  - Two or More Subjects Joined by *And*
  - Collective Nouns

Abbreviations *(Ab)*

Although most people are familiar with the kinds of abbreviations we encounter in everyday conversation and written material, from *mph* to *Mon.* to *Dr.*, technical abbreviations are something else. Each scientific and professional field generates hundreds of specialized terms, and many of these terms are often abbreviated for the sake of conciseness and simplicity.

Thus, before deciding to use technical abbreviations in an article or report, you must first consider your audience—laypersons, executives, experts, or technicians? Only readers with an appropriate background will be able to interpret the specialized shorthand for the field in question. When in doubt, avoid all but the most common abbreviations. If you must use a technical term, spell it out in full the first time it appears and include both the abbreviation and a definition in parentheses after it. You can then safely use the abbreviation if the term crops up again in your report.

Standard technical and scientific abbreviations include the following:

absolute	abs
acre or acres	acre or acres
alternating current (as adjective)	a-c
atomic weight	at. wt
barometer	bar.
Brinell hardness number	Bhn
British thermal units	Btu or B
meter	m
square meter	m^2
microwatt or microwatts	mu w or µW
miles per hour	mph
National Electric Code	NEC
per	per
revolutions per minute	rpm
rod	rod
ton	ton

The system implied by these illustrative abbreviations can be described by a brief set of rules.

1. Use the same (singular) form of abbreviation for both singular and plural terms:

cu ft	either cubic foot or cubic feet
cm	either centimeter or centimeters

But there are some common exceptions:

no.	number
nos.	numbers
p.	page
pp.	pages
ms.	manuscript
mss.	manuscripts

2. Use lowercase letters except for letters standing for proper nouns or proper adjectives:

ab	*but*	Btu or B
mph	*but*	Bhn

3. For technical terms, use periods only after abbreviations that spell complete words. For example, *in* is a word, and the abbreviation for inches could be confused with it. Therefore, use a period:

ft	*but*	in.
abs	*but*	bar.
cu ft	*but*	at. wt

4. Remember the hyphen in the abbreviations a-c and d-c when you use them as adjectives:

This a-c motor can be converted to 28 volts dc.

5. Spell out many short and common words:

acre rod per ton

6. In compound abbreviations, use internal spacing only if the first word is represented by more than its first letter:

rpm	*but*	cu ft
mph	*but*	mu w

7. With few exceptions, form the abbreviations of organization names without periods or spacing:

NEC ASA

8. Abbreviate terms of measurement only if they are preceded by an arabic expression of exact quantity:

55 mph *and* 20-lb anchor

But:

We will need an engine of greater horsepower.

Acronyms *(Acro)*

Acronyms are formed in two ways. In one way, the initial letters of each word in some phrase are combined to form a word. An example would be WYSIWYG, an acronym for the computer phrase "What you see is what you get." In a second way, some combination of initial letters or several letters of the words in the phrase are combined. An example would be *radar* for *ra*dio *d*etection *a*nd *r*anging.

Technical writing uses acronyms freely, as in this example from a description of a computer program that performs statistical analysis:

> It has good procedural capabilities, including some time-series-related plots and ARIMA forecasting, but it doesn't have depth in any one area. Although it has commands to create EDA displays, these graphics are static and are printed with characters rather than with lines[1]

Use acronyms without explanation only when you are absolutely sure your readers know them. If you have any doubts at all, at least provide the words from which the acronym stems. If you're unsure that the words are enough, provide a definition of the complete phrase. In the case of the paragraph just quoted, the computer magazine in which it was printed provided a glossary giving both the complete phrases and definitions:

> ARIMA (auto-regressive integrated moving-average): a model that characterizes changes in one variable over time. It is used in time-series analysis.
>
> EDA (exploratory data analysis): The use of graphically based tools, particularly in initial states of data analysis, to inspect data properties and to discover relationships among variables.[2]

Acronyms can be daunting to those unfamiliar with them. Even when you have an audience that knows their meaning, a too heavy use of acronyms can make your writing seem lumpish and uninviting.

Apostrophe *(Apos)*

The apostrophe has three chief uses: (1) to form the possessive, (2) to stand for missing letters or numbers, and (3) to form the plural of certain expressions.

Possessives

Add an apostrophe and an *s* to form the possessive of most singular nouns, including proper nouns, even when they already end in an *s* or another sibilant such as *x:*

man's
spectator's
jazz's
Marx's
Charles's

Exceptions to this rule occur when adding an apostrophe plus an *s* would result in an *s* or *z* sound that is difficult to pronounce. In such cases, usually just the apostrophe is added:

Xerxes'
Moses'
conscience'
appearance'

To understand this exception, pronounce *Marx's* and then a word like *Moses's* or *conscience's.*

To form plurals into the possessive case, add an apostrophe plus *s* to words that do not end in an *s* or other sibilant and an apostrophe only to those that do:

men's
data's
spectators'
agents'
witnesses'

To show joint possession, add the apostrophe and *s* to the last member of a compound or group; to show separate possession, add an apostrophe and *s* to each member:

Gregg and Klymer's experiment astounded the class.
Gregg's and Klymer's experiments were very similar.

Of the several classes of pronouns, only the indefinite pronouns use an apostrophe to form the possessive.

Possessive of Indefinite Pronouns	Possessive of Other Pronouns
anyone's	my (mine)
everyone's	your (yours)
everybody's	his, her (hers), its
nobody's	our (ours)
no one's	their (theirs)
other's	whose
neither's	

Missing Letters or Numbers

Use an apostrophe to stand for the missing letters in contractions and to stand for the missing letter or number in any word or set of numbers where for one reason or another a letter or number is omitted:

> can't, don't, o'clock, it's (it is), etc.
>
> We were movin' downriver, listenin' to the birds singin'.
>
> The class of '49 was Colgate's best class in years.

Plural Forms

An apostrophe is sometimes used to form the plural of letters and numbers, but this style is gradually dying, particularly with numbers.

> 6s and 7s (but also 6's and 7's)
>
> a's and b's

Brackets *(Brackets)*

Brackets are chiefly used when a clarifying word or comment is inserted into a quotation:

> "The result of this [disregard by the propulsion engineer] has been the neglect of the theoretical and mathematical mastery of the engine inlet problem."
> "An ideal outlet require [sic] a frictionless flow."
> "Last year [1993] saw a partial solution to the problem."

Sic, by the way, is Latin for *thus*. Inserted in a quotation, it means that the mistake found there is the original writer's, not yours. Use it with discretion.

Capitalization *(Cap)*

We provide the more important rules of capitalization. For a complete rundown, see your college dictionary.

Proper Nouns

Capitalize all proper nouns and their derivatives:

Places

America American Americanize Americanism

Days of the Week and Months

Monday Tuesday January February

But not the seasons:

winter spring summer fall

Organizations and Their Abbreviations

American Kennel Club (AKC)
United States Air Force (USAF)

Capitalize *geographic areas* when you refer to them as areas:

The Andersons toured the Southwest.

But do not capitalize words that merely indicate direction:

We flew west over the Pacific.

Capitalize the names of *studies* in a curriculum only if the names are already proper nouns or derivatives of proper nouns or if they are part of the official title of a department or course:

Department of Geology
English Literature 25
the study of literature
the study of English literature

Note: Many nouns (and their derivatives) that were originally proper have been so broadened in application and have become so familiar that they are no longer capitalized: *boycott, macadam, spoonerism, italicize, platonic, chinaware, quixotic.*

Literary Titles

Capitalize the first word, the last word, and every important word in literary titles:

But What's a Dictionary For
The Meaning of Ethics
How to Write and Be Read

Rank, Position, Family Relationships

Capitalize the titles of rank, position, and family relationship unless they are preceded by *my, his, their,* or similar possessive pronouns:

Professor J. E. Higgins
I visited Uncle Timothy.
I visited my uncle Timothy.
Dr. Milton Weller, Head, Department of Entomology

Colon *(Colon)*

The colon is chiefly used to introduce quotations, lists, or supporting statements. It is also used between clauses when the second clause is an example or amplification of the first and in certain conventional ways with numbers, correspondence, and bibliographical entries.

Introduction

Place a colon before a quotation, a list, or supporting statements and examples that are formally introduced:

> Mr. Smith says the following of wave generation:
>
> > The wind waves that are generated in the ocean and which later become swells as they pass out of the generating area are products of storms. The low pressure regions that occur during the polar winters of the Arctic and Antarctic produce many of these wave-generating storms.
>
> The various forms of engine that might be used would operate within the following ranges of Mach number:

M-0 to M-1.5	Turbojet with or without precooling
M-1.5 to M-7	Reheated turbojet, possibly with precooling
M-7 to M-10+	Ramjet with supersonic combustion

> Engineers are developing three new engines: turbojet, reheated turbojets, and ramjets.

Do not place a colon between a verb and its objects or a linking verb and the predicate nouns.

> *Objects*
>
> The engineers designed turbojets, reheated turbojets, and ramjets.
>
> *Predicate Nouns*
>
> The three engines the engineers are developing are turbojets, reheated turbojets, and ramjets.

Do not place a colon between a preposition and its objects:

> The plane landed at Detroit, Chicago, and Rochester.

Between Clauses

If the second clause consists of an example or an amplification of the first clause, then the colon may replace the comma, semicolon, or period:

> The docking phase involves the actual "soft" contact: the securing of lines, latches, and air locks.

> The difference between these two guidance systems is illustrated in Figure 2: The paths of the two vehicles are shown to the left and the motion of the ferry as viewed from the target station is shown to the right.

You may follow a colon with a capital or a small letter. Generally, a complete sentence beginning after a colon is given a capital.

Styling Conventions

Place a colon after a formal salutation in a letter, between hour and minute figures, between the elements of a double title, and between chapter and verse of the Bible:

> Dear Ms. Jones:
>
> at 7:15 p.m.
>
> Working Women: A Chartbook
>
> I Samuel 7:14–18

Comma *(C)*

The most used—and misused—mark of punctuation is the comma. Writers use commas to separate words, phrases, and clauses. Generally, commas correspond to the pauses we use in our speech to separate ideas and to avoid ambiguity. You will use the comma often: About two out of every three marks of punctuation you use will be commas. Sometimes your use of the comma will be essential for clarity; at other times you will be honoring grammatical conventions. (See also the entry for run-on sentences.)

Main Clauses

Place a comma before a coordinating conjunction *(and, but, or, nor, for, yet)* that joins two main (independent) clauses:

> During the first few weeks we felt a great deal of confusion, but as time passed we gradually fell into a routine.
>
> We could not be sure that the plumbing would escape frost damage, nor were we at all confident that the house could withstand the winds of almost hurricane force.

If the clauses are short, have little or no internal punctuation, or are closely related in meaning, then you may omit the comma before the coordinating conjunction:

> The wave becomes steeper but it does not tumble yet.

In much published writing there is a growing tendency to place two very short and closely related independent clauses (called contact clauses) side by side with only a comma between:

The wind starts to blow, the waves begin to develop.

Sentences consisting of *three* or more equal main clauses should be punctuated uniformly:

We explained how urgent the problem was, we outlined preliminary plans, and we arranged a time for discussion.

In general, identical marks are used to separate equal main clauses. If the equal clauses are short and uncomplicated, commas usually suffice. If the equal clauses are long or internally punctuated, or if their separateness is to be emphasized, semicolons are either preferable or necessary.

Clarification

Place a comma after an introductory word, phrase, or clause that might be over-read or that abnormally delays the main clause:

As soon as you have finished polishing, the car should be moved into the garage. (**Comma to prevent over-reading**)

Soon after, the winds began to moderate somewhat, and we were permitted to return to our rooms. (**First comma to prevent over-reading**)

If the Polar ice caps should someday mount in thickness and weight to the point that their combined weight exceeded the Equatorial bulge, the earth might suddenly flop ninety degrees. (**Introductory clause abnormally long**)

After a short introductory element (word, phrase, or clause) where there is no possibility for ambiguity, the use of the comma is optional. Generally, let the emphasis you desire guide you. A short introductory element set off by a comma will be more emphatic than one that is not.

Nonrestrictive Modifiers

Enclose or set off from the rest of the sentence every nonrestrictive modifier, whether a word, a phrase, or a clause. How can you tell a nonrestrictive modifier from a restrictive one? Look at these two examples:

Restrictive

A runway **that is not oriented with the prevailing wind** endangers the aircraft using it.

Nonrestrictive

The safety of any aircraft, **whether heavy or light**, is put in jeopardy when it is forced to take off or land in a crosswind.

The restrictive modifier is necessary to the meaning of the sentence. Not just any runway but "a runway that is not oriented with the prevailing wind" endangers aircraft. The writer has *restricted* the many kinds of runways he or she could talk about to one particular kind. In the nonrestrictive example, the modifier merely adds descriptive details. The writer does not restrict *aircraft* with the modifier but simply makes the meaning a little clearer.

Restrictive modifiers cannot be left out of the sentence if it is to have the meaning the writer intends; nonrestrictive modifiers can be left out.

Nonrestrictive Appositives

Set off or enclose every nonrestrictive appositive. As used here the term *appositive* means any element (word, phrase, or clause) that parallels and repeats the thought of a preceding element. According to this view, a verb may be coupled appositively with another verb, an adjective with another adjective, and so on. An appositive is usually more specific or more vivid than the element that it is an appositive to; an appositive makes explicit and precise something that has not been clearly implied.

Some appositives are restrictive and, therefore, are not set off or enclosed.

Nonrestrictive

A crosswind, **a wind perpendicular to the runway**, causes the pilot to make potentially dangerous corrections just before landing.

Restrictive

In some ways, Mr. Clinton **the President** had to behave differently from Mr. Clinton **the Governor**.

In the nonrestrictive example, the appositive merely adds a clarifying definition. The sentence makes sense without it. In the restrictive example, the appositives are essential to the meaning. Without them we would have, "In some ways, Mr. Clinton had to behave differently from Mr. Clinton."

Series

Use commas to separate members of a coordinate series of words, phrases, or clauses if *all* the elements are not joined by coordinating conjunctions:

Instructions on the label state clearly how to prepare the surfaces, how to apply the contents, and how to clean and polish the mended article.

To mold these lead figures you will need a hot flame, a two-pound block of lead, the molds themselves, a file or a rasp, and an awl.

Under the microscope the sensitive, filigree-like mold appeared luminous and transparent and faintly green.

Other Conventional Uses

Date

On August 24, 79 A.D., Mount Vesuvius erupted, covering Pompeii with 50 feet of ash and pumice.

Note: When you write the month and the year without the day, it is common practice to omit the comma between them—as in June 1993.

Geographical Expression

During World War II, Middletown, Pennsylvania, was the site of a huge military airport and supply depot.

Title after Proper Name

A card in yesterday's mail informed us that Penny Hutchinson, M.D., would soon open new offices in Hinsdale.

Noun of Direct Address

Lewis, do you suppose that we can find our way back to the cabin before nightfall?

Informal Salutation

Dear Jane,

Dangling Modifier *(DM)*

Many curious sentences result from the failure to provide the modifier something to modify:

Having finished the job, the tarpaulins were removed.

In this example it seems as though the tarpaulins have finished the job. As is so often the case, a passive voice construction has caused the problem (see pages 96–97). If we recast the sentence in active voice, we remove the problem:

Having finished the job, the workers removed the tarpaulins.

Dash *(Dash)*

In technical writing, you will use the dash almost exclusively to set off parenthetical statements. You may, of course, use commas or parentheses for the same function, but the dash is the most emphatic separator of the three. You may also use the dash to indicate a sharp transition. With typewriter print, you make the dash with two hyphens. You do not space between the words and the hyphens or between the hyphens themselves.

Typewriter

> The first phase in rendezvous--sighting and recognizing the target--is so vital that we will treat it at some length.

Typeset

> The target must emit or reflect light the pilot can see—but how bright must this light be?

Diction *(D)*

For good diction, choose words that are accurate, effective, and appropriate to the situation. Many different kinds of linguistic sins can cause faulty diction. Poor diction can involve a choice of words that are too heavy or pretentious: *utilize* for *use, finalize* for *finish, at this point in time* for *now,* and so forth. Tired old cliches are poor diction: *with respect to, with your permission, with reference to,* and many others. We talk about such language in Chapter 6, particularly in the section on pomposity (pages 100–103).

Sometimes the words chosen are simply too vague to be accurate: *inclement weather* for *rain, too hot* for *600° C.* See the section on specific words in Chapter 5 (pages 98–99) for more on this subject.

Poor diction can mean an overly casual use of language when some degree of formality is expected. One of the many synonyms for *intoxicated,* such as *bombed, stoned,* or *smashed,* might be appropriate in casual conversation but totally wrong in a police or laboratory report.

Poor diction can reflect a lack of sensitivity to language—to the way one group of words relates to another group. Someone who writes that "The airlines are beginning a crash program to solve their financial difficulties" is not paying attention to relationships. The person who writes that the "Steelworkers' Union representatives are getting down to brass tacks in the strike negotiations" has a tin ear, to say the least.

Make your language work for you, and make it appropriate to the situation.

Ellipsis *(Ell)*

Use three spaced periods to indicate words omitted within a quoted sentence, four spaced periods if the omission occurs at the end of the sentence:

> "As depth decreases, the circular orbits become elliptical and the orbital velocity . . . increases as the wave height increases."
>
> "As the ground swells move across the ocean, they are subject to headwinds or crosswinds. . . ."

You need not show an ellipsis if the context of the quotation makes it clear that it is not complete:

Wright said the accident had to be considered a "freak of nature."

Exclamation Point *(Exc)*

Place an exclamation point at the end of a startling or exclamatory sentence.

According to the Centers for Disease Control, every cigarette smoked shortens the smoker's life by seven minutes!

With the emphasis in technical writing on objectivity, you will seldom use the exclamation point.

Fragmentary Sentence *(Frag)*

Most fragmentary sentences are either verbal phrases or subordinate clauses that the writer mistakes for a complete sentence.

A verbal phrase has in the predicate position a participle, gerund, or infinitive, none of which functions as a complete verb:

Norton, **depicting** the electromagnetic heart. (participle)

The **timing** of this announcement about Triptycene. (gerund)

Braun, in order **to understand** tumor cell growth. (infinitive)

When your fragment is a verbal phrase, either change the verb to a complete verb or repunctuate the sentence so that the phrase is joined to the complete sentence of which it is actually a part.

Fragment

Norton, depicting the electromagnetic heart. She made a mockup of it.

Rewritten

Norton depicted the electromagnetic heart. She made a mockup of it.

Norton, depicting the electromagnetic heart, made a mockup of it.

Subordinate clauses are distinguishable from phrases in that they have complete subjects and complete verbs (rather than verbals) and are introduced by relative pronouns (*who, which, that*) or by subordinating conjunctions (*because, although, since, after, while*).

The presence of the relative pronoun or the subordinating conjunction is a signal that the clause is not independent but is part of a more complex sentence unit. Any independent clause can become a subordinate clause with the addition of a relative pronoun or subordinating conjunction.

Independent Clause

Women's unemployment rates were higher than men's.

Subordinate Clause

Although women's unemployment rates were higher than men's

Repunctuate subordinate clauses so that they are joined to the complex sentence of which they are a part.

Fragment

Although women's unemployment rates were higher than men's. Now the rates are similar.

Rewritten

Although women's unemployment rates were higher than men's, now the rates are similar.

Various kinds of elliptical sentences without a subject or a verb do exist in English, for example, "No!" "Oh?" "Good shot." "Ouch!" "Well, now." These constructions may occasionally be used for stylistic reasons, particularly to represent conversation, but they are seldom needed in technical writing. If you do use such constructions, use them sparingly. Remember that major deviations from normal sentence patterns will probably jar your readers and break their concentration on your report, the last thing that any writer wants.

Hyphen *(Hyphen)*

Hyphens are used to form various compound words and in breaking up a word that must be carried over to the next line.

Compound Numbers

See "Numbers."

Common Compound Words

Observe dictionary usage in using or omitting the hyphen in compound words.

governor-elect
ex-treasurer
Russo-Japanese
pro-American
court-martial
Croesus-like
drill-like
self-interest

But:

glasslike	wrist watch
neophyte	sweet corn
newspaper	weather map
newsstand	sun lamp
housewife	prize fight

Compound Words as Modifiers

Use the hyphen between words joined together to modify other words:

a half-spent bullet

an eight-cylinder engine

their too-little-and-too-late methods

Be particularly careful to hyphenate when omitting the hyphen may cause ambiguity:

two-hundred-gallon drums

two hundred-gallon drums

a pink-skinned hamster

Sometimes you have to carry a modifier over to a later word, creating what is called a *suspended hyphen:*

GM cars come with a choice of four-, six-, or eight-cylinder engines.

Word Division

Use a hyphen to break a word that must be carried over to the next line. Words compounded of two roots or a root and an affix are divided at the point of union:

self-/important	wind/jammer
cross-/pollination	desir/able
bladder/wort	anti/dote
summer/time	manage/able

Note: The first two words in this list are always spelled with a hyphen; the remaining words use the hyphen only when the word is divided at the end of a line.

In general, noncompound words of more than one syllable are divided between any two syllables but only between syllables:

soph/o/more	con/clu/sion
sat/is/fac/tion	sym/pa/thy

A syllable of one letter is never set down alone; a syllable of only two letters is seldom allowed to stand alone unless it is a prefix or a suffix, and then only if it is pronounced as spelled:

hello/	elec/tro/type
method/	de/mand
pilot/	ac/cept
many/	walked/
saga/	start/ed

If a consonant is doubled because a suffix is added, include the second consonant with the suffix:

spin/ner	slip/ping
stir/ring	slot/ted

But:

stopped/	pass/ing
lapped/	stall/ing

Because *-ped* is not pronounced as a syllable, it should not be carried over. In words like *passing* and *stalling*, both consonants belong to the root, and, therefore, only the suffix *-ing* is carried over.

Italicization *(Ital)*

Italic print is a distinctive typeface, like this sample: *Scientific American.* When you type or write, you represent italics by underlining, like this:

Scientific American

Foreign Words

Italicize foreign words that have not yet become a part of the English language:

We suspected him always of holding some *arrière pensée.*

Karl's everlasting *Weltschmerz* makes him a depressing companion.

Also italicize Latin scientific terms.

cichorium endivia (endive)

Percopsis omiscomaycus (trout-perch)

But do not italicize Latin abbreviations or foreign words that have become a part of the language:

etc.	bourgeois
vs.	status quo

Your collegiate dictionary will normally indicate which foreign words are still italicized and which are not.

Words, Letters, and Numbers Used as Such

The words **entrance** and **admission** are not perfectly interchangeable.

Don't forget the **k** in **picnicking**.

His **9**s and **7**s descended below the line of writing.

Titles

Italicize the titles of books, plays, magazines, newspapers, ships, and artistic works:

Webster's New World Dictionary
Othello
Scientific American
The Free Press
S.S. Pennsylvania
Mona Lisa

Misplaced Modifier *(MM)*

As in the case of dangling modifiers, curious sentences result from the modifier's not being placed next to the element modified:

An engine may crack when cold water is poured in unless it is running.

Probably, with a little effort, no one will misread this example, but, undeniably, it says that the engine will crack unless the water is running. Move the modifier to make the sentence clear:

Unless it is running, an engine may crack when cold water is poured in.

It should be apparent from the preceding examples that a modifier may be in the wrong position to convey one meaning but in the perfect position to convey a different meaning. In the next example, the placement of *for three years* is either right or wrong. It is in the right position to modify *to work* but in the wrong position to modify *have been trying*.

I have been trying to place him under contract to work here for three years. (three-year contract)

As the examples suggest, correct placement of modifiers sometimes amounts to more than mere nicety of expression. It can mean the difference between stating falsehood and truth, between saying what you mean and saying something else.

Numbers *(Num)*

There is a good deal of inconsistency in the rules for handling numbers. It is often a question as to whether you should write the number as a word or as a figure. We will give you the general rules. Your instructor or your organization may give you others. As in all matters of format, you must satisfy whomever you are working for at the moment. Do, however, be internally consistent within your reports. Do not handle numbers differently from page to page of a report.

Numbers as Words

Generally, you write out all numbers nine and under, and rounded-off large numbers, as words:

six generators
about a million dollars

However, when you are writing a series of numbers, do not mix up figures and words. Let the larger numbers determine the form used:

five boys and six girls

But:

It took us 6 months and 25 days to complete the experiment.

Numbers as Sentence Openers

Do not begin sentences with a figure. If you can, write the number as a word. If this would be cumbersome, write the sentence so as to get the figure out of the beginning position:

Fifteen months ago, we saw the new wheat for the first time.
We found 350 deficient steering systems.

Compound Number Adjectives

When you write two numbers together in a compound number adjective, spell out the first one or the shorter one to avoid confusing the reader:

Twenty 10-inch trout
100 twelve-volt batteries

Hyphens

Two-word numbers are hyphenated on the rare occasions when they are written out:

Eighty-five boxes

or:

Eighty-five should be enough.

Numbers as Figures

The general rule here is to write all exact numbers over nine as figures. This rule probably holds more true in technical writing, with its heavy reliance on numbers, than it does in general writing. However, as we noted, rounded-off numbers are commonly written as words. The precise figure could give the reader an impression of exactness that might not be called for.

Certain conventional uses call for figures at all times.

Dates, Exact Sums of Money, Time, Address

1 January 1994 or January 1, 1994

$3,422.67 **but** about three thousand dollars

1:57 P.M. **but** two o'clock

660 Fuller Road

Technical Units of Measurement

6 cu ft

4,000 rpm

Cross-References

See page 22.

Refer to Figure 2.

Fractions

When a fraction stands alone, write it as an unhyphenated compound:

two thirds

fifteen thousandths

When a fraction is used as an adjective, you may write it as a hyphenated compound. But if either the numerator or the denominator is hyphenated, do not hyphenate the compound. More commonly, fractions used as adjectives are written as figures.

two-thirds engine speed

twenty-five thousandths

3/4 rpm

Outlining

As illustrated in the accompanying sample outline, an outline has a title, purpose statement, audience statement, and body. We have annotated the sample outline to point out major outlining conventions. Following the sample outline, we provide other major outlining principles.

Desalination Methods for Air Force Use

Purpose: To choose a desalination method for Air Force bases located near large bodies of salt water
Audience: Senior officers

First level, use capital roman numeral

Second level, use capital letters

Third level, use arabic numerals

Fourth level, use lowercase letters

Capitalize only first letter of entry and proper nouns

Use no punctuation after entries

I. Statement of the problem
   A. Need for a choice
   B. Choices available
      1. Electrodialysis
      2. Reverse osmosis
   C. Sources of data
      1. Air Force manuals
      2. Expert opinion
         a. Journals
         b. Interviews
II. Explanation of criteria
   A. Cost
   B. Purity
   C. Quantity
III. Electrodialysis
   A. Theory of method
   B. Judgment of method
      1. Cost
      2. Purity
      3. Quantity
IV. Reverse osmosis
   A. Theory of method
   B. Judgment of method
      1. Cost
      2. Purity
      3. Quantity
V. Choice of method

- *Make all entries grammatically parallel.* (See the entry for parallelism.) Do not mix noun phrases with verb phrases, and so forth. A formal outline with a hodgepodge of different grammatical forms will seem to lack—and, in fact, may lack—logic and consistency.

Incorrect

I. The overall view
II. To understand the terminal phase
III. About the constant-bearing concept

Correct

I. The overall view
II. The terminal phase
III. The constant-bearing concept

- *Never have a single division.* Things divide into two or more; so obviously, if you have only one division, you have done no dividing. If you have a "I" you must have a "II." If you have an "A" you must have a "B," and so forth.

Incorrect

I. Visual capabilities
   A. Acquisition
II. Interception and closure rate
III. Braking

Correct

I. Visual capabilities
   A. Acquisition
   B. Interception and closure rate
II. Braking

- Do not have entries for your report's introduction or conclusion. Outline only the body of the report. Of course, the information in your purpose statement belongs in your introduction, and perhaps the information about audience belongs there as well.
- Use substantive statements in your outline entries. That is, use entries such as "Reverse osmosis" or "Judgment of method" that suggest the true substance of your information. Do not use cryptic expressions such as "Example 1" or "Minor premise."

Many word processing programs have an outlining feature. This feature has two advantages. First, you can choose the outlining scheme you want, for example, *I, A, B, 1, 2.* The program automatically writes the appropriate numbers or letters for you, and changes them when you change the outline. Second, you can write your text into the outline.

Parallelism *(Paral)*

When you link elements in a series, they must all be in the same grammatical form. Link an adjective with an adjective, a noun with a noun, a clause with a clause, and so forth. Look at the boldface portion of the sentence below:

> A good test would use **small amounts of plant material, require little time, simple to run, and accurate**.

The series begins with the verbs *use* and *require* and then abruptly switches to the adjectives *simple* and *accurate.* All four elements must be based on the same part of speech. In this case, it's simple to change the last two elements:

> A good test would use small amounts of plant material, require little time, **be simple to run**, and **be accurate**.

Always be careful when you are listing to keep all the elements of the list parallel. In the following example, the third item in the list is not parallel to the first two:

> The process has three stages: (1) the specimen is dried, (2) all potential pollutants are removed, and (3) atomization.

The error is easily corrected:

> The process has three stages: (1) the specimen is dried, (2) all potential pollutants are removed, and (3) the specimen is atomized.

When you start a series, keep track of what you are doing, and finish the series the same way you started it. Nonparallel sentences are at best awkward and off-key. At worst, they can lead to serious misunderstandings.

Parentheses *(Paren)*

Parentheses are used to enclose supplementary details inserted into a sentence. Commas and dashes may also be used in this role, but with some restrictions. You may enclose a complete sentence or several complete sentences within parentheses. But such enclosure would confuse the reader if only commas or dashes were used for the enclosure:

> The violence of these storms can scarcely be exaggerated. (Typhoons and hurricanes generate winds over 75 miles an hour and waves 50 feet high.) The study. . . .

Lists

Parentheses are also used to enclose numbers or letters used in listing:

> This general analysis consists of sections on (1) wave generation, (2) wave propagation, (3) wave action near a shoreline, and (4) wave energy.

Punctuation of Parentheses in Sentences

Within a sentence, place no mark of punctuation before the opening parenthesis. Place any marks needed in the sentence after the closing parenthesis:

> If a runway is regularly exposed to crosswinds of over 10 knots (11.6 mph), then the runway is considered unsafe.

Do not use any punctuation around parentheses when they come between sentences. Give the statement *inside* the parentheses any punctuation it needs.

Period *(Per)*

Periods have several conventional uses.

End Stop

Place a period at the end of any sentence that is not a question or an exclamation:

> Find maximum average daily temperature and maximum pressure altitude.

Abbreviations

Place a period after abbreviations:

M.D.	etc.
Ph.D.	Jr.

However, some style guides now call for no periods in academic abbreviations such as *BA*, *MD*, and *PhD*.

Decimal Point

Use the period with decimal fractions and as a decimal point between dollars and cents:

.4	$5.60
.05%	$450.23

Pronoun–Antecedent Agreement *(P/ag)*

Pronoun-antecedent agreement is closely related to verb-subject agreement. For example, the problem area concerning the use of collective nouns explained in "Verb-Subject Agreement" is closely related to the proper use of pronouns. When a collective noun is considered singular it takes a singular pronoun as well as a singular verb. Also, such antecedents as *each, everyone, either, neither, anybody, somebody, everybody,* and *no one* take singular pronouns as well as singular verbs:

Everyone had **his** assignment ready.

However, our sensitivity about using male pronouns exclusively when the reference may be to both men and women makes the choice of a suitable pronoun in this construction difficult. Many people object to the use of *his* as the pronoun in the preceding example. Do not choose to solve the problem by introducing a grammatical error, as in this example of incorrect usage:

Everyone had **their** assignment ready.

The use of male and female pronouns together is grammatically correct, if a bit awkward at times:

Everyone had **his or her** assignment ready.

Perhaps the best solution, one that is often applicable, is to use a plural antecedent that allows the use of a neutral plural pronoun, as in this example:

All the students had **their** assignments ready.

The same problem presents itself when we use such nouns as *student* or *human being* in their generic sense; that is, when we use them to stand for all students or all human beings. If used in the singular, such nouns must be followed by singular pronouns:

> The **student** seeking a loan must have **his or her** application in by 3 September.

Again, the best solution is to use a plural antecedent:

> **Students** seeking loans must have **their** applications in by 3 September.

See also the entry for sexist usage.

Pronoun Form *(Pron)*

Almost every adult can remember being constantly corrected by parents and elementary school teachers in regard to pronoun form. The common sequence is for the child to say, "Me and Johnny are going swimming," and for the teacher or parent to say patiently, "No, dear, 'Johnny and I are going swimming.'" As a result of this conditioning, **all** objective forms are automatically under suspicion in many adult minds, and the most common pronoun error is for the speaker or writer to use a subjective case pronoun such as *I, he,* or *she* when an objective case pronoun such as *me, him,* or *her* is called for.

Whenever a pronoun is the object of a verb or the object of a preposition, it must be in the objective case:

> It occurred to my colleagues and **me** to check the velocity data on the earthquake waves.
>
> Just between **you** and **me**, the news shook Mary and **him**.

However, use a subjective case pronoun in the predicate nominative position. This rule slightly complicates the use of pronouns after the verb. Normally, the pronoun position after the verb is thought of as objective pronoun territory, but when the verb is a linking verb (chiefly the verb *to be*), the pronoun is called a *predicate noun* rather than an object and is in the subjective case.

> It is she.
>
> It was he who discovered the mutated fruit fly.

Question Mark *(Ques)*

Place a question mark at the end of every sentence that asks a direct question:

What is the purpose of this report?

A request that you politely phrase as a question may be followed by either a period or a question mark:

Will you be sure to return the experimental results as soon as possible.

Will you be sure to return the experimental results as soon as possible?

When you have a question mark within quotation marks, you need no other mark of punctuation:

"Where am I?" he asked.

Quotation Marks *(Quot)*

Use quotation marks to set off short quotations and certain titles.

Short Quotations

Use quotation marks to enclose quotations that are short enough to work into your own text (normally, less than three lines):

According to Dr. Stockdale, "Ants, wonderful as they are, have many enemies."

When quotations are longer than three lines, set them off by single spacing and indenting them. See the entry for colon for an example of this style. Do not use quotation marks when quotations are set off and indented.

Titles

Place quotation marks around titles of articles from journals and periodicals:

Nihei's article "The Color of the Future" appeared in PC World.

Single Quotes

When you must use quotation marks within other quotation marks, use single marks (the apostrophe on your keyboard):

"Do you have the same trouble with the distinction between 'venal' and 'venial' that I do?" asked the copy editor.

Punctuation Conventions

The following are the conventions in the United States for using punctuation with quotation marks:

Commas and Periods Always place commas and periods inside the quotation marks. There are no exceptions to this rule:

> G. D. Brewer wrote "Manned Hypersonic Vehicles."

Semicolons and Colons Always place semicolons and colons outside the quotation marks. There are no exceptions to this rule:

> As Dr. Damron points out, "New technology has made photographs easy to fake"; therefore, they are no longer reliable as courtroom evidence.

Question Marks, Exclamation Points, and Dashes Place question marks, exclamation points, and dashes inside the quotation marks when they apply *to the quote only or to the quote and the entire sentence at the same time.* Place them outside the quotation marks when they apply to the entire sentence only.

Inside

> When are we going to find the answer to the question, "What causes clear air turbulence?"

Outside

> Did you read Minna Levine's "Business Statistics"?

Run-On Sentence *(Run-on)*

A run-on sentence is two independent clauses (that is, two complete sentences) put together with only a comma or no punctuation at all between them. Punctuate two independent clauses placed together with a period, semicolon, or a comma and a coordinating conjunction (*and, but, for, nor,* or *yet*). Infrequently, the colon or dash is used also. (There are some exceptions to these rules. See the entry for comma.) The following three examples are punctuated correctly, the first with a period, the second with a semicolon, the third with a comma and a coordinating conjunction:

> Check the hydraulic pressure. If it reads below normal, do not turn on the aileron boost.
>
> We will describe the new technology in greater detail; however, first we will say a few words about the principal devices found in electronic circuits.
>
> Ground contact with wood is particularly likely to cause decay, but wood buried far below the ground line will not decay because of a lack of sufficient oxygen.

If the example sentences had only commas or no punctuation at all between the independent clauses, they would be run-on sentences.

Writers most frequently write run-on sentences when they mistake conjunctive adverbs for coordinating conjunctions. The most common conjunctive adverbs are *also, anyhow, besides, consequently, furthermore, hence, however, moreover, nevertheless, therefore,* and *too.*

When a conjunctive adverb is used to join two independent clauses, the mark of punctuation most often used is a semicolon (a period is used infrequently), as in this correctly punctuated sentence:

> Ice fish are nearly invisible; however, they do have a few dark spots on their bodies.

Often the sentence will be more effective if it is rewritten completely, making one of the independent clauses a subordinate clause or a phrase.

> *Run-On Sentence*
>
> The students at the university are mostly young Californians, most of them are between the ages of 18 and 24.
>
> *Rewritten*
>
> The students at the university are mostly young Californians between the ages of 18 and 24.

Semicolon *(Semi)*

The semicolon lies between the comma and the period in force. Its use is quite restricted. (See also the entry for run-on sentences.)

Independent Clauses

Place a semicolon between two closely connected independent clauses that are not joined by a coordinating conjunction *(and, but, or, nor, for,* or *yet):*

> The expanding gases formed during burning drive the turbine; the gases are then exhausted through the nozzle.

When independent clauses joined by a coordinating conjunction have internal punctuation, then the comma before the coordinating conjunction may be increased to a semicolon:

> The front lawn has been planted with a Chinese Beauty Tree, a Bechtel Flowering Crab, a Mountain Ash, and assorted small shrubbery, including barberry and cameo roses; but so far nothing has been done to the rear beyond clearing and rough grading.

Series

When a series contains commas as internal punctuation within the parts, use semicolons between the parts:

Included in the experiment were Peter Moody, a freshman; Jesse Gatlin, a sophomore; Burrel Gambel, a junior; and Ralph Leone, a senior.

Sexist Usage *(Sexist)*

Conventional usages often discriminate against both men and women, but particularly against women. For example, a problem often arises when someone is talking about some group in general but refers to members of the group in the singular, as in the following passage:

> The modern secretary has to be an expert with electronic equipment. She has to be able to run a microcomputer and fix a fax machine. On the other hand, her boss still doodles letters on yellow pads. He has yet to come to grips with all the electronic gadgetry in today's office.

This paragraph makes two groundless assumptions: that all secretaries are female and all executives are male. Neither assumption, of course, is valid.

Similarly, in the past, letters began with "Dear Sir" or "Gentlemen." People who delivered mail were "mailmen" and those who protected our streets were "policemen." History books discussed "man's progress" and described how "man had conquered space."

However, of late we have recognized the unfairness of such discriminatory usages. Most organizations now make a real effort to avoid sexist usages in their documents. How can you avoid such usages once you understand the problem?

Titles of various kinds are fairly easy to deal with. *Mailmen* has become *mail carriers; policemen, police officers; chairmen, chairpersons* or simply *chairs;* and so forth. We no longer speak of "man's progress" but of "human progress."

The selection of pronouns when dealing with groups in general sometimes presents more of a problem. One way to deal with it is to move from the singular to the plural. You can speak of *secretaries/they* and *bosses/they,* avoiding the choice of either a male or female pronoun.

You can also write around the problem. You can convert a sentence like the following one from a sexist to a nonsexist statement by replacing the *he* clause with a verbal phrase such as an infinitive or a participle:

> The diver must close the mouthpiece shut-off valve before he runs the test.
>
> The diver must close the mouthpiece shut-off valve before running the test.

If you write instructions in a combination of the second person (you) and the imperative mood, you avoid the problem altogether:

> You must close the mouthpiece shut-off valve before you run the test.
>
> Close the mouthpiece shut-off valve before running the test.

At times, using plural forms or second person or writing around the problem simply won't work. In an insurance contract, for example, you might have to refer to the policyholder. It would be unclear to use a plural form because that might indicate two policyholders when only one is intended. When such is the case, writers have little recourse except to use such phrases as *he or she* or *he/she.* Both are a bit awkward, but they have the advantage of being both precise and nonsexist.

You can use the search program in your word processing program to find sexist language in your own work. Search for male and female pronouns and *man* and *men.* When you find them, check to see if you have used them in a sexist or nonsexist way. If you have used them in a sexist way, correct the problem, but be sure not to introduce inaccuracy or imprecision in doing so.

See also the entry for pronoun-antecedent agreement.

Spelling Error *(Sp)*

The condition of English spelling is chaotic and likely to remain so. George Bernard Shaw once illustrated this chaos by spelling *fish* as *ghoti.* To do so, he took the *gh* from *rough,* the *o* from *women,* and the *ti* from *condition.* If you have a spelling checker in your word processing program, it will help you avoid many spelling errors and typographical errors. Do remember, though, that a spelling checker will not catch the wrong word correctly spelled. That is, it won't warn you when you used *to* for *too.* You may obtain help from the spelling section in a collegiate dictionary where the common rules of spelling are explained. You can also buy, rather inexpensively, books that explain the various spelling rules and provide exercises to fix the rules in your mind.

To assist you here, we provide a list of common sound-alike words, each used correctly in a sentence.

I **accept** your gift.
Everyone went **except** Jerry.

His attorney gave him good **advice.**
His attorney **advised** him well.

Her cold **affected** her voice.
The **effect** was rather froglike.

He was **already** home by 9 P.M.
When her bag was packed, she was **all ready** to go.

The Senators stood **all together** on the issue.
Jim was **altogether** pleased with the result of the test.

He gave him **an** aardvark.
The aardvark **and** the anteater look somewhat alike.

The river **breached** the levee, letting the water through.
He loaded the cannon at the **breech.**

Springfield is the **capital** of Illinois.
Tourists were taking pictures of the **capitol** building.

Always **cite** your sources in a paper.
After the sun rose, we **sighted** the missing children.
She chose land near the river as the **site** for her house.

Burlap is a **coarse** cloth.
She was disappointed, of **course.**

His blue tie **complemented** his gray shirt.
I **complimented** him on his choice of ties.

Most cities have a governing body called a **council.**
The attorney's **counsel** was to remain quiet.

Being quiet, she said, was the **discreet** thing to do.
Each slice in a loaf of bread is **discrete** from the other slices.

"We must move **forward,**" the President said.
Many books have **forewords.**

Am I speaking so that you can **hear** me?
He was **here** just a minute ago.

It's obvious why he was here.
The sousaphone and **its** sound are both big and round.

Lead (Pb) has a melting point of 327.5°C.
Joan of Arc **led** the French troops to victory.

Our **principal** goal is to cut the deficit.
Hold to high ethical **principles.**

A thing at rest is **stationary.**
Choose white paper for your **stationery.**

A **straight** line is the shortest distance between two points.
The **Strait** of Gibraltar separates Europe from Africa.

I wonder when **they're** coming.
Are they bringing **their** luggage with them?
Put your luggage **there** in the corner.

He made a careful, **thorough** inspection.
He worked as **though** his life depended on it.
She **thought** until her head ached.

He **threw** the report on her desk.
His report cut **through** all the red tape.

Laurie moved **to** Trumansburg.
Gary moved to Trumansburg, **too.**
After one comes **two.**

We had two days of hot, sunny **weather.**
Whether he goes or not, I'm going.

Where **were** you on Monday?
The important thing is **we're** here today.
Where are you going tomorrow?

Whose house will you stay at?
Who's coming on the trip with us?

Is that **your** car you're driving?
You're right; it's my car.

Verb Form *(Vb)*

Improper verb form includes a wide variety of linguistic errors ranging from such nonstandard usages as "He seen the show" for "He saw the show" to such esoteric errors as "He was hung by the neck until dead" for "He was hanged by the neck until dead." Normally a few minutes spent with any collegiate dictionary will show you the correct verb form. College-level dictionaries list the principal parts of the verb after the verb entry.

Verb–Subject Agreement *(V/ag)*

Most of the time, verb–subject agreement presents no difficulty to the writer. For example, in the sentence, "He speaks for us all," only a child or a foreigner learning English might say, "He speak for us all." However, various constructions exist in English that do present agreement problems even for the adult, educated, native speaker of English. These troublesome constructions are examined in the following sections.

Words that Take Singular Verbs

The following words take singular verbs: *each, everyone, either, neither, anybody, somebody.*

Writers rarely have trouble with a sentence such as "No one is going to the game." Problems arise when, as is often the case, a prepositional phrase with a plural object is interposed between the simple subject and the verb, as in this sentence: "Each *of these disposal systems* is a possible contaminant." In this sentence the temptation is to let the object of the preposition, *systems,* govern the verb and write, "Each of these disposal systems *are* a possible contaminant."

Compound Subject Joined by Or or Nor

When a compound subject is joined by *or* or *nor,* the verb agrees with the closer noun or pronoun:

> Either the designer or the **builders are** in error.
>
> Either the builders or the **designer is** in error.

In informal and general usage, one might commonly hear, or see, the second sentence as "Either the builders or the designer are in error." In writing you should hold to the more formal usage of the example.

Parenthetical Expressions

Parenthetical expressions introduced by such words as *accompanied by, with, together with,* and *as well as* do not govern the verb:

> Mr. Roberts, **as well as** his two assistants, **is** working on the experiment.

Two or More Subjects Joined by And

Two or more subjects joined by *and* take a plural verb. Inverted word order does not affect this rule:

> Close to the Academy are Cathedral Rock and the Rampart Range.

Collective Nouns

Collective nouns such as *team, group, class, committee,* and many others take either plural or singular verbs, depending upon the meaning of the sentence. The writer must be sure that any subsequent pronouns agree with the subject and verb:

> The **team** is going to receive **its** championship trophy tonight.
>
> The **team** are going to receive **their** football letters tonight.

Note well: When the team was considered singular in the first example, the subsequent pronoun was *its.* In the second example the pronoun was *their.*

APPENDIX A TECHNICAL REFERENCE BOOKS AND GUIDES

Prepared by Donald J. Barrett
Chief Reference Librarian
United States Air Force Academy

Even an average library makes a staggering amount of technical information available to its users. You can gain ready access to that information through reference books and reference guides. A **reference book,** such as a dictionary, encyclopedia, or atlas, consolidates a good deal of technical information in one location. With a **reference guide** you can find the reference books, periodicals, and reports published in any specific field from agriculture to zoology. This appendix is a guide to field-specific reference books and field-specific reference guides. The preceding list of subjects covered in this appendix shows the extent of what is available to you and can help you locate the particular works you may need.

Book Guides

You can find information on books in science and technology in standard guides to book publication and general reviewing tools. A few tools devoted specifically to technical publications do exist, but most cover only a small portion of each year's book production, so, in this case, you must use the more general tools. Several of the major guides are listed here and will be found in most libraries.

Books in Print and ***Subject Guide to Books in Print.*** Annual.
Guide to book availability (in print) from 3,600 American publishers.
Books in Print: Author and title lists—give author, title, usually date of publication, edition, price, and publisher. *Subject Guide to Books in Print:* Books entered under Library of Congress subject heading, then by author, with title and other bibliographic data.

Book Review Digest. 1905–.
Index to selected book reviews, mostly from general periodicals. Gives publication data, brief descriptive notes, exact citations to reviews in about 80 periodicals. Subject and title indexes.

Cumulative Book Index. 1928–.
World list of all books published in the English language. Author, title, and subject listings arranged in a dictionary sequence. Main entry (fullest information) under author. Gives author, title, edition, series, pagination, price, publisher, date, Library of Congress card number.

Library of Congress Catalogs: Subject Catalog. 1950–.
Cumulative list by subject of works represented by Library of Congress cards for publications printed 1945 or later. Since 1983, issued in microfiche only.

New Technical Books. 1915–.
Selectively annotated titles by New York Public Library staff. Classed subject arrangement, includes table of contents, annotation for each book.

Proceedings in Print. 1964–.
Announcement journal for availability of proceedings of conferences. Full citation, price, subject, and agency indexes.

Technical Book Review Index. 1935–1988.
Very good guide to reviews appearing in scientific, technical, and trade journals, with bibliographic data and exact references to sources of reviews.

United Nations Documents Index. 1950–.
Contains checklists of documents and publications issued by various UN agencies, with subject and author indexes cumulated annually.

Reference Books

Each subject or academic field often has its own literature, often ranging from encyclopedias to indexes, dictionaries, biographical tools, and so on. The amount of literature specific to one subject may vary greatly. When you first approach a field, one principal guide, Sheehy's, is available to assist you in becoming familiar with its literature. The current edition of this guide should be found in every major collection:

Guide to Reference Books. Eugene P. Sheehy, ed. 10th ed. Chicago: American Library Association, 1986. Biennial Suppls.

Encyclopedias

The encyclopedia, although considered too general by some specialists, is often extremely useful to the researcher in getting under way and learning a field. Several of the encyclopedias for specific subjects are in fact quite detailed and scholarly. The editors and contributors to a well-written special encyclopedia will often be experts in their fields. Of course, the latest developments in a subject would be available only in periodical and report literature. Still, the general works are of great value toward understanding a subject, and they often include bibliographic citations to aid in further research.

Dictionary of Organic Compounds: The Constitution and Physical, Chemical, and Other Properties of the Principal Carbon Compounds and their Derivatives, Together with Relevant Literature References. Ian M. Heilbron, ed. 5th rev. ed. 7 vols. New York: Chapman and Hall, 1982. Plus annual suppls. 1985–.
Alphabetical list of compounds, large number of cross-references.

Encyclopaedic Dictionary of Physics. 9 vols, 5 suppls. New York: Pergamon, 1961–.
Scholarly work, alphabetically arranged, articles generally under 3,000 words, most with bibliographies. Includes articles on general, nuclear, solid state, molecular, chemical, metal, and vacuum physics. Index, plus a multilingual glossary in six languages.

Encyclopedia of the Biological Sciences. Peter Gray, ed. 2nd ed. New York: Van Nostrand Reinhold, 1970.
Contains 800 articles covering the broad field of the biological sciences as viewed by experts in their developmental, ecological, functional, genetic, structural, and taxonomic aspects. Bibliographies, biographical articles, illustrations, and diagrams are helpful features.

Encyclopedia of Chemical Technology. Raymond Kirk and Donald Othmer, eds. 25 vols, index, and suppl. 4th ed. New York: Wiley-Interscience, 1991–.
Main subject is chemical technology; about half the articles deal with chemical substances. There are also articles on industrial processes. A bibliography is

included for each product, as well as information on properties, sources, manufacture, and uses.

Encyclopedia of Materials Science and Engineering. Michael Bever, ed. 8 vols. Cambridge, MA: MIT Press, 1986.

Over 1,550 articles to assist in understanding design and development of new processes. Alphabetical topical arrangement. Index volume includes a systematic outline, author citation index, subject index, and materials information sources.

Encyclopedia of Physical Science and Technology. 2nd ed. 18 vols. San Diego: Academic Press, 1992. Supplemental yearbooks.

Covers all aspects of the physical sciences, including electronics, lasers, and optical technology. Over 500 articles, many illustrations, tables, and bibliographic citations. Separate index volume.

Encyclopedia of Polymer Science and Engineering. 2nd ed. 17 vols. New York: Wiley, 1985–1988. Plus supplement and index.

Articles designed to present a balanced account of all aspects of polymer science and technology, with bibliographies included.

McGraw-Hill Encyclopedia of Science and Technology. 7th ed. 20 vols, annual suppls. New York: McGraw-Hill, 1992.

Main set includes 7,500 articles, kept current by annual supplements. Covers the basic subject matter of all the sciences and their major applications in engineering, agriculture, and other technologies. Separate index volume. Has many diagrams and charts, and complicated subjects are treated in clear and readable language. Contributors identified in index volume.

Van Nostrand's Scientific Encyclopedia. 7th ed. 2 vols. New York: Van Nostrand Reinhold, 1989.

Includes articles on basic and applied sciences. Defines and explains over 17,000 terms, arranged alphabetically with extensive cross-references.

Subject Guides

Bibliographers and librarians have gathered research suggestions and bibliographies for many fields into published guides. It should be remembered that the rapidly changing literature in many subjects partially outdates any guide. Therefore, Sheehy's *Guide to Reference Books* and other tools listing current books and indexes should be consulted to supplement these guides.

Chemical Publications, Their Nature and Use. Melvin G. Mellon. 5th ed. New York: McGraw-Hill, 1982.

Describes publications by nature and sources. Identifies primary, secondary, and tertiary sources and evaluates their use by subject. Chapters on manual searching techniques and computer searching of data bases.

Geologic Reference Sources: A Subject and Regional Bibliography to Publications and Maps in the Geological Sciences. Dederick Ward and Marjorie Wheeler. 2nd ed. Metuchen, NJ: Scarecrow, 1981.

Subject bibliography, most items not annotated. Subject and geographic indexes. Section on geologic maps.

Guide to U.S. Government Scientific and Technical Resources. Rao Aluri. Littleton, CO: Libraries Unlimited, 1983.

Covers access to federally sponsored research, reports, patents, translations, and data bases.

How to Find Chemical Information: A Guide for Practicing Chemists, Educators, and Students. Robert E. Maizell. 2nd ed. New York: Wiley, 1986.
Extensive coverage on use of *Chemical Abstracts*, on-line data bases, patents, and standard literature.

Information Sources in Chemistry. 4th ed. R. T. Bottle. London: Bowker-Saur, 1993.
Subject arrangement, guides to resources and services evaluated.

Information Sources in Physics. Dennis F. Shaw, ed. 2nd ed. London: Butterworths, 1985.
Subject chapters, bibliographic essays, detailed examination of the literature.

Information Sources in Science and Engineering. C. D. Hurt. Littleton, CO: Libraries Unlimited, 1988.
Chapters on the literature of individual disciplines, annotated citations to over 2,000 titles.

Information Sources in Science and Technology. 2nd ed. Christopher Parker and Raymond Turley. London: Butterworths, 1986.
British emphasis.

Information Sources in the Life Sciences. H. V. Wyatt. 3rd ed. Boston: Butterworths, 1987.
General essays on searching, plus eight chapters on specific subjects. Worldwide literature coverage.

The Literature of Agricultural Engineering. Carl W. Hall and Wallace C. Olsen. Ithaca, NY: Cornell University Press, 1993,
Includes bibliographical references and index.

Science and Engineering Literature: A Guide to Reference Sources. Harold R. Malinowsky. 3rd ed. Littleton, CO: Libraries Unlimited, 1980.
Selected evaluative list of basic reference sources, arranged by major subjects such as physics, chemistry, astronomy, and the like.

Science and Technology: An Introduction to the Literature. Denis Grogan. 4th ed. London: Binbley, 1982.
Student guide to structure of the literature of science and technology.

Scientific and Technical Information Sources. Ching-Chih Chen. 2nd ed. Cambridge, MA: MIT Press, 1986.
Arranged by type of publication. Lists materials for each subject field. Annotated evaluative entries.

Technical Information Sources: A Guide to Patent Specifications, Standards, and Technical Reports Literature. Bernard Houghton. 2nd ed. Hamden, CT: Linnet, 1972.
British emphasis; covers use of patents as source of technical information, use of specifications and reports.

Use of Mathematical Literature. A. R. Dorling, ed. London: Butterworths, 1977.
Describes general literature and use of particular tools, then provides critical accounts by subject experts in their fields. Author and subject indexes.

Using the Biological Literature: A Practical Guide. Elisabeth B. Davis. New York: Marcel Dekker, 1981.
Broad subject chapters; bibliographic lists with brief subject content notes.

Using Science and Technology Information Sources. Ellis Mount and Beatrice Kovacs. Phoenix: Oryx Press, 1991.
Discusses nature and sources of information, textual and nontextual, with 35 chapters on types of sources.

Bibliographies

The literature of a field may often be compiled into bibliographies in connection with other publications, and in some cases as an indication of the work of an agency or company. A guide to bibliographies and examples of other types of compilations are given here.

Bibliographic Index: A Cumulative Bibliography of Bibliographies. 1937–.
Alphabetical subject list of separately published bibliographies and bibliographies appearing in books, pamphlets, and periodicals.
Bibliography of Agriculture. 1942–.
Classified bibliography of current literature received in the National Agricultural Library, with cumulative annual subject and author indexes.
Bibliography and Index of Geology. 1969–.
Index produced by Geological Society of America. Arranged in broad subject categories, with author and subject indexes. Formerly *Bibliography and Index of Geology Exclusive of North America* and *Bibliography of North American Geology.*
Chemical Titles. 1960–.
Author and key word indexes to titles from 700 journals in pure and applied chemistry. A computer-produced bibliography.
Dissertation Abstracts International: Abstracts of Dissertations Available on Microfilm or as Xerographic Reproductions. 1952–.
Compilation of abstracts of doctoral dissertations from most American universities. Since 1966, Part B has been devoted to the sciences and engineering.
Index of Selected Publications of the RAND Corporation. 1946–.
Coverage includes unclassified publications of the corporation. Abstracts, listed by subject and author, describe content and indexes.
Science Citation Index. 1961–.
Computer-produced index that provides access to related articles by indicating sources in which a known article by an author has been cited. Not ideally suited to subject searching.
Vertical File Index. 1932–.
Subject and title index to selected pamphlet material in all subjects of interest to the general library.

Biographies

Identification of authors and significant figures in the scientific and technical areas is frequently a problem. Some of the most notable biographical sources are commented on here. A check of Sheehy's *Guide to Reference Books* will reveal many more directories in almost every major subject field.

American Men and Women of Science: Physical and Biological Sciences. 17th ed. 8 vols. New York: R. R. Bowker, 1989–1990.
Standard biographical set for 130,000 people in the sciences.
Biography Index. 1946–.
Index to biographical material in books and magazines. Alphabetical, with index by profession and occupation.
Dictionary of Scientific Biography. 14 vols, suppl. and index. New York: Scribner's, 1970–1980.
Comprehensive: covers historical and current persons in science.
Who's Who in America: A Biographical Dictionary of Notable Living Men and Women. Chicago: Marquis, 1899–. Biennial.
The standard dictionary of contemporary biographical data. Regional volumes cover persons not of national prominence.

Who's Who in Frontiers of Science and Technology, 1985. 2nd ed. Chicago: Marquis Who's Who, 1985.
Covers scientists who have distinguished themselves by research in fields representing either new directions in traditional fields or research areas using advanced technologies.

Who's Who in Science in Europe: A New Reference Guide to West European Scientists. 5th ed. 3 vols. New York: Gale Research Service, 1987.
More than 40,000 entries including natural and physical sciences.

Dictionaries

Definition of terms for the student and scholar is a problem in the sciences, as in any field. A few general guides are available in addition to glossaries for a single field.

Academic Press Dictionary of Science and Technology. Christopher Morris, ed. San Diego: Academic Press, 1992.
Over 133, 000 entries in 124 fields of scientific knowledge.

Chambers Science and Technology Dictionary. Peter M. B. Walker, ed. 4th ed. New York: Cambridge UP, 1988.
Successor to *Chambers Technical Dictionary,* completely revised. Dictionary of concise definitions.

Concise Chemical and Technical Dictionary. Harry Bennett, ed. 4th ed. New York: Chemical Publishing, 1986.
Contains 50,000 brief definitions, including chemical formulas.

A Dictionary of Physical Sciences. John Daintith, ed. New York: Rowman, 1983.
Includes some diagrams and cross-references.

McGraw-Hill Dictionary of the Life Sciences. Daniel N. Lapedes, ed. New York: McGraw-Hill, 1976.
Provides vocabulary of the biological sciences and related disciplines. Over 20,000 terms, useful appendixes.

McGraw-Hill Dictionary of Physics and Mathematics. Daniel N. Lapedes, ed. New York: McGraw-Hill, 1978.
More than 20,000 terms, containing both basic vocabulary and current specialized terminology. Illustrated.

McGraw-Hill Dictionary of Scientific and Technical Terms. Sybil P. Parker, ed. 4th ed. New York: McGraw-Hill, 1989.
Gives almost 100,000 definitions, amplified by 2,800 illustrations. Each definition identified with the field of science in which it is primarily used.

Commercial Guides

Access to materials from companies working on a specific product can be facilitated by the use of product association and company address information. The standards are an example of tools that many industries must use to satisfy a contractor's requirements.

Annual Book of ASTM Standards, with Related Material. Philadelphia: American Society for Testing and Materials. Annual.
Approximately 50 parts including index. Contains 4,900 ASTM Standards and Tentatives in effect at the time of publication. An example of an essential reference book in industrial technology.

MacRae's Blue Book. 5 vols. Chicago: MacRae's Blue Book. Annual.
Buying directory for engineering products from over 50,000 companies.

Thomas Register of American Manufacturers and Thomas Register Catalog File. 26 vols. New York: Thomas. Annual.
Product lists, alphabetical listing of manufacturers, trade names, catalogs.

Periodicals

In almost any current research, the latest developments in a field will be published in the current periodicals and professional journals. Your first task as a researcher may be to determine what periodicals are published in a given field. Next, you may want to determine what indexing or abstracting services give access to a specific journal. The guides are the most significant in helping you to locate this type of information.

Ulrich's International Periodicals Directory: Now Including Irregular Serials and Annuals. 30th ed. 3 vols. New York: Bowker, 1991. Biennial, with supplement.
Alphabetical subject list of over 118,000 serials, published in all languages. Alphabetical title index. Indicates coverage of titles in periodical indexes and abstracting services.

World List of Scientific Periodicals Published in the Years 1900–1960. 4th ed. 3 vols. London: Butterworths, 1963–1965. Suppl., 1960–1968, 1970.
Lists more than 60,000 titles of periodicals concerned with the natural sciences and technology.

Periodical Indexes

A periodical index provides ready access to articles appearing in professional journals and general periodicals. Each index should be examined for inclusion principles, entry format, and any peculiarities unique to that index. The principal indexes for your consideration cover both general and specific fields, and the primary newspaper index to the *New York Times* is also listed.

Agricultural Index, Subject Index to a Selected List of Agricultural Periodicals and Bulletins. 1916–1964.
Detailed alphabetical subject index. Continued as *Biological and Agricultural Index.*

Air University Library Index to Military Periodicals. 1949–.
Subject index to significant articles in approximately 70 military and aeronautical periodicals not covered in readily available commercial indexes.

Applied Science and Technology Index. 1958–.
Subject index to periodicals in aeronautics, automation, chemistry, construction, electricity and electrical communication, engineering, geology and metallurgy, industrial and mechanical arts, machinery, physics, transportation, and related subjects. Formerly part of the *Industrial Arts Index.*

Art Index. 1929–.
Author and subject index to fine arts periodicals. Also includes coverage of architecture, graphic arts, industrial design, planning, and landscape design.

Biological and Agricultural Index. 1964–.
Detailed subject index to approximately 190 English language periodicals. Reports, bulletins, and other agricultural agency publications formerly covered in the *Agricultural Index* are no longer covered.

Business Periodicals Index. 1958–.
Subject index to business, financial, and management periodicals, and specific industry and trade journals. Formerly part of the *Industrial Arts Index.*

Current Technology Index. 1981–.
Subject guide to articles in British technical journals, with author index. Formerly *British Technology Index,* 1962–1980.

General Science Index. 1978–.
Subject index to 110 periodicals in fields including astronomy, atmospheric sciences, biological sciences, earth sciences, environment and conservation, genetics, oceanography, physics, physiology, and zoology.

Index to U.S. Government Periodicals. 1970–.
Computer-generated guide to 170 selected titles by author and subject.

Industrial Arts Index. 1913–1957.
Subject index, split into *Applied Science and Technology Index* and the *Business Periodicals Index.*

New York Times Index. 1851–.
Subject index with precise reference to date, page, and column for each article. Well cross-referenced, brief synopses of articles. Can serve as a guide to locating articles in other unindexed papers.

Public Affairs Information Service Bulletin (PAIS). 1915–.
Very useful index to government, economics, sociology, and so on, covering books, periodicals, documents, and reports. Includes selective indexing of over 1,000 periodicals.

Readers' Guide to Periodical Literature. 1900–.
Best-known periodical index. Covers U.S. periodicals of a broad, general nature in all subjects and scientific fields.

Social Sciences Index. 1974–.
Author and subject index to periodicals in fields including economics, environmental science, psychology, planning, and public administration. Covers 300 titles on a more scholarly level than the *Readers' Guide.* Formerly part of the *Social Sciences and Humanities Index* and the *International Index to Periodicals.*

Abstract Services

The abstracting journals assist access to periodical, book, and report literature, as does an index. The significant difference is the abstract itself, which frequently gives a better indication of article content and hence makes the abstracting journal significantly more useful than the periodical index.

Abstracts and Indexes in Science and Technology: A Descriptive Guide. 2nd ed. Dolores B. Owen. Metuchen, NJ: Scarecrow, 1985.
Describes approximately 220 abstract and indexing services by subject coverage, arrangement, indexes, abstracts, and other features. Includes information on data bases.

Abstracts of North American Geology. 1966–1971.
Abstracts of books, technical papers, maps on the geology of North America. Complements the *Bibliography of North American Geology.*

Air Pollution Abstracts. 1970–1976.
Produced by the Air Pollution Technical Information Center of the Environmental Protection Agency. Broad subject arrangement with specific author–subject indexes. Covers periodicals, books, proceedings, legislation, and standards.

ASM Review of Metal Literature. 1944–1967.
American Society for Metals abstracting journal of the world's literature concerned with the production, properties, fabrication, and application of metals, their alloys and compounds. Combined to form part of *Metals Abstracts.*

Biological Abstracts. 1926–.
Broad subject coverage of periodicals, books, and papers in all biological fields. Author, key word, and systematic indexes.

Ceramic Abstracts. 1922–.
Published in 17 subject sections with book, author, and subject indexes.

Chemical Abstracts. 1908–.
Covers chemical periodicals in all languages. Arranged in 80 subject sections; entry includes title, author, publication date, and abstract. Index sections by chemical substance, formula, numbered patent, patent concordance, author, and key word.

Energy Research Abstracts. 1976–.
Covers all scientific and technical reports, journal articles, conference papers and proceedings, books, patents, theses originated by the Department of Energy, its laboratories and contractors. Classed subject arrangement, with author, title, subject, report, and contract indexes. Succeeds *Nuclear Science Abstracts.*

Engineering Index. 1906–.
Abstracting journal includes coverage of serial publications, papers of conferences and symposia, separates and nonserial publications, some books. Excludes patents. Entries arranged by subject. Separate author index.

Geophysical Abstracts. 1929–1971.
Abstracts of current literature pertaining to the physics of the solid earth and to geophysical exploration. Annual author and subject indexes.

A Guide to the World's Abstracting and Indexing Services in Science and Technology. Washington, DC: National Federation of Science Abstracting and Indexing Services, 1963.
List of 1,855 titles originating in 40 countries. Covers the pure and applied sciences, including medicine and agriculture. Includes country and subject indexes.

Information Science Abstracts. 1966–.
Classified subject arrangement listing books and periodicals, international in scope. Formerly *Documentation Abstracts.*

International Aerospace Abstracts. 1961–.
Covers published literature in aeronautics and space science and technology. Companion publication to *Scientific and Technical Aerospace Reports.* Arranged in 75 subject categories, each entry gives accession number, title, author, source (book, conference, periodical, etc.), date, pagination, and abstract. Indexed by specific subject, personal author, contract number, meeting paper, and accession number.

Mathematical Reviews. 1940–.
Subject index to mathematical periodicals and books. Arranged by broad subject, with abstracts for most entries. Separate author index, subject classification.

Metallurgical Abstracts. 1934–1967.
British publication, combined to form part of *Metals Abstracts.*

Metals Abstracts. 1968–.
Covers all aspects of the science and practice of metallurgy and related fields. Classed subject arrangement with author index. Publication is a merger of *Metallurgical Abstracts* and the *ASM Review of Metal Literature.*

Meteorological and Geoastrophysical Abstracts. 1950–.
Includes foreign publications, arranged by Universal Decimal Classification subjects. Has separate author, subject, and geographical location indexes.

Mineralogical Abstracts. 1920–.
Classified list of abstracts covering current international literature, including books, periodicals, pamphlets, reports.

Nuclear Science Abstracts. 1947–1976.
Covers reports of the U.S. Energy Research and Development Administration (formerly the Atomic Energy Commission), government agencies, universities, industrial and independent research organizations, and worldwide book, journal, and patent literature dealing with nuclear science and technology. Arranged by subject field. Indexes by corporate author, personal author, subject, and report number. Continued by *Energy Research Abstracts.*

Oceanic Abstracts. 1964–.
Covers the worldwide book, periodical, and report literature on the oceans, including pollution, engineering, geology, and oceanography.

Pollution Abstracts. 1970–.
Classed subject listing of books, periodicals, reports, documents.

Psychological Abstracts. 1927–.
Covers books, periodicals, and reports, arranged by subject with full author and subject indexes, cumulative indexes for 1927–1980.

Science Abstracts. Section A—Physics Abstracts. 1898–. ***Section B—Electrical and Electronic Abstracts.*** 1898–. ***Section C—Computer and Control Abstracts.*** 1966–.
Covers books, periodicals, and papers in all languages; sections do not overlap. All sections arranged by subjects. Separate author and conference indexes.

Selected Water Resources Abstracts. 1968–.
Covers water in respect to quality, resources, engineering, and related aspects from books, journals, and reports. Subject and author indexes.

Report Literature

The publication of reports by academic, industrial, and government agencies has been a major development since World War II. Many contracts funded by government agencies have required publication of such reports. The report is often the most recent information on a subject. The bibliographical guides to this type of literature are only recently being developed.

Government Reports Announcements and Index. 1964–.
Formerly titled *Government Reports Announcements* (1971–1975), *U.S. Government Research and Development Reports* (1965–1971), and *U.S. Government Research Reports* (1954–1964). Covers new reports of U.S. government-sponsored research and development released by the Department of Defense and other federal agencies. Arranged by broad subject areas, each entry gives complete bibliographical citation, descriptors and availability data, and usually an abstract.

Government Reports Index. 1965–1975.
Continued as part of *Government Reports Announcements and Index.* Previously issued under other titles. Indexes reports by subject, author, corporate source, and report number.

Report Series Codes Dictionary. Eleanor J. Aronson, ed. 3rd ed. Detroit: Gale Research, 1986.

Identifies and explains over 20,000 letter and number codes used in issuing technical reports. Reference notes explain some of the systems used by various agencies. Alphabetical list of designations related to issuing agency, and alphabetical agency list to related series codes.

Scientific and Technical Aerospace Reports. 1963–.

Comprehensive abstracting journal covering worldwide report literature on the science and technology of space and aeronautics. Companion publication to *International Aerospace Abstracts.* Arranged in 74 subject categories. Indexed by subject, corporate source, individual author, contract number, report number, and accession number.

Technical Reports Awareness Circular. 1987–1989.

Guide to report literature acquired by the Defense Technical Information Center. Successor to the *Technical Abstract Bulletin* published from 1953–1986. Some recent issues of the former title are security classified. Well-indexed and good availability statements.

Use of Reports Literature. Charles P. Auger, ed. Hamden, CT: Archon, 1975.

Discusses report literature nature, control, and value. Evaluates sources.

Atlases and Statistical Guides

Basic data of interest to the technical researcher on many subjects are found in reliable and frequently updated standard guides. The quality atlas generally contains much more than maps of geographical locations. Statistical guides are of great reference importance to original research in economic, industrial, and social questions.

Commercial Atlas and Marketing Guide. Chicago: Rand McNally. Annual.

In addition to maps, contains much statistical data on trade, manufacturing, business, population, and transportation.

The National Atlas of the United States of America. Washington, DC: U.S. Department of the Interior Geological Survey, 1970.

Outstanding collection of 765 maps, many in color. Covers general reference and special subjects including landforms, geophysical forces, geology, marine features, soils, climate, water, history, and economic, sociocultural, and administrative data. Maps, data tables, and diagrams.

Statistical Abstract of the United States. 1878–. Washington, DC: Bureau of the Census. Annual.

Official government standard summary of statistics on the social, political, and economic organization of the United States. Excellent source citations, index.

The World Almanac and Book of Facts. 1868–. New York: Newspaper Enterprise Association. Annual.

The most comprehensive and generally useful almanac of miscellaneous information. Excellent statistical and news summary coverage.

Computerized Information Retrieval

Since the 1970s, bibliographic files have been made available for online interactive searching and information retrieval. In 1993 there were more than 7,900 data bases, with an estimated 4.5 billion citations available for searching. Normal access is from a local computer terminal to a firm or agency offering access to a data base or a system of data bases. Charges are calculated on the number of minutes a file is in use and the number of citations received. Citations are usually printed off-line and delivered by mail. Many services offer complete text document delivery for an additional charge.

A number of individual data base services are now accessible in libraries in compact disc, read-only-memory (CD-ROM) versions. With CD-ROM, there is no charge for line access or printing costs.

Three commercial services offer access to a wide spectrum of data bases:

- EPIC Service, OCLC Online Computer Library Center, Dublin, OH
- DIALOG, Dialog Information Services, Palo Alto, CA
- BRS, BRS Information Technologies, Latham, NY

From most academic and research libraries, subject searches of over 17 million book titles can also be done via the First Search service on the OCLC (Online Computer Library Center) System in Dublin, Ohio.

An average search can be expected to cost a minimum of $5–$50; costs can be minimized through careful planning of search terms and search strategy by an experienced operator. Many major colleges, universities, and information centers offer these services through service bureaus or libraries, with some libraries paying part or all of the costs. Individual access directly through originating firms is also possible. Custom searches of over 1.5 million reports from federal agencies and federally sponsored research are available from National Technical Information Service, U.S. Department of Commerce, Springfield, VA 22161. The cost is about $100 for up to 100 abstracts. Some of the current subject data bases available for on-line searching in the scientific and technical areas are the following:

Agriculture: AGRICOLA, developed by the National Library of Agriculture
Biological sciences: BIOSIS Previews, prepared by Biosciences Information Service
Business: ABI/INFORM, produced by Data Courier
Chemistry: CA Search, produced by Chemical Abstracts Service of the American Chemical Society
Education: ERIC (Educational Resources Information Center), developed by the National Institute of Education
Engineering: COMPENDEX, produced by Engineering Information
Environmental studies: ENVIROLINE, prepared by Environment Information Center

Geosciences: GEO-REF, produced by the American Geological Institute
Government research and development reports: NTIS, produced by the National Technical Information Service of the Department of Commerce
Mathematics: MATHSCI, produced by the American Mathematical Society
Mechanical engineering: ISMEC, prepared by Data Courier
Metallurgy: METADEX, prepared by the American Society of Metals
Physics: SPIN, Searchable Physics Information Notices, by American Institute of Physics
Pollution and environment: Pollution Abstracts, produced by Cambridge Scientific Abstracts
Psychology: PsycINFO, produced by the American Psychological Association
Science abstracts: INSPEC, produced by the Institution of Electrical Engineers
Science and technology: SCISEARCH, produced by the Institute for Scientific Information.

Many other data bases are available for searching, but some have special use restrictions, such as being limited to specific industrial group members.

U.S. Government Publications

Access to government publications, releases, and directives is possible only from a series of federally produced and commercially supplemented catalogs and indexes. The nature of the subject indexing is generally less specific than with nongovernmental periodical indexes and abstracting services. Therefore, more ingenuity on the part of the researcher is generally needed to find pertinent resources.

American Statistics Index and Abstracts, Annual and Retrospective Edition. A Comprehensive Guide to the Statistical Publications of the U.S. Government. Washington, DC: Congressional Information Service.

Aims to be a master guide and index to all federally produced statistical data. Does not contain the data but describes data and identifies sources.

Bureau of the Census Catalog of Publications, 1790–1972.

An example of a departmental catalog, kept up-to-date by frequent supplements. Many agencies issue such retrospective catalogs.

CIS/Index to Congressional Publications. Washington, DC: Congressional Information Service, 1970–.

Part one: Abstracts of congressional publications; Part two: Index of congressional publications and public laws; Part three: Legislative histories. Commercial index giving significant insight into contents of congressional publications, with a detailed subject index.

Code of Federal Regulations. 1949–.
Contains codifications of general and permanent administrative rules and regulations of general applicability and future effect.

Cumulative Subject Index to the Monthly Catalog of United States Government Publications. 1900–1971. 15 vols. Washington, DC: Carrollton Press, 1973–1975.
Covers about 800,000 publications. Be sure to read the introduction for exclusions and entry policies.

Federal Register. 1936–.
Daily publication of executive orders, presidential proclamations, and announcements of important rules and regulations of the federal government. Indexed by agency and significant subjects.

Monthly Catalog of United States Government Publications. 1895–.
List by agency of publications, printed and processed, issued each month. Subject index only until 1973, then added author and title indexes. Entry gives title, author, publication data, price or availability indication. Superintendent of Documents classification number.

Monthly Checklist of State Publications.
Records the documents and publications issued by the various states and received in the Library of Congress.

Subject Bibliographies. 1975–.
Lists of publications available in specified subject areas from the Government Printing Office.

United States Government Manual. 1935–. Annual. Washington, DC: U.S. GPO.
The official handbook of the federal government; describes the purposes and programs of most official and quasi-official agencies, with addresses and lists of current officials.

Weekly Compilation of Presidential Documents. 1965–.
Makes available transcripts of the president's news conferences, messages to Congress, public speeches and statements, and other presidential materials. Indexed.

APPENDIX B A SELECTED BIBLIOGRAPHY

Technical Writing

Anderson, Paul V. 1990. *Technical Writing: A Reader Centered Approach.* 2nd ed. San Diego: Harcourt.

Blicq, Ron S. 1992. *Technically Write!* 4th ed. Englewood Cliffs, NJ: Prentice Hall.

Brusaw, Charles, Gerald J. Alred, and Walter E. Oliu. 1992. *The Handbook of Technical Writing.* 4th ed. New York: St. Martin's.

Burnett, Rebecca E. 1994. *Technical Communications.* 3rd ed. Belmont, CA: Wadsworth.

Cohen, Gerald, and Donald H. Cunningham. 1984. *Creating Technical Manuals.* New York: McGraw-Hill.

Day, Robert A. 1994. *How to Write and Publish a Scientific Paper.* 4th ed. Phoenix: Oryx.

Felker, Daniel B., et al. 1981. *Guidelines for Document Designers.* Washington, DC: American Institutes for Research.

Horton, William K. 1990. *Designing and Writing Online Documentation.* New York: Wiley.

Huckin, Thomas N. 1991. *Technical Writing and Professional Communication for Non-Native Speakers.* 2nd ed. New York: McGraw-Hill.

Jordan, Stello, Joseph M. Kleinman, and H. Lee Shimberg, eds. 1971. *Handbook of Technical Writing Practices.* 2 vols. New York: Wiley.

Keene, Michael L. 1993. *Effective Professional and Technical Writing.* 2nd ed. Lexington, MA: Heath.

Killingsworth, M. Jimmie, and Michael K. Gilbertson. 1992. *Signs, Genres, and Communities in Technical Communication.* Amityville, NY: Baywood Publishing.

Markel, Michael H. 1992. *Technical Writing: Situations and Strategies.* 3rd ed. New York: St. Martin's.

Michaelson, Herbert B. 1990. *How to Write and Publish Engineering Papers and Reports.* 3rd ed. Philadelphia: ISI.
Pickett, Nell A., and Ann A. Laster. 1989. *Technical English: Writing, Reading, and Speaking.* 5th ed. New York: Harper.
Souther, James W., and Myron L. White. 1977. *Technical Report Writing.* 2nd ed. New York: Wiley.
Warren, Thomas L. 1985. *Technical Writing.* Belmont, CA: Wadsworth.

Business Communication

Andrews, Deborah C., and William D. Andrews. 1991. *Business Communication.* 2nd ed. New York: Macmillan.
Brusaw, Charles T., et al. 1993. *The Business Writer's Handbook.* 4th ed. New York: St. Martin's.
Kolin, Phillip C. 1990. *Successful Writing at Work.* 3rd ed. Lexington, MA: Heath.
Pearsall, Thomas E., and Donald H. Cunningham. 1994. *How to Write for the World of Work.* 5th ed. Fort Worth: Harcourt.
Sigband, Norman B., and Arthur H. Bell. 1989. *Communication for Management and Business.* 5th ed. Glenview, IL: Scott.
Tebeaux, Elizabeth. 1990. *Design of Business Communications: The Process and the Product.* New York: Macmillan.
Treece, Malra. 1990. *Successful Communication for Business and the Professions.* 4th ed. Boston: Allyn & Bacon.

Writing in General

Flower, Linda. 1993. *Problem Solving Strategies for Writing.* 4th ed. San Diego: Harcourt.
Kinneavy, James L. 1980. *A Theory of Discourse.* New York: Norton.
Lambuth, David, et al. 1976. *The Golden Book on Writing.* New York: Penguin.
Quiller-Couch, Sir Arthur. 1916. *On the Art of Writing.* New York: Putnam's.
Pearsall, Thomas E., and Donald H. Cunningham. 1988. *The Fundamentals of Good Writing.* New York: Macmillan.
Strunk, William, and E. B. White. 1979. *Elements of Style.* New York: Macmillan.
Van Buren, Robert, and Mary Fran Buehler. 1980. *The Levels of Edit.* 2nd ed. Pasadena, CA: JPL, California Institute of Technology.
Williams, Joseph M. 1988. *Style: Ten Lessons in Clarity and Grace.* 3rd ed. New York: HarperCollins.

Usage

Corbett, Edward P. 1987. *The Little English Handbook.* 5th ed. Glenview, IL: Scott.
Ebbitt, Wilma R., and David R. Ebbitt. 1990. *Index to English.* 8th ed. New York: Oxford.
Follett, Wilson. 1966. *Modern American Usage.* Edited and completed by Jacques Barzun. New York: Hill and Wang.
Fowler, Henry W. 1987. *A Dictionary of Modern English Usage.* ed. Ernest Gowers. 2nd rev. ed. New York: Oxford UP.

O'Hare, Frank, and Edward A. Kline. 1992. *The Modern Writer's Handbook.* 3rd ed. New York: Macmillan.

Style Manuals

Achtert, Walter S., and Joseph Gibaldi. 1988. *MLA Handbook for Writers of Research Papers,* 3rd ed. New York: MLA.
CBE Style Manual Committee. 1983. *Council of Biology Editors Style Manual.* 5th ed. Arlington, VA: Council of Biology Editors.
Chicago Manual of Style. 1993. 14th ed. Chicago: U of Chicago P.
Publications Manual of the American Psychological Association. 1983. 3rd ed. Washington, DC: American Psychological Association.
U.S. Government Printing Office Style Manual. 1984. Washington, DC: GPO.

Speech

Becker, Paula B., and Dennis Becker. 1992. *Business Speaking.* Burr Ridge, IL: Irwin.
Beebe, Steven A., and Susan Beebe. 1990. *Public Speaking: An Audience Centered Approach.* Englewood Cliffs, NJ: Prentice Hall.
DeVito, Joseph A. 1990. *The Elements of Public Speaking.* 4th ed. New York: Harper Collins.
Hunt, Gary T. 1987. *Public Speaking.* 2nd ed. Englewood Cliffs, NJ: Prentice Hall
Lucas, Stephen E. 1989. *The Art of Public Speaking.* 3rd ed. New York: Random House.

Logic

Copi, Irving M., and Keith Burgess. 1991. *Informal Logic.* 2nd ed. New York: Macmillan.
Copi, Irving M., and Carl Cohen. 1990. *Introduction to Logic.* 8th ed. New York: Macmillan.
Toulmin, Steven, et al. 1984. *An Introduction to Reasoning.* 2nd ed. New York: Macmillan.

Graphics

Conover, Theodore E. 1991. *Graphic Communications Today.* 2nd ed. St. Paul, MN: West.
Duff, Jon M. 1993. *Technical Illustration with Computer Applications.* Englewood Cliffs, NJ: Prentice Hall.
Hill, Francis S. 1990. *Computer Graphics.* New York: Macmillan.
Hoffman, E. Kenneth. 1990. *Computer Graphics Applications.* Belmont, CA: Wadsworth.
Lefferts, Robert. 1982. *How to Prepare Charts and Graphs for Effective Reports.* New York: Barnes & Noble.
White, Jan V. 1988. *Graphic Design for the Electronic Age.* New York: Watson-Guptill.
_____. 1990. *Color for the Electronic Age.* New York: Watson-Guptill.

Library Research

Barzun, Jacques, and Graff, Henry F. 1992. *The Modern Researcher.* 5th ed. Boston: Houghton Mifflin.

Hurt, C. D. 1988. *Information Sources in Science and Technology.* Englewood, CO: Libraries Unlimited.

Mann, Thomas. 1990. *A Guide to Library Research Methods.* New York: Oxford UP.

McCormick, Mona. 1986. *The New York Times Guide to Reference Materials.* New York: NAL.

CHAPTER NOTES

Chapter 1: An Overview of Technical Writing

[1] Paul V. Anderson, "What Survey Research Tells Us about Writing at Work," *Writing in Nonacademic Settings*, eds. Lee Odell and Dixie Goswami (New York: Guilford, 1985) 30.

[2] Anderson 40.

[3] Anderson 54.

[4] Philip W. Swain, "Giving Power to Words," *American Journal of Physics* 13 (1945): 320.

Chapter 2: Composing

[1] Fred L. Luconi, "Artificial Intelligence," *Vital Speeches of the Day* 52 (1986): 605.

[2] Terry Winograd of Stanford University, as quoted in Bob Ryan, "AI in Identity Crisis," *Byte* (June 1991): 241–42.

[3] Stephen S. Hall, "Aplysia and Hermissenda," *Science* 85 (May 1985): 33.

[4] Lester Faigley and Thomas P. Miller, "What We Learn from Writing on the Job," *College English* 44 (1982): 562–63.

[5] Lee Odell, Dixie Goswami, Anne Herrington, and Doris Quick, "Studying Writing in Non-Academic Settings," *New Essays in Technical and Scientific Communication: Research, Theory, Practice*, eds. Paul V. Anderson, R. John Brockmann, and Carolyn R. Miller (Farmingdale, NY: Baywood, 1983) 27–28.

[6] For this concept we are indebted to Victoria M. Winkler, "The Role of Models in Technical and Scientific Writing," *New Essays in Technical and Scientific Communication: Research, Theory, Practice*, eds. Paul V. Anderson, R. John Brockmann, and Carolyn R. Miller (Farmingdale, NY: Baywood, 1983) 111–22.

[7] Blaine McKee, "Do Professional Writers Use an Outline When They Write?" *Technical Communication* (1st Quarter 1972): 10–13.

[8] Lillian Bridwell and Ann Duin, "Looking In-Depth at Writers: Computers as Writing Medium and Research Tool," *Writing On-Line: Using Computers in the Teaching of Writing,* eds. J. L. Collins and E. A. Sommers (Montclair, NJ: Boyton/Cook, 1985) 119.

[9] Eric Brown, "Word Processing and the Three Bears," *PC World* (Dec. 1985): 197.

[10] In writing this section on ethical considerations, we have drawn upon the following books and articles: John Bryan, "Down the Slippery Slope: Ethics and the Technical Writer as Marketer," *Technical Communication Quarterly* (Winter 1992): 73–88; William K. Franken, *Ethics* (Englewood Cliffs, NJ: Prentice-Hall, 1963); Dean G. Hall and Bonnie A. Nelson, "Integrating Professional Ethics into the Technical Writing Course," *Journal of Technical Writing and Communication* 17 (1987): 45–61; Mike Markel, "A Basic Unit on Ethics for Technical Communicators," *Journal of Technical Writing and Communication* 21 (1991): 327–50; H. Lee Shimberg, "Ethics and Rhetoric in Technical Writing," *Technical Communication* (4th Quarter 1978): 16–18; and Arthur E. Walzer, "The Ethics of False Implicature in Technical and Professional Writing," *Journal of Technical Writing and Communication* 19 (1989): 149–60.

[11] The information about radon in this exercise came from United States Environmental Protection Agency, *Consumer's Guide to Radon Reduction: How to Reduce Radon Levels in Your Home* (Washington, DC: U.S. GPO, 1992).

Chapter 3: Writing Collaboratively

[1] In this chapter we are indebted to the following: Rebecca E. Burnett, "Substantive Conflict in a Cooperative Context: A Way to Improve the Collaborative Planning of Workplace Documents," *Technical Communication* 38 (1991): 532–39; Mary Beth Debs, "Collaborative Writing in Industry," *Technical Writing: Theory and Practice,* eds. Bertie E. Fearing and W. Keats Sparrow (New York: Modern Language Association, 1989) 33–42; and *Collaborative Writing in Industry: Investigations in Theory and Practice,* eds. Mary M. Lay and William M. Karis (Amityville, NY: Baywood, 1991). In *Collaborative Writing in Industry* we are particularly indebted to David K. Farkas, "Collaborative Writing, Software Development, and the Universe of Collaborative Activity," 13–30; James R. Weber, "The Construction of Multi-Authored Texts in One Laboratory Setting," 49–64; Barbara Couture and Jone Rymer, "Discourse Interaction between Writer and Supervisor: A Primary Collaboration in Workplace Writing," 87–108; Ann Hill Duin, Linda A. Jorn, and Mark S. DeBower, "Collaborative Writing: Courseware and Telecommunications," 146–69; and William Van Pelt and Alice Gillam, "Peer Collaboration and the Computer-Assisted Classroom: Bridging the Gap between Academia and the Workplace," 170–206.

[2] For more on this, see Edgar R. Thompson, "Ensuring the Success of Peer Revision Groups," *Focus on Collaborative Learning,* ed. Jeff Golub (Urbana, IL: NCTE, 1988) 109–16.

Chapter 4: Writing for Your Readers

[1] U. S. Bureau of the Census, *Statistical Abstract of the United States: 1992,* 112th ed. (Washington, DC: GPO, 1992) Table 219.

[2] See "Education Report Finds U.S. Reading and Writing Skills Worsen," *Publishers Weekly* (8 Nov 1991):15 and "When Johnny's Whole Family Can't Read," (20 July 1992): 68.

[3] Mary B. Coney, "The Use of the Reader in Technical Writing," *Journal of Technical Writing and Communication* 8 (1978): 104.

[4] Coney 104.

[5] U.S. Environmental Protection Agency, *Manual of Individual and Non-Public Water Supply Systems* (Washington, DC: GPO, 1991) 7.

[6] Michael Ryan and James Tankard, Jr., "Problem Areas in Science News Writing," *Journal of Technical Writing and Communication* 4 (1974): 230.

[7] J. J. Degan, "Microwave Resonance Isolators," *Bell Laboratories Record* (Apr 1966): 123.

[8] For material in this section we are indebted to Janice C. Redish, "Understanding Readers," *Techniques for Technical Communicators*, eds. Carol M. Barnum and Saul Carliner (New York: Macmillan, 1993) 14–41.

[9] U.S. Geological Survey, *Our Changing Continent* (Washington, DC: GPO, 1991) 6–7.

[10] Thomas N. Huckin, "A Cognitive Approach to Readability," *New Essays in Technical and Scientific Communication*, eds. Paul V. Anderson, R. John Brockmann, and Carolyn R. Miller (Farmingdale, NY: Baywood, 1983) 99.

[11] Susan Huseonica, "Archaeology from Above," *NASA Magazine* (Summer 1993): 17.

[12] James W. Souther, "Identifying the Informational Needs of Readers: A Management Responsibility," *IEEE Transactions on Professional Communication* PC-28 (1985): 10.

[13] Souther 10.

[14] Souther 10.

[15] Souther 10.

[16] Thomas E. Pinelli, Virginia M. Cordle, and Raymond F. Vondran, "The Function of Report Components in the Screening and Reading of Technical Reports," *Journal of Technical Writing and Communication* 14 (1984): 89.

[17] National Institutes of Health, *Inside the Cell* (Washington, DC: U.S. Department of Health and Human Services, 1990) 19.

[18] Richard Conniff, "Eye on the Storm," *Raytheon Magazine* (Fall 1982): 21.

[19] Allen Hammond, "Limits of the Medium," *SIPIscope* 11. 2 (1983): 6–7.

[20] See Ann Hill Duin, "How People Read: Implications for Writers," and Wayne Slater, "Current Theory and Research on What Constitutes Readable Expository Text," *The Technical Writing Teacher* 15 (1988): 185–93, 195–206.

[21] Ryan and Tankard 233.

[22] David F. Cope, "Nuclear Power: A Basic Briefing," *Mechanical Engineering* 89 (1967): 50.

[23] Junko Torii et al., "Effect of Time of Day on Adaptive Response to a 4-Week Aerobic Exercise Program," *The Journal of Sports Medicine and Physical Fitness* (Dec. 1992): 350–51.

[24] Lester Faigley, "Nonacademic Writing: The Social Perspective," *Writing in Nonacademic Settings*, eds. Lee Odell and Dixie Goswami (New York: Guilford, 1985) 238.

[25] The material about cultural differences in this section is based upon William B. Gudykunst, *Bridging Differences: Effective Intergroup Communication* (Newbury Park, CA: Sage, 1991) 42–55.

Chapter 5: Achieving a Readable Style

[1] Raymond K. Neff, "Computing in the University: The Implications of New Technologies," *Perspectives in Computing* (Fall 1987): 15.

[2] U.S. Geological Survey, *Our Changing Continent* (Washington, DC: GPO, 1991) 3–4.

[3] Janice C. Redish and Jack Selzer, "The Place of Readability Formulas in Technical Communication," *Technical Communication* (4th Quarter 1985): 49.

[4] Francis Christensen, "Notes Toward a New Rhetoric," *College English* (Oct. 1963): 7–18.

[5] Daniel B. Felker et al., *Guidelines for Document Designers* (Washington, DC: American Institutes for Research, 1981) 47–48.

[6] As quoted in Felker et al. 64.
[7] As revised in Felker et al. 65.
[8] As quoted in Janice C. Redish, *The Language of Bureaucracy* (Washington, DC: American Institutes for Research, 1981) 1.
[9] CBE Style Manual Committee, *CBE Style Manual,* 5th ed. (Bethesda, MD: Council of Biology Editors, Inc., 1983) 38.
[10] *CBE Style Manual* 38.
[11] For a perceptive essay on elegant variation, see H. W. Fowler, *A Dictionary of Modern English Usage* (New York: Oxford UP, 1950) 130–33.
[12] The excerpts from the St. Paul Fire and Marine Insurance Company's old and new Personal Liability Catastrophe Policy are reprinted with the permission of the St. Paul Companies, St. Paul, MN 55102.
[13] *CBE Style Manual* 36–37. Reproduced with permission from *CBE Style Manual,* 5th ed. Copyright 1983, CBE Style Manual Committee, Council of Biology Editors, Inc.
[14] "Planners Outlaw Jargon," *Plain English* (Apr. 1981): 1.
[15] National Institutes of Health, *Inside the Cell* (Washington, DC: U.S. Department of Health and Human Services, 1990) 48.

Chapter 6: Informing

[1] Robert I. Tilling, Christina Heliker, and Thomas Wright, *Eruptions of Hawaiian Volcanoes: Past, Present, and Future* (Washington, DC: U.S. Geological Survey, 1987) 16–17.
[2] Tony Capaccio, "Bosnia Air Drop," *Air Force Magazine* (July 1993): 52–53. Reprinted by permission.
[3] U.S. National Aeronautics and Space Administration, *Voyager at Neptune: 1989* (Washington, DC: GPO, 1989) 16.
[4] This quotation and the next are from U.S. Geological Survey, *Safety and Survival in an Earthquake* (Washington, DC: GPO, 1991) 3–5.
[5] William B. Gudykunst, *Bridging Differences: Effective Intergroup Communication* (Newbury Park, CA: Sage Publications, 1991) 44.
[6] Stephan Wilkinson, "Tiny Keys to Our Electronic Future," *Raytheon Magazine* (Winter 1985): 4.
[7] U.S. National Aeronautics and Space Administration, *Exploring the Universe with the Hubble Space Telescope* (Washington, DC: GPO, n.d.) 11.
[8] "If All Our Reporters Were Laid End to End," *The Wall Street Journal* 4 (April 1983): 10.
[9] Sir James Jeans, *Stars in Their Courses* (Cambridge, UK: Cambridge U P, 1931) 23–24. Copyright 1931. Reprinted with the permission of Cambridge University Press.
[10] Raymond H. Beal, Joe K. Mauldin, and Susan C. Jones, *Subterranean Termites: Their Prevention and Control* (Washington, DC: U.S. Department of Agriculture, 1989).
[11] Stephan Wilkinson, "Earth Shakers," *Raytheon Magazine* (Winter 1987): 12–15.

Chapter 7: Defining and Describing

[1] Centers for Disease Control, *The Public Health Consequences of Disasters* (Washington, DC: U.S. Department of Health and Human Services, 1989) 33.
[2] U.S. Congress, Office of Technology Assessment, *Biological Effects of Power Frequency Electric and Magnetic Fields* (Washington, DC: GPO, 1989) 19.
[3] U.S. Environmental Protection Agency, *Manual of Individual and Non-Public Water Supply Systems* (Washington, DC: GPO, 1991) 29.
[4] National Institutes of Health, *Inside the Cell* (Washington, DC: U.S. Department of Health and Human Services, 1990) 21.

[5] Robert I. Tilling, Christina Heliker, and Thomas Wright, *Eruptions of Hawaiian Volcanoes: Past, Present, and Future* (Washington, DC: U.S. Geological Survey, 1987) 17.

[6] Robin Birley, "A Frontier Post in Roman Britain," *Scientific American* (Feb. 1977): 39.

[7] U.S. National Aeronautics and Space Administration, *The Voyager Flights to Jupiter and Saturn* (Washington, DC: GPO, 1982) 11.

[8] *The Plow and the Hearth Fall/Winter Catalog* (1986): 27.

[9] U.S. National Aeronautics and Space Administration, *Planetary Exploration Through Year 2000* (Washington, DC: GPO, 1988) 30–31.

[10] David L. Wallace and John R. Hayes, "Redefining Revision for Freshmen," *Research in the Teaching of English* 25 (1991): 58–59.

[11] Stephan Wilkinson, "Tiny Keys to Our Electronic Future," *Raytheon Magazine* (Winter 1985): 4.

Chapter 8: Arguing

[1] U.S. National Aeronautics and Space Administration, *Viking: The Exploration of Mars* (Washington, DC: GPO, 1984) 6.

[2] Joel Gurin, "In the Beginning," *Science* 80 (July/Aug. 1980): 50.

[3] For a more detailed explanation of Toulmin logic, see Steven Toulmin, Richard Rieke, and Allan Janik, *An Introduction to Reasoning*, 2nd ed. (New York: Macmillan, 1984).

[4] The material in this example is adapted from Sharon Begley, Mary Hager, and Larry Wilson, "Is It All Just Hot Air?" *Newsweek* 20 (Nov. 1989): 64–66; Warren T. Brookes, "The Global Warming Panic," *Forbes* 25 (Dec. 1989): 96–100; Gregg Easterbrook, "A House of Cards," *Newsweek* 1 (June 1992): 24–33; Gregg Easterbrook, "Green Cassandras," *The New Republic* 6 (July 1992): 23–25; and Fred Krupp, "A Global Bargain for Global Warming," *Vital Speeches of the Day* 59 (1992): 90–93.

[5] Dale Blumenthal, "Food Irradiation: Toxic to Bacteria, Safe for Humans," *FDA Consumer Magazine* (Nov. 1990): As reprinted in DHHS Publication No. (FDA) 91-2241.

Chapter 9: Document Design

[1] Daniel B. Felker et al., *Guidelines for Document Designers* (Washington, DC: American Institutes for Research, 1981). This collection of research-based guidelines provides useful insights on most of the principles discussed in this chapter. For a rationale for teaching design, see Stephen A. Bernhardt, "Seeing the Text," *College Composition and Communication* 37 (1986): 66–78.

[2] Thomas N. Huckin, "A Cognitive Approach to Readability," *New Essays in Technical and Scientific Communication: Research, Theory, and Practice*, eds. Paul V. Anderson, R. John Brockmann, and Carolyn R. Miller (Farmingdale, NY: Baywood, 1983) 90–101; Janice C. Redish, Robin M. Battison, and Edward S. Gold, "Making Information Accessible to Readers," *Writing in Nonacademic Settings*, eds. Lee Odell and Dixie Goswami (New York: Guilford, 1985) 129–53.

[3] Felker et al. 79–80.

[4] Jan V. White, *Graphic Design for the Electronic Age* (New York: Watson-Guptill, 1988) 20; Felker et al. 85–86.

[5] Rolf F. Rehe, *Typography: How to Make It Most Legible*, 3rd ed. (Carmel, IN: Design Research International) 34.

[6] M. Gregory and E. C. Poulton, "Even Versus Uneven Right-Margins and the Rate of Comprehension in Reading," *Ergonomics* 13 (1970): 427–34; Rehe.

[7] Huckin; Redish et al.

[8] M. A. Tinker, *Legibility of Print* (Ames: Iowa State UP, 1963); M. A. Tinker, *Bases for Effective Reading* (Minneapolis: U of Minnesota P, 1965); Felker et al. 77–78. Felker et al. recommend 8- to 10-point type. Although that guideline works well for many fonts in typeset documents where the resolution from the printer is excellent, 8-point type from most word processors and desktop printers is not large enough to be read easily.

[9] Philippa J. Benson, "Writing Visually: Design Considerations in Technical Publications," *Technical Communication* (4th Quarter 1985): 35–39. Felker et al. 73–76.

[10] Felker et al. 87–88; Rehe 35–36.

[11] J. Foster and P. Coles, "An Experimental Study of Typographical Cuing in Printed Text," *Ergonomics* 20 (1977): 57–66.

[12] Linda Flower, John R. Hayes, and Heidi Swarts, "Revising Functional Documents: The Scenario Principle," *New Essays in Technical and Scientific Communication: Research, Theory, Practice*, eds. Paul V. Anderson, R. John Brockmann, and Carolyn R. Miller (Farmingdale, NY: Baywood, 1983) 41–58.

[13] Jeanne W. Halpern, "An Electronic Odyssey," *Writing in Nonacademic Settings*, eds. Lee Odell and Dixie Goswami (New York: Guilford, 1985) 157.

Chapter 10: Design Elements of Reports

[1] M. Jimmie Killingsworth and Betsy G. Jones, "Division of Labor or Integrated Teams: A Crux in the Management of Technical Communication?" *Technical Communication* 36 (1989): 210.

[2] Bill Heald, "Putting Out Fires with Privateers," *Code One* (July 1991): 17–18. Reprinted by permission.

[3] U.S. Department of Labor, *Job Search Guide: Strategies for Professionals* (Washington, DC: GPO, 1993) 2.

[4] Lenore S. Ridgway, Roger A. Grice, and Emilie Gould, "I'm OK; You're Only a User: A Transactional Analysis of Computer–Human Dialogs," *Technical Communication* 39 (1992): 39.

[5] *Phthalates in Food* (Chicago: Institute of Food Technologists, 1974) 1.

[6] David R. Russell, "The Ethics of Teaching Ethics in Professional Communication: The Case of Engineering Publicity at MIT in the 1920s," *Journal of Business and Technical Communication* 7 (1993): 84–85. Reprinted by permission.

[7] *Phthalates in Food* 2.

[8] Russell 107.

[9] J. J. Bull, R. C. Vogt, and C. J. McCoy, "Sex Determining Temperatures in Turtles: A Geographic Comparison," *Evolution* 36 (1982): 331.

[10] Walter S. Achtert and Joseph Gibaldi, *MLA Handbook for Writers of Research Papers*, 3rd ed. (New York: MLA, 1988).

[11] *Publication Manual of the American Psychological Association*, 3rd ed. (Washington, DC: APA, 1983).

[12] *The Chicago Manual of Style*, 13th ed. (Chicago: U of Chicago P, 1982) 107–28.

[13] National Institutes of Health, *Smoking and Your Digestive System* (Washington, DC: U.S. Department of Health and Human Services, 1991) 1–3.

Chapter 13: The Strategies and Communications of the Job Hunt

[1] In writing this chapter we have drawn upon two books from the U.S. Department of Labor: *Job Search Guide: Strategies for Professionals* (Washington, DC: GPO, 1993) and *Tips for Finding the Right Job* (Washington, DC: GPO, 1992).

[2] *Job Search Guide* 22.

[3] Jane L. Anton, Michael L. Russell, and the Research Committee of the Western College Placement Association, *Employer Attitudes and Opinions Regarding Potential College Graduate Employees* (Hayward, CA: Western College Placement Association, 1974) 10.

[4] For example, see Rosemary Ullrich, "The Power of a Positive Job Search," *Business World Women* (Fall 1977): 11; Tom Jackson, "10 Musts for a Powerful Resume," *CPC Annual,* 30th ed. (Bethlehem, PA: College Placement Council, 1986) 3: 19–28; and Sandra Grundfest, "A Cover Letter and Resume Guide," *Business Week's Guide to Careers* (New York: McGraw-Hill, 1986) 8–13.

[5] Avery Comarow, "Tracking the Elusive Job," *The Graduate* (Knoxville: Approach 13-30 Corporation, 1977) 42.

Chapter 16: Instructions

[1] G. B. Harrison, *Profession of English* (New York: Harcourt, 1962) 149.

[2] The questions posed here are based on questions presented by Janice C. Redish, Robbin M. Battison, and Edward S. Gold, "Making Information Accessible to Readers," *Writing in Nonacademic Settings,* eds. Lee Odell and Dixie Goswami (New York: Guilford, 1985) 139–43.

[3] Redish, Battison, and Gold 134.

[4] U.S. Department of Health and Human Services, *Eating to Lower Your High Blood Cholesterol* (Washington, DC: GPO, 1989) 1.

[5] U.S. General Services Administration, *Paint and Painting* (Washington, DC: GPO, 1977) 15.

[6] Shoebox for Sebago Docksides.

[7] Kathryn Coonrod, "Why Should Technical Communicators Care About Product Liability?" *STC Intercom* (Feb. 1993): 5.

[8] Charles H. Sides, *How to Write Papers and Reports About Computer Technology* (Philadelphia: ISI, 1984) 70.

[9] U.S. Department of Agriculture, *Simple Home Repairs: Inside* (Washington, DC: GPO, 1986) 7–8.

[10] *Simple Home Repairs* 16.

[11] Protocol analysis material furnished by Professor Victoria Mikelonis, U of Minnesota, St. Paul, MN.

Chapter 18: Oral Reports

[1] These introspections were compiled at a session of the National Training Laboratory in Group Development that one of the authors attended in Bethel, Maine.

[2] The material on visual aids has been especially prepared for this chapter by Professor James Connolly of the University of Minnesota.

Part V: Handbook

[1] Minna Levine, "Business Statistics," *MacUser* (April 1990): 128.

[2] Levine 120.

INDEX

Page numbers followed by *f* indicate figures.

MARKING SYMBOLS

This list of marking symbols refers you to Part V, the "Handbook," where you can find discussions of style and usage. The list furnishes you with a heading and a page reference.